PROCESS INTEGRATION FOR RESOURCE CONSERVATION

GREEN CHEMISTRY AND CHEMICAL ENGINEERING

Series Editor: Sunggyu Lee
Ohio University, Athens, Ohio, USA

Proton Exchange Membrane Fuel Cells: Contamination and Mitigation Strategies
Hui Li, Shanna Knights, Zheng Shi, John W. Van Zee, and Jiujun Zhang

Proton Exchange Membrane Fuel Cells: Materials Properties and Performance
David P. Wilkinson, Jiujun Zhang, Rob Hui, Jeffrey Fergus, and Xianguo Li

Solid Oxide Fuel Cells: Materials Properties and Performance
Jeffrey Fergus, Rob Hui, Xianguo Li, David P. Wilkinson, and Jiujun Zhang

Efficiency and Sustainability in the Energy and Chemical Industries: Scientific Principles and Case Studies, Second Edition
Krishnan Sankaranarayanan, Jakob de Swaan Arons, and Hedzer van der Kooi

Nuclear Hydrogen Production Handbook
Xing L. Yan and Ryutaro Hino

Magneto Luminous Chemical Vapor Deposition
Hirotsugu Yasuda

Carbon-Neutral Fuels and Energy Carriers
Nazim Z. Muradov and T. Nejat Veziroğlu

Oxide Semiconductors for Solar Energy Conversion: Titanium Dioxide
Janusz Nowotny

Lithium-Ion Batteries: Advanced Materials and Technologies
Xianxia Yuan, Hansan Liu, and Jiujun Zhang

Process Integration for Resource Conservation
Dominic C. Y. Foo

GREEN CHEMISTRY AND CHEMICAL ENGINEERING

PROCESS INTEGRATION FOR RESOURCE CONSERVATION

Dominic C. Y. Foo

CRC Press
Taylor & Francis Group
Boca Raton   London   New York

CRC Press is an imprint of the
Taylor & Francis Group, an **informa** business

Contents

Part II   Mathematical Optimization Techniques

Series Preface

The subjects and disciplines of chemistry and chemical engineering have encountered a new landmark in the way of thinking about developing and designing chemical products and processes. This revolutionary philosophy, termed green chemistry and chemical engineering, focuses on the designs of products and processes that are conducive to reducing or eliminating the use and/or generation of hazardous substances. In dealing with hazardous or potentially hazardous substances, there may be some overlaps and interrelationships between environmental chemistry and green chemistry. Whereas environmental chemistry is the chemistry of the natural environment and the pollutant chemicals in nature, green chemistry proactively aims to reduce and prevent pollution at its very source. In essence, the philosophies of green chemistry and chemical engineering tend to focus more on industrial application and practice rather than on academic principles and phenomenological science. However, as both a chemistry and chemical engineering philosophy, green chemistry and chemical engineering derive from and build on organic chemistry, inorganic chemistry, polymer chemistry, fuel chemistry, biochemistry, analytical chemistry, physical chemistry, environmental chemistry, thermodynamics, chemical reaction engineering, transport phenomena, chemical process design, separation technology, automatic process control, and more. In short, green chemistry and chemical engineering advocate the rigorous use of chemistry and chemical engineering to prevent pollution and protect the environment.

The Pollution Prevention Act of 1990 in the United States established a national policy to prevent or reduce pollution at its source whenever feasible. And adhering to the spirit of this policy, the Environmental Protection Agency (EPA) launched its Green Chemistry Program in order to promote innovative chemical technologies that reduce or eliminate the use or generation of hazardous substances in the design, manufacture, and use of chemical products. Global efforts in green chemistry and chemical engineering have recently gained substantial support from the international community of science and engineering as well as from academia, industry, and government in all phases and aspects.

Some of the successful examples and key technological developments include the use of supercritical carbon dioxide as a green solvent in separation technologies; application of supercritical water oxidation for destruction of harmful substances; process integration with carbon dioxide sequestration steps; solvent-free synthesis of chemicals and polymeric materials; exploitation of biologically degradable materials; use of aqueous hydrogen peroxide for efficient oxidation; development of hydrogen proton exchange membrane (PEM) fuel cells for a variety of power generation needs; advanced biofuel production; devulcanization of spent tire rubber; avoidance of the use of chemicals and processes causing generation of volatile organic compounds (VOCs); replacement of traditional petrochemical processes by microorganism-based bioengineering processes; replacement of chlorofluorocarbons (CFCs) with nonhazardous alternatives; advances in the design of energy-efficient processes; use of clean, alternative, and renewable energy sources in manufacturing; and much more. This list, even though it is only a partial compilation, is undoubtedly growing exponentially.

This book series on Green Chemistry and Chemical Engineering by CRC Press/Taylor & Francis Group is designed to meet the challenges of the twenty-first century in the fields of chemistry and chemical engineering by publishing books based on cutting-edge research and development to the effect of reducing adverse impacts on the environment by chemical enterprise. In achieving this, the series will detail the development of alternative sustainable technologies that will minimize the hazard and maximize the efficiency of any chemical choice. The series aims at delivering to the readers in academia and industry an authoritative source of information in the field of green chemistry and chemical engineering. The publisher and its series editor are fully aware of the rapidly evolving nature of the subject and its long-lasting impact on the quality of human life both in the present and in future. As such, the team is committed to making this series the most comprehensive and accurate literary source in the field of green chemistry and chemical engineering.

Sunggyu Lee

Foreword

The ubiquitous use of natural resources throughout the process industries is one of the major hurdles limiting the extent of sustainability worldwide. Tremendous amounts of raw materials and energy are used in manufacturing chemical products. As industry endeavors to conserve natural resources, two critical questions need to be answered:

1. What are the targets for minimum consumption of mass and energy?
2. How to methodically achieve these targets in a cost-effective manner?

This book provides an integrated set of frameworks, methodologies, and tools to answer these two questions.

The author is a renowned leader in the area of resource conservation through process integration, and I had the pleasure of closely observing his professional growth and remarkable contributions that have been documented through excellent publications. In this book, Dr. Foo manages to elegantly transform the theories and concepts into effective educational tools, exciting reading materials, and very useful applications. Various techniques are used, including graphical, algebraic, and mathematical tools. The addressed problems span a wide range, including direct recycle, in-plant modifications, and waste treatment for continuous and batch systems. Numerical examples are used to further explain the tools and to demonstrate their effectiveness.

Overall, this is an excellent contribution that will benefit numerous researchers, students, and process engineers and will serve the cause of sustainability worldwide.

Mahmoud El-Halwagi, PhD
Professor and Holder of McFerrin Professorship
Texas A&M University
College Station, Texas

It is no exaggeration to say that the integration of environmental considerations is one of the most significant changes to occur in basic chemical engineering curriculum in recent decades. In fact, the same can probably be said for other engineering disciplines as well. The major environmental issues of the twenty-first century, such as climate change, air and water pollution, and the gradual but inevitable depletion of natural resources, are now routinely discussed by lecturers in university classrooms worldwide.

The state-of-the-art approach combines so-called end-of-pipe treatment technologies with pollution prevention strategies to ensure that engineering systems are designed to be sustainable, legally compliant, and economically viable. While end-of-pipe technologies such as gas scrubbers and wastewater treatment plants make use of physical and chemical processes to reduce the environmental hazards of existing waste streams, pollution prevention approaches make use of a broader and less well-defined set of technologies—both "hard" and "soft"—to achieve reductions in environmental impacts.

This book deals with the efficient use of utilities in industrial systems through systematic recycle and reuse of process streams. Many applications deal with efficient water utilization, which is fast becoming a critical aspect of process plant operations in many parts of the world, and is likely to become even more so as the effects of global climate change make themselves felt at the local level. However, many of the principles described here are readily extended toward problems involving efficient use of other resources, including utility gases and industrial solvents. It is quite instructive how analogous structures can be solved by common techniques even though the physical nature of the underlying industrial processes is fundamentally different. Furthermore, efficient use of purchased utilities yields economic benefits that may offset some, or all, of the costs incurred in implementing these pollution prevention strategies. Finally, an added benefit is that enhancing the efficiency of an industrial system generally reduces the quantity of waste streams that need to be treated after the fact. The main contribution of this textbook is that it brings together a family of *systematic design tools* that can be used to determine the most cost-effective measures to implement recycle and reuse of process streams in industrial plants. These techniques are drawn from the state-of-the-art in process integration literature, mostly from the author's own research over the past decade, and presents it in a form suitable for a wide range of audiences, from advanced undergraduate students to practicing engineers from the process industries.

It has been some years since Dr. Dominic C. Y. Foo, then an ambitious, freshly minted PhD at the beginning of his academic career, described to me his idea for this book at a conference banquet dinner. I thought at the time that the idea was promising, but needed time to mature. Today, with Dr. Foo having established himself as one of the world's most renowned young researchers in the area of process integration, with an extensive body of work and experience compressed into a relatively brief (by academic standards) career to date (and, meanwhile, as the ever-growing public concern about sustainability, in general, and the consequences of climate change, in particular, continues to induce changes in the chemical engineering curriculum), this excellent book comes at just the right time to teach the next generation of process designers how to "save the planet" more systematically and intelligently.

Raymond R. Tan, PhD
Full Professor and University Fellow
Chemical Engineering Department
Center for Engineering and Sustainable Development Research
De La Salle University-Manila, Philippines

Preface

Resource conservation and *waste minimization* have been major concerns in the process industry in the past two decades. This could be due to several reasons. Among these, the increase of public awareness toward environmental sustainability is probably the most influential factor. Hence, ever-stringent emission regulations are observed in many developed and developing countries. To maintain business sustainability and to fulfill social responsibilities, many industrial sectors have taken initiatives to improve emission quality by reducing their emission load. Besides, industrial sectors also enjoy economic benefits from this initiative, since reduced waste generation means more efficient use of resources, which in turn helps reduce overall production costs.

Concurrently, the development of various systematic design tools within the *process systems engineering* (PSE) community has seen significant achievements in the past decades. It should also be noted that most recent scholarly activities in the PSE community have been treating chemical processes as an integrated system rather than focusing on the individual units as in the past. This is indeed a good move, as it is now generally recognized that an optimum unit operation designed independently does not necessary lead to an optimum overall process. Hence, taking a holistic approach toward the synthesis of an optimum process flowsheet is indeed important. In this aspect, *process integration* technique that emphasizes the unity of the process has much to offer. The technique has its roots in energy recovery system design due to the first world oil crisis back in the 1970s. It was then extended to various waste recovery systems in the late 1980s. Since the mid-1990s, various process integration tools were developed, focusing on the recovery of material resources, such as water, utility gas, solvent, and solid wastes. However, most of these techniques are published in technical papers, targeted for academic researchers in the PSE community. Hence, this book is meant to respond to the need to disseminate the state-of-the-art techniques developed in the past one-and-a-half decades to a wider audience, especially those looking for cost-effective techniques for various resource conservation problems. Both *pinch analysis* and *mathematical optimization* techniques are covered.

The book aims to bridge the gap between academic and industrial practitioners, with the use of various industrial examples. Hence, the book is useful for upper level undergraduate or postgraduate students in chemical, process, or environmental engineering, who will soon face the "real world" upon graduation. On the other hand, industrial practitioners will also find the various industrial examples as good reference points in their efforts toward implementing resource conservation initiatives in their plants. The main emphasis of the book is *to set performance targets ahead of detailed design,* following the philosophy of process integration. With the performance targets identified, one will get away with the question of "Is there a better design?" that is always raised in any technical project.

One of the main aims of this book is to enable readers self-learning. Hence, most of the examples are explained in a methodical manner to facilitate independent reading. Besides, worksheets and calculation files for selected examples and problems are also made available on publisher's website (www. crcpress.com/product/ISBN/9781439860489) for the ease of use for readers (please look for the icon). Also made available on the book website and the demo version of LINGO and What's*Best!* softwares (from LINDO Systems Inc.), as well as the prototype software of my research group, i.e. RCN-Net (with user guide).

I hope you enjoy reading the book and have fun in setting targets!

Acknowledgments

I am indebted to many organizations and institutions for the completion of this book. I would like to express my gratitude to the various funding agencies that supported my scholarly activities in the past decade. These include the Ministry of Science, Technology and Innovation (MOSTI) of Malaysia, World Federation of Scientists, as well as the University of Nottingham Research Committee.

Special thanks are due to Allison Shatkin and Jennifer Ahringer, editors at CRC Press/Taylor & Francis Group, who have been very helpful in supporting the making of this book. I would also like to thank Robert Sims, project editor at Taylor & Francis Group, and Arunkumar Aranganathan, project manager at SPi Global, Puducherry, India, for their help.

I am grateful to LINDO Systems Inc., who have provided the demo versions of LINGO and What'sBest! software that accompany this book.

I am also indebted to several people who have played important roles in my life. Without them, I would not have been able to achieve success in my professional career. I am especially grateful to my high school teacher, Boon Nam Khoo, who taught me O-level physics and inspired me to pursue my professional career as a chemical engineer. I should also be grateful to all my formal teachers in Chi Wen Primary and Secondary Schools (where I had the elementary and middle school education); as well as Datuk Mansor Secondary School (where I completed high school education).

I am also grateful to all my lecturers at Universiti Teknologi Malaysia, where I spent 10 years for my undergraduate and graduate studies in Chemical Engineering. Special thanks are due to Professor Ramlan Abdul Aziz at the Institute of Bioproduct Development (formerly known as Chemical Engineering Pilot Plant), who inspired me tremendously during my early days as a postgraduate student. His dedication to work and the motivation to help others are the core values that I have tried to emulate. He has also given me a lot of career and self development opportunities since the initial stages of my career (and even until now!). Professor Zainuddin Abdul Manan, who was my thesis advisor and my professional writing trainer, also provided constant support throughout my career as a researcher and author.

I am fortunate to have been able to work with a group of helpful and supportive colleagues in the Department of Chemical and Environmental Engineering, University of Nottingham Malaysia Campus (UNMC). During the academic year 2010–2011, my colleagues helped me by temporarily relieving me of my academic load in terms of teaching and project supervision as I wrote this book. Without their help, this book would not have been a reality. Special thanks are due to Dr. Denny K.S. Ng, my first PhD student and now a colleague in the department, who helped me in teaching the various design courses in the past two years. With his help, I have been able to develop several assignments for students, which later become examples and problems in this book. I should also mention my second PhD student, Dr. Irene M.L. Chew (now with Monash University Sunway Campus, Malaysia), whose PhD work had also contributed significantly to the content of this book. Also thanks to the undergraduate and graduate students in our department, who contributed in "spot-checking" for errors and generated solutions to some of the exercises in this book (through their assignments)!

I am honored to be able to work with a group of collaborators who became very close friends over the years. Professor Raymond Tan of De La Salle University, Philippines, is one those in the top of the list. We started our collaboration as early as when I was still a PhD student back in 2004. The idea of writing a book was also nurtured over a conference dinner during our early days of friendship (see the foreword by Professor Tan). Throughout my professional career, he has been a good mentor. I should also mention that he has been the main reviewer for most of the chapters in this book (along with Denny and Irene). Besides, I have been very fortunate to work with Professor Mahmoud El-Halwagi of Texas A&M University since 2003. He is an inspiring mentor who once told me "contribution will always prevail" during the early, difficult years of my research career. To this day, I am still using his quote to tell the novices in the academic and research community that "your detractors cannot stop you; they can only slow you down." I should also mention that his first two process integration books have been a great source of inspiration in the writing of this book.

I would also like to acknowledge my other close collaborators for their continuous support. These include Professor Cheng-Liang Chen (National Taiwan University, Taiwan), Professor Santanu Bandyopadhyay (Indian Institute of Technology, Bombay, India), Professor Thokozani Majozi (University of Pretoria, South Africa), Professor Jiří Klemeš (University of Pannonia, Hungary), Professor Feng Xiao (China University of Petroleum, formally with Xi'an Jiaotong University, China), Professor Robin Smith (University of Manchester, United Kingdom), the late Professor Jacek M. Jeżowski (Rzeszow University of Technology, Poland), Professor Ramlan Abdul Aziz (Universiti Teknologi Malaysia, Malaysia), Professor Paul Stuart (École Polytechnique, Montreal, Canada), Professor Rosli Yunus (Universiti Malaysia Pahang, Malaysia), Professor Valentin Pleşu (University Politehnica of Bucharest, Romania), Professor Vasile Lavric (University Politehnica of Bucharest, Romania), Dr. Vasiliki Kazantzi (Technological Educational Institute of Larissa, Greece), Dr. Sivakumar Kumaresan (Universiti Malaysia Sabah, Malaysia), Dr. Chung Lim Law (UNMC), Dr. Petar Varbanov (University of Pannonia, Hungary), Dr. Alberto Alva-Argaez (Process Ecology, formally with Canmet ENERGY, Canada), Dr. Abeer Shoaib (Suez-Canal University, Egypt), Dr. Nick Hallale (Essar Oil Stanlow Refinery, United Kingdom), Yin Ling Tan (Curtin University of Technology Sarawak Campus, Malaysia), Dr. Hon Loong Lam (UNMC), Dr. Mimi Haryani Hashim (Universiti Teknologi

Malaysia), Dr. ChangKyoo Yoo (KyungHee University, Korea), Dr. Demetri Petrides (Intelligen, Inc., the United States), Dr. Alexandros Koulouris (Technological Educational Institute of Thessaloniki, Greece), Mustafa Kamal (Universiti Teknologi Malaysia, Malaysia), Cheng Seong Khor (Universiti Teknologi PETRONAS, Malaysia), Mike Boon Lee Ooi (NGLTech, Malaysia), Jully Tan (UCSI University, Malaysia), and Chiang Choon Lai (UNMC).

I am also very fortunate to have been working with a group of excellent young researchers throughout my academic career. Some of them came to my research group as short-term visiting scholars. They include Dr. Seingheng Hul (Institute of Technology of Cambodia, Cambodia), Dr. Lee Jui-Yuan (National Taiwan University, Taiwan), Dr. Chun Deng (China University of Petroleum, China), Tianhua Wang (Central Research Institute of China Chemical Science and Technology, China), Sophanna Nun (National Cleaner Production Office, Cambodia), Dr. Gopal Chandra Sahu (Forbes Marshall), Choon Hock Pau (ESI Sampling, Malaysia), Edward Li Zhian Yong (Perfection Enterprise, Canada). Besides, the undergraduate and graduate students whom I have guided for their projects are worth mentioning. They include Mahendran Subramaniam (Shell, Malaysia), Noor Zuraihan Mohamad Noor (Procter & Gamble, Malaysia), Raymond Eam Hooi Ooi (Worley Parsons, Malaysia), Choon Keat Kuan (MTBE Petronas, Malaysia), Jun Hoa Chan (Koch Glitsch, Singapore), Dr. Douglas Han Shin Tay (Global Green Synergy, Malaysia), Joseph Song Hok Lim (Adirondack, Malaysia), Omar Sulhi (Laurentian University, Canada), Arunasalam Athimulam (Sarawak Shell, Malaysia), Sin Cherng Lee (IOI Group, Malaysia), Wendy Tjan (Saturn Electronic Parts & CCTV System, Indonesia), Wai Hong Wong (ExxonMobil, Malaysia), Shin Yin Saw (SBM Offshore Group, Malaysia), Liangming Lee (Global Process Systems, Malaysia), Melwynn Kuok Yauu Leong (Sulzer Chemtech, Ming Hann Lim (Citibank, Malaysia) Singapore), Sinthi Laya Thillaivarrna (Shell Refinery Company, Malaysia), S. Sadish Kumar (Pan Century Edible Oils, Malaysia), Victor Francis Obialor Onyedim (Halliburton, Houston, Texas), Diban Pitchaimuthu (Global Process Systems, Malaysia), Ee Ling Toh (Worley Parsons Services, Malaysia), Sarah Seen Teng Soo (Exxon Mobil, Malaysia), Kevin Kar Keng Yap (Muda Paper Mills, Malaysia). My academic career would not have been as exciting without the contributions of these young people!

Lastly, I am very grateful to my wonderful family. I am indebted to my wife, Cecilia Choon Shiuan Cheah, for her constant support. Her dedication and love toward our family and kids, Irene Xian Hui Foo and Jessica Xin Hui Foo, have been crucial in supporting me to continuously excel in my professional career. I would also like to mention the invaluable support I have received from my parents, Tian Juan Foo and Mathilda Ah Nooi Siaw, and from my siblings, Agnes Hui Hwen Foo and Jude Chwan Woei Foo.

Author

Ir. Dr. Dominic C.Y. Foo is a Professor of Process Design and Integration at the University of Nottingham, Malaysia Campus, and is the Founding Director for the Centre of Excellence for Green Technologies. He is a professional engineer registered with the Board of Engineers Malaysia (BEM). He is also a world-renowned researcher in process integration for resource conservation. Dr. Foo has collaborated with researchers from various countries in Asia, Europe, America, and Africa. He is an active author, with more than 70 journal papers, and has made more than 120 conference presentations (with more than 10 plenary/keynote presentations). He served as international scientific committee for several important conferences in process design (PRES, FOCAPD, ESCAPE, PSE, etc.). He is the winner of the Innovator of the Year Award 2009 of the Institution of Chemical Engineers United Kingdom (IChemE) as well as the 2010 Young Engineer Award of the Institution of Engineers Malaysia (IEM). He is an active member in both IEM and IChemE Malaysia Branch. He also actively conducts professional training for practicing engineers.

How to Make Use of This Book

Instructors for Process Design Courses

Incorporating sustainability concepts into the undergraduate syllabus has been a recent trend in chemical engineering. Hence, topics such as environmentally benign design, heat recovery system, and resource conservation are among those that have attracted the attention of academics who conduct process design courses at undergraduate and graduate levels. This may also be partially due to the influence of industrial practitioners.

Nevertheless, balancing the topics to be covered in an undergraduate process design course is indeed a challenge. In most universities, process design is taught at the final year (senior—year 4 level) of an undergraduate chemical engineering program along with the capstone design project. In some cases, only limited lecture series are carried out throughout the design project (i.e., no formal taught course elements). Hence, the course instructors will have to balance between the necessary materials needed in a typical design project (e.g., process synthesis, equipment design, process simulation, and economic evaluation) while adding some extra elements where they feel relevant.

Tables 0.1 and 0.2 show a list of suggested topics to be included in a typical undergraduate process design course, along with sections that may be excluded. For an introductory level (Table 0.1), it is suggested to cover the basic concept on resource conservation network (RCN) with direct water reuse/recycle alone. This typically requires a duration of one to two weeks (assuming three contact hours per week), depending on how much time is being spent for the working sessions. For an intermediate level course (three to five weeks), most sections in Chapters 2 through 8 may be covered (Table 0.2). Students will be trained to synthesize an RCN with reuse/recycle, regeneration, and waste treatment schemes (waste treatment can be optional) in this intermediate level course. If the instructor feels that the students should learn some basic optimization techniques, it would be appropriate to also include Chapter 12. Note again that there are sections that may be excluded due to time constraint.

In some cases, the instructors may want to offer specialized/optional modules (e.g., for graduate level) that run throughout the entire semester. For such a case, three different types of topics may be selected from the suggested list in Table 0.3. For instructors who wish to focus on the insight-based techniques,

TABLE 0.1

Suggested Topics for Undergraduate Chemical Engineering Process Design Course (Introductory Level)

Chapters (Ch)	Sections That May Be Excluded
Ch 2: Data extraction	Sections 2.4 (refinery hydrogen network) and 2.5 (property integration), if focus is on water recovery
Ch 3: Graphical targeting for direct reuse/recycle	Sections 3.3 (multiple resources), 3.4 (threshold problems), and 3.5 (property integration)
Ch 7: Network design and evolution	Sections 7.3 (design of material regeneration network) and 7.4 (network evolution techniques)

TABLE 0.2

Suggested Topics for Undergraduate Chemical Engineering Process Design Course (Intermediate Level)

Chapters (Ch)	Sections That May Be Excluded
Ch 2: Data extraction	—
Ch 3: Graphical targeting for direct reuse/recycle	Section 3.4 (threshold problems)
Ch 4: Algebraic targeting for direct reuse/recycle	Section 4.3 (threshold problems)
Ch 5: Process changes[a]	—
Ch 6: Material regeneration network	Section 6.3 (mass exchanger as regeneration unit)
Ch 7: Network design and evolution	Section 7.4 (network evolution techniques)
Ch 8: Total material network[a]	—
Ch 12: Superstructural approach[a]	Section 12.6 (interplant resource conservation networks)

[a] These few chapters are a matter of choice, based on the preference of instructors.

topics in Chapters 2 through 11 should be covered (column 1). The material in Chapters 12 through 17, where superstructural and automated targeting techniques are covered, should be relevant for those who want to focus on mathematical optimization (column 2). One may also opt for a hybrid way, i.e. a combination of insight-based and optimization techniques (column).

Instructors for Professional Courses

As resource conservation is a good means to achieve sustainability for the process industry, it has attracted lots of attention among industrial practitioners over the last decades. I have myself conducted many successful

TABLE 0.3

Suggested Topics for Optional Courses

Insight-Based Techniques	Mathematical Optimization	Hybrid
Ch 2; Data extraction	Ch 2: Data extraction	Ch 2: Data extraction
Ch 3; Graphical targeting for direct reuse/recycle	Ch 4: Algebraic targeting for direct reuse/recycle[a]	Ch 3: Graphical targeting for direct reuse/recycle
Ch 4: Algebraic targeting for direct reuse/recycle	Ch 6: Material regeneration network[a]	Ch 4: Algebraic targeting for direct reuse/recycle
Ch 5: Process changes	Ch 12: Superstructural approach	Ch 5: Process changes[a]
Ch 6: Material regeneration network	Ch 13: ATM for direct reuse/recycle	Ch 6: Material regeneration network
Ch 7: Network design and evolution	Ch 14: ATM for material regeneration network	Ch 7: Network design and evolution
Ch 8: Total material network	Ch 15: ATM for total material network	Ch 12: Superstructural approach[a]
Ch 9: Pre-treatment networks	Ch 16: ATM for batch material networks	Ch 13: ATM for direct reuse/recycle
Ch 10: Batch material networks[a]	Ch 17: ATM for IPRCN	Ch 14: ATM for material interception network
Ch 11: Interplant resource conservation network[a]		Ch 15: ATM for total material network[a]
Ch 12: Superstructural approach[a]		

[a] These few chapters are a matter of choice, based on the preference of instructors. Refer to Tables 0.1 and 0.2 for sections that may be excluded for each chapter.

workshops in various countries on resource conservation. The typical suggested topics are listed in Table 0.4 for different levels of professional courses.

Advice for Instructors

Based on my experience, the most effective way to teach the various resource conservation topics found in this book is the "hands-on" approach, i.e., practical/workshop sessions. These are needed in order for the students to get to experience the "know-how" in solving an RCN problem.

The instructors are also encouraged to make use of the various teaching materials in conducting the course/workshop, which consists of PowerPoint slides (with worksheets), calculation files, etc.

TABLE 0.4

Suggested Topics for Different Levels of Professional Courses

Introductory (Half Day)	Basic (1–2 Days)	Advanced (4–5 Days)
Ch 2: Data extraction	Ch 2: Data extraction	Ch 2: Data extraction
Ch 3: Graphical targeting for direct reuse/recycle	Ch 3: Graphical targeting for direct reuse/recycle	Ch 3: Graphical targeting for direct reuse/recycle
Ch 4: Algebraic targeting[a] for direct reuse/recycle	Ch 4: Algebraic targeting for direct reuse/recycle	Ch 4: Algebraic targeting for direct reuse/recycle
Ch 7: Network design[a] and evolution	Ch 5:Process changes	Ch 5: Process changes[a]
	Ch 6: Material regeneration network	Ch 6: Material regeneration network
	Ch 7: Network design and evolution	Ch 7: Network design and evolution
	Ch 8: Total material network[a]	Ch 8: Total material network[a]
	Ch 9: Pre-treatment networks[a]	Ch 12: Superstructural approach
	Ch 12: Superstructural approach[a]	Ch 13: ATM for direct reuse/recycle
		Ch 14: ATM for material regeneration network
		Ch 15: ATM for total material network[a]

[a] These few chapters are a matter of choice, based on the preference of instructors. Refer to Tables 0.1 and 0.2 for sections that may be excluded for each chapter.

1

Introduction

Resource conservation is one of the most important elements for establishing sustainability in modern society. The growth of global population, overextraction, mismanagement of natural resources, climate change, and global water availability are some of the factors that need serious consideration to achieve the goal of sustainability. Climate change and global water availability are also being emphasized in the *planetary boundaries* as proposed by Rockstrom et al. (2009). To this end, we are seeking a generic family of techniques that address the various resource conservation problems of similar nature.

1.1 Motivating Examples

Many regard resource conservation as a simplistic problem, with a straightforward solution. Some plant operation personnel tend to leave this problem to the consultants in the belief that they will provide them with quick and sensible solutions. However, it is always good to have a benchmark in order to compare with the solution proposed by an external consultant.

Let us begin this discussion with several motivating examples. Figure 1.1 shows an acrylonitrile (AN) plant where the wastewater treatment facility is operated at full hydraulic capacity (El-Halwagi, 1997, 2006). Since water is a by-product of the reactor, any increase in AN production will lead to larger wastewater flowrate and hence overload the existing wastewater treatment facility (see detailed description in Example 2.1.).

To debottleneck the wastewater treatment facility of the AN plant, different parties may adopt different solution strategies. If the problem is approached by a wastewater treatment consultant, a typical solution would be to construct a larger wastewater treatment plant to handle the desired wastewater throughput. However, this involves substantial capital investment and operating cost, which may not even justify the desired production expansion.

On the other hand, a process or utility engineer of the AN plant who has been instructed by the plant manager to look into the debottlenecking problem may opt for different options. He or she may want to carry out wastewater recycling in order to reduce the wastewater flowrate to the

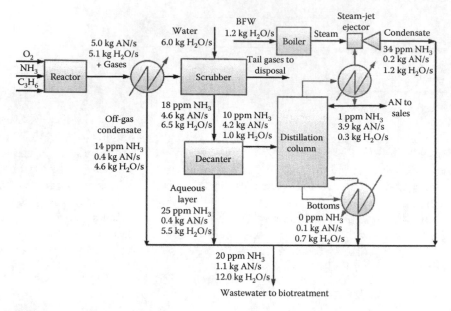

FIGURE 1.1
AN production plant. (From El-Halwagi, M.M., *Pollution Prevention through Process Integration: Systematic Design Tools*, Academic Press, San Diego, CA, 1997. With permission.)

treatment facility. He or she may then call up various suppliers to look for a good filtration unit (or something similar) to purify the terminal waste-water stream for recycling purposes. This seems to be the most common industrial practice, especially for those who claim it to be a straightforward solution. However, this may not be as straightforward as it seems to be. One would certainly need to worry about the various contaminants present in the terminal wastewater stream, since it is contributed by various waste-water streams from different sections of the plant. Next, when the engineer presents the proposal to the plant manager, the plant manager may in turn question the engineer on the choice of the filtration unit, as there could be other types of purification units that might be cheaper and worth exploring. He or she may also question if any of the individual wastewater streams could be directly recovered before any treatment; and if yes, what would be the maximum recovery for the plant? The engineer has more questions in his or her mind. He or she will have to sort out the various suggestions to figure out an economically viable solution. In fact, if one were to explore the various options possible for water recycling in this problem, there are easily a dozen of them, ranging from various direct water recovery options as well as those associated with the use of purification units (see El-Halwagi [2006] for details). Note that each of these solutions is unique and needs detailed investigation. Obviously, industrial practitioners have no such luxury of time to inspect each solution in detail.

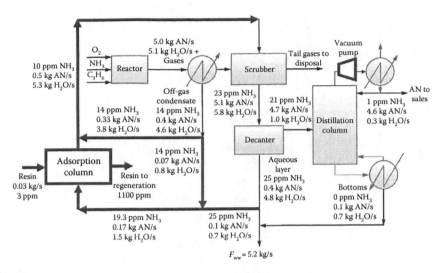

FIGURE 1.2
Optimum water recovery option for AN case.

The optimum solution for this case is given in Figure 1.2. The solution includes an installation of a resin adsorption unit that purifies the effluent from the decanter and off-gas condensate, before it is recovered to the scrubber. Note also that a big portion of the off-gas condensate is also recovered to the scrubber directly. Besides, the boiler that runs the steam-jet ejector of the distillation column is replaced by a vacuum pump, which eliminates the condensate from the steam-jet ejector. This solution is generated by a systematic approach* and is not immediately obvious to the engineer through an ad hoc approach.

Let us now look at a second case study, where solvent is utilized in a metal degreasing process (Kazantzi and El-Halwagi, 2005). In order to reduce the solvent waste of the process, suggestions are made to recover the solvent from the condensate streams (Figure 1.3).

The optimum solution for this case is given in Figure 1.4. For this case, the operating condition of the thermal processing unit is modified to allow direct recovery of the condensate streams (without any purification unit). Similar to the AN case, the solution for this case is generated via a systematic design tool* and is not obvious to even an experienced engineer.

For a third example, let us look at Figure 1.5 showing a refinery hydrogen network. It is desirable to reduce both the fresh hydrogen feed and the waste gas that is sent purged as fuel.

The minimum cost solution for the hydrogen recovery network is given in Figure 1.6. As shown, apart from direct recovery of hydrogen sources, a pressure swing adsorption (PSA) unit is used to purify some hydrogen sources

---

* See the generation of solution for this example in Chapters 2 through 7.

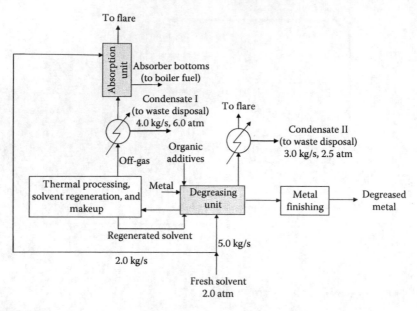

FIGURE 1.3
A metal degreasing process. (From Kazantzi, V. and El-Halwagi, M.M., *Chem. Eng. Prog.*, 101(8), 28, 2005. Reproduced with permission. Copyright © 2005. American Institute of Chemical Engineers (AIChE)).

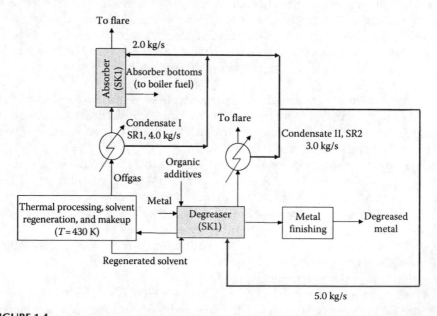

FIGURE 1.4
Solvent recovery scheme for a metal degreasing process. (From Kazantzi, V. and El-Halwagi, M.M., *Chem. Eng. Prog.*, 101(8), 28, 2005. Reproduced with permission. Copyright © 2005. American Institute of Chemical Engineers (AIChE)).

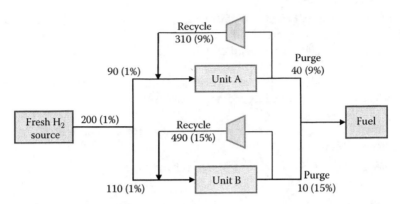

FIGURE 1.5
A refinery hydrogen network (stream flowrate given in million standard cubic feet/day-MMscfd; its purity is given in parenthesis). (From *Adv. Environ. Res.*, 6, Hallale, N. and Liu, F., Refinery hydrogen management for clean fuels production, 81–98, Copyright 2001, with permission from Elsevier.)

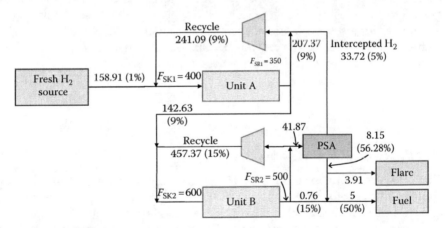

FIGURE 1.6
Hydrogen recovery for a refinery hydrogen network (stream flowrate given in million standard cubic feet/day-MMscfd; its purity is given in parenthesis).

for recovery. Again, this does not seem to be an obvious solution, even for an experienced process engineer. However, the recovery scheme in Figure 1.6 may be generated by a series of systematic design steps.*

From the earlier examples, it is obvious that developing a material recovery system may not be as straightforward as commonly thought, even for relatively simple processes. The question "is there a better design?" often arises among process engineers. Furthermore, we realize that material recovery

---

* See the generation of solution for this example in Chapters 2 and 13 through 15.

may involve various types of resources, e.g., water, utility gases, and solvent. Hence, a systematic approach is needed to develop an optimum flowsheet for such a system. We are in need of a "quick kill" solution that addresses this problem based on some generic principles common to all material recovery systems. In a nutshell, the approach should answer the following questions that often arise in a material recovery problem:

- What is the maximum recovery potential for a given material recovery problem?
- What is the minimum waste generation for a process?
- If a purification unit is to be used, what is the minimum capacity for the unit?
- What is the minimum cost solution for a given material recovery problem?

Is there such a technique that can answer the questions given earlier? Fortunately, there is. With the development of various systematic process design techniques in the past decades, it is now possible to set rigorous performance targets for a material recovery problem. One of such promising techniques is known as *process integration*, which consists of two important elements, i.e., *process synthesis* and *process analysis* (El-Halwagi, 1997).

1.2 Process Synthesis and Analysis

Until recently, many regarded process design exercise as performing a process simulation study for a given flowsheet. This is indeed not true. A typical process design exercise should consist of both elements of *process synthesis* and *process analysis*. The term "process synthesis" was first proposed in the late 1960s by Rudd (1968). The definition for the term has evolved over the years, and one commonly acceptable definition takes the form of "the discrete decision-making activities of conjecturing (1) which of the many available component parts one should use, and (2) how they should be interconnected to structure the optimal solution to a given design problem" (Westerberg, 1987, p. 128). In a nutshell, we are aiming to produce an optimum (unknown) process flowsheet, when the process input and output streams are provided (Figure 1.7).

On the other hand, in process analysis, we aim to predict how a defined process would actually behave under a given set of operating conditions. This involves the decomposition of the process into its constituent elements (e.g., units) for performance analysis. In most cases, once a process flowsheet is synthesized, its detailed characteristics (e.g., temperature, pressure, and flow-rates) are predicted using various analytical techniques. In other words, we

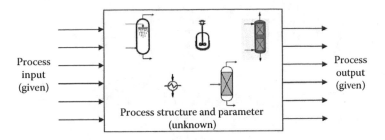

FIGURE 1.7
A process synthesis problem. (From *Process Integration*, El-Halwagi, M.M., Copyright 2006, with permission from Elsevier Inc.)

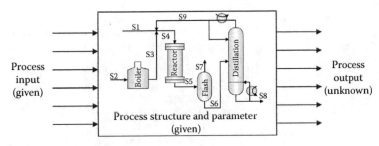

FIGURE 1.8
A process analysis problem. (From *Process Integration*, El-Halwagi, M.M., Copyright 2006, with permission from Elsevier Inc.)

aim to predict the process outputs for given process inputs and their flowsheet (Figure 1.8). Most often, various commercial process simulation software may be used, for both continuous (e.g., Aspen HYSYS, Aspen Plus, UniSim Design, and DESIGN II for Windows) and batch processes (e.g., SuperPro Designer).

Process synthesis and process analysis supplement each other in a process design exercise. Through process analysis, the basic conditions of a process flowsheet (e.g., mass and energy balances) may be obtained. We can then make use of the process synthesis technique to develop an optimum flowsheet. Note that in some instances, we may also reverse the task sequence of synthesis and analysis. For such a case, a process flowsheet is first constructed with the process synthesis technique; it is then analyzed to obtain its process conditions. The interactions of process synthesis and analysis in a design exercise are depicted in Figure 1.9.

Figure 1.10 shows the various mass and energy input and output streams for a processing unit. With process synthesis and analysis, the mass–energy dimensions for a chemical process are better understood. This is essential in setting performance targets for a material recovery problem.

We next discuss the main technique that can be used to set rigorous performance targets, i.e., *process integration*.

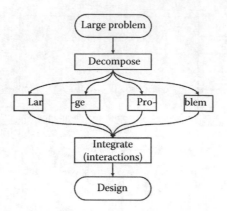

FIGURE 1.9
Interactions of process synthesis and analysis. (From Westerberg, A.W./Liu, Y.A., McGee, H.A., and Epperly, W.R. (eds.): Process synthesis: A morphology review. In: *Recent Developments in Chemical Process and Plant Design.* 1987. Copyright Wiley-VCH Verlag GmbH & Co. KGaA. With permission.)

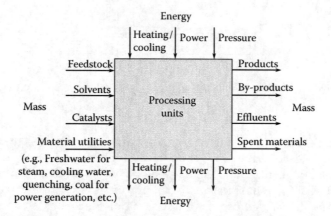

FIGURE 1.10
Mass–energy dimensions for a processing unit. (From El-Halwagi, M.M., *Pollution Prevention through Process Integration: Systematic Design Tools*, Academic Press, San Diego, CA, 1997. With permission.)

1.3 Process Integration: A Brief Overview

El-Halwagi (1997, 2006) defined process integration as *a holistic approach to process design, retrofitting, and operation, which emphasizes the unity of the process.* To date, three main branches of process integration problems have been established, i.e., *heat*, *mass*, and *property integration.* They are briefly described as follows.

The first trend in process integration was dedicated to *heat integration* problems. It was first introduced in the early 1970s and gained great attention during the first oil crisis a few years back then. The technique was first established for the optimal synthesis of a heat exchanger network (HEN, Hohmann, 1971; Linnhoff and Flower, 1978), followed by the efficient design of various energy intensive processes, e.g., distillation and utility system. With extensive development of heat integration techniques in the 1980s and 1990s, this trend has reached a mature stage in recent years. Many reviews (e.g., Gundersen and Naess, 1988; Linnhoff, 1993; Furman and Sahinidis, 2002) and textbook materials (Linnhoff et al., 1982, 1994; Douglas, 1988; Shenoy, 1995; Smith, 1995, 2005; Biegler et al., 1997; Turton et al., 1998; Seider et al., 2003; Kemp, 2007; Klemeš et al., 2010; El-Halwagi, 2011) are readily available.

Following the analogy between heat and mass transfer, *mass integration* was then established in the late 1980s for the synthesis of a mass exchange network (MEN, El-Halwagi and Manousiouthakis, 1989). The technique was further extended into other subcases of MEN synthesis problems, e.g., reactive MEN, combined heat and reactive MEN, and waste-interception network. References for this trend are available in review articles (El-Halwagi, 1998; El-Halwagi and Spriggs, 1998; Dunn and El-Halwagi, 2003) and textbooks (El-Halwagi, 1997, 2006, 2011; Seider et al., 2003). It is worth mentioning that the main driving force for the development of these techniques was the waste minimization initiative.

Apart from mass integration problems that are concentration based, *property integration* has been established as a new trend in the last decade. The technique is based on property and functionality tracking of stream quality, since many chemical processes are characterized by chemical or physical properties rather than composition (Shelley and El-Halwagi, 2000; El-Halwagi, 2006, 2011).

Within the domain of mass and property integration, there exist some special cases of subareas that focus on material recovery, which has been initiated since the mid-1990s. These include the introduction of *water minimization* (Wang and Smith, 1994) and refinery hydrogen network problems (Towler et al., 1996). At a later stage, the techniques were extended to cover the recovery of various other materials, such as solvent, solids, etc. (see Kazantzi and El-Halwagi, 2005; Foo et al., 2006). In recent years, process integration researchers have collectively termed this special area of interest as *resource conservation network* (RCN) synthesis, which addresses the optimum use of various material resources, such as water, utility gases (e.g., hydrogen and nitrogen), solvent, paper, plastic waste, etc. Some recent archives have been made available in review articles (Bagajewicz, 2000; Foo, 2009; Gouws et al., 2010; Jeżowski, 2010) and textbook materials too (Mann and Liu, 1999; Smith, 2005; El-Halwagi, 2006, 2011; Klemeš et al., 2010). This is also the main subject of this book, which aims to compile the most state-of-the-art techniques developed in the past decade in RCN synthesis.

Besides, there are also other emerging trends of process integration that are worth mentioning, e.g., energy sector planning with environmental constraints (Tan and Foo, 2007; Foo et al., 2008a; Lee et al., 2009; Tan et al., 2009), supply chain production planning (Singhvi and Shenoy, 2002; Singhvi et al., 2004;

Foo et al., 2008b; Ludwig et al., 2009), process scheduling (Foo et al., 2007), human resource planning (Foo et al., 2010), and financial analysis (Zhelev, 2005).

We might now wonder how the aforementioned process integration problems are interconnected. In fact, there exist some similarities in all these problems. In summary, all these problems may be described in terms of stream *quantity* and *quality* aspects. For instance, in HEN and other heat integration problems, we rely on the minimum approach temperature (quality) for feasible heat transfer (quantity). On the other hand, for MEN and concentration-based RCN problems, the impurity concentration (quality) will ensure that feasible mass transfer (quantity) takes place. The same principle applies to other problems, as summarized in Table 1.1 (Sahu and Bandyopadhyay, 2011).

It is also worth noting that within the process integration community, there exist two distinct approaches in addressing the various RCN problems, i.e., *insight-based pinch analysis* and *mathematical optimization* techniques. Both of these techniques have their strengths and limitations. The insight-based techniques are well accepted in providing good insights for the designers. They can be used to set performance targets for a given problem before detailed design of an RCN. However, they can only handle a single quality constraint at one time. In contrast, mathematical optimization techniques do not provide such good insights for designers, but they are powerful in solving complex cases (e.g., process constraints, detailed stream matching, etc.). Comparisons of these approaches are summarized in Table 1.2.

TABLE 1.1

Quantity and Quality Aspects of Various Process Integration Problems

Quantity	Quality	Examples/Problems	References
Heat	Temperature	Heat exchange network	Linnhoff et al. (1982, 1994)
		Heat integration	Smith (1995, 2005)
Mass	Concentration	Mass exchange network	El-Halwagi and Manousiouthakis (1989)
		Water minimization	Wang and Smith (1994)
		Refinery hydrogen network	Towler et al. (1996)
Mass	Properties	Property-based RCNs	Kazantzi and El-Halwagi (2005)
Steam	Pressure	Cogeneration	Dhole and Linnhoff (1993)
Energy	CO_2	Carbon-constrained energy planning	Tan and Foo (2007)
Mass	Time	Supply chain management	Singhvi and Shenoy (2002)
Time	Time	Process scheduling	Foo et al. (2007)
		Human resource planning	Foo et al. (2010)
Energy	Time	Isolated power system	Arun et al. (2007)

Source: From Sahu, G. C. and Bandyopadhyay, S. Holistic approach for resource conservation, *Chemical Engineering World*, Dec 2011, 104–108, 2011.

TABLE 1.2

Comparison for Insight-Based and Mathematical Optimization Techniques

Approaches	Strengths	Limitations
Insight-based pinch analysis techniques	• Provide good insights for designers • Ability to set performance targets ahead of design	• Handle single quality constraint at one time
Mathematical optimization techniques	• Handle multiple quality constraints simultaneously • Handle complex cases (e.g., process constraints and detailed stream matching) relatively well	• Poor insights for designers • Global optimality may not be guaranteed for certain types of problems

Besides, for the insight-based techniques, one may also categorize the nature of the techniques as graphical or algebraic methods. While both methods essentially obtain the same results, graphical techniques are more powerful in providing good insights for designers as compared with the algebraic techniques. However, the latter are normally more computationally effective, which facilitates the calculation steps, especially for large and complex problems. Besides, the numerical nature of the algebraic techniques also makes them a good platform for software interactions (e.g., process simulation). A summary of comparison for both methods is given in Table 1.3.

Note that the insight-based technique is a two-step approach. In Step 1, performance targets for a given problem are first identified based on the quality and quantity aspects as discussed earlier. For the case of RCNs, the performance targets correspond to their minimum flowrates of fresh resources and waste discharge. In Step 2, the network structure that achieves the performance targets in Step 1 is then identified; this is commonly termed as the *network design* stage.

It is also worth nothing that in recent years, the trend to combine both insight-based and mathematical optimization approaches is becoming more apparent (Smith, 2000). This is different from the early 1980s

TABLE 1.3

Comparison for Insight-Based and Mathematical Optimization Techniques

Techniques	Strengths	Limitations
Graphical	• Better insights for designers	• Tedious solution for complex problems • Inaccuracy problems
Algebraic	• Computational effectiveness • Ease for large and complex problems • Interaction with other softwares, e.g., process simulators, spreadsheets	• Less insights for designers

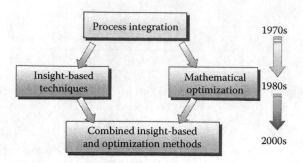

FIGURE 1.11
Development in process integration research in recent years. (Courtesy of Professor Robin Smith, University of Manchester).

where developments were mainly focused on the individual approaches (Figure 1.11). As a result, some hybrid approaches (e.g., Alva-Argáez et al., 1998, 1999; Ng et al., 2009a,b,c) have been developed in recent years.

All insight-based, optimization and hybrid techniques for RCN synthesis are found in this book.

1.4 Strategies for Material Recovery and Types of RCNs

Several terminologies are now widely accepted for RCN synthesis in the process integration community. Following the definition of Wang and Smith (1994, 1995), when an effluent stream of a process is sent to other operations and does not reenter its original process, it is termed as *direct reuse* scheme (Figure 1.12a). On the other hand, if the effluent is permitted to reenter the operation where it was generated, the scheme is termed as *direct recycle* scheme (Figure 1.12b). When the maximum potential for direct reuse/recycle is exhausted, the effluent may be sent for *regeneration* in an *interception unit* (any purification process that improves its quality) for further reuse (Figure 1.12c) or recycle (Figure 1.12d).

When the strategies of reuse, recycle, or regeneration are implemented within an RCN, we term them as *in-plant direct reuse/recycle* or *material regeneration networks*. On the other hand, we may also incorporate a *distributed waste treatment* system into an RCN, in which the overall scheme is known as the *total material network* (TMN, Kuo and Smith, 1998; see Figure 1.13). Note also that in some cases, both regeneration and waste treatment may be handled by the same interception unit. In other cases, a separate interception unit is needed to purify fresh resources before it is utilized in a process. This is commonly known as a pretreatment network (Figure 1.13c).

Upon the exhaustion of in-plant material recovery potential, we may then explore the potential of material recovery across different RCNs (Figure 1.14).

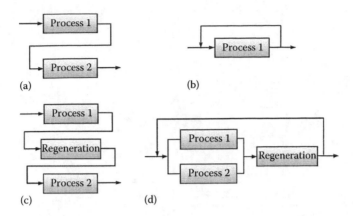

FIGURE 1.12
Strategies for in-plant resource conservation: (a) direct reuse, (b) direct recycle, (c) regeneration reuse, and (d) regeneration recycling. (From Wang, Y P. and Smith, R., *Chem. Eng. Sci.*, 49, 981, 1994. With permission.)

In practice, these RCNs may be part of a large RCN that are geographically segregated into different sections, or different process plants that operate with each other through *industrial symbiosis*. We shall collectively term this category of RCN problems as *inter-plant resource conservation networks* (IPRCNs). For an IPRCN, various reuse/recycle and regeneration strategies may be implemented for each individual RCNs, before material recovery is carried out among them.

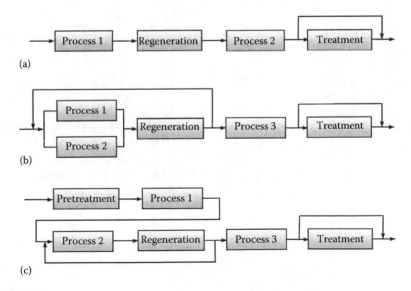

FIGURE 1.13
Other types of RCNs: (a) TMN with regeneration–reuse scheme, (b) TMN with regeneration–recycle scheme, and (c) TMN with pretreatment and regeneration scheme.

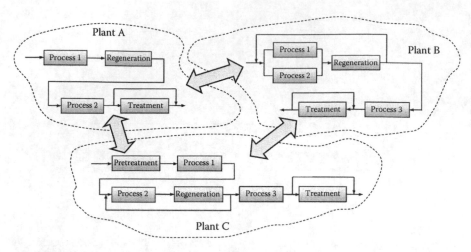

FIGURE 1.14
IPRCN where material recovery is carried out among different RCNs.

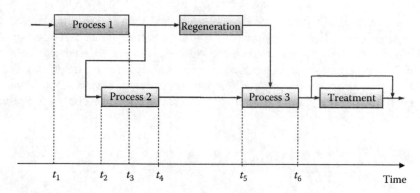

FIGURE 1.15
A batch material network.

Apart from the various RCN problems that are operated in continuous mode, there are also RCNs that are operated in batch process mode. Hence, the time dimension of this *batch material network* (BMN) is of important consideration in the analysis. Similar to the IPRCN problem, reuse/recycle and regeneration strategies may be implemented for a BMN, as shown in Figure 1.15.

1.5 Problem Statements

To better understand an RCN problem, we should first understand the concept of *process sink* and *source*. For a given process where a resource is found

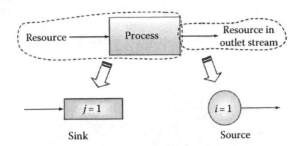

FIGURE 1.16
A material sink and source. (From Foo, D.C.Y., *Ind. Eng. Chem. Res.*, 48(11), 5125, 2009. With permission.)

in its outlet streams and needs to be recovered, the outlet stream is termed as a process source. On the other hand, a process sink refers to a unit where a resource is consumed (Figure 1.16).

Hence, the formal problem statement for a reuse/recycle network can be stated as follows:

- Given a number of resource-consuming units in the process, designated as process sink, or $SK = \{j = 1, 2, 3, ..., N_{SK}\}$, each sink requires a fixed flow/flowrate requirement of material (F_{SKj}) with a targeted quality index (q_{SKj}). The quality index should satisfy the following constraint:

$$q_{SKj}^{min} \leq q_{SKj} \leq q_{SKj}^{max} \tag{1.1}$$

where q_{SKj}^{min} and q_{SKj}^{max} are the lowest and highest limits of the targeted quality. Note that the quality index of an RCN varies from one case to another, which may take the form of impurity concentration of a concentration-based RCN (e.g., ppm, mass ratio, and percentage) or property operator values for a property-based RCN (see detailed discussion in Chapter 2).

- Given a number of resource-generating units/streams, designated as process source, or $SR = \{i = 1, 2, 3, ..., N_{SR}\}$, each source has a given flow/flowrate (F_{SRi}) and a quality index (q_{SRi}). Each source can be sent for reuse/recycle to the sinks. The unutilized source(s) are sent for environmental discharge (treated as a sink).

- When the process source is insufficient for use by the sinks, external fresh resources (treated as source) may be purchased to satisfy the requirement of sinks.

The superstructural representation for this problem is given in Figure 1.17.

On the other hand, for material regeneration and pretreatment networks, as well as TMNs, interception units are used for purifying process sources, before they are recovered to the sinks and/or

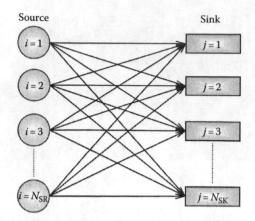

FIGURE 1.17
Superstructural representation for reuse/recycle scheme.

sent for final waste discharge. In this case, the superstructural representation is given in Figure 1.17, and the following is added to the problem statement:

- Given interception units of known performance, the process source is to be purified for recovery in the sinks and for meeting the environmental discharge limit (Figure 1.18).

For an IPRCN, apart from in-plant reuse/recycle/regeneration, material recovery is also carried out across different RCNs. Hence, the superstructural

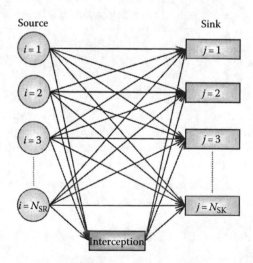

FIGURE 1.18
Superstructural representation for RCN with interception unit.

representations take the form as in Figure 1.19. Note that we may differentiate them as *direct* and *indirect* integration schemes. In the latter schemes, material recovery is carried out through a *centralized utility facility* (CUF), which acts as a hub through which streams are collected before reuse/recycle. In some cases, the CUF supplies the individual RCNs with fresh resources and provides various interception units for regeneration and waste treatment facilities.

For a BMN, where process sinks and sources are available in different *time intervals*, reuse/recycle, interception has to be carried out within the same time interval. Besides, we may also make use of storage tanks to recover process sources across different time intervals. The superstructural representation for the BMN is shown in Figure 1.20.

For all RCN problems described earlier, the objective of the RCN problem may be set to minimize the requirement of the external fresh resources and minimum cost. Note that minimizing the fresh resources also leads to

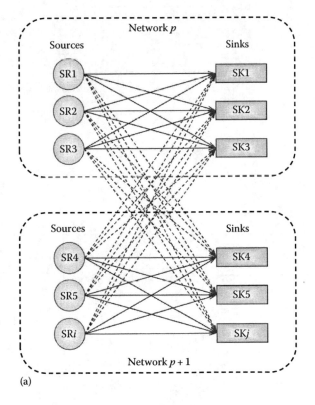

(a)

FIGURE 1.19
Superstructural model for IPRCN: (a) direct integration and (b) indirect integration schemes. (From Chew, I., et al., *Industrial & Engineering Chemistry Research*, 47, 9485, 2008.)

(*continued*)

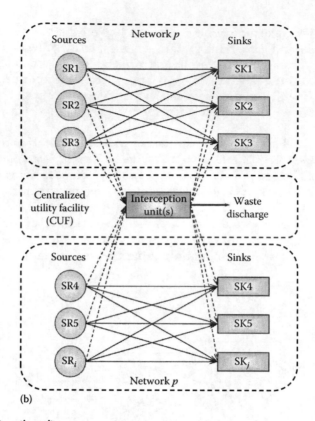

(b)

FIGURE 1.19 (continued)

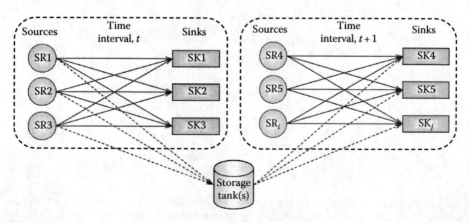

FIGURE 1.20
Superstructural model for BMN.

minimum waste generation for an RCN. This achieves the waste minimization objective of an RCN.

This book covers all types of RCN problems described earlier.

1.6 Structure of the Book

This book has 17 chapters. Chapter 1 discusses the general aspects of process synthesis and process integration. We then discuss the various data extraction principles in performing resource conservation activities in Chapter 2. Correct data extraction is the most critical step in ensuring that an optimum RCN is synthesized.

The remaining chapters of the book are divided into two separate parts, classified based on the approach used. Chapters 3 through 11 in Part I present various insight-based pinch analysis techniques for RCN synthesis, covering both stages on targeting and network design. Chapters 12 through 17 in Part II, on the other hand, present mathematical optimization techniques for RCN synthesis. Note that techniques in both parts cover the most commonly found RCN problems, which covers direct material reuse/recycle, regeneration, pretreatment, TMN, IPRCNs, and BMNs. The reader is advised to read the "Notes for Readers" that precede the respective parts for an overview of the various techniques.

Readers who are interested in industrial applications should read the Appendix, where the various hypothetical and industrial problems are tracked throughout the book.

References

Alva-Argáez, A., Kokossis, A. C., and Smith, R. 1998. Wastewater minimisation of industrial systems using an integrated approach. *Computers and Chemical Engineering*, 22, S741–744.

Alva-Argáez, A., Vallianatos, A., and Kokossis, A. 1999. A multi-contaminant transhipment model for mass exchange network and wastewater minimisation problems. *Computers and Chemical Engineering*, 23, 1439–1453.

Arun, P., Banerjee, R., Bandyopadhyay, S. 2007. Sizing curve for design of isolated power systems, *Energy for Sustainable Development*, 11(4), 21–28.

Bagajewicz, M. 2000. A review of recent design procedures for water networks in refineries and process plants. *Computers and Chemical Engineering*, 24, 2093–2113.

Bandyopadhyay, S. 2010. Design and optimization of isolated energy systems through pinch analysis. *Asia-Pacific Journal of Chemical Engineering Journal*, 6(3), 518–526.

Biegler, L. T., Grossman, E. I., and Westerberg, A. W. 1997. *Systematic Methods of Chemical Engineering and Process Design.* Upper Saddle River, NJ: Prentice Hall.

Chew, I. M. L., Ng, D. K. S., Foo, D. C. Y., Tan, R. R., Majozi, T., and Gouws, J. 2008. Synthesis of direct and indirect inter-plant water nerwork. *Industrial & Engineering Chemistry Research*, 47, 9485–9496.

Dhole, V. R. and Linnhoff, B. 1993. Total site targets for fuel, co-generation, emissions, and cooling. *Computers and Chemical Engineering*, 17 (suppl.), Sl01–S109.

Douglas, J. M. 1988. *Conceptual Design of Chemical Processes.* New York: McGraw Hill.

Dunn, R. F. and El-Halwagi, M. M. 2003. Process integration technology review: Background and applications in the chemical process industry. *Journal of Chemical Technology and Biotechnology*, 78, 1011–1021.

El-Halwagi, M. M. 1997. *Pollution Prevention through Process Integration: Systematic Design Tools.* San Diego, CA: Academic Press.

El-Halwagi, M. M. 1998. Pollution prevention through process integration. *Clean Product and Processes*, 1, 5–19.

El-Halwagi, M. M. 2006. *Process Integration.* San Diego, CA: Elsevier Inc.

El-Halwagi, M. M. 2011. *Sustainable Design through Process Integration,* Butterworth-Heinemann.

El-Halwagi, M. M. and Manousiouthakis, V. 1989. Synthesis of mass exchange networks. *AIChE Journal*, 35(8), 1233–1244.

El-Halwagi, M. M. and Spriggs, H. D. 1998. Solve design puzzle with mass integration. *Chemical Engineering Progress*, 94(8), 25–44.

Foo, D. C. Y. 2009. A state-of-the-art review of pinch analysis techniques for water network synthesis. *Industrial and Engineering Chemistry Research*, 48(11), 5125–5159.

Foo, D. C. Y., Hallale, N., and Tan, R. R. 2007. Pinch analysis approach to short-term scheduling of batch reactors in multi-purpose plants. *International Journal of Chemical Reactor Engineering*, 5, A94.

Foo, D. C. Y., Hallale, N., and Tan, R. R. 2010. Optimize shift scheduling using pinch analysis. *Chemical Engineering*, 2010, 48–52.

Foo, D. C. Y., Kazantzi, V., El-Halwagi, M. M. and Manan, Z. A. 2006. Cascade analysis for targeting property-based material reuse networks, *Chemical Engineering Science*, 61(8), 2626–2642.

Foo, D. C. Y., Tan, R. R., and Ng, D. K. S. 2008a. Carbon and footprint-constrained energy sector planning using cascade analysis technique. *Energy*, 33(10), 1480–1488.

Foo, D. C. Y., Ooi, M. B. L., Tan, R. R., and Tan, J. S. 2008b. A heuristic-based algebraic targeting technique for aggregate planning in supply chains. *Computers and Chemical Engineering*, 32(10), 2217–2232.

Furman, K. C. and Sahinidis, N. V. 2002. A critical review and annotated bibliography for heat exchanger network synthesis in the 20th century. *Industrial and Engineering Chemistry Research*, 41(10), 2335–2370.

Gouws, J., Majozi, T., Foo, D. C. Y., Chen, C. L., and Lee, J.-Y. 2010. Water minimisation techniques for batch processes. *Industrial and Engineering Chemistry Research*, 49(19), 8877–8893.

Gundersen, T. and Naess, L. 1988. The synthesis of cost optimal heat exchange networks—An industrial review of the state of the art. *Computers and Chemical Engineering*, 6, 503–530.

Hallale, N. and Liu, F. 2001. Refinery hydrogen management for clean fuels production. *Advances in Environmental Research*, 6, 81–98.

Hohmann, E. C. 1971. *Optimum Networks for Heat Exchange*, PhD thesis. University of Southern California, Los Angeles, CA.

Jeżowski, J. 2010. Review of water network design methods with literature annotations. *Industrial and Engineering Chemistry Research*, 49, 4475–4516.

Kazantzi, V. and El-Halwagi, M. M. 2005. Targeting material reuse via property integration. *Chemical Engineering Progress*, 101(8), 28–37.

Kemp, I. C. 2007. *Pinch Analysis and Process Integration: A User Guide on Process Integration for the Efficient Use of Energy*. Amsterdam, the Netherlands: Elsevier, Butterworth-Heinemann.

Klemeš, J., Friedler, F., Bulatov, I., and Varbanov, P. 2010. *Sustainability in the Process Industry: Integration and Optimization*. New York: McGraw-Hill.

Kuo, W. C. J. and Smith R. 1998. Designing for the interactions between water-use and effluent treatment. *Chemical Engineering Research and Design*, 76, 287–301.

Lee, S. C., Ng, D. K. S., Foo, D. C. Y., and Tan, R. R. 2009. Extended pinch targeting techniques for carbon-constrained energy sector planning. *Applied Energy*, 86(1), 60–67.

Linnhoff, B. 1993. Pinch analysis: A state-of-art overview. *Chemical Engineering Research and Design*, 71, 503–522.

Linnhoff, B. and Flower J. R. 1978. Synthesis of heat exchanger networks. *AIChE Journal*, 24, 633–642.

Linnhoff, B., Townsend, D. W., Boland, D., Hewitt, G. F., Thomas, B. E. A., Guy, A. R., and Marshall, R. H. 1982, 1994. *A User Guide on Process Integration for the Efficient Use of Energy*. Rugby, U.K.: IChemE.

Ludwig, J., Treitz, M., Rentz, O., and Geldermann, J. 2009. Production planning by pinch analysis for biomass use in dynamic and seasonal markets. *International Journal of Production Research*, 47(8), 2079–2090.

Mann, J. G. and Liu, Y. A. 1999. *Industrial Water Reuse and Wastewater Minimization*. New York: McGraw Hill.

Ng, D. K. S., Foo, D. C. Y., and Tan, R. R. 2009a. Automated targeting technique for single-component resource conservation networks—Part 1: Direct reuse/recycle. *Industrial and Engineering Chemistry Research*, 48(16), 7637–7646.

Ng, D. K. S., Foo, D. C. Y., and Tan, R. R. 2009b. Automated targeting technique for single-component resource conservation networks—Part 2: Single pass and partitioning waste interception systems. *Industrial and Engineering Chemistry Research*, 48(16), 7647–7661.

Ng, D. K. S., Foo, D. C. Y., Tan, R. R., Pau, C. H., and Tan, Y. L. 2009c. Automated targeting for conventional and bilateral property-based resource conservation network. *Chemical Engineering Journal*, 149, 87–101.

Rockstrom, J., Steffen, W., Noone, K., Persson, A., Chapin, F. S., Lambin, E. F., Lendon, T. M et al., 2009. A safe operating space for humanity. *Nature*, 461, 472–475.

Rudd, D. F. 1968. The synthesis of system designs, I. Elementary decomposition theory. *AIChE Journal*, 14(2), 343–349.

Seider, W. D., Seader, J. D., and Lewin, D. R. 2003. *Product and Process Design Principles: Synthesis, Analysis and Evaluation*. New York: John Wiley & Sons.

Sahu, G. C. and Bandyopadhyay, S. 2011. Holistic approach for resource conservation, *Chemical Engineering World*, Dec 2011, 104–108.

Shelley, M. D. and El-Halwagi, M. M. 2000. Componentless design of recovery and allocation systems: A functionality-based clustering approach. *Computers and Chemical Engineering*, 24, 2081–2091.

Shenoy, U. V. 1995. *Heat Exchanger Network Synthesis: Process Optimization by Energy and Resource Analysis*. Houston, TX: Gulf Publishing Company.

Singhvi, A., Madhavan, K. P., and Shenoy, U. V. 2004. Pinch analysis for aggregated production planning in supply chains. *Computers and Chemical Engineering*, 28, 993–999.

Singhvi, A. and Shenoy, U. V. 2002. Aggregate planning in supply chains by pinch analysis. *Transactions of the Institute of Chemical Engineers, Part A*, 80, 597–605.

Smith, R. 1995. *Chemical Process Design*. New York: McGraw-Hill.

Smith, R. 2000. State of the art in process integration. *Applied Thermal Engineering*, 20, 1337–1345.

Smith, R. 2005. *Chemical Process: Design and Integration*. New York: John Wiley & Sons.

Tan, R. R. and Foo, D. C. Y. 2007. Pinch analysis approach to carbon-constrained energy sector planning. *Energy*, 32(8), 1422–1429.

Tan, R. R., Ng, D. K. S., and Foo, D. C. Y. 2009. Pinch analysis approach to carbon-constrained planning for sustainable power generation. *Journal of Cleaner Production*, 17, 940–944.

Towler, G. P., Mann, R., Serriere, A. J.-L., and Gabaude, C. M. D. 1996. Refinery hydrogen management: Cost analysis of chemically integrated facilities. *Industrial and Engineering Chemistry Research*, 35(7), 2378–2388.

Turton, R., Bailie, R. C., Whiting, W. B., and Shaeiwitz, J. A. 1998. *Analysis, Synthesis and Design of Chemical Processes*. Upper Saddle River, NJ: Prentice Hall.

Wang, Y. P. and Smith, R. 1994. Wastewater minimisation. *Chemical Engineering Science*, 49, 981–1006.

Wang, Y. P. and Smith, R. 1995. Wastewater minimization with flowrate constraints. *Chemical Engineering Research and Design*, 73, 889–904.

Westerberg, A. W. 1987. Process synthesis: A morphology review. In: Liu, Y. A., McGee, H. A. and Epperly, W. R. eds., *Recent Developments in Chemical Process and Plant Design*. New York: John Wiley & Sons.

Zhelev, T. K. 2005. On the integrated management of industrial resources incorporating finances. *Journal of Cleaner Production*, 13, 469–474.

2

Data Extraction for Resource Conservation

Data extraction is the most crucial step in any resource conservation activity. If the data for carrying out an analysis are wrongly extracted, we might end up synthesizing a suboptimum resource conservation network (RCN). Hence, the data extraction strategies in this chapter supersede various insight-based and optimization approaches for the rest of the chapters in this book.

As outlined in the problem statements in Chapter 1, there exist several material *sinks* in a process where the material *sources* may be recovered to (i.e., via direct reuse/recycle and/or regeneration). In this chapter, we focus on some basic principles on how to identify these process sinks and sources correctly, along with their limiting data for synthesizing an optimum RCN. Note that for an RCN that is based on mass integration, e.g., water and gas networks, the quality index for maximum recovery corresponds to the impurity concentration in the process sinks and sources. On the other hand, the corresponding quality index for property integration is the linearized operator for the property mixing rule.

2.1 Segregation for Material Sources

To maximize the potential for material recovery, it is of utmost importance to keep the individual source candidates segregated. However, this is not a common practice for a manufacturing process in general. In most cases, many material sources in the plant are normally mixed before being sent to the next processing unit. For instance, effluent streams from units that contain contaminated water are normally mixed in the equalizing tank before being fed to downstream wastewater treatment units. Note that some of these effluent streams may only contain a small amount of impurities (while some streams are more contaminated). Having all effluent streams mixed in the equalizing tank thus degrades the potential for water recovery. Hence, we have the following heuristic:

Heuristic 1—Keep material sources segregated to maximize their recovery potential.

Example 2.1 Data extraction for water minimization in acrylonitrile (AN) production

Figure 2.1 shows a process flow diagram (PFD) for AN production (El-Halwagi, 1997), one of the motivating examples in Chapter 1. AN is produced in a fluidized-bed reactor via vapor-phase ammoxidation of propylene. The reaction is single pass, with almost complete conversion of propylene. The reactor is operated at 2 atm and 450°C, with the stoichiometry shown in Equation 2.1:

$$C_3H_6 + NH_3 + 1.5O_2 \xrightarrow{\text{catalyst}} C_3H_3N + 3H_2O \tag{2.1}$$

Effluent from the reactor is cooled and partially condensed. The off-gas from the condenser is sent to a scrubber for purification, while its condensate to the biotreatment facility. Freshwater is used as the scrubbing agent in the scrubber, before the tail gas may be sent for disposal. The bottom product from the scrubber is sent to a decanter, where it is separated into aqueous and organic layers. The aqueous phase is sent to the biotreatment facility, while the organic phase (which contains most AN products) is fractionated in a mild vacuumed distillation column. The latter is induced by a steam-jet ejector, where steam condensate is produced (this is sent for biotreatment too). Besides, the distillation column bottom also produces wastewater, which is sent to the biotreatment facility. Detailed material balances for the process are given in Figure 2.1.

Due to increased customer demand, the plant authority is exploring opportunities to increase the plant's overall AN throughput. However,

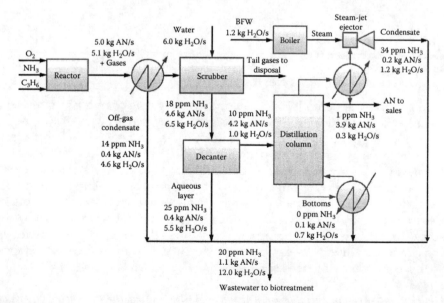

FIGURE 2.1
PFD for AN production. (From El-Halwagi, M.M., *Pollution Prevention through Process Integration: Systematic Design Tools*, Academic Press, San Diego, CA, 1997, p. 87. With permission.)

Equation 2.1 reveals that water is a by-product of AN production. Hence, an increased AN production also leads to increased wastewater flowrate to the biotreatment facility. Furthermore, the latter is currently operated at full hydraulic capacity. Hence, any increase in production capacity is only possible upon the debottlenecking of the biotreatment facility, either by reducing the total wastewater flowrate or by installing another biotreatment unit. The latter strategy is reported to be an expensive one (El-Halwagi, 1997). Thus, it has been decided to approach the problem by reducing its total wastewater flowrate, by carrying out water recovery between its process sinks and sources.

Identify the water sinks and sources to carry out a water recovery scheme for the process. Note that for this process, ammonia (NH_3) concentration is the main concern for water recovery (i.e., the quality index in this case).

Solution

From the process description, it is estimated that the two units that require water intake are boiler feed water (BFW) and scrubber. So they are designated as the *water sinks* if water recovery is to be carried out.

On the other hand, it is a common practice to take the terminal wastewater stream that is sent to the wastewater facility as a candidate for water reuse/recycle, i.e., *water source*. Note, however, that doing this does not result in the maximum water recovery potential for the process. As shown in Figure 2.1, the terminal wastewater stream has an NH_3 concentration of 20 ppm, which is actually an average concentration after the mixing of four individual wastewater streams, i.e., off-gas condensate (from condenser), aqueous layer from the decanter, distillation bottom, as well as the steam condensate from the steam-jet ejector. Among four wastewater streams, distillation bottom is the cleanest source, as it is free from ammonia content. Hence, this stream can be fully recovered. However, when the stream is mixed with other wastewater streams (and reaches a concentration of 20 ppm), its potential for recovery is degraded.

Following Heuristic 1, the individual streams that contribute to the terminal wastewater stream are segregated as water sources for recovery:

1. Distillation bottom (0.1 kg/s; 0 ppm)
2. Off-gas condensate (5.0 kg/s; 14 ppm)
3. Aqueous layer from the decanter (5.9 kg/s; 25 ppm)
4. Steam condensate from steam-jet ejector (1.4 kg/s; 34 ppm)

2.2 Extraction of Limiting Data for Material Sink for Concentration-Based RCN

To maximize the material recovery potential for a concentration-based RCN problem (e.g., water minimization), the following heuristics are useful for material sinks:

Heuristic 2—Minimize the flowrate of a sink to reduce the overall fresh resource intake.

Heuristic 3—Maximize the inlet concentration of a sink to maximize material recovery.

It is a common fact that during plant operation, processing units are normally operated within a given range for their operating parameters. For an RCN, flowrate and inlet concentration are two important parameters for any resource-consuming processes. Similar to other operating parameters, operating flowrate and concentration are also maintained within the range that is bound by their upper and lower limits, such as that shown in Figure 2.2. For the operating flowrate, the lower limit is important to ensure that the unit (e.g., scrubber, which utilizes water as scrubbing agent) is performing according to its design specification (e.g., to purify the sour gas content to a given quality). In contrast, the upper limit is always decided based on the designed capacity of the unit (e.g., to avoid flooding in the column, etc.). A similar situation also applies for inlet concentration for the processes. In this case, the lower and upper limits correspond to the desired operating concentration range that ensures normal operating conditions for the resource-consuming units.

Note, however, that experience shows that the establishment of the operating range for flowrate and concentration are often not directly measurable. One will have to rely on experience or some empirical models to predict the lower and upper limits of these variables. It is always useful to work with experienced plant personnel to identify these limits. Besides, the safety factor (10%–20% of the most severe case) should also be incorporated in choosing the operating limits. For instance, a vessel that can tolerate 100 kg/h of water throughput is always operated at its maximum limit of 90 kg/h, i.e.,

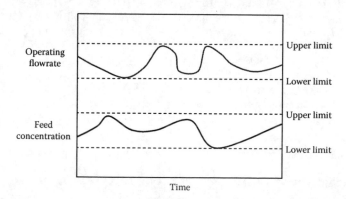

FIGURE 2.2
Historical data for plant operation. (From El-Halwagi, M.M., *Process Integration*, Academic Press, San Diego, CA, 2006, p. 42. With permission.)

90% of its design capacity. Hence, even if the upper limit has been taken for a material recovery scheme, it is still within the operating range that has incorporated the safety factor.

Following Heuristics 2 and 3 means that the lower flowrate and upper concentration limits of the process sinks are taken for material recovery. Since the flowrate and concentration values are always taken based on their operating limits, they are often known as the *limiting data*, which will enable the maximum potential to be explored for a material recovery scheme.

Example 2.2 Water minimization for AN production (revisited)

The AN process in Example 2.1 is revisited. To ensure a feasible water recovery scheme, the following process constraints on flowrate and concentration are to be complied with:

i. Scrubber

$$5.8 \leq \text{Flowrate of wash feed (kg/s)} \leq 6.2 \tag{2.2}$$

$$0.0 \leq NH_3 \text{ content of wash feed (ppm)} \leq 10.0 \tag{2.3}$$

ii. BFW

$$NH_3 \text{ content} = 0.0 \text{ ppm} \tag{2.4}$$

$$AN \text{ content} = 0.0 \text{ ppm} \tag{2.5}$$

iii. Decanter

$$10.6 \leq \text{Feed flowrate (kg/s)} \leq 11.1 \tag{2.6}$$

iv. Distillation column

$$5.2 \leq \text{Feed flowrate (kg/s)} \leq 5.7 \tag{2.7}$$

$$0.0 \leq NH_3 \text{ content of feed (ppm)} \leq 30.0 \tag{2.8}$$

$$80.0 \leq AN \text{ content of feed (wt\%)} \leq 100.0 \tag{2.9}$$

Identify the limiting data for the process sinks in the AN process to carry out a water minimization study.

Solution

Example 2.1 identifies that the two process sinks for a water minimization study are BFW and the scrubber inlet. Hence, we focus only on the

TABLE 2.1

Limiting Water Data for AN Production

j	Water Sinks, SK$_j$ Stream	Flowrate F_{SKj} (kg/s)	Concentration C_{SKj} (ppm)
1	BFW	1.2	0
2	Scrubber inlet	5.8	10

i	Water Sources, SR$_i$ Stream	Flowrate F_{SRi} (kg/s)	Concentration C_{SRi} (ppm)
1	Distillation bottoms	0.8	0
2	Off-gas condensate	5	14
3	Aqueous layer	5.9	25
4	Ejector condensate	1.4	34

constraints that are related to these sinks (i.e., decanter and distillation column will not be considered). For the BFW, Equation 2.4 indicates that no NH_3 content is tolerated in this sink; hence, an impurity of 0 ppm is taken as the limiting concentration of the latter. Note that its flowrate may be extracted from the PFD in Figure 2.1. On the other hand, Equations 2.2 and 2.3 indicate the operating range for the flowrate and concentration in this sink. Applying Heuristics 2 and 3 leads to the lower flowrate (5.8 kg/s) and upper concentration (10 ppm) limits being identified as the limiting data for this process sink. The limiting data for both sinks are summarized in Table 2.1.

Besides, Example 2.1 also identifies four water sources for the water minimization study, i.e., off-gas condensate, aqueous layer from the decanter, distillation bottom, and steam condensate from the steam-jet ejector. The limiting flowrate and concentration for these sources are also summarized in Table 2.1.

2.3 Data Extraction for Mass Exchange Processes

Many unit operations in the process industry may be categorized as mass exchange processes, e.g., washing, absorption, extraction, etc. The typical characteristic of these operations is that, mass separating agents (MSAs, e.g., water, solvent, etc.) are used to selectively remove certain components (e.g., impurities, pollutants, by-products, products) from a component-laden stream (El-Halwagi, 1997, 2006). The latter is normally termed as the *rich stream*, while the MSA is termed as the *lean stream*. Figure 2.3 shows a typical product washing operation, where water is used as an MSA (lean stream) to partially remove the impurity content from the product (rich) stream.

We may represent the product washing in Figure 2.3a as a countercurrent operation in Figure 2.3b. For most mass exchange processes, the inlet

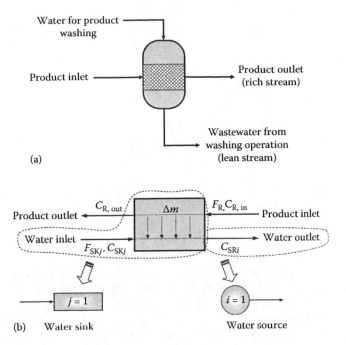

FIGURE 2.3
A product washing operation (a) for the abvoe and (b) for the below.

and outlet flowrates of the operation are normally nearly the same (i.e., with negligible flowrate losses). Hence, for a product stream with flowrate F_R, to be washed in order to reduce its impurity content from inlet ($C_{R,in}$) to outlet concentrations ($C_{R,out}$), the following mass balance equation is used to determine the impurity load (Δm) to be removed from the rich stream:

$$\Delta m = F_R (C_{R,in} - C_{R,out}). \tag{2.10}$$

Since water is used as an MSA for this operation, the inlet of the operation is hence designated as the water sink, while its outlet as the water source (see Figure 2.3b). Similarly, due to the negligible flowrate losses for the water sink and source, their flowrates are normally kept uniform (i.e., $F_{SKj} = F_{SRi}$). A material balance for the MSA to account for the impurity load (Δm) that is transferred from the rich stream is hence given as

$$\Delta m = F_{SKj}(C_{SRi} - C_{SKj}). \tag{2.11}$$

A convenient way of analyzing the mass exchange processes is the concentration versus load diagram (Wang and Smith, 1994). Such a diagram for a single washing operation in Figure 2.3 is shown in Figure 2.4. As shown, inlet and

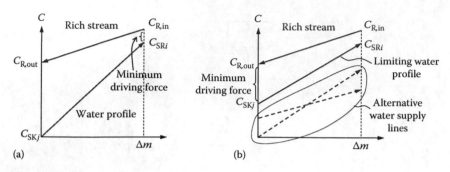

FIGURE 2.4
A product washing operation represented in concentration versus load diagram. (From *Chem. Eng. Sci.*, 49, Wang, Y.P. and Smith, R., Wastewater minimisation, 981–1006, 1994. Copyright 1994, with permission from Elsevier.)

outlet concentrations of the rich stream as well as those of the water stream (known as the *water profile*) are plotted versus the impurity load transferred between them. If pure freshwater (i.e., zero impurity) is used in this operation, the water profile is drawn from the origin, such as that in Figure 2.4a. To minimize the freshwater flowrate, the slope of the water profile is kept to the maximum, so that the water stream will exit at the highest possible outlet concentration. Note that the latter value is normally separated from the inlet concentration of the rich stream by the minimum driving force, to ensure feasible mass transfer between the two streams. This is shown in Figure 2.4a.

However, doing this only ensures the use of the minimum freshwater flowrate in the washing operation but does not include the consideration of water reuse/recycle, since the inlet concentration is set at zero. Reduction of freshwater is only possible when the washing operation allows the recycling of effluent from the same unit or the reuse of effluent from other units (when there is more than one water-using process). For such a case, the inlet concentration of the washing process should be raised to a higher value, such as that shown in Figure 2.4b. Similar to the outlet concentration, the inlet concentration should take the maximum possible value that is bounded by the minimum mass transfer driving force at the lean end of the concentration versus load diagram to ensure feasible mass transfer, as shown in Figure 2.4b. Note also that this corresponds to Heuristic 3 discussed earlier, i.e., *maximize the inlet concentration of a sink to maximize material recovery*.

When inlet and outlet concentrations of the water profile are both set to their limiting (maximum) values, the latter is termed as the *limiting water profile*. As the name implies, the limiting water profile dictates that the washing operation stream is operated at its minimum flowrate for its maximum inlet and outlet concentrations. Note that since the limiting water profile represents the boundary of the operating concentrations, any alternative water supply lines below the limiting water profile are also feasible in operating the washing process (see Figure 2.4b). However, only when the process is operated at the condition of the limiting water profile (i.e., maximum inlet and outlet concentrations), the maximum water recovery potential for the overall process will be ensured.*

It is also worth noting that the earlier discussion is not restricted to washing operations alone. The same principle applies to all mass exchange processes where MSA is employed, such as absorption, extraction, etc. The limiting profile can be extended to identify the minimum flowrate needed for the MSA(s) in these operations.

Example 2.3 Data extraction for extraction

Figure 2.5 shows an extraction process where an organic solvent (lean stream) is used to purify a product stream that contains an unwanted impurity (rich stream). It may be assumed that there is a negligible flowrate loss in both streams. For all the following scenarios, determine the missing parameters (F_{SK}, C_{SK}, and/or C_{SR}) for the profile of the organic solvent stream:

1. Scenario 1
   - $F_{SK} = 10\,\text{kg/s}$
   - $C_{SK} = 0\ \text{wt\%}$ (i.e., fresh solvent is used)
2. Scenario 2
   - $F_{SK} = 10\,\text{kg/s}$
   - $0\ \text{wt\%} \leq C_{SK} \leq 0.005\ \text{wt\%}$
3. Scenario 3
   - $0\ \text{wt\%} \leq C_{SK} \leq 0.005\ \text{wt\%}$
   - $0\ \text{wt\%} \leq C_{SR} \leq 0.030\ \text{wt\%}$

Solution

For all scenarios, the impurity load is first determined using Equation 2.10 as follows:

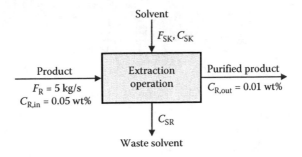

FIGURE 2.5
An extraction operation.

---

* This condition applies to RCN of single impurity case, and also to one of the impurity for a multiple impurity RCN, see discussion in Savelski and Bagajewicz (2000, 2003).

$$5 \text{ kg/s} (0.0005 - 0.0001) = 0.0020 \text{ kg/s}$$

For Scenario 1, since the inlet flowrate and concentration are fixed, we shall determine the outlet concentration of the waste solvent (C_{SR}) using Equation 2.11:

$$0.0020 \text{ kg/s} = 10 \text{ kg/s} (C_{SR} - 0)$$

$$C_{SR} = 0.0002 = 0.02 \text{ wt}\%$$

For Scenario 2, the earlier discussion indicates that the inlet concentration of an extraction operation should be taken at a higher value for maximizing solvent recovery. Hence, for an inlet concentration of 0.005 wt%, the outlet concentration of the waste solvent (C_{SR}) is then determined using Equation 2.11 as

$$0.0020 \text{ kg/s} = 10 \text{ kg/s} (C_{SR} - 0.00005)$$

$$C_{SR} = 0.00025 = 0.025 \text{ wt}\%$$

In Scenario 3, the inlet and outlet concentrations of the solvent stream are set at the limiting values, and the corresponding limiting solvent flowrate can then be determined as

$$0.0020 \text{ kg/s} = F_{SK} (0.00030 - 0.00005)$$

$$F_{SK} = 8 \text{ kg/s}$$

The concentration versus load diagrams for all scenarios are shown in Figure 2.6. For Scenarios 1 and 2, since the solvent flowrate is fixed, we can only alter its inlet and outlet concentrations. Comparing with Scenario 1,

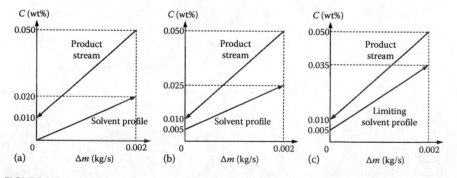

FIGURE 2.6
Concentration versus load diagrams for Example 2.3: (a) Scenario 1, (b) Scenario 2, and (c) Scenario 3—limiting solvent profile.

the outlet concentration in Scenario 2 is increased by 0.005 wt%. We will have to ensure that it is within the maximum possible value for feasible mass transfer, as discussed in Figure 2.4. For Scenario 3, we obtain the *limiting solvent profile*, where the solvent flowrate is minimized at its limiting inlet and outlet concentrations. This ensures that the solvent flowrate through the process is minimum (i.e. with steepest solvent profile).

2.4 Data Extraction for Hydrogen-Consuming Units in Refinery

Significant amounts of hydrogen are utilized in refinery operations. Typical hydrogen-consuming units are hydrotreating and hydrocracking processes. In these processes, hydrogen gas is used to remove impurities (e.g., sulfur, nitrogen, etc.) from the crude oil stream. Since there are many hydrogen-consuming units in treating various crude products within a refinery, we may explore the opportunity of hydrogen recovery among these units. However, the extraction of data for hydrogen recovery is not as straightforward as compared with the previously discussed resource-consuming processes. In this section, we shall focus on the general operation of a typical hydrogen-consuming unit and show how data extraction could be carried out.

Figure 2.7 shows a simplified PFD for a hydrogen-consuming unit (Hallale and Liu, 2001; Alves and Towler, 2002). A liquid hydrocarbon feed is mixed with a hydrogen-rich gas stream, before being heated and sent to the hydrotreating/hydrocracking reactor. In the reactor, some hydrogen is consumed, while other light hydrocarbon (e.g., methane, ethane, etc.) and

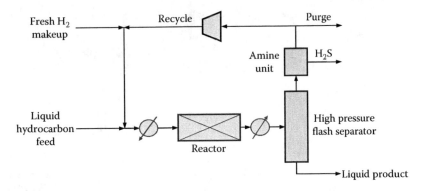

FIGURE 2.7
A simplified PFD for a hydrogen-consuming unit. (Reprinted with permission from Alves, J.J. and Towler, G.P., Analysis of refinery hydrogen distribution systems, *Ind. Eng. Chem. Res.*, 41, 5759–5769, 2002. Copyright 2002 American Chemical Society.)

gaseous components (e.g., H_2S, NH_3) are formed. The reactor effluent is cooled and sent to a high-pressure flash separator. The bottom liquid product of the separator is normally sent to a low-pressure separator for further purification. The top stream from the flash separator contains a significant amount of hydrogen, which is always sent for recycling. Prior to recycling, the gas stream is treated with an amine unit for H_2S removal. A small portion of the treated gas stream is then purged (normally to fuel or flare systems) to prevent the buildup of the contaminant, while the remainder of the gas stream is recompressed and recycled to the reactor, with some fresh hydrogen makeup.

In analyzing the hydrogen-consuming unit, we may mistakenly think that the purge stream in the recycle loop is the hydrogen source (to be sent for reuse/recycle), while the hydrogen sink corresponds to the fresh hydrogen makeup. However, this is not the case. The actual sink for a hydrogen-consuming unit is its net requirement of hydrogen that is fed to the hydrotreating/hydrocracking reactor. In other words, this corresponds to the hydrogen gas stream that is mixed with the liquid hydrocarbon feed (Figure 2.8). Since this stream is a mixture of fresh hydrogen makeup and recycle streams, we can perform simple mass balance to determine its flowrate and concentration, once the flowrates and concentrations of the makeup and recycle streams are known.

Note that the individual streams in a hydrogen-consuming unit are normally characterized in terms of its hydrogen purity (mol% or vol%). However, it is often more convenient to carry out the analysis by characterizing its impurity concentration, which is the collective impurity content in the stream. For the given flowrates and impurity concentrations of the makeup and recycle streams, the flowrate and impurity concentration of the hydrogen sink may be determined using Equations 2.12 and 2.13, respectively.

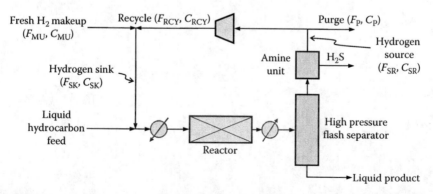

FIGURE 2.8
Identification of hydrogen sink and source. (From *Adv. Environ. Res.*, 6, Hallale, N. and Liu, F., Refinery hydrogen management for clean fuels production, 81–98, 2001. Copyright 2001, with permission from Elsevier.)

$$F_{SK} = F_M + F_R \tag{2.12}$$

$$C_{SK} = \frac{F_M C_M + F_R C_R}{F_M + F_R} \tag{2.13}$$

where F_{SK}, F_{MU}, and F_{RCY} are the respective flowrates of the hydrogen sink, makeup, and recycle streams, with their respective impurity concentrations of C_{SK}, C_{MU}, and C_{RCY}.

On the other hand, the actual source for a hydrogen-consuming unit is the stream that emits from the amine unit (i.e., after being treated for H_2S removal; see Figure 2.8). This stream is taken as the source as it contains the entire flowrate of hydrogen that may be considered for potential reuse/recycle. In other words, the summation of the recycle and purge stream flowrates gives the total flowrate of the hydrogen source (Equation 2.14), while the impurity concentrations of all these streams are essentially the same, given as in Equation 2.15:

$$F_{SK} = F_{RCY} + F_P \tag{2.14}$$

$$C_{SR} = C_{RCY} = C_P \tag{2.15}$$

Example 2.4 Data extraction for a refinery hydrogen network

Figure 2.9 shows a PFD for a refinery hydrogen network (Hallale and Liu, 2001), one of the motivating example in Chapter 1. At present, two

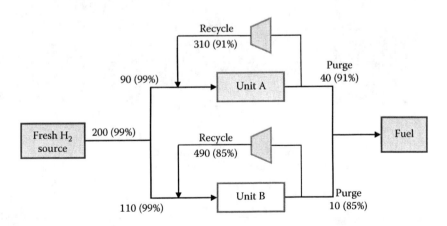

FIGURE 2.9
PFD of a refinery hydrogen network (stream flowrate given in million standard cubic feet/day—MMscfd; its hydrogen purity is given in parenthesis). (From *Adv. Environ. Res.*, 6, Hallale, N. and Liu, F., Refinery hydrogen management for clean fuels production, 81–98, 2001. Copyright 2001, with permission from Elsevier.)

TABLE 2.2

Data for Example 2.4

	Makeup Stream	Recycle Stream	Purge Stream	Hydrogen Sink	Hydrogen Source
Unit A					
Flowrate (MMscfd)	90	310	40	400	350
Concentration (%)	1	9	9	7.20	9
Unit B					
Flowrate (MMscfd)	110	490	10	600	500
Concentration (%)	1	15	15	12.43	15

hydrogen-consuming processes, i.e. Units A and B are fed by a fresh hydrogen source; and purge a significant amount of unused hydrogen as fuel. Note that both units have an existing internal hydrogen recycle stream.

Extract the data for this case to carry out a hydrogen recovery analysis.

Solution

In the first step, all hydrogen purity in the makeup, recycle, and purge streams of both Units A and B are converted into impurity concentration, given in columns 2–4 in Table 2.2.

To identify the flowrate ($F_{SK,A}$) and concentration ($C_{SK,A}$) for the hydrogen sink of Unit A, Equations 2.12 and 2.13 are utilized, i.e.,

$$F_{SK,A} = 90 + 310 = 400 \text{ MMscfd}$$

$$C_{SK,A} = \frac{90(1) + 310(9)}{90 + 310} = 7.2\%$$

Equations 2.14 and 2.15 are then utilized to calculate the flowrate ($F_{SR,A}$) and concentration ($C_{SR,A}$) for hydrogen source of Unit A, i.e.,

$$F_{SR,A} = 310 + 40 = 350 \text{ MMscfd}$$

$$C_{SR,A} = 9\%$$

Similarly, flowrates and concentrations for the hydrogen sinks and sources for Unit B are also calculated using Equations 2.12 through 2.15. Data for both sinks are summarized in the last two columns of Table 2.2.

2.5 Data Extraction for Property Integration

Unlike water minimization and gas recovery, which are both concentration based, property integration involves properties or functionalities of streams

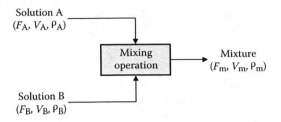

FIGURE 2.10
Mixing of solutions with different density.

that are not driven by their chemical constituents. Hence, to extract the correct limiting data for a property integration study, the basic background of the property mixing rule needs to be discussed. This is shown with an example in Figure 2.10, which involves the mixing of solutions A and B, with respective density ρ_A and ρ_B, mass flowrate F_A and F_B, and volumetric flowrate V_A and V_B. The resulting mixture consists of mass (F_M) and volumetric flowrates (V_M) and a mean density of ρ_M.

Assuming that the volumetric flowrate of the mixture is the sum of the individual solution flowrates, i.e.,

$$V_M = V_A + V_B. \tag{2.16}$$

Equation 2.16 may be written as

$$\frac{F_M}{\rho_M} = \frac{F_A}{\rho_A} + \frac{F_B}{\rho_B} \tag{2.17a}$$

and rearranged as

$$\frac{1}{\rho_M} = x_A \frac{1}{\rho_A} + x_B \frac{1}{\rho_B} \tag{2.17b}$$

where x_A and x_B are the mass fraction of solutions A and B in the mixture, respectively. A more generic form of this *mixing rule* is given as

$$\frac{1}{\rho_M} = \sum_i \frac{x_i}{\rho_i} \tag{2.17c}$$

where ρ_i is the density of solution i.

Note that Equation 2.17c is a nonlinear equation, which may lead to a complication in calculation (especially when the equation becomes highly nonlinear). A more convenient way of representation is its linearized form, given as

$$\psi_M (\rho_M) = \sum_i x_i \psi_i \left(\frac{1}{\rho_i} \right) \tag{2.18}$$

where $\psi_M(\rho_M)$ and $\psi_i(1/\rho_i)$ are the linearizing operators for the density of mixture and solution i, respectively. In other words, the nonlinear mixing rule in Equation 2.17c is transformed into a linear mixing rule given as Equation 2.18.

The mixing example in Figure 2.10 is revisited to verify that both Equations 2.19 and 2.20 are equivalent. Assume that both solutions A and B have a mass flow of 1 kg each. The density values for these solvents are given as 1000 and 1500 kg/m³, respectively. We can determine from Equation 2.17c or 2.18 that the mixture density is equivalent to 1200 kg/m³ (= 2/[1/1000 + 1/1500]). Alternatively, we can also calculate the density operator of solvents A and B as 0.001 (= 1/1000 kg/m³) and 0.00067 m³/kg (= 1/1500 kg/m³), respectively. Note that in this case, the operators become the specific gravity of the solvents. However, in some other property integration problems, the operator values may be meaningless (e.g. Examples 2.5 and 2.6). From Equation 2.19, the density operator (specific gravity) of the mixture is calculated as 0.00083 m³/kg (= 0.5(0.001) + 0.5(0.00067)) which is equivalent to the density value of 1200 kg/m³ (= 1/0.00083 kg/m³).

Similarly, we may apply the same linearizing principle for a wide range of properties P for any given mixing rule patterns. A generic form of the linearized mixing rule is given as

$$\psi_M (P_M) = \sum_i x_i \psi_i (P_i) \tag{2.19}$$

where $\psi_M(P_M)$ and $\psi_i(P_i)$ are the linearizing operators for property P in the mixture and individual stream i, respectively.

It is interesting to note that Equation 2.19 is also applicable for concentration-based water minimization and gas recovery problems. For instance, when two water streams of equal flowrate but with respective impurity concentrations of 20 and 40 ppm are mixed, Equation 2.19 determines that the resulting mixture will have a concentration (C_M) of

$$C_M = 0.5(20) + 0.5(40) = 30 \text{ ppm.}$$

In this case, the property operator of $\psi(C)$, is equivalent to concentration. For simplicity, $\psi(P)$ will be denoted as ψ in this book. Operators for some of the common nonlinear properties are summarized in Table 2.3. Note that these operators may be determined empirically, or from first principles.

Apart from identifying the linearized mixing rule for properties, we need to define the correct limiting flowrate for the material sink in a property integration problem. This relates to the property operator of the fresh resource in the given problem. To maximize material recovery in a property integration problem, Heuristic 3 is modified as follows.

TABLE 2.3

Operator Expressions for Nonlinear Properties

Property of Mixture, P_M	Mixing Rule	Operator
Density, ρ_M	$\dfrac{1}{\rho_M} = \sum_i \dfrac{x_i}{\rho_i}$	$\psi = \dfrac{1}{\rho_i}$
Reid vapor pressure, RVP_M	$RVP_M = \sum_i x_i RVP_i^{1.44}$	$\psi = RVP_i^{1.44}$
Electric resistivity, R_M	$\dfrac{1}{R_M} = \sum_i \dfrac{x_i}{R_i}$	$\psi = \dfrac{1}{R_i}$
Viscosity, μ_M	$\log(\mu_M) = \sum_i x_i \log(\mu_i)$	$\psi = \log(\mu_i)$
Paper reflectivity, $\bar{R}_\infty$	$R_{\infty,M} = \sum_i x_i R_{\infty,i}^{5.92}$	$\psi = R_{\infty,i}^{5.92}$

Source: From *Chem. Eng. Sci.*, 61(8), Foo, D.C.Y., Kazantzi, V., El-Halwagi, M.M., and Manan, Z.A., Cascade analysis for targeting property-based material reuse networks, 2626–2642, 2006b. Copyright 2006, with permission from Elsevier.

Heuristic 4—Choose the property operator of the sink that is furthest from the operator of the fresh resource.

The main reason behind Heuristic 4 is that, there exist two distinct cases for the property integration problems with respect to the property operator value of the fresh resource. In the first case, the fresh resource has an operator value that is *superior* among all process sources. This is similar to the concentration-based water minimization and gas recovery problems, where freshwater/gas resource is always available at the lowest concentration among all sources. Hence, the limiting inlet property operator for the material sinks is set to the upper bound for a given operating range. In contrast, when a fresh resource has an operator value that is *inferior* among all sources, the limiting inlet operator for the material sink will take the lower bound of the given operating range. Doing this ensures the maximum recovery of process sources or the minimum usage of fresh resource(s).

Example 2.5 Data extraction for metal degreasing process

Figure 2.11 shows the PFD of the metal degreasing process (motivating example in Chapter 1) where fresh organic solvent is used in the degreasing and absorption units (Kazantzi and El-Halwagi, 2005; Foo et al., 2006a). The treated metal from the degreasing unit is sent to metal finishing to produce the final product. On the other hand, the used solvent from the degreasing unit is sent for reactive thermal processing for regeneration, by decomposing its grease and organic additives content. The regenerated solvent exits as the liquid product from the thermal processing unit and is then recycled to the degreasing unit, while the off-gas from the former

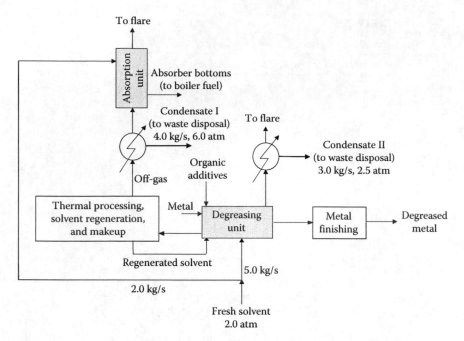

FIGURE 2.11
PFD of a metal degreasing process. (Reproduced from Kazantzi, V. and El-Halwagi, M.M., *Chem. Eng. Prog.*, 101(8), 28, 2005. With permission. Copyright 2005 American Institute of Chemical Engineers (AIChE).)

(which contains solvent) is treated in a condenser and then an absorption unit before being sent for flare. Besides, another off-gas stream leaving the degreasing unit is also passed through a condenser before it is flared. Two generated condensate streams are observed to contain some solvent (source), which may be recovered to the process sinks (i.e., degreasing and absorption units) to reduce the consumption of fresh solvent.

The main property of the solvent that dictates its extent of recovery is the Reid vapor pressure (RVP), which is important in characterizing the volatility (and, indirectly, the composition) of the solvent. The general mixing rule for the RVP is given in Equation 2.20:

$$RVP_M = \sum_i x_i RVP_i^{1.44} \qquad (2.20)$$

The fresh solvent has an RVP value of 2.0 atm; the RVP values for the condensate streams are given in Figure 2.11. The following process constraints on flowrate and RVP values are to be complied with for the process sinks, i.e.,

i. Degreasing unit

Flowrate of solvent = 5.0 kg/s   (2.21)

$$2.0 \leq \text{RVP of solvent (atm)} \leq 3.0 \qquad (2.22)$$

ii. Absorption unit

$$\text{Flowrate of solvent} = 2.0 \text{ kg/s} \qquad (2.23)$$

$$2.0 \leq \text{RVP of solvent (atm)} \leq 4.0 \qquad (2.24)$$

Solution

Table 2.3 shows that the property operator of the RVP is given as

$$\psi = \text{RVP}_i^{1.44} \qquad (2.25)$$

Equation 2.25 is first used to calculate the operator values of the fresh solvent, as well as Condensates I and II, i.e., 2.71, 13.20, and 3.74 atm$^{1.44}$, respectively.* It is observed that the fresh solvent has a superior operator value among all process sources. Similarly, Equation 2.25 is used to convert the operating boundary of the RVP values of the process sinks. Hence, the revised form of Equations 2.22 and 2.24 are given in their operator values as follows:

$$2.71 \leq \psi \quad \text{for degreasing unit (atm}^{1.44}) \leq 4.86 \qquad (2.26)$$

$$2.71 \leq \psi \quad \text{for absorption unit (atm}^{1.44}) \leq 7.36 \qquad (2.27)$$

Following Heuristic 4, the upper limit of the property operator operating range of both sinks is taken as the limiting data for the problem, i.e., 4.86 (Condensate I) and 7.36 atm$^{1.44}$ (Condensate II), respectively. A summary of the limiting data for all process sinks and sources, along with the fresh solvent, is given in Table 2.4.

TABLE 2.4

Limiting Data for Example 2.5

Sink	F_{SKj} (kg/s)	ψ_{SKj} (atm$^{1.44}$)
Degreasing unit	5	4.86
Absorption unit	2	7.36
Source	F_{SRi} (kg/s)	ψ_{SRi} (atm$^{1.44}$)
Condensate I	4	13.20
Condensate II	3	3.74
Fresh solvent	To be determined	2.71

* This is a factitious unit without any physical meaning.

Example 2.6 Data extraction for papermaking process

Figure 2.12 shows the PFD of a Kraft papermaking process (Kazantzi and El-Halwagi, 2005; El-Halwagi, 2006; Foo et al., 2006a). Wood chips are mixed with white liquor in the Kraft digester, where they are digested and treated chemically. The black liquor from the digester is recycled, while the produced pulp is then sent to the bleaching section. Next, the bleached fiber is supplied to Paper Machines I and II to be converted into two grades of paper products. An external fresh fiber source is also fed to both paper machines to supplement its fiber need. A portion of the off-spec product is rejected (termed as *reject*). This latter is sent for treatment in the hydro pulper and hydro sieve units. Two effluents are produced from these units, i.e., an underflow and an overflow waste that is termed as *broke*. The former is normally sent for burning, while the broke is sent for waste treatment. It is also observed that the broke contains some fiber content, which may be recovered to the paper machine.

A paper product property known as *reflectivity* (R_∞) is the quality of concern in determining the extent of recovery for the broke. Reflectivity is defined as the reflectance of an infinitely thick material compared with an absolute standard, i.e., magnesium oxide, and is dimensionless. The mixing rule for reflectivity R_∞ is given as

$$R_{\infty,M} = \sum_i x_i R_{\infty,i}^{5.92} \tag{2.28}$$

Reflectivity index for the three fiber sources, i.e., fresh fiber, bleached fiber, and broke, is given as 0.95, 0.88, and 0.75, respectively. The process sources, i.e., bleached fiber and broke, are maximized before the fresh fiber is used. The present operating flowrates of these fiber sources are shown in Figure 2.12.

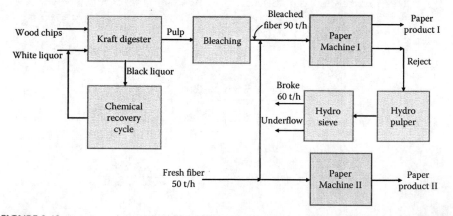

FIGURE 2.12
PFD of a papermaking process. (Reproduced from Kazantzi, V. and El-Halwagi, M.M., *Chem. Eng. Prog.*, 101(8), 28, 2005. With permission. Copyright 2005 American Institute of Chemical Engineers (AIChE).)

On the other hand, the following process constraints on flowrate and fiber reflectivity are to be complied with for process sinks in recovering fiber, i.e.,

   i. Paper Machine I

     Fiber requirement = 100 t/h$\qquad$(2.29)

     $0.85 \leq$ Fiber reflectivity (dimensionless) $\leq 0.95$$\qquad$(2.30)

  ii. Paper Machine II

     Fiber requirement = 40 t/h$\qquad$(2.31)

     $0.90 \leq$ Fiber reflectivity (dimensionless) $\leq 0.95$$\qquad$(2.32)

Solution

Table 2.3 shows that the property operator of reflectivity is given as

$$\psi = R_{\infty,i}^{5.92} \qquad (2.33)$$

Operator values of the fresh and bleached fiber, as well as the broke, are determined using Equation 2.33 as 0.74, 0.47, and 0.18, respectively. It is observed that the fresh fiber has an inferior operator value among all process sources. Similarly, the operating boundary of reflectivity of the process sinks are converted into their respective operator values using Equation 2.33 as follows:

   $0.38 \leq \psi$ for Paper Machine I (dimensionless) $\leq 0.74$$\qquad$(2.34)

   $0.54 \leq \psi$ for Paper Machine II (dimensionless) $\leq 0.74$$\qquad$(2.35)

Following Heuristic 4, the lower limit of the property operator operating range of both sinks is taken as the limiting data for the problem, i.e., 0.38 (Paper Machine I) and 0.54 atm$^{1.44}$ (Paper Machine II), respectively. A summary of the limiting data for all process sinks and sources, along with the fresh solvent, is given in Table 2.5.

TABLE 2.5

Limiting Data for Example 2.6

Sink	F_{SKj} (t/h)	ψ_{SKj}
Machine I	100	0.382
Machine II	40	0.536

Source	F_{SRi} (t/h)	ψ_{SRi}
Bleached fiber	90	0.469
Broke	60	0.182
Fresh fiber	To be determined	0.738

2.6 Additional Readings

An illustration of the impact of wrongly extracted limiting data is found in Foo et al. (2006b), focusing on water network synthesis of the fixed flowrate problems.* In essence, the work shows that when the wrong limit of the operating range is taken as the limiting data, the synthesized RCN will experience higher consumption of fresh resource as well as higher waste discharge.

Problems

Water Minimization Problems

2.1   Figure 2.13 shows a tire-to-fuel process, where scrap tires are converted into energy (El-Halwagi, 1997; Noureldin and El-Halwagi, 1999). Scrap tires are shred by a high-pressure water-jet before they are fed to the pyrolysis reactor. The spent water from the shredding unit is then filtered, recompressed, and recycled. Wet cake from the filtration unit is sent to the solid waste handling section. To compensate water loss in the filtered wet cake, freshwater is added to the recycled water loop at the water-jet compression station.

   In the pyrolysis reactor, the shredded tires are turned into oils and gaseous fuels. The oils are sent for separation and finishing to

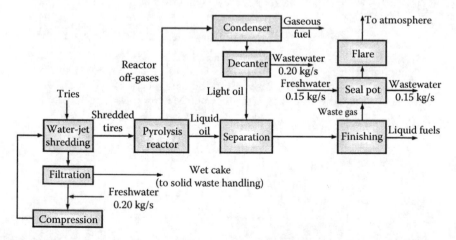

FIGURE 2.13
Tire-to-fuel process. (From El-Halwagi, M.M., *Pollution Prevention through Process Integration: Systematic Design Tools*, Academic Press, San Diego, CA, 1997, p. 97. With permission.)

---

* Reader may also solve Problems 4.1 and 4.2 to see the consequence of wrong data extraction.

produce liquid transportation fuels. Off-gases from the reactor are sent to a condenser to separate its light oil content from the gaseous fuel. The condensate is sent to a decanter where a two-phase liquid is produced. The organic layer from the decanter is mixed with the liquid oil from the reactor for separation and finishing, while the aqueous layer forms the wastewater effluent. On the other hand, a waste gas emits from the finishing section and is sent for flare. A seal pot is used to prevent the back-propagation of fire from the flare, by passing through a water stream to form a buffer zone between the fire and the source of the flare gas.

Two wastewater streams are observed in the process, i.e., effluent from the decanter and seal pot. These sources may be recovered to the process water sinks, i.e., seal pot and water-jet compression station, in order to reduce freshwater consumption. To evaluate the potential of water recovery, heavy organic content of the wastewater streams is the impurity that is of main concern. For the water sinks, the following constraints on feed flowrate and impurity concentration should be satisfied:

i. Seal pot:

$$0.10 \leq \text{Flowrate of feed water (kg/s)} \leq 0.20 \tag{2.36}$$

$$0 \leq \text{Impurity concentration of feed water (ppm)} \leq 500 \tag{2.37}$$

ii. Makeup to water-jet compression station:

$$0.18 \leq \text{Flowrate of feed water (kg/s)} \leq 0.20 \tag{2.38}$$

$$0 \leq \text{Impurity concentration of makeup water (ppm)} \leq 50 \tag{2.39}$$

On the other hand, the water sources have the following operating condition:

i. Decanter:

$$\text{Wastewater flowrate} = 0.20 \text{ kg/s} \tag{2.40}$$

$$\text{Impurity concentration} = 500 \text{ ppm} \tag{2.41}$$

ii. Seal pot:

$$\text{Wastewater flowrate} = \text{flowrate of its water sink}$$
$$\text{(given as in Equation 2.36)} \tag{2.42}$$

$$\text{Impurity concentration} = 200 \text{ ppm} \tag{2.43}$$

Note that Equation 2.42 indicates that the water sink and source of the seal pot will always have uniform flowrates. Identify the limiting data for water recovery for this process (see answers in Problem 3.1).

2.2   Figure 2.14 shows a Kraft pulping process where paper product is produced from wood chips (El-Halwagi, 1997). The wood chips (with 50% water content) are fed with steam in both chip bin and presteaming unit. The steaming process will assist the impregnation of chips with chemicals at the latter processing steps. The mixture from presteaming unit is then sent to a digester where white liquor (a chemical mixture) is used to solubilize lignin in the wood chips, which is commonly termed as the *cooking* operation. By-products from the cooking operation include methanol, turpentine, etc. The cooked chips are then sent for washing in a countercurrent multistage washer, where cooking chemicals are removed. The washed product is then sent for concentration and the finishing units to produce the paper product. In the latter units, wastewater streams (W13 and W14) are formed.

Extraction liquor from the digester is sent for two-stage flashing operations. Steams generated from these flash units are recovered to the chip bin and presteaming, respectively. The liquid portion from the second flash tank is cooled and sent to the multistage evaporation unit, where a large amount of combined wastewater stream (W12) is produced from the condensate of each effect of the evaporation unit.

Vapor stream from the first flash unit is sent to the primary condenser, along with the steam from the presteaming unit. The noncondensable material from the primary condenser is sent to the secondary condenser, where most gases are condensed. The condensate is sent to the turpentine decanter, where turpentine is recovered. The leftover noncondensable material from the secondary condenser is finally condensed by cooling water in the direct contact condenser. The condensates from these units generate several wastewater streams, i.e., W6 and W7.

Due to the use of a significant amount of water in the Kraft pulping process, water recovery is considered to reduce the overall process freshwater and wastewater flowrates. For this process, methanol concentration is the main concern. The flowrates and methanol concentrations of the individual wastewater streams are shown in Figure 2.14. These water sources may be recovered to the water sinks, i.e., countercurrent multistage washing, chemical recovery, as well as the direct contact condenser.

The following constraints on feed flowrate and methanol concentration should be satisfied for the water sinks while considering water recovery:

i.   Countercurrent multistage washing:

Flowrate of wash water = 467 t/h   (2.44)

Methanol concentration in wash water ≤ 20 ppm   (2.45)

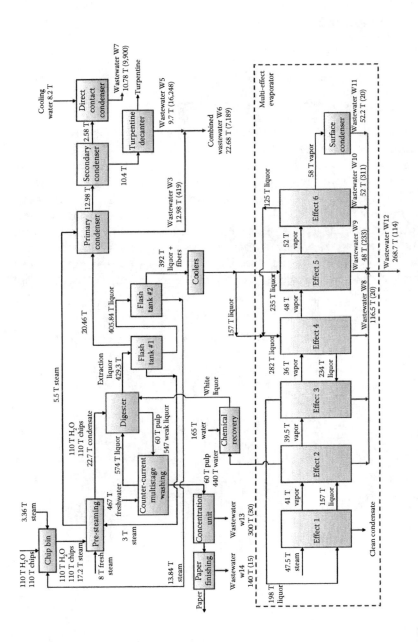

FIGURE 2.14

A Kraft pulping process (basis: 1 h; T refers to ton; values in parenthesis indicate methanol concentration in ppm). (From Hamad, A.A. et al., Systematic integration of source reduction and recycle reuse for cost-effective compliance with the cluster rules, in *AIChE Annual Meeting, Miami, FL*, 1995. With permission.)

ii. Chemical recovery

$$165 \leq \text{Flowrate of feed water (t/h)} \leq 180 \tag{2.46}$$

$$\text{Methanol concentration in feed water} \leq 20 \text{ ppm} \tag{2.47}$$

iii. Direct contact condenser

$$\text{Flowrate of feed water} = 8.2 \text{ t/h} \tag{2.48}$$

$$\text{Methanol concentration in feed water} \leq 10 \text{ ppm} \tag{2.49}$$

Identify the limiting data for water recovery for this process (see answer in Problem 4.1).

2.3 Figure 2.15 shows a manufacturing process where tricresyl phosphate is produced from cresol and phosphorous oxychloride (El-Halwagi, 1997). The reactants are mixed and heated before being fed to the reactor. Inert gas is added as diluent in the reactor. Effluent from the reactor is sent to a flash vessel. Liquid product from the flash is cooled and sent to a two-stage washing operation to remove the cresol content in the product stream. On the other hand, the vapor stream from the flash unit is sent to a two-stage scrubbing operation for purification, before

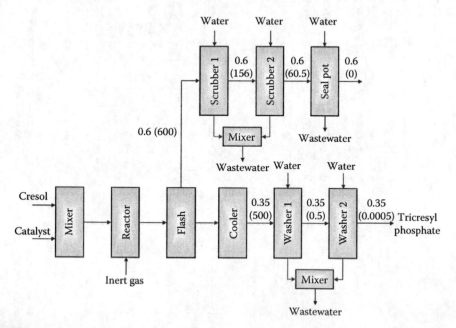

FIGURE 2.15
Tricresyl phosphate manufacturing process (stream flowrates given in kg/s; values in parenthesis indicate cresol concentration in ppm). (From Hamad, A.A. et al., Optimal design of hybrid separation systems for in-plant waste reduction, in *Proceedings of the Fifth World Congress of Chemical Engineering*, San Diego, CA, Vol. III, pp. 453–458, 1996. With permission.)

it is sent for flare. A seal pot is used between scrubbing operation and flare to prevent backward propagation of fire.

It is observed that a significant amount of water is used in the water-using processes, i.e., washing and scrubbing operations, as well as in the seal port. Hence, water recovery may be considered in order to reduce the freshwater consumption of the process. In considering water recovery, cresol concentration is the impurity that is of main concern for the process sinks. Since all water-using processes may be modeled as mass transfer operations, it is appropriate to take the inlet streams of these units as water sinks and their outlet streams as sources. For the water sinks, the following constraints are to be complied during water recovery:

i. Washer 1

$$\text{Flowrate of wash water} = 2.45 \text{ kg/s} \tag{2.50}$$

$$0 \leq \text{Cresol concentration in wash water (ppm)} \leq 5 \tag{2.51}$$

ii. Washer 2

$$\text{Flowrate of wash water} = 2.45 \text{ kg/s} \tag{2.52}$$

$$\text{Cresol concentration in wash water} = 0 \text{ ppm} \tag{2.53}$$

iii. Scrubber 1

$$0.7 \leq \text{Flowrate of feed water (kg/s)} \leq 0.84 \tag{2.54}$$

$$0 \leq \text{Cresol concentration in feed water (ppm)} \leq 30 \tag{2.55}$$

iv. Scrubber 2

$$0.5 \leq \text{Flowrate of feed water (kg/s)} \leq 0.6 \tag{2.56}$$

$$0 \leq \text{Cresol concentration in feed water (ppm)} \leq 30 \tag{2.57}$$

v. Seal pot

$$0.2 \leq \text{Flowrate of feed water (kg/s)} \leq 0.25 \tag{2.58}$$

$$0 \leq \text{Cresol concentration in feed water (ppm)} \leq 100 \tag{2.59}$$

On the other hand, the source flowrate of each unit is to be kept uniform with that of the sink. Besides, the cresol concentration of the water sources is to be determined through mass balance, i.e., based on the cresol content of the product stream/exhaust gas that is sent to the water-using processes (hint: Equations 2.10 and 2.11 are to be used here).

Identify the limiting data for water recovery for this process (see answer in Problem 4.2).

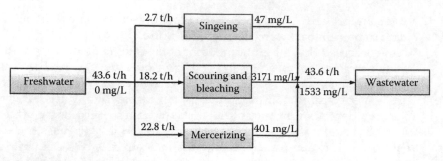

FIGURE 2.16
Water-using scheme for the bleaching section in a textile plant. (From Ujang, Z. et al., *Water Sci. Technol.*, 46, 77, 2002. With permission.)

2.4 Figure 2.16 shows a water-using scheme for the bleaching section in a textile plant before carrying out water minimization study (Ujang et al., 2002). As shown, freshwater (0 mg/L) is utilized in singeing, scouring, and bleaching, as well as in mercerizing processes. These processes are treated as mass exchange processes, as water is used as MSA to reduce the impurity load from the process streams. Hence, these processes have uniform inlet and outlet flowrates. Chemical oxygen demand (COD) concentration is taken as the main consideration for water recovery, as it is the most significant parameter for process operations. Identify the limiting data for water recovery for this process. Note that the limiting inlet concentration for the three processes are given as 0 (singeing), 47 (scouring and bleaching) and 75 ppm (mercerizing) respectively (hint: Equations 2.10 and 2.11 may be used; see answer in Problem 3.2).

Utility Gas Recovery Problems

2.5 A magnetic tape manufacturing process is shown in Figure 2.17 (El-Halwagi, 1997). Coating ingredients are first dissolved in an organic solvent to form a slurry mixture. The slurry is then suspended with

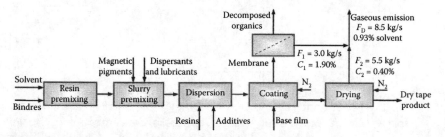

FIGURE 2.17
A magnetic tape process. (From Dunn, R.F. et al., Selection of organic solvent blends for environmental compliance in the coating industries, in Griffith, E.D., Kahn, H., and Cousins, M.C. eds., *Proceedings of the First International Plant Operations and Design Conference*, Vol. III, pp. 83–107, AIChE, New York, 1995. With permission.)

resin binders and special additives, which are then deposited on the base film. The coated film is then sent to the drying operation where the dry tape product is made Note that the limiting inlet concentration for the three processes are given as o (singing), 47 (scouring and bleaching) and 75 ppm (mercerzing) respectively (hint: Equations 2.10 and 2.11 may be used; see answer in Problem 3.2).

In the coating chamber, nitrogen gas is used to induce the solvent evaporation rate that suits the deposition operation. During the operation, a small portion of the solvent is decomposed into organic species. A membrane unit is then used to purify the chamber exhaust gas. The membrane reject stream mainly consists of nitrogen gas and some leftover organic solvent. Besides, nitrogen gas is also used in the drying operation to evaporate the remaining solvent. The dryer exhaust gas is then mixed with the membrane reject stream and sent for disposal. To reduce nitrogen consumption of the process, the solvent-laden exhaust gases (source) may be recovered to the process sinks, i.e., coating and drying processes. Solvent content is the primary concern in considering nitrogen recovery. Besides, the following process constraints of the sinks are to be complied with for any nitrogen recovery scheme:

i. Coating

$$3.0 \leq \text{Flowrate of gaseous feed (kg/s)} \leq 3.2 \qquad (2.60)$$

$$0.0 \leq \text{wt\% of solvent} \leq 0.2 \qquad (2.61)$$

ii. Drying

$$5.5 \leq \text{Flowrate of gaseous feed (kg/s)} \leq 6.0 \qquad (2.62)$$

$$0.0 \leq \text{wt\% of solvent} \leq 0.1 \qquad (2.63)$$

The flowrate and concentration of the nitrogen sources are shown in Figure 2.17. It may be assumed that the characteristics of these sources are independent of the feed gas. Note that the fresh nitrogen is solvent free, i.e., with solvent concentration of 0 wt%.

Identify the limiting data for nitrogen recovery in the magnetic tape manufacturing process (see answers in Problem 3.8).

2.6 Figure 2.18 shows a refinery hydrogen network (Alves and Towler, 2002), where hydrogen recovery is considered. Four hydrogen-consuming processes are found in the network, i.e., hydrocracking unit (HCU), naphtha hydrotreater (NHT), cracked naphtha hydrotreater (CNHT), and diesel hydrotreater (DHT). The internal recycled hydrogen stream data for these processes are given in Table 2.6.

Two in-plant hydrogen-producing facilities are also found, i.e., catalytic reforming unit (CRU) and steam reforming unit (SRU). These internal hydrogen sources are to be maximized before the

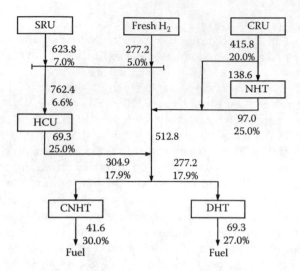

FIGURE 2.18
A refinery hydrogen network (numbers represent the total gas flowrate in mol/s and impurity concentration in mol%). (Reprinted with permission from Alves, J.J. and Towler, G.P., Analysis of refinery hydrogen distribution systems, *Ind. Eng. Chem. Res.*, 41, 5759 , 2002. Copyright 2002 American Chemical Society.)

TABLE 2.6

Flowrate and Concentration of Recycle Stream in Figure 2.18

Units	F_{RCY} (mol/s)	C_{RCY} (mol%)
HCU	1732.6	25.0
NHT	41.6	25.0
CNHT	415.8	30.0
DHT	277.2	27.0

purchase of fresh hydrogen. The fresh hydrogen feed has an impurity content of 5%.

Identify the limiting data in order to carry out a hydrogen recovery analysis (see answer in Problem 4.15).

Property Integration Problems

2.7 Figure 2.19 shows a wafer fabrication process where a large amount of ultrapure water (UPW) is consumed (Ng et al., 2009). Four UPW-using processes are the wet processing section (Wet), lithography, combined chemical and mechanical processing (CMP), and miscellaneous operations (Etc.). A total of six wastewater streams are generated from these processes, with some processes (Wet and CMP) generating two wastewater streams of different quality.

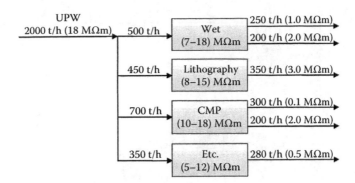

FIGURE 2.19
Usage of UPW in a wafer fabrication process.

In order to reduce the UPW and wastewater flowrates, the wastewaters are considered for recovery to the water-using processes. The most significant water quality factor in considering water recovery is *resistivity* (R), which is an index that reflects the total ionic content in the aqueous streams, with the property mixing rule of

$$\frac{1}{R_M} = \sum_i \frac{x_i}{R_i} \tag{2.64}$$

The sink and source of the water-using processes are given as fixed flowrates, as shown in Figure 2.19. Besides, it may be assumed that the characteristics of sources are independent from those of the sinks. The sinks may accept water with a given range of resistivity values, as shown in parenthesis in Figure 2.19.

Identify the limiting data for water recovery for this process (see answer in Problem 4.19).

2.8 Recovery of acetic acid (AA) is being considered in a vinyl acetate manufacturing process, where vinyl acetate monomer (VAM) is produced from the reactants of ethylene, oxygen, and AA (El-Halwagi, 2006). Note that an equimolar of water impurity is also formed as a reaction by-product. As shown in the PFD in Figure 2.20, AA is first evaporated before it is fed to the reactor. The reactor effluent is then sent to Absorber I, where AA is used to absorb VAM. The bottom stream of Absorber I is sent to the distillation tower, where VAM is recovered as top product; the latter is sent for final finishing. On the other hand, the top product stream of Absorber I, which contains unreacted ethylene and oxygen, is sent to Absorber II, where water is used as a scrubbing agent to absorb the remaining AA content. Ethylene and oxygen gases leave Absorber II in the top stream and are purified before being recycled to the reactor. The bottom product streams of Absorber II

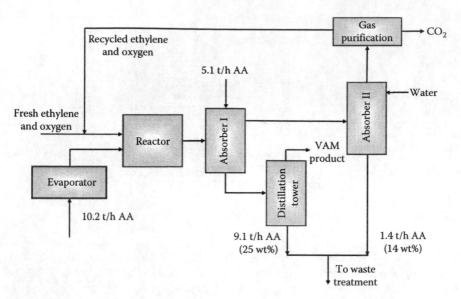

FIGURE 2.20
Vinyl acetate manufacturing process (numbers represent AA stream flowrate; values in parenthesis represent water concentrations in AA streams). (From El-Halwagi, M.M., *Process Integration*, Academic Press, San Diego, CA, 2006, p. 56. With permission.)

and distillation tower are sent for waste treatment. The latter consists of neutralization and biological treatment units. In considering the recovery of AA, water content is the impurity that is of main concern. Besides, the following technical constraints are to be observed in evaluating options for AA recovery.

i. Neutralization unit (in waste treatment)

$$0 \leq \text{Feed flowrate (t/h)} \leq 11.0 \tag{2.65}$$

$$0 \leq \text{AA content in feed stream (wt\%)} \leq 85 \tag{2.66}$$

ii. Evaporator

$$10.2 \leq \text{Feed flowrate (t/h)} \leq 11.2 \tag{2.67}$$

$$0.0 \leq \text{Water content in feed stream (wt\%)} \leq 10.0 \tag{2.68}$$

iii. Absorber I

$$5.1 \leq \text{Feed flowrate (t/h)} \leq 6.0 \tag{2.69}$$

$$0.0 \leq \text{Water content in feed stream (wt\%)} \leq 5.0 \tag{2.70}$$

Identify the limiting data for AA recovery for this process (see answers in Problem 3.10).

References

Alves, J. J. and Towler, G. P. 2002. Analysis of refinery hydrogen distribution systems. *Industrial and Engineering Chemistry Research*, 41, 5759–5769.

Dunn, R. F., El-Halwagi, M. M., Lakin, J., and Serageldin, M. 1995. Selection of organic solvent blends for environmental compliance in the coating industries. In: Griffith, E. D., Kahn, H., and Cousins, M. C. eds., *Proceedings of the First International Plant Operations and Design Conference*, Vol. III, pp. 83–107. New York: AIChE.

El-Halwagi, M. M. 1997. *Pollution Prevention through Process Integration: Systematic Design Tools.* San Diego, CA: Academic Press.

El-Halwagi, M. M. 2006. *Process Integration.* San Diego, CA: Academic Press.

Foo, D. C. Y., Kazantzi, V., El-Halwagi, M. M., and Manan, Z. A. 2006a. Cascade analysis for targeting property-based material reuse networks. *Chemical Engineering Science*, 61(8), 2626–2642.

Foo, D. C. Y., Manan, Z. A., and El-Halwagi, M. M. 2006b. Correct identification of limiting water data for water network synthesis. *Clean Technologies and Environmental Policy*, 8(2), 96–104.

Hallale, N. and Liu, F. 2001. Refinery hydrogen management for clean fuels production. *Advances in Environmental Research*, 6, 81–98.

Hamad, A. A., Crabtree, E. W., Garrison, G. W., and El-Halwagi, M. M. 1996. Optimal design of hybrid separation systems for in-plant waste reduction. In: *Proceedings of the Fifth World Congress of Chemical Engineering*, San Diego, CA, Vol III, pp. 453–458.

Hamad, A. A., Varma, V., El-Halwagi, M. M., and Krishnagopalan, G. 1995. Systematic integration of source reduction and recycle reuse for cost-effective compliance with the cluster rules. In: *AIChE Annual Meeting*, Miami, FL.

Kazantzi, V. and El-Halwagi, M. M. 2005. Targeting material reuse via property integration. *Chemical Engineering Progress*, 101(8), 28–37.

Ng, D. K. S., Foo, D. C. Y., Tan, R. R., Pau, C. H., and Tan, Y. L. 2009. Automated targeting for conventional and bilateral property-based resource conservation network. *Chemical Engineering Journal*, 149, 87–101.

Noureldin, M. B. and El-Halwagi, M. M. 1999. Interval-based targeting for pollution prevention via mass integration. *Computers and Chemical Engineering*, 23, 1527–1543.

Savelski, M. and Bagajewicz, M. 2000. On the optimality conditions of water utilization systems in process plants with single contaminants. *Chemical Engineering Science*, 55, 5035–5048.

Savelski, M. and Bagajewicz, M. 2003. On the necessary conditions of optimality of water utilization systems in process plants with multiple contaminants. *Chemical Engineering Science*, 58, 5349–5362.

Ujang, Z., Wong, C. L., and Manan, Z. A. 2002. Industrial wastewater minimization using water pinch analysis: A case study on an old textile plant. *Water Science and Technology*, 46, 77–84.

Wang, Y. P. and Smith, R. 1994. Wastewater minimisation. *Chemical Engineering Science*, 49, 981–1006.

Part I

Insight-Based Pinch Analysis Techniques

In Part I of this book, the readers will learn several well-established insight-based *pinch analysis* techniques for the synthesis of a resource conservation network (RCN). A typical pinch analysis technique consists of two stages. In stage 1, the performance targets for an RCN are first identified based on first principles (i.e., mass balance and thermodynamic constraints). In most cases, the minimum fresh resource(s) needed for an RCN should be determined, since it translates into minimum operating cost (and minimum fixed cost indirectly) of an RCN. Chapters 3 through 6 and 8 cover various graphical or algebraic techniques that can be used to determine the performance targets. This stage is commonly termed as *targeting* and is performed prior to the determination of actual structure (interconnection) of an RCN. The latter is regarded as stage 2, which is commonly termed as *network design*, is covered in Chapter 7. Chapters 3 through 8 should be read sequentially as they cover basic techniques for direct material reuse/recycle, regeneration, and waste treatment.

Upon learning the techniques discussed in Chapters 1 through 8. the readers can move to advanced topics such as pre-treatment network (Chapter 9), inter-plant networks (Chapter 10), and batch material networks (Chapter 11), which require the techniques discussed in the earlier chapters. Chapters 9 through 11 stand alone, with no interconnection among them.

To facilitate self-learning, some drawing and spreadsheet calculation files of selected examples are also made available to the readers on publisher website (please look for the icon 🖥). However, it is advisable that the readers redo the calculation (for those examples with calculation files) in order to have a better understanding of the techniques.

3

Graphical Targeting Techniques for Direct Reuse/Recycle

Graphical techniques are well recognized for their advantage in providing good insights for problem analysis. In this chapter, a commonly used graphical technique for the targeting of various resource conservation networks (RCNs), is introduced i.e., *material recovery pinch diagram* (MRPD), developed independent by El-Halwagi et al. (2003) as well as Prakash and Shenoy (2005). This chapter focuses on the targeting of a direct reuse/recycle RCN for both single and multiple fresh resources. Besides, targeting for multiple fresh resources and threshold problems are also discussed.

3.1 Material Recovery Pinch Diagram

For an RCN with single fresh resource with superior quality (highest purity or lowest impurity concentration among all sources), steps to construct the MRPD are given as follows (El-Halwagi et al., 2003; Prakash and Shenoy, 2005):

1. Arrange the process sinks in descending order of quality levels. Do the same for process sources.

2. Calculate the maximum load acceptable by the sinks (m_{SKj}) and the load possessed by the sources (m_{SRi}), which are given as the product of the sink (F_{SKj}) or source (F_{SRi}) flowrate with its corresponding quality level (q_{SKj} or q_{SRi}). These load values are given by Equations 3.1 and 3.2, respectively:

$$m_{SKj} = F_{SKj} \, q_{SKj} \tag{3.1}$$

$$m_{SRi} = F_{SRi} \, q_{SRi} \tag{3.2}$$

3. All process sinks are plotted from the origin on a load-versus-flowrate diagram, with descending order of their quality levels to form the *sink composite curve*. The individual sink segments are connected by linking the tail of a latter to the arrowhead of an earlier segment (Figure 3.1).

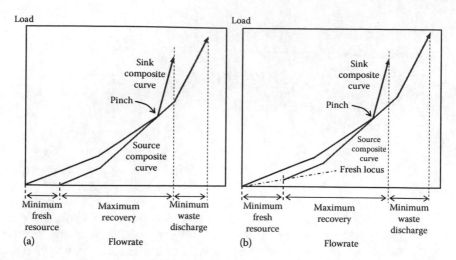

FIGURE 3.1
The MRPD for (a) pure fresh resource and (b) impure fresh resource. (From El-Halwagi, M.M. et al., *Ind. Eng. Chem. Res.*, 42, 4319, 2003; Prakash, R. and Shenoy, U.V., *Chem. Eng. Sci.*, 60(1), 255, 2005.)

4. Step 3 is repeated for all process sources to form the *source composite curve* on the same load-versus-flowrate diagram.

5. For the RCN with pure (zero impurity content) fresh resources, the source composite curve is moved horizontally to the right until it just touches the sink composite curve, with the latter lying above and to the left of the source composite curve (Figure 3.1a). For an RCN with impure fresh resources, a fresh locus is to be plotted on the MRPD, with its slope corresponding to its quality level (q_R). The source composite curve is slid on the locus until it just touches the sink composite curve and to the right and below the latter (Figure 3.1b).

Several *targets* are identified from the MRPD. First, the overlap region between the two composite curves gives the maximum recovery target among all process sinks and sources of the RCN. When sources are insufficient to satisfy the requirement of the sinks, fresh resources (to be purchased) are needed for the RCN. The minimum requirement of the fresh resource is given by the horizontal distance where the sink composite curve extends to the left of the source composite curve. On the other hand, the horizontal distance where the source composite curve extends to the right of the sink composite curve represents the minimum waste generated from the RCN, which will be sent for final treatment before environment discharge. The point where the two composite curves touch is termed as the *pinch*, which is the overall bottleneck for maximum recovery among all sinks and sources in the RCN. Next, the use of MRPD is demonstrated on several examples that follow.

Example 3.1 Water minimization for acrylonitrile (AN) production

The AN production case study in Example 2.1 is revisited here. The original limiting data for the problem are given in Table 2.1 (also found in the first three columns of Table 3.1). Determine the following targets for the problem using the MRPD:

1. Minimum fresh resource (water) for the RCN
2. Maximum water recovery between process sinks and sources
3. Minimum wastewater discharge from the RCN

Solution

The five-step procedure is followed to plot the MRPD for the AN case study:

1. *Arrange the process sinks in descending order of their quality level. Do the same for process sources.*

    For the AN production case study, the quality level corresponds to the concentration of ammonia (see Example 2.1). As shown in the third column of Table 3.1, all sinks and sources are arranged in ascending order of their impurity concentrations (i.e., descending order of quality).

2. *Calculate the maximum load acceptable by the sinks (m_{SKj}) and the load possessed by the sources (m_{SRi}).*

    The load values for all sinks or sources are calculated using Equations 3.1 and 3.2, respectively, and are shown in the fourth column of Table 3.1. Note that in this case, the loads correspond to the flowrates of ammonia load in the sinks and sources, respectively.

3. *All process sinks are plotted from the origin on a load-versus-flowrate diagram, with descending order of their quality levels to form the sink composite curve. The individual sink segments are connected by linking the tail of a latter to the arrowhead of an earlier segment.*

TABLE 3.1

Limiting Water Data for Plotting the MRPD for AN Case Study

Sinks, SK$_j$	F_{SKj} (kg/s)	C_{SKj} (ppm)	m_{SKj} (mg/s)	Cum. F_{SKj} (kg/s)	Cum. m_{SKj} (mg/s)
BFW (SK1)	1.2	0	0	1.2	0
Scrubber inlet (SK2)	5.8	10	58.0	7.0	58.0

Sources, SR$_i$	F_{SRi} (kg/s)	C_{SRi} (ppm)	m_{SRi} (mg/s)	Cum. F_{SRi} (kg/s)	Cum. m_{SRi} (mg/s)
Distillation bottoms (SR1)	0.8	0	0	0.8	0
Off-gas condensate (SR2)	5	14	70.0	5.8	70.0
Aqueous layer (SR3)	5.9	25	147.5	11.7	217.5
Ejector condensate (SR4)	1.4	34	47.6	13.1	265.1

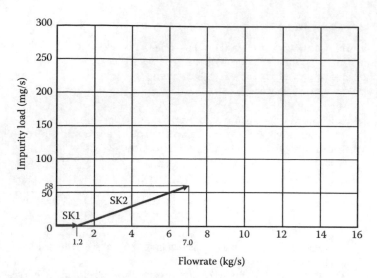

FIGURE 3.2
Plotting of sink composite curve for Example 3.1. (From *Process Integration*, El-Halwagi, M.M., Copyright 2006, with permission from Elsevier.)

Since the flowrate and load values of the sink composite curve are cumulative in nature, cumulative flowrates (Cum. F_{SKj}) and cumulative loads (Cum. m_{SKj}) for all sinks are first calculated, given in the last two columns of Table 3.1. These values are then used to plot the sink composite curve for the AN case on the load-versus-flowrate diagram, as shown in Figure 3.2. Note that the slope of each segment corresponds to its quality level, i.e., ammonia concentration in this case. Hence, the sink composite curve indicates that a total flowrate of 7 kg/s is needed by sinks SK1 and SK2, and they can only accept a total impurity load of 58 mg/s.

4. *Step 3 is repeated for all process sources to form the source composite curve on the same load-versus-flowrate diagram.*

Similar to the plotting of the sink composite curve (Step 3), the cumulative flowrates (Cum. F_{SRi}) and cumulative loads (Cum. m_{SRi}) for all sources are first calculated in the last two columns of Table 3.1. These values are used to plot the source composite curve in Figure 3.3. Similar to the sink composite curve, the slope of each segment in the source composite curve corresponds to the ammonia concentration of these sources. The source composite curve in Figure 3.3 indicates that the sources have a total water flowrate of 13.1 kg/s and posses a total impurity load of 265.1 mg/s.

Before proceeding to Step 5, the sink and source composite curves in Figure 3.3 are first examined for their feasibility. Note that there are two criteria that a feasible MRPD needs to fulfill, i.e., flowrate and load constraints. The flowrate constraint has been fulfilled in this case, as the total source flowrate (13.1 kg/s)

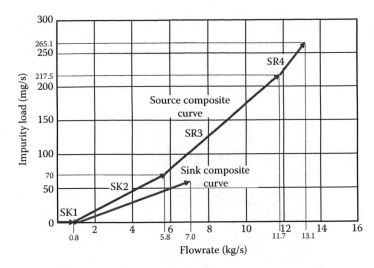

FIGURE 3.3
Plotting of source composite curve for Example 3.1 (From *Process Integration*, El-Halwagi, M.M., Copyright 2006, with permission form Elsevier.)

exceeds that required by the sinks (7.0 kg/s). Nevertheless, the sinks can only tolerate a maximum total load limit of 58 mg/s impurity. However, if one were to feed the available sources to fulfill the total flowrate requirement of the sinks, the maximum total load limit of the sinks will be exceeded. This can be easily observed from Figure 3.3, where the source composite curve lies almost completely above the sink composite curve. In other words, the available sources are "too concentrated" to be fed to the sinks. In order to "dilute" the water sources for recovery, freshwater is required for the process. The minimum extent of the freshwater flowrate is to be determined in Step 5.

5. *For the RCN with pure (zero impurity content) fresh resources, the source composite curve is moved horizontally to the right until it just touches the sink composite curve, with the latter lying above and to the left of the source composite curve.*

   Fresh resource for the AN case is freshwater feed, which has zero ammonia content. Hence, the source composite curve is moved horizontally to the right until it just touches the sink composite curve and lies completely below and to the right of the latter. This forms a feasible MRPD as shown in Figure 3.4, where both the flowrate and load constraints of the RCN are fulfilled.

   Several network targets are readily identified from the feasible MRPD in Figure 3.4. The minimum fresh resource for the AN case, i.e., freshwater flowrate (F_{FW}), is identified as 2.1 kg/s. This represents the minimum freshwater needed by the RCN after the water sources are fully recovered to the sinks. The extent of recovery (F_{REC}), which is given by the overlapping region between the sink and source composite

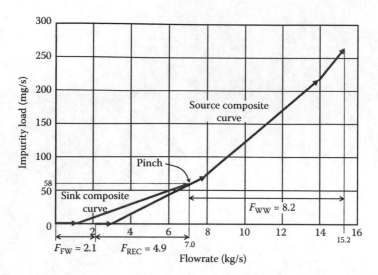

FIGURE 3.4
MRPD for Example 3.1. (From *Process Integration*, El-Halwagi, M.M., Copyright 2006, with permission from Elsevier.)

curves, is identified as 4.9 kg/s (= 7.0 – 2.1 kg/s) from Figure 3.4. Besides, the excess water sources (after recovery to the sinks) emit as wastewater from the network, with its total flowrate (F_{WW}) identified as 8.2 kg/s. Figure 3.4 also shows the pinch where the two composite curves touch, which corresponds to a cumulative flowrate of 7.0 kg/s and a cumulative load of 58 mg/s. The significance of the pinch is discussed in the following section.

3.2 Significance of the Pinch and Insights from MRPD

As discussed earlier, the sink and source composite curves of a feasible MRPD touch each other at the pinch, with the latter representing the overall bottleneck for maximum material recovery in an RCN. In other words, the identified fresh resource target is the minimum requirement of a particular resource in the RCN. If one were to utilize x amount of additional resources in this RCN, the same amount of process sources (which is supposed to be recovered to the sinks) would end up being discharged as waste. This leads to an x amount of additional waste being generated from the RCN, as shown in Figure 3.5a. In contrast, if one were to utilize less than the targeted resource, the RCN would experience infeasibility in part of the network (see Figure 3.5b).

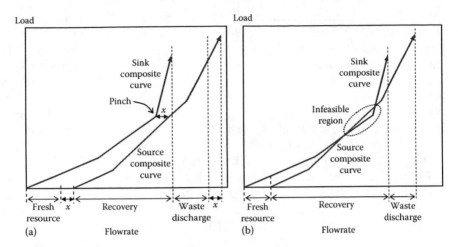

FIGURE 3.5
RCN with (a) additional and (b) insufficient resources. (From *Process Integration*, El-Halwagi, M.M., Copyright 2006, with permission from Elsevier.)

Note that the pinch always occurs at one of the source quality levels. The source where the pinch occurs is termed as the *pinch-causing source* (Manan et al., 2004; Foo et al., 2006). As shown in Figure 3.6, the pinch divides the overall network into two separate regions with respect to the pinch quality, i.e., *higher quality region* (HQR, below the pinch) and *lower quality region*

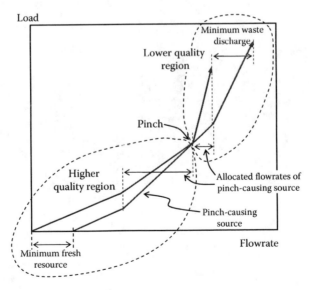

FIGURE 3.6
HQR and LQR.

(LQR, above the pinch). With respect to the pinch quality, we can easily classify to which region the individual sinks and sources belong.

Several remarks are worth mentioning for HQR and LQR. For the HQR, the total flowrate of the process sources is always less than that of the process sinks. Hence, the flowrate deficit is supplemented by the fresh resource. In other words, this region is the most constrained part of the RCN, as it controls the overall fresh resource flowrate for an RCN. Besides, it is noticed that the sinks in the HQR will always accept the maximum load sent from the sources (including fresh sources).

For the LQR, the total flowrate of the process sources is always more than that of the process sinks. This means that this region has a surplus of process sources, which will emit as waste from the RCN.

A few "golden rules" can then be derived for the HQR and LQR (Hallale, 2002; Manan et al., 2004; El-Halwagi, 2006):

1. Sources in the HQR (including fresh resource) should not be fed to sink in the LQR and may not be mixed with sources in the latter region. If a source that belongs to the HQR is being used in the LQR, such as that in Figure 3.7a, an additional x amount of fresh resource is needed in the RCN. This also results in the same amount of additional waste discharge from the network.

2. The pinch-causing source is an exception for Rule 1, as it belongs to both HQR and LQR. The horizontal distance of the pinch-causing source segment in the HQR and LQR represents its allocated flowrate in the respective regions (see labeling in Figure 3.6).

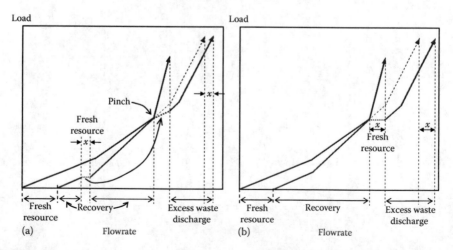

FIGURE 3.7
Cases with higher fresh resource and waste discharge: (a) the use of source in the HQR for sink in the LQR and (b) the use of fresh resource in the LQR.

3. Fresh resource(s) shall only be used in the HQR, as this region experiences flowrate deficit. Using this resource in the LQR (where flowrate surplus is observed) will result in higher waste discharge. This is shown in Figure 3.7b. As x amount of additional fresh resource is used in the LQR, the waste discharge from the RCN also increases with the same magnitude.

Example 3.2 Water minimization for AN production (revisited)

Make use of the MRPD for the AN case study in Figure 3.4 to carry out the following tasks:

1. Identify the HQR and LQR where the process sinks and sources belong.
2. Identify the pinch-causing source and its allocated flowrates to the HQR and LQR.
3. Identify a reuse/recycle scheme for the case study based on the insights.

Solution

With respect to the pinch concentration, the RCN may be divided into HQR (lower concentration) and LQR (higher concentration), as shown in Figure 3.8. Both sinks SK1 and SK2 in the sink composite curve, along with the freshwater feed ($F_{FW} = 2.1$ kg/s), belong to the HQR. Besides, source SR1 and a big portion of SR2 also belong to the same region. On the other hand, a small portion of SR2, as well as SR3 and SR4 belong to

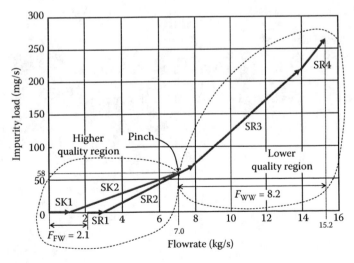

FIGURE 3.8
HQR and LQR for Example 3.2.

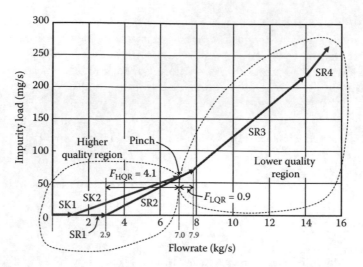

FIGURE 3.9
Pinch-causing source and its allocated flowrates for Example 3.2.

the LQR, while no sink presents in this region. Hence, all sources will emit as wastewater (F_{WW} = 8.2 kg/s) from the LQR.

Note that the sinks in the LQR accept the maximum impurity load sent from the sources. For this case, this corresponds to 58 mg/s ammonia content for the AN case, as shown in Figure 3.8.

It is observed from Figure 3.8 that the pinch occurs at SR2. The slope of SR2, i.e., 14 ppm, hence corresponds to the pinch concentration of the network, while SR2 (F_{SR2} = 5.0 kg/s) is identified as the pinch-causing source. As shown in Figure 3.9, the allocated flowrates of the pinch-causing source to the HQR (F_{HQR}) and LQR (F_{LQR}) are determined as 4.1 and 0.9 kg/s, respectively.

Based on these insights, a water reuse/recycle network for the AN case may be developed. In the HQR, the allocated flowrate of the pinch-causing source, i.e. SR2 to this region (F_{HQR}) will be used along with SR1 and the freshwater feed to satisfy sinks SK1 and SK2. Since SK1 (boiler feed water—BFW) does not tolerate any impurity (see Table 3.1), it is fed with freshwater. The remaining water sources, i.e., freshwater feed and SR1, and the remaining flowrate of SR2 are mixed to be fed to SK2 (scrubber inlet). As mentioned earlier, in the LQR, the unutilized sources SR3 and SR4 as well as the remaining flowrate of SR2 are discharged as wastewater. The resulting water reuse/recycle network is shown in Figure 3.10. However, note that the development of an RCN based on targeting insights will only work for simple cases.* For a more complicated problem, the detailed network design procedure should be used (see Chapter 7 for details).

---

* To develop an RCN based on targeting insights, there should only be maximum one sink with nonzero quality index in the HQR.

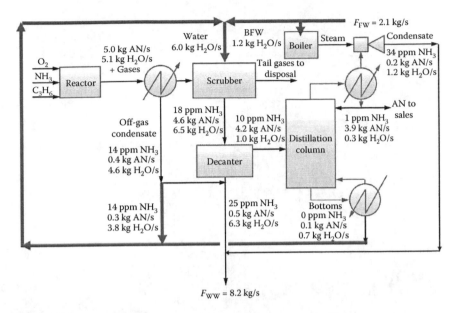

FIGURE 3.10

Water reuse/recycle network for AN case study. Note: Detailed mass balance has not been revised yet. (From El-Halwagi, M.M., *Pollution Prevention through Process Integration: Systematic Design Tools*, Academic Press, San Diego, CA, 1997.)

Example 3.3 Hydrogen reuse/recycle in a refinery hydrogen network

The refinery hydrogen network case study in Example 2.4 (Hallale and Liu, 2001) is revisited here. The original limiting data needed for the case study are shown in Table 2.2 (also in the first three columns of Table 3.2). Determine the following targets for this RCN:

1. Minimum fresh resource (hydrogen) and purge flowrates, given that the fresh hydrogen feed has 1% impurity
2. Maximum recovery target between process sinks and sources

TABLE 3.2

Limiting Data for Example 3.3

Sinks, SK_j	F_{SKj} (MMscfd)	C_{SKj} (%)	m_{SKj} (MMscfd)	Cum. F_{SKj} (MMscfd)	Cum. m_{SKj} (MMscfd)
A inlet (SK1)	400	7.20	28.8	400	28.8
B inlet (SK2)	600	12.43	74.58	1000	103.38
Sources, SR_i	F_{SRi} (MMscfd)	C_{SRi} (%)	m_{SRi} (MMscfd)	Cum. F_{SRi} (MMscfd)	Cum. m_{SRi} (MMscfd)
A outlet (SR1)	350	9.00	31.5	350	31.5
B outlet (SR2)	500	15.00	75	850	106.5

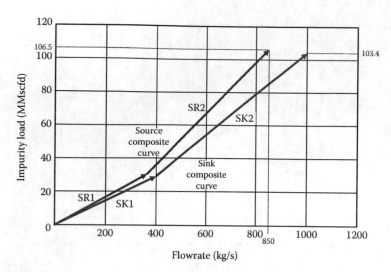

FIGURE 3.11
Plotting of composite curves for Example 3.3.

3. Pinch concentration that limits the maximum recovery between process sinks and sources
4. The pinch-causing source and its allocated flowrate to the HQR and LQR

Solution

Following Steps 1–3 of the MRPD procedure, the cumulative flowrates and loads of the process sinks and sources are shown in the last two columns of Table 3.2. We then construct the MRPD following Steps 3 and 4, resulting in the sink and source composite curves in Figure 3.11.

As shown in Figure 3.11, the source composite curve is found to be on the left and above the sink composite curve. This means that all available sources have a higher concentration than the maximum acceptable limit of the sinks. Besides, the total flowrate (850 MMscfd) of all sources is also insufficient to be used by the sinks (1000 MMscfd). This means that the MRPD is infeasible for both flowrate and load constraints. Hence, fresh hydrogen is needed to restore the flowrate and load feasibilities of the RCN.

Since fresh hydrogen feed has 1% impurity, a fresh locus is plotted on the MRPD with a slope that corresponds to its impurity concentration. The source composite curve is then slid on this locus until it just touches the sink composite curve and is to the right and below the latter (Figure 3.12).

From Figure 3.12, the horizontal distance of the sink composite curve that extends to the left of the source composite curve gives the minimum fresh hydrogen feed (F_H) needed by the network as 183 MMscfd. On the other hand, the horizontal distance where the source composite curve extends to the right of the sink composite curve dictates the minimum hydrogen that is purged from the network (F_P), which is identified as

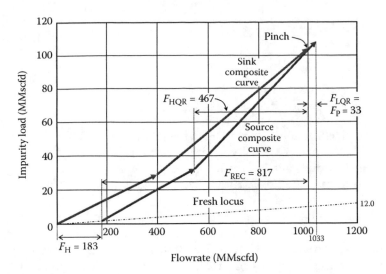

FIGURE 3.12
MRPD for Example 3.3.

33 MMscfd. The overlapping region between the two composite curves then gives the maximum recovery target (F_{REC}), which is identified as 817 MMscfd (= 1000 – 183 MMscfd). The pinch corresponds to the concentration of the second segment of the source composite curve, i.e., 15% of SR2. The latter is hence identified as the pinch-causing source for the problem, with 467 MMscfd (500 – 33 MMscfd) of its flowrate being allocated to the HQR (F_{HQR}), while the rest to the LQR (F_{LQR}). Since there is no process sink in the LQR, the allocated flowrate to this region is purged. In other words, the purge stream for this case is originated from SR2.

3.3 Targeting for Multiple Resources

In earlier sections, minimum flowrate targeting procedure is dedicated to RCN with single fresh resource of superior quality among all sources. There are also cases where RCN is supplied with multiple fresh resources of different quality levels, which call for a revised targeting procedure. Note that fresh resources should only be utilized in the HQR of an RCN, since it is the area that experiences resource deficit.

Since there is more than one fresh resource in the RCN, it is always useful to compare the relative costs among the available fresh resources. However, since these fresh resources have different impurity levels (which lead to different minimum flowrates), comparing them based on their unit cost alone is insufficient. A useful index termed as the *prioritized cost* (Shenoy and Bandyopadhyay, 2007) may be used, which is given in Equation 3.3:

$$CTP_R = \frac{CT_R}{q_R - q_P} \tag{3.3}$$

where
  CT_R and CTP_R are the unit cost and prioritized costs for fresh resource
  q_R and q_P are the quality indices of fresh resource and the pinch

Note that pinch quality is obtained when the RCN is supplied by a single resource. When a lower quality resource has a prioritized cost that is lower than that of a high-quality resource (i.e., $CTP_{R1} < CTP_{R2}$ and $q_{R1} > q_{R2}$), the use of the former should be maximized; otherwise, only the high-quality resource should be used.

For the case where the lower quality resource is to be maximized prior to the higher quality resource, a two-step targeting procedure is used (Wan Alwi and Manan, 2007). Figure 3.13 demonstrates the case where two fresh resources are available for use in an RCN, i.e., a pure primary resource that has superior quality among all sources and secondary resource (with intermediate quality level). Step 1 involves the targeting of primary resource. As shown in Figure 3.13a, a partial section of the source composite curves is first plotted, with only the source segments of better quality than of the secondary fresh resource. A locus of the secondary fresh resource (with slope corresponding to its quality level) is added at the end of the earlier source segments. Next, the source segments along with the impure fresh resource locus are then slid horizontally until they stay below and to the right of the sink composite curve, and touch the latter at the pinch point (Figure 3.13a). The overhang of the sink

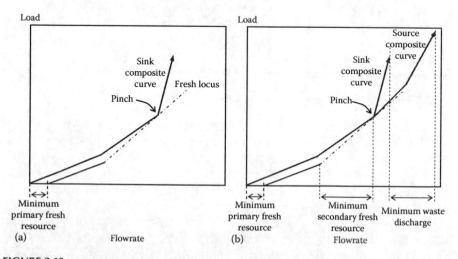

FIGURE 3.13
Minimum flowrate targeting for multiple fresh resource: (a) targeting for primary fresh resource and (b) targeting for secondary fresh resource. (From Foo, D.C.Y., *Ind. Eng. Chem. Res.*, 48(11), 5125, 2009. With permission.)

composite curve gives the minimum flowrate of the pure fresh resource. We then proceed to step 2 to determine the flowrate target of the secondary fresh resource. The rest of the source segments (of lower quality) are connected and slide them along the locus of the impure fresh resource until they touch the sink composite curve. The horizontal distance of the fresh locus gives the minimum flowrate needed for the secondary fresh resource (Figure 3.13b). Similar to earlier cases, the overhang of the source composite curve gives the minimum flowrate of waste discharge from the RCN.

Note that apart from in-plant recovery the same targeting procedure may also be utilized for *inter-plant resource conservation network*, where resources are shared by different RCNs. For such a case, the targeting procedure may be used to set the minimum cross-plant flowrate (virtually no cost needed), which in turn leads to reduced overall fresh resource and waste flowrates (see Chapter 10 for details).

Example 3.4 Targeting for multiple freshwater sources

The AN production case study in Examples 3.1 and 3.2 is revisited. In this case, an impure freshwater source with 5 ppm ammonia content (secondary fresh resource) is available for use besides the pure freshwater source (primary fresh resource). Unit costs for the pure (CT_{FW1}) and impure freshwater sources (CT_{FW2}) are given as \$0.001/kg and \$0.0005/kg, respectively. Determine the minimum flowrates for both freshwater sources as well as that for wastewater. Determine also the operating costs for freshwater sources of the RCN.

Solution

In Example 3.2, the targeting procedure identifies that the pinch concentration for this problem corresponds to 14 ppm. Hence, the prioritized costs for pure (CTP_{FW1}) and impure freshwater sources (CTP_{FW2}) are determined using Equation 3.3 as follows:

$$CTP_{FW1} = \frac{\$0.001/kg}{(14-0)\,ppm} = \frac{\$0.071}{g\,NH_3}$$

$$CTP_{FW2} = \frac{0.0005/kg}{(14-5)\,ppm} = \frac{\$0.056}{g\,NH_3}$$

Since the prioritized cost of the impure freshwater source is much lower than that of the pure freshwater source, the use of the former should be maximized prior to the latter. Following step 1 of the targeting procedure, the source composite curve should only consist of a single segment at this stage, i.e., source SR1, which has a lower concentration (0 ppm) than that of impure freshwater source (5 ppm). Locus of the latter is then connected to the SR1 segment. Both SK1 segment and the impure

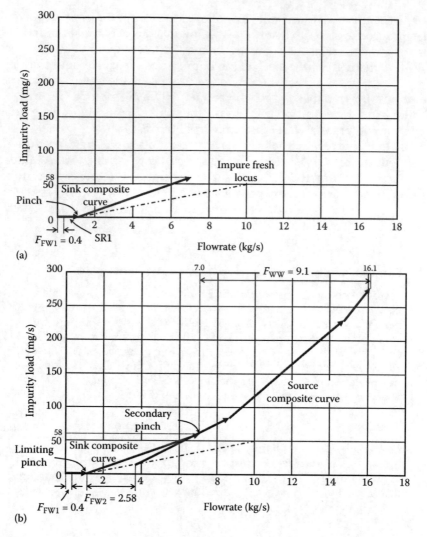

FIGURE 3.14
MRPD for Example 3.4: (a) targeting for pure freshwater source and (b) targeting for impure freshwater source.

freshwater locus are then slid horizontally until they stay below and to the right of the sink composite curve. Figure 3.14a shows that a pinch (5 ppm) is formed between the impure fresh locus and the sink composite curve. The overhang of the sink composite curve dictates the minimum flowrate of pure freshwater source (F_{FW1}) as 0.4 kg/s.

Next, the remaining water source segments (SR2 – SR4) are then connected and slid along the locus of the impure freshwater locus until they touch the sink composite curve (step 2). For this case, a new pinch point

TABLE 3.3

Comparison of Various Cost Scenarios for Example 3.4

Scenarios	Pure Freshwater Flowrate (kg/s)	Impure Freshwater Flowrate (kg/s)	Freshwater Cost of RCN ($/s)
Pure freshwater source alone, with $CT_{FW2} = \$0.001/\text{kg}$ (Example 3.1)	2.1	0	0.0021
With pure and impure freshwater sources ($CT_{FW2} = \$0.0005/\text{kg}$)	0.4	2.58	0.0017
With pure and impure freshwater sources ($CT_{FW2} = \$0.0007/\text{kg}$)	0.4	2.58	0.0022

is formed at the end of the sink composite curve. The source segment where the pinch is formed indicates that the new pinch concentration corresponds to 14 ppm (i.e., identical to the case where a single pure freshwater source is used—Examples 3.1 and 3.2). The minimum flowrate of the impure freshwater source (F_{FW2}) is identified from the horizontal distance of the impure fresh locus, i.e., 2.58 kg/s (Figure 3.14b). Note that for this kind of *multiple pinch* cases, the highest quality pinch (i.e., 5 ppm for this case) is known as the *limiting pinch*, in which it controls the overall fresh resource flowrates (Hallale, 2002; Manan et al., 2004). Finally, the overhang of the source composite curve gives the minimum wastewater flowrate of this RCN, i.e., 9.1 kg/s.

We next determine the operating cost for freshwater sources of the RCN. With unit costs of $1/t and $0.5/t for pure ($CT_{FW1}$) and impure freshwater sources (CT_{FW2}), the freshwater cost of the RCN is determined as $0.0017/s (=0.4 kg/s × $0.001/kg + 2.58 kg/s × $0.0005/kg). On the other hand, if the impure freshwater source is not utilized, i.e., 2.1 kg/s of pure freshwater source will have to be used (see Examples 3.1 and 3.2), the freshwater cost rises to $0.0021/s (=2.1 kg/s × $0.001/kg). Nevertheless, when the unit cost of impure freshwater source is raised to $0.0007/kg, the RCN experiences higher freshwater cost, $0.0022/s (=0.4 kg/s × $0.001/kg + 2.58 kg/s × $0.0007/kg). Hence, in this latter case, it will be more sensible to make use of the freshwater source alone (with lower cost). A comparison of these scenarios is given in Table 3.3.

3.4 Targeting for Threshold Problems

Examples that we have discussed so far involve the use of fresh resources and also the generation of waste. There are also cases where neither fresh resource is needed nor waste is generated for an RCN. These special cases are known as the *threshold problems* (Foo, 2008). These are illustrated in the following examples.

Example 3.5 Targeting for threshold problem—zero fresh resource

Table 3.4 shows the limiting data for a water minimization case study. Target the minimum freshwater and wastewater flowrates for the water network.

Solution

Following Steps 1–4 of the procedure, the MRPD for the example is plotted in Figure 3.15. However, before the source composite curve is being shifted (Step 5 of the MRPD procedure), it is already below and to the right of the sink composite curve. In other words, it fulfils both the flowrate and load constraints of the problem. Hence, no freshwater is needed for this case. Besides, the MRPD dictates that a wastewater flowrate (F_{WW}) of 20 t/h is generated from the network. Since the two composite

TABLE 3.4

Limiting Water Data for Example 3.5

SK_j	F_{SK_j} (t/h)	C_{SK_j} (ppm)	m_{SK_j} (kg/h)	Cum. F_{SK_j} (t/h)	Cum. m_{SK_j} (kg/h)
SK1	40	125	5	40	5
SK2	20	500	10	60	15
SK3	40	1125	45	100	60
SR_i	F_{SR_i} (t/h)	C_{SR_i} (ppm)	m_{SR_i} (kg/h)	Cum. F_{SR_i} (t/h)	Cum. m_{SR_i} (kg/h)
SR1	50	100	5	50	5
SR2	50	300	15	100	20
SR3	20	1000	20	120	40

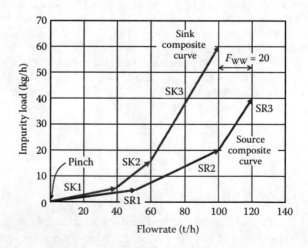

FIGURE 3.15
MRPD for Example 3.5. (From *J. Environ. Manag.*, 88(2), Foo, D.C.Y., Flowrate targeting for threshold problems and plant-wide integration for water network synthesis, 253–274, Copyright 2008, with permission from Elsevier.)

TABLE 3.5

Limiting Water Data for Example 3.6

SK_j	F_{SK_j} (t/h)	C_{SK_j} (ppm)	m_{SK_j} (kg/h)	Cum. F_{SK_j} (t/h)	Cum. m_{SK_j} (kg/h)
SK1	50	20	1	50	1
SK2	20	50	1	70	2
SK3	100	400	40	170	42

SR_i	F_{SR_i} (t/h)	C_{SR_i} (ppm)	m_{SR_i} (kg/h)	Cum. F_{SR_i} (t/h)	Cum. m_{SR_i} (kg/h)
SR1	20	20	0.4	20	0.4
SR2	50	100	5	70	5.4
SR3	40	250	10	110	15.4

curves touch each other at the lowest concentration end, the pinch for this case corresponds to the concentration of SR1, i.e., 100 ppm.

Example 3.6 Targeting for threshold problem—zero waste discharge

Table 3.5 shows the limiting water data for a water minimization case study taken from Foo (2008). Use the MRPD to target the minimum freshwater and wastewater flowrates of the network.

Solution

Following the five-step procedure, the MRPD for this case study is plotted in Figure 3.16. However, as shown in Figure 3.16, even though the

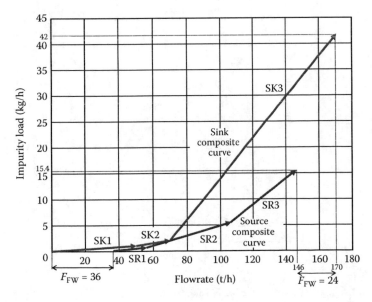

FIGURE 3.16
MRPD for Example 3.6.

source composite curve lies completely below and to the right of the sink composite curve, the total available flowrate of the sources is still insufficient for the sinks. Hence, freshwater is required to supplement the flowrate requirement in the last segment of the sink composite curve. Summing the freshwater flowrates (F_{FW}) at both ends of the MRPD gives a total of 60 t/h. For better representation, the source composite curve is further shifted to the right so that the total freshwater requirement is collected on the left of the MRPD, as shown in Figure 3.17. Since the source composite curve does not extend further from the sink composite curve, this is a zero discharge RCN. Foo (2008) defined the concentration of the last segment of the source composite curve as the *threshold concentration* of the problem.

Note that the threshold problem exists in all types of RCN problems, apart from the water minimization examples shown here.

3.5 Targeting for Property Integration

For an RCN with resource at the superior property operator level (see Section 2.5 for discussion of superior and inferior property operators), the same MRPD procedure discussed in earlier section may be used. However, for the special

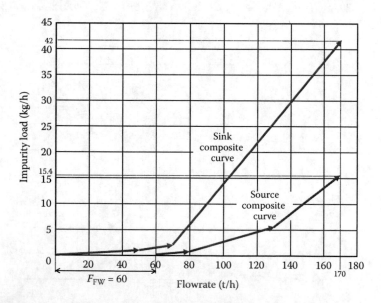

FIGURE 3.17
MRPD for Example 3.6 (with collection of total freshwater flowrates). (From *J. Environ. Manag.*, 88(2), Foo, D.C.Y., Flowrate targeting for threshold problems and plant-wide integration for water network synthesis, 253–274, Copyright 2008, with permission from Elsevier.)

case where property operator of the fresh resource exists at the *inferior* level, Step 5 of the MRPD procedure is modified slightly (see modification in italics):

5. For the RCN with zero operator fresh resource of inferior value, *the sink composite curve is moved horizontally to the right until it just touches the source composite curve, with the latter lying above and to the left of the sink composite curve.* For RCN with nonzero operator fresh resource, a fresh locus is plotted on the MRPD, with its slope corresponding to the quality level of the fresh resource (q_R). The source composite curve is slid on the locus until it just touches the sink composite curve and is to the *left and above the latter.*

Two examples on property-based RCNs with fresh resources of superior and inferior property operators next shown in the following for illustration.

Example 3.7 Solvent recovery for metal degreasing process

The metal degreasing process in Example 2.5 is revisited here. The main property for the evaluation of a reuse/recycle scheme is the Reid vapor pressure (RVP), which is converted to the corresponding operator values for the process sinks and sources in Section 2.5 and summarized in Table 2.3. Similarly, the RVP value of the fresh resource (solvent), i.e., 2.0 atm, is converted to its operator of $2.713\,\text{atm}^{1.44}$ ($=2^{1.44}\,\text{atm}^{1.44}$) using Equation 2.25. Identify the minimum fresh solvent and solvent waste targets for the RCN of this process.

Solution

Since the fresh solvent has the lowest operator value among all sources (superior level), the earlier discussed procedure for MRPD plotting is followed to identify the RCN targets. Following Steps 1–3, the property loads as well as the cumulative flowrates and loads for the sinks or sources are determined and summarized in Table 3.6. Note that in Table 3.6, SR2 with lower operator value (higher quality) is placed before SR1, following Step 1 of the MRPD procedure. Note also that unlike the earlier examples on concentration-based RCN (e.g., water minimization, hydrogen network), the loads for these property-based composite curves

TABLE 3.6

Limiting Data for Example 3.7

Sink, SK$_j$	F_{SKj} (kg/s)	ψ_{SKj} (atm$^{1.44}$)	m_{SKj} (kg atm$^{1.44}$/s)	Cum. F_{SKj} (kg/s)	Cum. m_{SKj} (kg atm$^{1.44}$/s)
Degreaser (SK1)	5.0	4.87	24.32	5.0	24.32
Absorber (SK2)	2.0	7.36	14.72	7.0	39.04

Source, SR$_i$	F_{SRi} (kg/s)	ψ_{SRi} (atm$^{1.44}$)	m_{SRi} (kg atm$^{1.44}$/s)	Cum. F_{SRi} (kg/s)	Cum. m_{SRi} (kg atm$^{1.44}$/s)
Condensate II (SR2)	3.0	3.74	11.22	3.0	11.22
Condensate I (SR1)	4.0	13.20	52.80	7.0	64.02

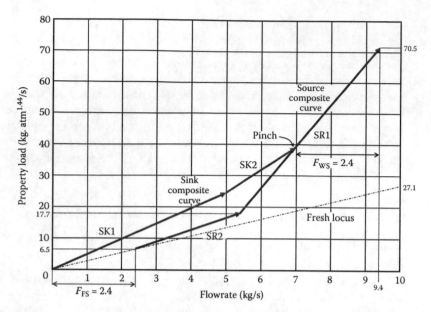

FIGURE 3.18
MRPD for Example 3.7. (From Kazantzi, V. and El-Halwagi, M.M., *Chem. Eng. Prog.*, 101(8), 28, 2005. Reproduced with permission. Copyright © 2005. American Institute of Chemical Engineers (AIChE)).

(represented by the y-axis) do not posses any physical meaning. They are merely used for calculation purposes.

Since the fresh solvent consists of nonzero operator value, a fresh locus is added in the MRPD, with its slope being its operator value. Following Steps 3–5, the MRPD is plotted and is shown in Figure 3.18. As shown, the minimum fresh solvent (F_{FS}) needed for the problem is targeted as 2.4 kg/s. The same amount of solvent waste (F_{WS}) is generated from network. Besides, SR1 is identified as the pinch-causing source, with its property operator being the pinch of the RCN.

Example 3.8 Fiber recovery for paper making process

The paper making process in Example 2.6 is revisited here. The property that is of concern for reuse/recycle is reflectivity, which may be converted to the corresponding operator value (see Example 2.6). The fresh resource use is fresh fiber, which has a reflectivity of 0.95 (dimensionless) or an operator of 0.738 ($=0.95^{5.92}$). Identify the minimum fresh fiber and waste targets for the RCN of this process.

Solution

Since the fresh fiber has the highest operator value among all sources, it is considered to possess an inferior property operator value. Following Steps 1–2, the cumulative flowrates and loads for plotting the MRPD are calculated and summarized in Table 3.7.

TABLE 3.7

Limiting Data for Example 3.8

Sink, SK_j	F_{SKj} (t/h)	ψ_{SKj} (Dimensionless)	m_{SKj} (t/h)	Cum. F_{SKj} (t/h)	Cum. m_{SKj} (t/h)
Machine II (SK2)	40	0.536	21.44	40	21.44
Machine I (SK1)	100	0.382	38.20	140	59.64
Source, SR_i	F_{SRi} (t/h)	ψ_{SRi} (Dimensionless)	m_{SRi} (t/h)	Cum. F_{SRi} (t/h)	Cum. m_{SRi} (t/h)
Bleached Fiber (SR1)	90	0.469	42.23	90	42.23
Broke (SR2)	60	0.182	10.93	150	53.16

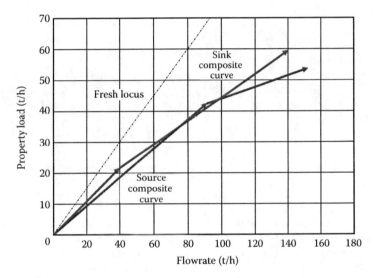

FIGURE 3.19

Plotting of sink and source composite curve for Example 3.8. (From Kazantzi, V. and El-Halwagi, M.M., *Chem. Eng. Prog.*, 101(8), 28, 2005. Reproduced with permission. Copyright © 2005. American Institute of Chemical Engineers (AIChE)).

Following Steps 3–4, the MRPD for the example is given in Figure 3.19. Note that the individual segments of the sink and source composite curves appear in descending slope (with descending order of quality), which is different from earlier examples. A fresh locus is also added since the fresh fiber has a nonzero operator value, with the locus slope being the property operator of the fresh fiber.

Similar to earlier cases, to ensure a feasible RCN, the two feasibility criteria on flowrate and load constraints need to be fulfilled. From Figure 3.19, it is observed that the source composite curve has a flowrate that is larger than that of the sink composite curve. This means that the flowrate constraint has been fulfilled. However, the load constraint has not been fulfilled as the two composite curves intersect each other.

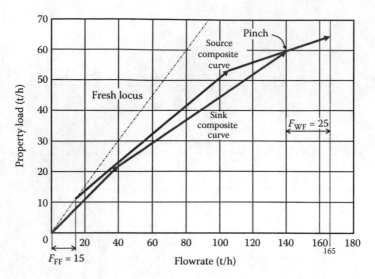

FIGURE 3.20
MRPD for Example 3.8. (From Kazantzi, V. and El-Halwagi, M.M., *Chem. Eng. Prog.*, 101(8), 28, 2005. Reproduced with permission. Copyright © 2005. American Institute of Chemical Engineers (AIChE)).

This can be accomplished by adding fresh resources (fiber) to the net-work. Following the revised Step 5 of the MRPD procedure, the source composite curve is slid along the fresh locus until it is above and to the left of the sink composite curve (Figure 3.20).

The interpretation for the MRPD is the same as in earlier examples. As shown in Figure 3.20, the minimum fresh fiber (F_{FF}) needed by the RCN is identified from the horizontal distance where the sink composite curve extends to the left of the source composite curve, i.e., 15 t/h. On the other hand, the minimum waste discharge (F_{WF}) is identified from the section where the source composite curve extends to the right of the sink composite curve, i.e., 25 t/h.

3.6 Additional Readings

Apart from the MRPD, many different graphical techniques have also been developed for the targeting of material reuse/recycle in the RCN in the past decade. These include the *limiting composite curve* (Wang and Smith, 1994, 1995), *surplus diagram* (Alves and Towler, 2002; Hallale, 2002), source com-posite curve (Bandyopadhyay, 2006; Bandyopadhyay et al., 2006), etc. Each of these methods has its own strength and limitations. A review that compares these techniques may be found in Foo (2009).

Problems

Water Minimization Problems

3.1 For the tire-to-fuel process in Problem 2.1, determine the minimum flowrates of freshwater (0 ppm) and wastewater. Use the limiting data in Table 3.8 (see answer in Problem 7.3).

3.2 For the textile plant in Problem 2.4, determine the minimum flowrates of freshwater (0 ppm) and wastewater for the process. The limiting data is given in Table 3.9.

3.3 Table 3.10 summarizes the limiting data for the classical water minimization example from Wang and Smith (1994). Identify the minimum freshwater and wastewater flowrates when the RCN is fed with single pure freshwater resource (see solution in Example 8.1).

3.4 Table 3.11 summarizes the limiting data for a water minimization case study taken from Polley and Polley (2000). Identify the minimum freshwater and wastewater flowrates for this case for the following scenarios:

TABLE 3.8

Limiting Data for Problem 3.1

j	Sinks, SK_j	F_{SKj} (kg/s)	C_{SKj} (ppm)	i	Sources, SR_i	F_{SRi} (kg/s)	C_{SRi} (ppm)
1	Seal pot	0.1	500	1	Decanter	0.2	500
2	Water jet makeup	0.18	50	2	Seal-pot	0.1	200

TABLE 3.9

Limiting Data for Problem 3.2

j	SK_j	F_{SK_i} (t/h)	C_{SK_i} (ppm)	i	SR_i	F_{SR_i} (t/h)	C_{SR_i} (ppm)
1	Singeing	2.67	0	1	Singeing	2.67	47
2	Mercerizing	22.77	75	2	Mercerizing	22.77	476
3	Scouring and bleaching	18.17	47	3	Scouring and bleaching	18.17	3218

TABLE 3.10

Limiting Data for Problem 3.3

Sinks, SK_j	F_{SKj} (t/h)	C_{SKj} (ppm)	Sources, SR_i	F_{SRi} (t/h)	C_{SRi} (ppm)
SK1	20	0	SR1	20	100
SK2	100	50	SR2	100	100
SK3	40	50	SR3	40	800
SK4	10	400	SR4	10	800

TABLE 3.11

Limiting Data for Problem 3.4

j	Sinks, SK_j	F_{SK_j} (t/h)	C_{SK_j} (ppm)	i	Sources, SR_i	F_{SR_i} (t/h)	C_{SR_i} (ppm)
1	SK1	50	20	1	SR1	50	50
2	SK2	100	50	2	SR2	100	100
3	SK3	80	100	3	SR3	70	150
4	SK4	70	200	4	SR4	60	250

    a. Single pure freshwater feed, i.e., zero impurity content (solution is found in Example 7.1).

    b. Single impure freshwater feed, with 10 ppm impurity content (see answer in Problem 7.1).

    c. Two freshwater feeds, i.e., pure (0 ppm, $1/ton) and impure freshwater feeds (80 ppm, $0.2/ton). Determine also the operating cost for freshwater feed (see answers in Problem 7.1).

3.5 Figure 3.21 shows a process flow diagram for a bulk chemical production (Foo, 2011). Two gas phase reactants are fed to the reactor to be converted into the desired product with the presence of water (as reaction carrier). In the separation unit, the unconverted reactant is recovered to the reactor, while the separated products are sent to the downstream

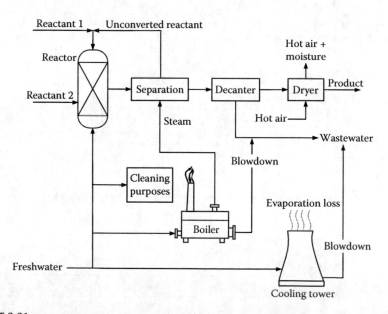

FIGURE 3.21

Process flow diagram for a bulk chemical production. (From Foo, D.C.Y., Resource conservation through pinch analysis, in Foo, D.C.Y., El-Halwagi, M.M., and Tan, R.R. eds., *Recent Advances in Sustainable Process Design and Optimisation*, World Scientific, in press, 2011.)

TABLE 3.12

Limiting Data for Problem 3.5

j	SK_j	F_{SKj} (t/h)	C_{SKj} (ppm)	i	SR_i	F_{SRi} (t/h)	C_{SRi} (ppm)
1	Reactor	10.5	0	1	Decanter	13.6	20
2	Boiler makeup	0.6	0	2	Cooling tower blowdown	1.0	400
3	Cooling tower makeup	9.3	50	3	Boiler blowdown	0.2	400
4	Cleaning	3.5	100	4			

purification units, which consist of decanter and dryer. Wastewater is generated from the decanter due to the huge amount of water being used as reaction carrier as well as the steam condensate used in the separation unit. Besides, a significant amount of water is also utilized in the utility section of the plant. This includes the makeup water for boiler and cooling tower (with a large evaporation loss) as well as water for general plant and vessel cleaning. Note that due to piping configuration, water for vessel cleaning also contributes to the decanter effluent. Apart from the decanter effluent, two other water sources are also found in the utility section, i.e., blowdown streams from the cooling tower and boiler.

Limiting data for water minimization are summarized in Table 3.12. Suspended solid is taken as the primary impurity for water recovery. Determine the minimum flowrates of freshwater and wastewater for the process. Freshwater feed is assumed to have zero impurity (see answers in Problem 7.6).

3.6 Figure 3.22 shows an organic chemical production (Hall, 1997; Foo, 2008). A large amount of water is consumed as reactor feed, as well as for utility and washing water (in washers and for hosepipe). Besides, a large amount of wastewater is also produced from the various processing units. The limiting flowrates and concentrations for the water sinks and sources are given in Figure 3.22. Determine the minimum flowrates of freshwater and wastewater for the process. Freshwater feed is assumed to have zero impurity (see answers in Problem 7.11).

3.7 Use the insights from the MRPD to develop the reuse/recycle scheme for the threshold problems in Examples 3.5 and 3.6. Show the RCN in a superstructural representation similar to that in Figure 1.17.

Utility Gas Recovery Problems

3.8 For the magnetic tape process in Problem 2.5 (El-Halwagi, 1997), identify the minimum fresh nitrogen feed and purge gas flowrate targets for its RCN. The limiting data are given in Table 3.13 (note: beware of the impurity concentration when plotting the source composite curve; see answer in Problem 7.18).

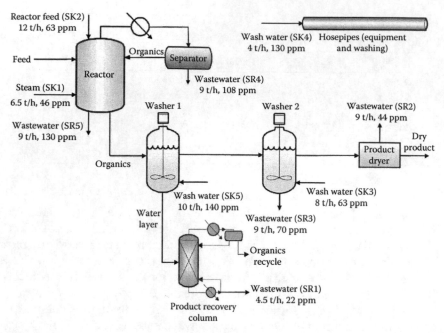

FIGURE 3.22
An organic chemical production. (From *J. Environ. Manag.*, 88(2), Foo, D.C.Y., Flowrate targeting for threshold problems and plant-wide integration for water network synthesis, 253–274, Copyright 2008, with permission from Elsevier.)

TABLE 3.13

Limiting Data for Problem 3.8

j	Sinks, SK_j	F_{SK_j} (kg/s)	C_{SK_j} (wt%)	i	Sources, SR_i	F_{SR_i} (kg/s)	C_{SR_i} (wt%)
1	Drying	5.5	0.1	1	Membrane retentate	3	1.9
2	Coating	3	0.2	2	Drying	5.5	0.4

3.9 Figure 3.23 shows a process where a large amount of oxygen gas is utilized (Foo and Manan, 2006). As shown, oxygen is used in both the oxidation processes as well as to enhance the combustion system and the aerobic section of the wastewater treatment plant. The purge oxygen streams from the oxidation processes are still of good quality and may be considered for reuse/recycling. Limiting data for the process sinks and sources are given in Table 3.14. The fresh oxygen supply has a supply of 5% impurity. Determine the minimum fresh oxygen gas and purge flowrate targets for the RCN (see answer in Problem 7.19).

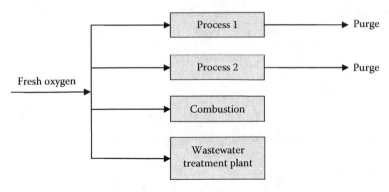

FIGURE 3.23
An oxygen recovery problem. (From Foo, D.C.Y. and Manan, Z.A., Setting the minimum flow-rate targets for utility gases using cascade analysis technique, *Ind. Eng. Chem. Res.*, 45(17), 5986. Copyright 2006 American Chemical Society. With permission.)

TABLE 3.14

Limiting Data for Problem 3.9

i	Sinks	F_{SKj} (kg/s)	C_{SKj} (wt%)	j	Sources	F_{SRi} (kg/s)	C_{SRi} (wt%)
1	Process 1	25	10	1	Process 1	10	40
2	Process 2	10	10	2	Process 2	5	50
3	Combustion	4	65				
4	Wastewater treatment	2	65				

Property Integration Problems

3.10 For the vinyl acetate manufacturing process in Problem 2.8, identify the minimum fresh and waste acetic acid flowrate targets for its RCN. The limiting data are given in Table 3.15. Assume that the fresh acetic acid does not poses any water content.

3.11 Figure 3.24 shows a microelectronics manufacturing facility where resource conservation is considered (Gabriel et al., 2003; El-Halwagi, 2006). Wafer fabrication (Wafer Fab) and combined chemical and mechanical processing (CMP) are identified as both sinks and sources for the reuse/recycle problem. Both units accept the ultrapure water (UPW) as feed and produce some rejects that may be recovered, i.e., 50% and 100% spent rinse streams. The property that is of concern for water recovery is resistivity (R), with the mixing rule given in Equation 2.64. The limiting data for the process are shown in Table 3.16. Determine the minimum flowrate targets for UPW as well as the reject streams for this process. The UPW has a resistivity of $18,000\,k\Omega/cm$ (see answer in Problem 7.25).

TABLE 3.15

Limiting Data for Problem 3.10

j	Sinks, SK_j	F_{SKj} (t/h)	ψ_{SKj} (wt%)	i	Sources, SR_i	F_{SRi} (t/h)	ψ_{SRi} (wt%)
1	Evaporator	10.2	10	1	Absorber II bottom	1.4	14
2	Absorber I	5.1	5	2	Distillation bottom	9.1	25

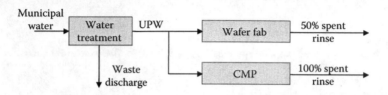

FIGURE 3.24

Microelectronics manufacturing facility. (From *Process Integration*, El-Halwagi, M.M., Copyright 2006, with permission from Elsevier.)

TABLE 3.16

Limiting Data for Problem 3.11

j	Sink	F_{SKj} (gal/min)	R_{SKj} (kΩ/cm)	ψ_{SKj} (cm/kΩ)
1	Wafer Fab	800	16,000	0.0000625
2	CMP	700	10,000	0.0001000

i	Source	F_{SRi} (gal/min)	R_{SRi} (kΩ/cm)	ψ_{SRi} (cm/kΩ)
1	50% spent rinse	1,000	8,000	0.000125
2	100% spent rinse	1,000	2,000	0.000500

TABLE 3.17

Limiting Data for Problem 3.12

j	Sinks	F_{SKj} (t/year)	ψ_{SKj} (Water Fraction)	i	Sources	F_{SRi} (t/year)	ψ_{SRi} (Water Fraction)
1	Magazine	400	0.038	1	Magazine	336	0.135
2	Manila card	400	0.049	2	Manila card	2.8	0.144

3.12 Table 3.17 shows the limiting data for a paper recovery study (Wan Alwi et al., 2010). The property that is of concern for paper reuse/recycle is the water fraction in the paper material. Fresh fiber with zero water fraction may be used to supplement the process sources. Use the MRPD to determine the minimum amount of fresh and waste fiber for the study.

References

Alves, J. J. and Towler, G. P. 2002. Analysis of refinery hydrogen distribution systems. *Industrial and Engineering Chemistry Research*, 41, 5759–5769.

Bandyopadhyay, S. 2006. Source composite curve for waste reduction. *Chemical Engineering Journal*, 125, 99–110.

Bandyopadhyay, S., Ghanekar, M. D., and Pillai, H. K. 2006. Process water management. *Industrial and Engineering Chemistry Research*, 45, 5287–5297.

El-Halwagi, M. M. 1997. *Pollution Prevention through Process Integration: Systematic Design Tools*. San Diego, CA: Academic Press.

El-Halwagi, M. M. 2006. *Process Integration*. San Diego, CA: Elsevier Inc.

El-Halwagi, M. M., Gabriel, F., and Harell, D. 2003. Rigorous graphical targeting for resource conservation via material recycle/reuse networks. *Industrial and Engineering Chemistry Research*, 42, 4319–4328.

Foo, D. C. Y. 2008. Flowrate targeting for threshold problems and plant-wide integration for water network synthesis. *Journal of Environmental Management*, 88(2), 253–274.

Foo, D. C. Y. 2009. A state-of-the-art review of pinch analysis techniques for water network synthesis. *Industrial and Engineering Chemistry Research*, 48(11), 5125–5159.

Foo, D. C. Y. 2012. Resource conservation through pinch analysis. In Foo, D. C. Y., El-Halwagi, M. M. and Tan, R. R. eds., *Recent Advances in Sustainable Process Design and Optimisation*. World Scientific.

Foo, D. C. Y. and Manan, Z. A. 2006. Setting the minimum flowrate targets for utility gases using cascade analysis technique. *Industrial and Engineering Chemistry Research*, 45(17), 5986–5995.

Foo, D. C. Y., Manan, Z. A., and Tan, Y. L. 2006. Use cascade analysis to optimize water networks. *Chemical Engineering Progress*, 102(7), 45–52.

Gabriel, F. B., Harell, D. A., Dozal, E., and El-Halwagi, M. M. 2003. Pollution targeting for functionality tracking. *AIChE Spring Meeting*, New Orleans, LA, March 31–April 4, 2003.

Hall, S. G. 1997. Water and effluent minimisation. Institution of Chemical Engineers, North Western Branch Papers, No. 4, November 12, 1997.

Hallale, N. 2002. A new graphical targeting method for water minimisation. *Advances in Environmental Research*, 6(3), 377–390.

Hallale, N. and Liu, F. 2001. Refinery hydrogen management for clean fuels production. *Advances in Environmental Research*, 6, 81–98.

Kazantzi, V. and El-Halwagi, M. M. 2005. Targeting material reuse via property integration. *Chemical Engineering Progress*, 101(8), 28–37.

Manan, Z. A., Tan, Y. L., and Foo, D. C. Y. 2004. Targeting the minimum water flowrate using water cascade analysis technique. *AIChE Journal*, 50(12), 3169–3183.

Ng, D. K. S., Foo, D. C. Y., Tan, R. R., Pau, C. H., and Tan, Y. L. 2009. Automated targeting for conventional and bilateral property-based resource conservation network. *Chemical Engineering Journal*, 149, 87–101.

Polley, G. T. and Polley, H. L. 2000. Design better water networks. *Chemical Engineering Progress*, 96(2), 47–52.

Prakash, R. and Shenoy, U. V. 2005. Targeting and design of water networks for fixed flowrate and fixed contaminant load operations. *Chemical Engineering Science*, 60(1), 255–268.

Shenoy, U. V. and Bandyopadhyay, S. 2007. Targeting for multiple resources. *Industrial and Engineering Chemistry Research*, 46, 3698–3708.

Wan Alwi, S. R. and Manan, Z. A. 2007. Targeting multiple water utilities using composite curves. *Industrial and Engineering Chemistry Research*, 46, 5968–5976.

Wan Alwi, S. R., Soh, G. K., and Manan, Z. A. 2010. A graphical approach for design of a maximum paper recycling network. *The 5th International Symposium on Design, Operation and Control of Chemical Processes* (PSE Asia 2010), Singapore, 25–28 July, 2010.

Wang, Y. P. and Smith, R. 1994. Wastewater minimisation. *Chemical Engineering Science*, 49, 981–1006.

Wang, Y. P. and Smith, R. 1995. Wastewater minimization with flowrate constraints. *Chemical Engineering Research and Design*, 73, 889–904.

4

*Algebraic Targeting Techniques for Direct Reuse/Recycle**

In Chapter 3, graphical technique that provides various useful insights for a resource conservation network (RCN) prior to detailed network design has been demonstrated. However, all graphical techniques are associated with several common pitfalls, such as low accuracy, as well as cumbersome drawing exercises. In this chapter, a well-established algebraic technique *known as material cascade analysis* (MCA) is introduced. The technique can be used to determine various RCN targets for direct reuse/recycle scheme, for both single and multiple fresh resources. Targeting for threshold problems is also discussed. Depending on the application, the MCA technique may be further categorized as follows: *water cascade analysis* (WCA; see Manan et al., 2004; Foo et al., 2006b; Foo, 2007, 2008), *gas cascade analysis* (GCA, Foo and Manan, 2006), and *property cascade analysis* (PCA, Foo et al., 2006a). However, all of these possess similar structure.

4.1 Generic Procedure for Material Cascade Analysis Technique

Steps to construct the MCA are given as follows:

1. Quality levels (q_k) of all process sinks and sources are arranged in descending order. Quality level for the fresh resource is added, if it does not already exist for the process sinks and sources. An arbitrary value is also added for the lowest quality level.

2. Flowrates of the process sinks (F_{SKj}) and sources (F_{SRi}) are located at their quality levels in the columns 2 and 3 of Table 4.1, respectively. When there is more than one sink and source at the same quality level, their individual flowrates should be added.

3. At each quality level k, the total flowrate of the process sink(s) is deducted from that of the process source(s), with the net flowrate given (see column 4, Table 4.1). For each level, a positive net flowrate indicates the presence of excess process source(s), while negative

---

* All examples of this chapter are available in spreadsheet, downloadable from book website.

TABLE 4.1

MCA for Minimum Resource Targeting

q_k	$\Sigma_j F_{SKj}$	$\Sigma_i F_{SRi}$	$\Sigma_i F_{SRi} - \Sigma_j F_{SKj}$	$F_{C,k}$	Δm_k	Cum. Δm_k	$F_{R,k}$
				F_R			
q_k	$(\Sigma_j F_{SKj})_1$	$(\Sigma_i F_{SRi})_1$	$(\Sigma_i F_{SRi} - \Sigma_j F_{SKj})_1$	⇩			
				$F_{C,k}$	Δm_k		
q_{k+1}	$(\Sigma_j F_{SKj})_{k+1}$	$(\Sigma_i F_{SRi})_{k+1}$	$(\Sigma_i F_{SRi} - \Sigma_j F_{SKj})_{k+1}$	⇩		Cum. Δm_{k+1}	$F_{R,k+1}$
				$F_{C,k+1}$	Δm_{k+1}	⇩	
⋮	⋮	⋮	⋮	⋮	⋮	⋮	⋮
⋮	⋮	⋮	⋮	⋮	⋮	⋮	⋮
⋮	⋮	⋮	⋮	⋮	⋮	⋮	⋮
q_{n-2}	$(\Sigma_j F_{SKj})_{n-2}$	$(\Sigma_i F_{SRi})_{n-2}$	$(\Sigma_i F_{SRi} - \Sigma_j F_{SKj})_{n-2}$	⇩		Cum. Δm_{n-2}	
				$F_{C,n-2}$	Δm_{n-2}	⇩	
q_{n-1}	$(\Sigma_j F_{SKj})_{n-1}$	$(\Sigma_i F_{SRi})_{n-1}$	$(\Sigma_i F_{SRi} - \Sigma_j F_{SKj})_{n-1}$	⇩		Cum. Δm_{n-1}	$F_{R,n-1}$
				$F_{C,n-1} = F_D$	Δm_{n-1}	⇩	
q_n						Cum. Δm_n	$F_{R,n}$

Source: From *Chem. Eng. Res. Des.*, 85(A8), Foo, D. C. Y., Water cascade analysis for single and multiple impure fresh water feed, 1169–1177, Copyright 2007, with permission from Elsevier.

value indicates insufficient process source(s) in fulfilling the requirement of the process sink(s).

4. The net material flowrate is cascaded down the quality levels to yield the cumulative flowrate ($F_{C,k}$) in column 5. The first entry of this column corresponds to the fresh resource consumption for the RCN (F_R). However, at this stage, it is assumed that no fresh resource is used, i.e., $F_R = 0$.

5. In column 6, the impurity/property load in each quality interval (Δm_k) is calculated. The load value is given by the product of the cumulative flowrate ($F_{C,k}$, column 5), combined with the difference across two quality levels ($q_{k+1} - q_k$), where $F_{C,k}$ is located.

6. The load values are cascaded down the quality levels to yield the cumulative load (Cum. Δm_k) in column 7. If negative Cum. Δm_k values are observed, Steps 7 and 8 are followed.

7. The *interval fresh resource flowrate* ($F_{R,k}$) is calculated for each quality level in column 8. This flowrate value is determined by dividing the cumulative loads (Cum. Δm_k, column 7) by the difference between the quality levels of interest (q_k) and of the fresh resource (q_R). This is given by Equation 4.1:

$$F_{R,k} = \frac{\Delta m_k}{(q_k - q_R)} \tag{4.1}$$

8. The absolute value of the largest negative $F_{R,k}$ in column 8 is identified as the *minimum fresh resource consumption* (F_R) of the network. Steps 4 to 6 are repeated by using the newly obtained F_R value in the first entry of column 5.

For convenience, it is always better to show the results of the MCA in tabulated form. When negative Cum. Δm_k values are observed in Step 7 (column 7), the result is known as the *infeasible cascade table*. After all negative Cum. Δm_k values in column 7 are removed (by carrying out Step 8), a *feasible cascade table* is created.

Several RCN targets are readily obtained from the feasible cascade table, similar to the graphical techniques described in Chapter 3. Apart from the fresh resource target (F_R) identified in Step 8, the final entry of column 5 in the feasible cascade table indicates the minimum waste discharge (F_D) from the RCN. Besides, the material recovery pinch is found at the quality level where zero Cum. Δm_k value (column 7) is observed. Process source that exists at the pinch is known as the *pinch-causing source*, where a portion of this source is allocated to the higher quality region (HQR) above the pinch, and the rest to the lower quality region (LQR) below the pinch. These flowrate allocation targets (F_{HQR} and F_{LQR} for HQR and LQR respectively) are observed in column 5 (among the $F_{C,k}$ values) in the intervals just above and just below the pinch quality level (Manan et al., 2004; Foo et al., 2006a,b).

Example 4.1 Water minimization for acrylonitrile (AN) production

Water minimization case study of the AN process given in Examples 2.1 and 3.1 is revisited here. The limiting water data for the problem are given in Table 4.2 (reproduced from Table 2.1). Assuming pure freshwater (i.e., no impurity) is used, determine the following RCN targets using the WCA technique:

TABLE 4.2

Limiting Water Data for AN Production

Water Sinks, SK_j		Flowrate	NH_3 Concentration
j	Stream	F_{SKj} (kg/s)	C_{SKj} (ppm)
1	BFW	1.2	0
2	Scrubber inlet	5.8	10

Water Sources, SR_i		Flowrate	NH_3 Concentration
i	Stream	F_{SRi} (kg/s)	C_{SRi} (ppm)
1	Distillation bottoms	0.8	0
2	Off-gas condensate	5	14
3	Aqueous layer	5.9	25
4	Ejector condensate	1.4	34

1. Minimum freshwater flowrate
2. Minimum wastewater flowrate
3. The flowrate allocation targets of the pinch-causing source to the HQR and LQR
4. Develop the RCN structure on the basis of insights gained from WCA

Solution

The eight-step procedure is followed to carry out WCA for the AN case study:

1. *Quality levels (q_k) of all process sinks and sources are arranged in descending order. Quality level for the fresh resource is also added, if it does not already exist for the process sinks and sources. An arbitrary value is also added for the highest quality level.*

   As mentioned in Examples 2.1 and 3.1, the quality level in this case corresponds to the concentration of ammonia (NH_3) in the water sinks and sources. We then arrange the concentrations in the final column of Table 4.2 in ascending order, i.e., 0, 10, 14, 25, and 34 ppm (i.e. descending order of quality). Note that the freshwater concentration (0 ppm) is already present among the concentration levels of the water sinks and sources. For the highest quality level, it is appropriate to set the highest concentration of the unit used in this case, i.e., 1,000,000 ppm. All quality levels are shown in the first column (C_k) of the cascade table in Table 4.3.

2. *Flowrates of the process sinks (F_{SKj}) and sources (F_{SRi}) are located at their quality levels in columns 2 and 3 of the cascade table.*

3. *At each quality level k, the total flowrate of the process sink(s) is deducted from that of the process source(s), with the net flowrate given in column 4.*

TABLE 4.3

Preliminary Cascade Table for Steps 2–4

C_k (ppm)	$\Sigma_j F_{SKj}$ (kg/s)	$\Sigma_i F_{SRi}$ (kg/s)	$\Sigma_i F_{SRi} - \Sigma_j F_{SKj}$ (kg/s)	$F_{C,k}$ (kg/s)
				$F_{FW} = 0$
0	1.2	0.8	−0.4	
				−0.4
10	5.8		−5.8	
				−6.2
14		5	5	
				−1.2
25		5.9	5.9	
				4.7
34		1.4	1.4	
				6.1
1,000,000				

4. *The net material flowrate is cascaded down the quality levels to yield the cumulative flowrate ($F_{C,k}$) in column 5. The first entry of this column corresponds to the fresh resource consumption for the RCN (F_R). However, at this stage, it is assumed that no fresh resource is used, i.e., $F_R = 0$.*

   Following Steps 2–4, a preliminary cascade table is constructed in Table 4.3. In the final column, fresh resource usage (corresponds to freshwater feed, F_{FW} in this case) is taken as zero.

5. *In column 6, the impurity/property load in each quality interval (Δm_k) is calculated.*

6. *The load values are cascaded down the quality levels to yield the cumulative load (Cum. Δm_k) in column 7. If negative Cum. Δm_k values are observed, Steps 7 and 8 are carried out.*

7. *The interval fresh resource flowrate ($F_{R,k}$) is calculated for each quality level in column 8.*

   Following Steps 5–7, an infeasible cascade table is constructed (Table 4.4). In column 7, negative Cum. Δm_k values are observed in the concentration levels of 10, 14, and 25 ppm (Step 6), which means that the RCN experiences load infeasibility. Hence, the interval freshwater flowrate ($F_{FW,k}$) for each concentration level is calculated and shown in column 8 of Table 4.4.

8. *The absolute value of the largest negative $F_{R,k}$ in column 8 is identified as the minimum fresh resource consumption (F_R) of the network. Steps 4–6 are repeated by using the newly obtained F_R value in the first entry of column 5.*

   In column 8 of Table 4.4, the largest negative $F_{FW,k}$ value is identified as –2.06 kg/s. Next, its absolute value is identified as the minimum freshwater consumption (F_{FW}) of the network. Repeating Steps 4–6 with $F_{FW} = 2.06$ kg/s results in the feasible cascade table in Table 4.5. In Table 4.5, the final entry of column 5 indicates the minimum waste (wastewater) discharge from

TABLE 4.4

Infeasible Cascade Table

C_k (ppm)	$\Sigma_j F_{SKj}$ (kg/s)	$\Sigma_i F_{SRi}$ (kg/s)	$\Sigma_i F_{SRi} - \Sigma_j F_{SKj}$ (kg/s)	$F_{C,k}$ (kg/s)	Δm_k (mg/s)	Cum. Δm_k (mg/s)	$F_{FW,k}$ (kg/s)
				$F_{FW}=0$			
0	1.2	0.8	–0.4				
				–0.4	–4		
10	5.8		–5.8			–4	–0.40
				–6.2	–24.8		
14		5	5			–28.8	–2.06
				–1.2	–13.2		
25		5.9	5.9			–42	–1.68
				4.7	42.3		
34		1.4	1.4			0.3	0.01
				6.1	6,099,792.6		
1,000,000						6,099,792.9	6.10

TABLE 4.5

Feasible Cascade Table for AN Case

C_k (ppm)	$\Sigma_j F_{SKj}$ (kg/s)	$\Sigma_i F_{SRi}$ (kg/s)	$\Sigma_i F_{SRi} - \Sigma_j F_{SKj}$ (kg/s)	$F_{C,k}$ (kg/s)	Δm_k (mg/s)	Cum. Δm_k (mg/s)
				$F_{FW} = 2.06$		
0	1.2	0.8	−0.4			
				1.66	16.57	
10	5.8		−5.8			16.57
				−4.14	−16.57	
14		5	5			**0.00**
				0.86	9.43	**(PINCH)**
25		5.9	5.9			9.43
				6.76	60.81	
34		1.4	1.4			70.24
				$F_{WW} = 8.16$	8,156,865.51	
100,0000						8,156,935.76

the RCN, with a flowrate (F_{WW}) of 8.16 kg/s. The material recovery pinch is found at the concentration level where zero Cum. Δm_k value is observed in column 7, i.e., 14 ppm.

The pinch divides the network into HQR and LQR. Hence, both process sinks and sources with concentration lower than 14 ppm, i.e., SK1, SK2, and SR1 (identified from Table 4.2), belong to the HQR (area shaded with vertical lines in Table 4.6), while those with concentration higher than 14 ppm, i.e., SR3 and SR4, belong to the LQR (area shaded with horizontal lines in Table 4.6). However, process source at 14 ppm, i.e., the pinch-causing

TABLE 4.6

HQR and LQR for AN Case

C_k (ppm)	$\Sigma_j F_{SKj}$ (kg/s)	$\Sigma_i F_{SRi}$ (kg/s)	$\Sigma_i F_{SRi} - \Sigma_j F_{SKj}$ (kg/s)	$F_{C,k}$ (kg/s)	Δm_k (mg/s)	Cum. Δm_k (mg/s)
				$F_{FW} = 2.06$		
0	1.2	0.8	−0.4			
				1.66	16.57	
10	5.8		−5.8			16.57
				$F_{HQR} = -4.14$	−16.57	
14		5	5			0.00
				$F_{LQR} = 0.86$	9.43	(PINCH)
25		5.9	5.9			9.43
				6.76	60.81	
34		1.4	1.4			70.24
				$F_{WW} = 8.16$	8,156,865.51	
1,000,000						8,156,935.76

source (SR2), belongs to both regions. As shown in column 5 of Table 4.6, 4.14 kg/s of the pinch-causing source is sent to the HQR (F_{HQR}, interval between 10 and 14 ppm, negative sign indicates water is sent from lower to higher quality levels), while 0.86 kg/s to the LQR (F_{LQR}, interval between 14 and 25 ppm).

For a simple case such as this, we can develop the RCN structure based on the flowrate allocation insights of the cascade table. As shown in column 5 of Table 4.6, 2.06 kg/s of freshwater is fed to the RCN. A portion of this freshwater flowrate (0.4 kg/s) is consumed at 0 ppm level, where negative net flowrate is observed. Since this level only consists of SK1 (boiler feed water—BFW) and SR1 (distillation bottoms), both freshwater feed and SR1 are used to fulfill the flowrate requirement of SK1.

At 10 ppm level, where SK2 (scrubber inlet) is found, the remaining freshwater flowrate (1.66 kg/s) is consumed by this sink. However, SK2 has a flowrate requirement of 5.8 kg/s. Hence, 4.14 kg/s of water flowrate is supplied by the allocation flowrate of the pinch-causing source, SR2 (off-gas condensate, 14 ppm), to the HQR (F_{HQR}). With the flowrate allocation to SK2, the pinch-causing source SR2 has its flowrate reduced to 0.86 kg/s, which then leaves as allocation flowrate to the LQR (F_{LQR}). This allocation flowrate is then cascaded to lower concentration levels. In the absence of process sink at 25 and 34 ppm levels, the allocation flowrate of SR2 is mixed with other unutilized sources (SR3 and SR4) to be discharged as wastewater stream from the RCN.

With the earlier-discussed insights, we can develop the structure for the RCN in Figure 4.1, which is essentially the same as that in Figure 3.10. However, note that the development of an RCN based on targeting insights will only work for simple cases such as this example.* For more complicated problems, a detailed network design procedure should be used (see Chapter 7 for details).

4.2 Targeting for Multiple Fresh Resources

When an RCN is served by several fresh resources of different quality, the MCA procedure is to be revised in order to determine the minimum resource flowrates. In principle, the revised procedure consists of three steps (Foo, 2007):

1. Flowrate targeting for lower quality resource—this step determines the minimum flowrate for lower quality resource in the absence of higher quality resource.

2. Flowrate targeting for higher quality resource—minimum higher quality resource flowrate needed to supplement the lower quality resource is determined.

---

* There should be maximum one sink with non-zero quality index in the HQR.

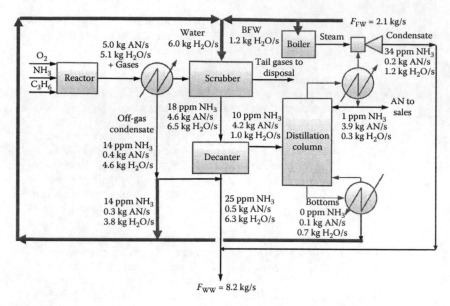

FIGURE 4.1
RCN structure for AN case study (Example 4.1). Note: Detailed mass balance has not been revised yet.

3. Flowrate adjustment of lower quality resource—the lower quality resource flowrate is reduced, as higher quality resource is used for the RCN.

Note that flowrate targeting/adjustment for each step should follow the MCA procedure as described in previous section.

Example 4.2 Targeting for multiple freshwater sources

The AN production case study in Example 4.1 is revisited. Similar to Example 3.4, a secondary impure freshwater source with 5 ppm ammonia content is available for use apart from the primary pure freshwater source. Unit costs for the primary (CT_{FW1}) and secondary freshwater sources (CT_{FW2}) are given as \$0.001/kg and \$0.0005/kg, respectively. Determine the minimum flowrates for both freshwater sources and wastewater stream, as well as the operating cost for freshwater sources of the RCN.

Solution

Since the unit costs of both freshwater sources are identical to those in Example 3.4, their prioritized costs remain identical. Hence the use of the secondary impure freshwater is maximized before the primary freshwater source is utilized. A three-step procedure is followed to carry out the targeting task:

1. *Flowrate targeting for lower quality resource.*

    Following the previously described MCA procedure, an infeasible cascade for targeting the minimum flowrate of the secondary freshwater source (5 ppm) is given in Table 4.7. As shown in the last column of Table 4.7, the minimum flowrate for this water source is determined as 3.2 kg/s.

2. *Flowrate targeting for higher quality resource.*

    We then include the targeted flowrate of secondary freshwater source ($F_{FW2} = 3.2$ kg/s) into the column for water sources in the cascade table. A pinch concentration is observed at 14 ppm, where zero cumulative load is found (see Table 4.8). However, it is also observed that a negative cumulative load value is found at 5 ppm. In other words, the RCN experiences load infeasibility at this level. Hence, we next proceed to determine the minimum flowrate of primary freshwater source (0 ppm) needed to restore the feasibility of the RCN. The interval flowrates for the primary freshwater source are then determined for each concentration level. As shown in the final column of Table 4.8, the minimum flowrate for this water source is determined as 0.4 kg/s.

3. *Flowrate adjustment of lower quality resource.*

    The targeted flowrate for the primary water source is included in the first level of column 5 in the cascade table (Table 4.9). However, we then observe that a new pinch concentration has emerged at 5 ppm and that the original pinch concentration at 14 ppm (due to the use of secondary freshwater source) has disappeared. It means that the secondary freshwater source has a larger flowrate than its minimum requirement, due to

TABLE 4.7

Targeting for Secondary Freshwater Source: Infeasible Cascade

C_k (ppm)	$\Sigma_j F_{SKj}$ (kg/s)	$\Sigma_i F_{SRi}$ (kg/s)	$\Sigma_i F_{SRi} - \Sigma_j F_{SKj}$ (kg/s)	$F_{C,k}$ (kg/s)	Δm_k (mg/s)	Cum. Δm_k (mg/s)	$F_{FW2,k}$ (kg/s)
0	1.2	0.8	−0.4				
				−0.4	−2		
5			0			−2.00	
				−0.4	−2		
10	5.8		−5.8			−4.00	−0.80
				−6.2	−24.8		
14		5	5			−28.80	−3.20
				−1.2	−13.2		
25		5.9	5.9			−42.00	−2.10
				4.7	42.3		
34		1.4	1.4			0.30	0.01
				6.1	6,099,792.6		
1,000,000						6,099,792.90	6.10

TABLE 4.8

Targeting for Primary Freshwater Source: Infeasible Cascade

C_k (ppm)	$\Sigma_j F_{SKj}$ (kg/s)	$\Sigma_i F_{SRi}$ (kg/s)	$\Sigma_i F_{SRi} - \Sigma_j F_{SKj}$ (kg/s)	$F_{C,k}$ (kg/s)	Δm_k (mg/s)	Cum. Δm_k (mg/s)	$F_{FW1,k}$ (kg/s)
0	1.2	0.8	−0.4				
				−0.40	−2.00		
5		$F_{FW2}=3.2$	3.2			−2.00	−0.40
				2.80	14.00		
10	5.8		−5.8			12.00	1.20
				−3.00	−12.00		
14		5	5			0.00	0.00
				2.00	22.00	(PINCH)	
25		5.9	5.9			22.00	0.88
				7.90	71.10		
34		1.4	1.4			93.10	2.74
				9.30	9,299,683.80		
1,000,000						9,299,776.90	9.30

TABLE 4.9

Flowrate Adjustment for Secondary Freshwater Source

C_k (ppm)	$\Sigma_j F_{SKj}$ (kg/s)	$\Sigma_i F_{SRi}$ (kg/s)	$\Sigma_i F_{SRi} - \Sigma_j F_{SKj}$ (kg/s)	$F_{C,k}$ (kg/s)	Δm_k (mg/s)	Cum. Δm_k (mg/s)	$F_{FW2,k}$ (kg/s)
				$F_{FW1}=0.40$			
0	1.2	0.8	−0.4				
				0.00	0.00		
5		$F_{FW2}=3.2$	3.2			0.00	
				3.20	16.00	(PINCH)	
10	5.8		−5.8			16.00	3.20
				−2.60	−10.40		
14		5	5			5.60	0.62
				2.40	26.40		
25		5.9	5.9			32.00	1.60
				8.30	74.70		
34		1.4	1.4			106.70	3.68
				$F_{WW}=9.70$	9,699,670.20		
1,000,000						9,699,776.90	9.70

the use of the primary freshwater source. Hence, we proceed to reduce the flowrate of the secondary freshwater source. The last column in Table 4.9 tabulates the *excessive interval flowrates* supplied by the secondary freshwater source to each concentration level, which is determined using Equation 4.1. For instance, the excessive interval flowrate at 10 ppm is determined as follows:

TABLE 4.10

Minimum Flowrates for Primary and Secondary Freshwater Sources

C_k (ppm)	$\Sigma_j F_{SKj}$ (kg/s)	$\Sigma_i F_{SRi}$ (kg/s)	$\Sigma_i F_{SRi} - \Sigma_j F_{SKj}$ (kg/s)	$F_{C,k}$ (kg/s)	Δm_k (mg/s)	Cum. Δm_k (mg/s)
				$F_{FW1} = 0.40$		
0	1.2	0.8	−0.4			
					0.00	0.00
5		$F_{FW2} = 2.58$	3.2			0.00
					2.58	12.89 (PINCH)
10	5.8		−5.8			12.89
					−3.22	−12.89
14		5	5			0.00
					1.78	19.56 (PINCH)
25		5.9	5.9			19.56
					7.68	69.10
34		1.4	1.4			88.66
				$F_{WW} = 9.08$	9,077,469.13	
1,000,000						9,077,557.79

$$\frac{16\,\text{mg/s}}{(10-5)\text{ppm}} = 3.2\,\text{kg/s}$$

From the last column of Table 4.9, it is observed that the minimum excessive interval flowrate is found at 14 ppm, i.e., 0.62 kg/s. In other words, 0.62 kg/s excessive flowrate of the secondary freshwater source has been supplied to the RCN. Hence, this excessive flowrate is to be deducted from the earlier targeted secondary freshwater source flowrate (in Step 1), i.e., 3.2 − 0.62 = 2.58 kg/s. Including this reduced flowrate in the cascade table restores the original pinch concentration at 14 ppm (see Table 4.10). As discussed in Example 3.4, for a *multiple pinch* RCN, the higher quality pinch (5 ppm) is termed as the *limiting pinch*, which controls the overall fresh resource flowrates of the RCN. Note that since both freshwater sources have unit costs identical to those given in Example 3.4, the operating cost of freshwater is essentially equal to that given in Example 3.4, i.e., $0.0017/s (= 0.4 kg/s × $0.001/kg + 2.58 kg/s × $0.0005/kg).

4.3 Targeting for Threshold Problems

As discussed in Chapter 3, there exist two distinct cases of threshold problems, i.e., RCN with zero fresh resource and/or zero waste discharge. For

RCN with zero fresh resource, the procedure described earlier for MCA may be followed without modification. Note, however, that only Steps 1–6 of the procedure are needed to establish the RCN targets, while Steps 7 and 8 are omitted.

On the other hand, for RCN with zero waste discharge, performing Steps 1–8 of the MCA procedure will result in negative waste discharge. Hence, Step 9 is added to establish the rigorous targets for this kind of RCN, which is described as follows:

9. The absolute value of the negative waste discharge flowrate is added to the fresh resource consumption determined in Step 8 to obtain the rigorous fresh resource consumption (F_R) for the threshold case. Steps 4–6 are repeated by using the newly obtained F_R value in the first entry of column 5.

Example 4.3 Targeting for threshold problem—zero fresh resource

Table 4.11 shows the limiting data for a water minimization case study described in Example 3.5. Determine the following RCN targets using the WCA technique:

1. Minimum freshwater flowrate
2. Minimum wastewater flowrate
3. Pinch concentration for the RCN

Solution

Following Steps 1–6 of the procedure, the cascade table for the example is constructed in Table 4.12. Since none of the Cum. Δm_k values in column 7 is negative, the WCA procedure is completed. The first and final entries in column 5 indicate that the RCN requires no freshwater feed ($F_{FW}=0$) but will generate a wastewater flowrate (F_{WW}) of 20 t/h. The pinch for this problem is located at the highest concentration level, i.e., 0 ppm. This means that all water sinks and sources are located in the LQR.

Example 4.4 Targeting for threshold problem—zero waste discharge

Table 4.13 shows the limiting data for a water minimization case study described in Example 3.6. Determine the minimum freshwater and wastewater flowrates using the WCA technique.

TABLE 4.11

Limiting Water Data for Example 4.3

SK_j	F_{SKj} (t/h)	C_{SKj} (ppm)	SR_i	F_{SRi} (t/h)	C_{SRi} (ppm)
SK1	40	125	SR1	50	100
SK2	20	500	SR2	50	300
SK3	40	1125	SR3	20	1000

TABLE 4.12

Cascade Table for Example 4.3

C_k (ppm)	$\Sigma_j F_{SKj}$ (t/h)	$\Sigma_i F_{SRi}$ (t/h)	$\Sigma_i F_{SRi} - \Sigma_j F_{SKj}$ (t/h)	$F_{C,k}$ (t/h)	Δm_k (kg/h)	Cum. Δm_k (kg/h)
				$F_{FW}=0$		
0			0			(PINCH)
				0	0	
100		50	50			0
				50	1.25	
125	40		−40			1.25
				10	1.75	
300		50	50			3
				60	12	
500	20		−20			15
				40	20	
1000		20	20			35
				60	7.5	
1125	40		−40			42.5
				$F_{WW}=20$	19,977.5	
1,000,000			0			20,020

TABLE 4.13

Limiting Water Data for Example 4.4

SK_j	F_{SKj} (t/h)	C_{SKj} (ppm)	SR_i	F_{SRi} (t/h)	C_{SRi} (ppm)
SK1	50	20	SR1	20	20
SK2	20	50	SR2	50	100
SK3	100	400	SR3	40	250

Solution

Following Steps 1–6 of the procedure, the infeasible cascade table for the example is constructed in Table 4.14. Since negative Cum. Δm_k values are observed in column 7, the interval freshwater flowrates are determined for each concentration level (Step 7), with the results shown in the final column of Table 4.14. The largest negative value for this column is observed at the lowest concentration level (1,000,000 ppm), i.e., 59.97 t/h. This means that 59.97 t/h of freshwater is needed at this concentration level.

Following Step 8 of the procedure, 59.97 t/h freshwater is added at the first entry of column 5, with the cascade result shown in Table 4.15. However, the wastewater flowrate is observed to have a negative value at the last entry of column 5. Obviously, this is an infeasible solution, which means that the freshwater flowrate is insufficient to supplement the requirement of the water sink for the network.

TABLE 4.14

Infeasible Cascade Table for Example 4.4

C_k (ppm)	$\Sigma_j F_{SKj}$ (t/h)	$\Sigma_i F_{SRi}$ (t/h)	$\Sigma_i F_{SRi} - \Sigma_j F_{SKj}$ (t/h)	$F_{C,k}$ (t/h)	Δm_k (kg/h)	Cum. Δm_k (kg/h)	$F_{FW,k}$ (t/h)
0				0.00			
					0.00	0.00	
20	50	20	−30			0.00	0.00
				−30.00	−0.90		
50	20		−20			−0.90	−18.00
				−50.00	−2.50		
100		50	50			−3.40	−34.00
				0.00	0.00		
250		40	40			−3.40	−13.60
				40.00	6.00		
400	100		−100			2.60	6.50
				−60.00	−59,976.00		
1,000,000						−59,973.40	−59.97

TABLE 4.15

Cascade Table with Infeasible Wastewater Flowrate

C_k (ppm)	$\Sigma_j F_{SKj}$ (t/h)	$\Sigma_i F_{SRi}$ (t/h)	$\Sigma_i F_{SRi} - \Sigma_j F_{SKj}$ (t/h)	$F_{C,k}$ (t/h)	Δm_k (kg/h)	Cum. Δm_k (kg/h)
				$F_{FW} = 59.97$		
0						
				59.97	1.20	
20	50	20	−30			1.20
				29.97	0.90	
50	20		−20			2.10
				9.97	0.50	
100		50	50			2.60
				59.97	9.00	
250		40	40			11.59
				99.97	15.00	
400	100		−100			26.59
				$F_{WW} = -0.03$	−26.59	
1,000,000						0.00 (PINCH)

Following Step 9 of the procedure, absolute value of the negative wastewater discharge is added to the freshwater flowrate, which results in a total feed of 60 t/h. A revised cascade table with 60 t/h freshwater feed is constructed in Table 4.16. It is observed that the RCN has zero wastewater discharge ($F_{WW} = 0$). The concentration where the pinch was formed earlier (Table 4.15) is now termed as the *threshold point* of the RCN (Foo, 2008).

TABLE 4.16

Feasible Cascade Table for Example 4.4

C_k (ppm)	$\Sigma_j \Gamma_{SKj}$ (t/h)	$\Sigma_i F_{SRi}$ (t/h)	$\Sigma_i F_{SRi} - \Sigma_j F_{SKj}$ (t/h)	$F_{C,k}$ (t/h)	Δm_k (kg/h)	Cum. Δm_k (kg/h)
				$F_{FW} = 60.00$		
0						
				60.00	1.20	
20	50	20	−30			1.20
				30.00	0.90	
50	20		−20			2.10
				10.00	0.50	
100		50	50			2.60
				60.00	9.00	
250		40	40			11.60
				100.00	15.00	
400	100		−100			26.60
				$F_{WW} = 0.00$	0.00	
1,000,000						26.60
						(THRESHOLD)

Note again that apart from water minimization examples presented here, threshold problems also exist in other.

4.4 Targeting for Property Integration with Inferior Property Operator Level

Procedure for carrying out MCA for most property integration problems (i.e., PCA) remains the same as described earlier. However, the cummulative load in column 7 is now termed as the property load (ΔL_k). Besides, in special case where the property operator of the fresh resource exists at the inferior level, modified Steps 1 and 7 are required (see modification in italic case):

1. *Property operator levels* (Ψ_k) of all process sinks and sources are arranged in decending order. Operator level for the fresh resource is added, if it does not already exist for the process sinks and sources. *Zero value is also added for the lowest quality level.*

7. The interval fresh resource flowrate ($F_{R,k}$) is calculated for each *operator* level in column 8. This flowrate value is determined by dividing the cumulative loads (Cum. ΔL_k, column 7) by the difference between the *quality level of the fresh resource* (q_R) *and the quality level of interest* (q_k). This is given by Equation 4.2:

$$F_{R,k} = \frac{\Delta L_k}{(q_R - q_k)} \tag{4.2}$$

TABLE 4.17

Limiting Data for Example 4.5

Sink, SK$_j$	F_{SKj} (t/h)	ψ_{SKj} (Dimensionless)	Source, SR$_i$	F_{SRi} (t/h)	ψ_{SRi} (Dimensionless)
Machine I (SK1)	100	0.382	Bleached fiber (SR1)	90	0.469
Machine II (SK2)	40	0.536	Broke (SR2)	60	0.182
			Fresh fiber (FF)	To be determined	0.738

Example 4.5 Targeting for property integration—fresh resource with inferior operator value

Table 4.17 shows the limiting data for the fiber recovery case study described in Examples 2.6 and 3.8. Determine the minimum fresh and waste fiber targets using the PCA technique.

Solution

Following the revised Step 1, the property operator levels are arranged in descending order, as shown in column 1 of Table 4.18. A zero-quality level is also added. Next, following Steps 2–6 of the MCA procedure, an infeasible cascade table for the example is constructed in Table 4.18. Since negative Cum. ΔL_k values are observed in column 7, the interval fresh fiber flowrates ($F_{FF,k}$) are calculated for each operator level in column 8 (Step 7).

Finally, a feasible cascade is constructed in Table 4.19, following Step 8 of the MCA procedure. The minimum fresh fiber requirement (F_{FF}) is

TABLE 4.18

Infeasible Cascade Table for Example 4.5

ψ_k	$\Sigma_j F_{SKj}$ (t/h)	$\Sigma_i F_{SRi}$ (t/h)	$\Sigma_i F_{SRi} - \Sigma_j F_{SKj}$ (t/h)	$F_{C,k}$ (t/h)	Δm_k (t/h)	Cum. ΔL_k (t/h)	$F_{FF,k}$ (t/h)
0.738							
				0	0		
0.536	40		−40			0	0
				−40	−2.68		
0.469		90	90			−2.68	−9.96
				50	4.35		
0.382	100		−100			1.67	4.69
				−50	−10		
0.182		60	60			−8.33	−14.98
				10	1.82		
0.000						−6.51	−8.82

TABLE 4.19

Feasible Cascade Table for Example 4.5

ψ_k	$\Sigma_j F_{SKj}$ (t/h)	$\Sigma_i F_{SRi}$ (t/h)	$\Sigma_i F_{SRi} - \Sigma_j F_{SKj}$ (t/h)	$F_{C,k}$ (t/h)	ΔL_k (t/h)	Cum. Δm_k (t/h)
				$F_{FF}=14.98$		
0.738						
				14.98	3.03	
0.536	40		−40			3.03
				−25.02	−1.68	
0.469		90	90			1.35
				64.98	5.65	
0.382	100		−100			7.00
				−35.02	−7.00	
0.182		60	60			0
				$F_{WF}=24.98$	4.55	(PINCH)
0.000						4.55

targeted at 14.95 t/h, while the waste fiber target (F_{WF}) is identified at 24.95 t/h. These targets match those obtained via graphical targeting technique described in Chapter 3 (Figure 3.20).

Problems

Water Minimization Problems

4.1 For the Kraft pulping process described in Problem 2.2, determine the minimum flowrates of freshwater and wastewater for the RCN using WCA for the following cases. Limiting data for the RCN are given in Table 4.20. Assume pure freshwater is used.

   a. Correctly extracted limiting data—sources are segregated (use data in Table 4.20a) (answers are found in Problem 7.15).

   b. Incorrectly extracted limiting data—source are not segregated (use data in Table 4.20b).*

   c. Determine the additional operating cost incurred with the use of incorrect limiting data in (b) as compared to the case with correct limiting data in (a). Assume an annual operating time of 8000 h and unit costs for freshwater and wastewater as $1.5/ton and $10/ton, respectively.

---

* See Section 2.1 for discussion on segregation of material sources.

TABLE 4.20

Limiting Water Data for Problem 4.2

j	Sinks, SK$_j$	F_{SKj} (t/h)	C_{SKj} (ppm)	i	Sources, SR$_i$	F_{SRi} (t/h)	C_{SRi} (ppm)
(a) Correctly extracted limiting data							
1	Pulp washing	467	20	1	W3	12.98	419
2	Chemical recovery	165	20	2	W5	9.7	16,248
3	Condenser	8.2	10	3	W7	10.78	9,900
				4	W8	116.5	20
				5	W9	48	233
				6	W10	52	311
				7	W11	52.2	20
				8	W13	300	30
				9	W14	140	15
(b) Incorrectly extracted limiting data							
1	Pulp washing	467	20	1	W6	22.68	7189
2	Chemical recovery	165	20	2	W7	10.78	9900
3	Condenser	8.2	10	3	W12	268.7	114
				4	W13	300	30
				5	W14	140	15

4.2 For the tricresyl phosphate manufacturing process in Problem 2.3, determine the minimum flowrates of freshwater (0 ppm) and wastewater for the RCN for the following cases.

(a) Incorrectly extracted limiting data—flowrates for sources SR3, SR4, and SR5 are taken as the largest values (use data in Table 4.21a).

(b) Correctly extracted limiting data—flowrates for sources SR3, SR4, and SR5 are taken as the smallest values (use data in Table 4.21b).*

4.3 Table 4.22 shows the limiting data for a water minimization case study (Jacob et al., 2002). Determine the minimum flowrates of freshwater (0 ppm) and wastewater for the RCN (see answer in Problem 7.10).

4.4 Limiting data for a water network with 10 pairs of water sinks and sources are summarized in Table 4.23 (Savelski and Bagajewicz, 2001). Use WCA to identify the minimum freshwater and wastewater flowrates for this RCN. Assume pure freshwater is used (see answer in Problem 7.17).

4.5 An existing water network for a paper milling process is shown in Figure 4.2 (Foo et al., 2006b). The feedstock for the paper mill (mainly consists of old newspapers and magazines) is first blended with dilution water and chemicals to form pulp slurry called *stock*. The stock is sent to the forming section of the paper machine to form paper sheets. In the paper machine, freshwater is fed to the forming and pressing sections to

---

* See Section 2.2 for discussion on data extraction principles. Refer to Equations 2.54, 2.56, and 2.58 for operating flowrates of process sources.

TABLE 4.21

Limiting Water Data for Problem 4.2

j	Sinks, SK_j	F_{SKj} (kg/s)	C_{SKj} (ppm)	i	Sources, SR_i	F_{SRi} (kg/s)	C_{SRi} (ppm)
(a) Incorrectly extracted limiting data							
1	Washing I	2.45	5	1	Washing I	2.45	76.36
2	Washing II	2.45	0	2	Washing II	2.45	0.07
3	Scrubber I	0.84	30	3	Scrubber I	0.84	410.57
4	Scrubber II	0.60	30	4	Scrubber II	0.6	144
5	Flare seal pot	0.25	100	5	Flare seal pot	0.25	281.5
(b) Correctly extracted limiting data							
1	Washing I	2.45	5	1	Washing I	2.45	76.36
2	Washing II	2.45	0	2	Washing II	2.45	0.07
3	Scrubber I	0.70	30	3	Scrubber I	0.70	410.57
4	Scrubber II	0.50	30	4	Scrubber II	0.50	144
5	Flare seal pot	0.20	100	5	Flare seal pot	0.20	281.5

TABLE 4.22

Limiting Water Data for Problem 4.3

SK_j	F_{SKj} (t/h)	C_{SKj} (ppm)	SR_i	F_{SRi} (t/h)	C_{SRi} (ppm)
SK1	1200	120	SR1	500	100
SK2	800	105	SR2	2000	110
SK3	500	80	SR3	400	110
			SR4	300	60

TABLE 4.23

Limiting Water Data for Problem 4.4

Sinks, SK_j	F_{SKj} (t/h)	C_{SKj} (ppm)	Sources, SR_i	F_{SRi} (t/h)	C_{SRi} (ppm)
1	36.364	25	1	36.364	80
2	44.308	25	2	44.308	90
3	22.857	25	3	22.857	200
4	60.000	50	4	60.000	100
5	40.000	50	5	40.000	800
6	12.500	400	6	12.500	800
7	5.000	200	7	5.000	600
8	10.000	0	8	10.000	100
9	80.000	50	9	80.000	300
10	43.333	150	10	43.333	300

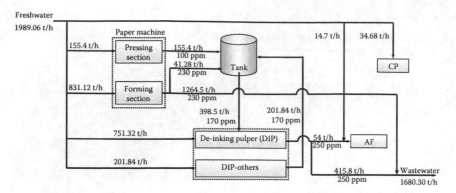

FIGURE 4.2
Water network of a paper mill. (From Foo, D. C. Y. et al., *Chem. Eng. Prog.*, 102(7), 45, 2006b. Reproduced with permission. Copyright © 2006. American Institute of Chemical Engineers (AIChE)).

remove debris. At the same time, wastewater is removed from the stock during paper sheet formation. A big portion of this wastewater stream is sent for waste treatment, while the rest to the water storage tank for reuse.

In the de-inking pulper (DIP) and its associated processes ("DIP-Others"), a mixture of freshwater and spent water (from the water storage) is used to remove printing ink from the main stock. A big portion of the DIP effluent is sent for waste treatment, whereas others are mixed with freshwater to be reused in the approach flow (AF) system. Besides, the effluent from "DIP-Others" is sent to the water storage tank for reuse/recycle. Freshwater is also consumed in the chemical preparation (CP) unit to dilute the de-inking chemicals.

It is believed that further water recovery is possible with this process. In considering water recovery, total suspended solid (TSS) is the most significant water quality factor that is of concern. Table 4.24 summarizes the limiting water data for the case study. Use WCA to identify the minimum freshwater and wastewater flowrate targets for the process. Assume pure freshwater is used.

TABLE 4.24

Limiting Water Data for Problem 4.5

j	Sinks, SK_j	F_{SKj} (t/h)	C_{SKj} (ppm)	i	Sources, SR_i	F_{SRi} (t/h)	C_{SRi} (ppm)
1	Pressing section	155.40	20	1	Pressing section	155.40	100
2	Forming section	831.12	80	2	Forming section	1305.78	230
3	DIP-Others	201.84	100	3	DIP-Others	201.84	170
4	DIP	1149.84	200	4	DIP	469.80	250
5	Chemical preparation (CP)	34.68	20				
6	Approach flow (AF)	68.70	200				

4.6 Limiting data for water minimization for a specialty chemical production process (Wang and Smith, 1995) are shown in Table 4.25. Freshwater in this case has zero impurity. Determine the minimum flowrates of freshwater and wastewater for the process (see answer in Problem 7.7).

4.7 Develop the RCN structure for the water minimization case studies in Examples 4.3 and 4.4, using the targeting insights from WCA.

4.8 A palm oil mill (Chungsiriporn et al., 2006) has undergone water minimization study. The limiting water data are shown in Table 4.26. Freshwater in this case is free from impurity. Determine the minimum flowrates of freshwater and wastewater for the water network of the mill (see answer in Problem 7.4). Develop the RCN structure using the targeting insights.

4.9 A water minimization study was carried out in a steel plant (Tian et al., 2008). The limiting water data are shown in Table 4.27. Determine the minimum freshwater and wastewater flowrates for the water network. Assume freshwater has an impurity (chlorine) content of 20 ppm (see answer in Problem 7.5).

4.10 In a Kraft pulping process, water minimization study is carried out to analyze its water-saving potential (Parthasarathy and Krishnagopalan, 2001). Impurity that is of concern for water recovery is the chloride content in the water sources. Determine the minimum freshwater and wastewater flowrates for the RCN. The freshwater source available for service has a chloride content of 4.2 ppm. Limiting data for the problem are summarized in Table 4.28.

TABLE 4.25

Limiting Water Data for Problem 4.6

Sinks	F_{SKj} (t/h)	C_{SKj} (ppm)	Sources	F_{SRi} (t/h)	C_{SRi} (ppm)
Reactor	80	100	Reactor	20	1000
Cyclone	50	200	Cyclone	50	700
Filtration	10	0	Filtration	40	100
Steam system	10	0	Steam system	10	10
Cooling system	15	10	Cooling system	5	100

TABLE 4.26

Limiting Water Data for Problem 4.8

j	Sinks, SK_j	F_{SKj} (t/h)	C_{SKj} (wt/wt)	i	Sources, SR_i	F_{SRi} (t/h)	C_{SRi} (wt/wt)
1	Mixing water	14.0	0.02	1	Separator outlet	24	0.093
2	Blending water	2.5	0.12	2	Sterilizer condensate	6	0.010
3	Phase balancing	7.5	0.093	3	Cooling water	4	0

TABLE 4.27

Limiting Water Data for Problem 4.9

j	Sinks, SK_j	F_{SKj} (t/day)	C_{SKj} (mg/L)	i	Sources, SR_i	F_{SRi} (t/day)	C_{SRi} (mg/L)
1	Hot air furnace	1680	90	1	Hot air furnace	960	400
2	Blast furnace	1920	75	2	Blast furnace	1440	100
3	Power plant	1680	40	3	Power plant	1320	45

TABLE 4.28

Limiting Water Data for Problem 4.10

j	Sinks, SK_j	F_{SKj} (t/d)	C_{SKj} (ppm)	i	Sources, SR_i	F_{SRi} (t/d)	C_{SRi} (ppm)
1	Washer	7,000	110	1	Evaporation condensate	7,971	0
2	Recovery furnace	4,200	5.5	2	Washer purge	945	275
3	Washers filters	2,000	38.5	3	Acid bleach effluent	14,520	235
4	ClO_2 stage	12,000	10.12	4	Alkali bleach effluent	13,750	504
5	Alkaline extraction	11,200	13.2				

4.11 Rework Scenarios (a)–(c) in Problem 3.4 by using WCA. Identify the minimum freshwater and wastewater flowrates for the RCN.* In addition, determine also the minimum flowrates for the following scenarios. In all cases, assume that the lower quality freshwater source(s) has lower prioritized cost as compared to the higher quality source(s) (solutions are found in Examples 6.1, 7.1 and Problem 7.1).

  d. Three freshwater sources are available for use: one pure (0 ppm) and two impure freshwater sources (20 and 50 ppm, respectively; solution is found in Problem 7.1).

  e. Same freshwater sources are available for use as in (d). However, the 20 ppm freshwater source has limited flowrate of 40 t/h.

4.12 Limiting data for literature example (Wang and Smith, 1995) are shown in Table 4.29. Determine the minimum flowrates of freshwater and wastewater for the following scenarios. In all cases, assume that the lower quality freshwater source(s) has lower prioritized cost as compared to the higher quality source(s).

  a. Two freshwater sources are available for use: pure (0 ppm) and impure freshwater sources (25 ppm).

  b. Two freshwater sources are available for use: pure (0 ppm) and impure freshwater sources (40 ppm).

* Answer for Scenario (a)—single pure fresh water resource is available in Example 6.1.

TABLE 4.29

Limiting Water Data for Problem 4.6

Sinks, SK$_j$	F_{SKj} (t/h)	C_{SKj} (ppm)	Sources, SR$_i$	F_{SRi} (t/h)	C_{SRi} (ppm)
1	20	0	1	20	100
2	100	50	2	100	100
3	40	50	3	40	800
4	10	80			

c. Three freshwater sources are available for use: one pure (0 ppm) and two impure freshwater sources (25 and 50 ppm).

  What conclusions can we make by observing the earlier-discussed scenarios?

Utility Gas Recovery Problems

4.13 For the refinery hydrogen network in Examples 2.4 and 3.3, use GCA to identify the minimum flowrates of the fresh hydrogen feed and purge stream for the RCN (for fresh hydrogen feed with 1% impurity). See limiting data in Table 2.2 (see answers in Problem 7.20). Identify also which hydrogen source(s) is being purged from the RCN.

4.14 Use GCA to identify the minimum flowrate targets of fresh nitrogen feed and purge gas stream for the magnetic tape process in Problems 2.5 and 3.8 (use limiting data in Table 3.13). Fresh nitrogen is free of impurity (see answer in Problem 7.18).

4.15 For the refinery hydrogen network in Problem 2.6, identify the minimum flowrates of fresh hydrogen feed (5% impurity content) and purge stream for the RCN using GCA technique. Limiting data are given in Table 4.30. The fresh hydrogen feed has an impurity content of 5%. Identify also which hydrogen source(s) is being purged from the RCN (see answer in Problem 7.21).

4.16 Data for a refinery hydrogen network are shown in Table 4.31 (Foo and Manan, 2006); determine the minimum flowrates of the fresh hydrogen

TABLE 4.30

Limiting Data for Hydrogen Network in Problem 4.15

j	Sinks, SK$_j$	F_{SKj} (mol/s)	C_{SKi} (mol%)	i	Sources, SR$_i$	F_{SRj} (mol/s)	C_{SRi} (mol%)
1	HCU	2495.0	19.39	1	HCU	1801.9	25.0
2	NHT	180.2	21.15	2	NHT	138.6	25.0
3	CNHT	720.7	24.86	3	CNHT	457.4	30.0
4	DHT	554.4	22.43	4	DHT	346.5	27.0
				5	SRU	623.8	7.0
				6	CRU	415.8	20.0

TABLE 4.31

Limiting Data for Hydrogen Network in Problem 4.16

Sinks, SK$_j$	F_{SKj} (mol/s)	C_{SKi} (mol%)	Sources, SR$_i$	F_{SRj} (mol/s)	C_{SRi} (mol%)
1	120.00	0.10	1	80.00	1.70
2	27.80	1.40	2	75.00	15.00
3	80.00	2.50	3	28.55	4.00
4	60.00	2.50	4	80.00	5.00
5	100.00	3.00	5	120.00	10.00
6	150.00	10.00	6	40.00	1.70
			7	80.00	2.50

feed and purge stream from the RCN using GCA technique. The fresh hydrogen feed has an impurity of 0.1% (note that this is multiple-pinch problem; see answer in Problem 7.22).

Property Integration Problems

4.17 Use PCA to identify the minimum fresh solvent and solvent waste targets for the metal degreasing process in Examples 2.5 and 3.7. Verify the targets with those identified using the graphical targeting technique described in Example 3.7. Also, identify the source that is discharged as waste from the RCN.

4.18 Identify the minimum fresh and solvent waste targets for the vinyl acetate–manufacturing process in Problems 2.8 and 3.10. Limiting data for the problem are given in Table 3.14. Verify the targets with those identified using the graphical targeting technique in Problem 3.10. Identify also the sources that are discharged as waste from the RCN.

4.19 For the wafer fabrication problem in Problem 2.7. Carry out flowrate targeting for the following scenarios (answers are available in Problem 7.26):

a. Identify the minimum ultrapure water (UPW—fresh resource) and wastewater flowrates for direct reuse/recycle scheme in the wafer fabrication process (FAB), utilizing the limiting data in Table 4.32a. Identify also the sources that are discharged as waste from the RCN.

b. The overall plant for wafer fabrication is given in Figure 4.3. Apart from the FAB section, three other process units are also consuming large amount of municipal freshwater. Identify the minimum flowrates for the fresh resources (UPW and municipal freshwater) and wastewater for direct reuse/recycle scheme for the entire plant. The limiting data for this scenario are given in Table 4.32b. The municipal freshwater has a resistivity value of 0.02 MΩ m.

TABLE 4.32

Limiting Data for Problem 4.19

j	Sinks, SK_j	F_{SKj} (t/h)	ψ_{SKj} ([MΩ m]$^{-1}$)	i	Sources, SR_i	F_{SRi} (t/h)	ψ_{SRi} ([MΩ m]$^{-1}$)
(a) FAB section							
1	Wet	500	0.143	1	Wet I	250	1
2	Lithography	450	0.125	2	Wet II	200	0.5
3	CMP	700	0.100	3	Lithography	350	0.333
4	Etc	350	0.200	4	CMP I	300	10
				5	CMP II	200	0.5
				6	Etc	280	2
(b) Entire plant							
1	Wet	500	0.143	1	Wet I	250	1
2	Lithography	450	0.125	2	Wet II	200	0.5
3	CMP	700	0.100	3	Lithography	350	0.333
4	Etc	350	0.200	4	CMP I	300	10
5	Cleaning	200	125.000	5	CMP II	200	0.5
6	Cooling tower makeup	450	50.000	6	Etc	280	2
7	Scrubber	300	100.000	7	Cleaning	180	500
				8	Scrubber	300	200
				9	UF reject	30% of UF inlet	100
				10	RO reject	30% of RO inlet	200

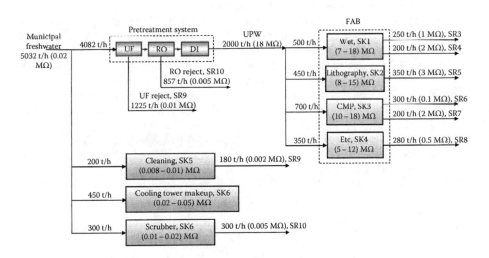

FIGURE 4.3

A wafer fabrication process. (From *Chem. Eng. J.*, 149, Ng, D. K. S., Foo, D. C. Y., Tan, R. R., Pau, C. H., and Tan, Y. L., Automated targeting for conventional and bilateral property-based resource conservation network, 87–101, Copyright 2009, with permission from Elsevier.)

TABLE 4.33

Limiting Data for Problem 4.20

Sinks	F_{SKj} (t/year)	ψ_{SKj} (Water Fraction)	Sources	F_{SRi} (t/year)	ψ_{SRi} (Water Fraction)
Newspaper	400	0.004	Newspaper	383.36	0.104
Tissue	400	0.006	Tissue	60.74	0.106
A4 paper	400	0.939	A4 paper	221.02	0.155
Coagulated box	600	0.839	Coagulated box	160.50	0.245

Note that the reject streams of the reverse osmosis (RO) and ultra-filtration (UF) units will have to be adjusted based on the targeted UPW flowrate.

4.20 Table 4.33 shows the limiting data for a paper recovery study (Wan Alwi et al., 2010). The property that is of concern for paper reuse/recycle is the water fraction in the paper material. Fresh fiber with zero water fraction may be used to supplement the process sources. Determine the minimum amount of fresh and waste fiber for the study.

References

Alves, J. J. and Towler, G. P. 2002. Analysis of refinery hydrogen distribution systems. *Industrial and Engineering Chemistry Research*, 41, 5759–5769.

Chungsiriporn, J., Prasertsan, S., and Bunyakan, C. 2006. Minimization of water consumption and process optimization of palm oil mills. *Clean Technologies and Environmental Policy*, 8, 151–158.

Foo, D. C. Y. 2007. Water cascade analysis for single and multiple impure fresh water feed. *Chemical Engineering Research and Design*, 85(A8), 1169–1177.

Foo, D. C. Y. 2008. Flowrate targeting for threshold problems and plant-wide integration for water network synthesis. *Journal of Environmental Management*, 88(2), 253–274.

Foo, D. C. Y., Kazantzi, V., El-Halwagi, M. M., and Manan, Z. A. 2006a. Cascade analysis for targeting property-based material reuse networks. *Chemical Engineering Science*, 61(8), 2626–2642.

Foo, D. C. Y. and Manan, Z. A. 2006. Setting the minimum flowrate targets for utility gases using cascade analysis technique. *Industrial and Engineering Chemistry Research*, 45(17), 5986–5995.

Foo, D. C. Y., Manan, Z. A., and Tan, Y. L. 2006b. Use cascade analysis to optimize water networks. *Chemical Engineering Progress*, 102(7), 45–52.

Jacob, J., Kaipe, H., Couderc, F., and Paris, J. 2002. Water network analysis in pulp and paper processes by pinch and linear programming techniques. *Chemical Engineering Communications*, 189(2), 184–206.

Manan, Z. A., Tan, Y. L., and Foo, D. C. Y. 2004. Targeting the minimum water flowrate using water cascade analysis technique. *AIChE Journal*, 50(12), 3169–3183.

Ng, D. K. S., Foo, D. C. Y., Tan, R. R., Pau, C. H., and Tan, Y. L. 2009. Automated targeting for conventional and bilateral property-based resource conservation network. *Chemical Engineering Journal*, 149, 87–101.

Parthasarathy, G. and Krishnagopalan, G. 2001. Systematic reallocation of aqueous resources using mass integration in a typical pulp mill. *Advances in Environmental Research*, 5, 61–79.

Savelski, M. J. and Bagajewicz, M. J. 2001. Algorithmic procedure to design water utilization systems featuring a single contaminant in process plants. *Chemical Engineering Science*, 56, 1897–1911.

Tian, J. R., Zhou, P. J., and Lv, B. 2008. A process integration approach to industrial water conservation: A case study for a Chinese steel plant. *Journal of Environmental Management*, 86, 682–687.

Wan Alwi, S. R., Soh, G. K., and Manan, Z. A. 2010. A graphical approach for design of a maximum paper recycling network. *The 5th International Symposium on Design, Operation and Control of Chemical Processes* (PSE Asia 2010), Singapore, July 25–28, 2010.

Wang, Y. P. and Smith, R. 1995. Wastewater minimization with flowrate constraints. *Chemical Engineering Research and Design*, 73, 889–904.

5

Process Changes for Resource Conservation Networks

In the previous chapters, we have seen how graphical and algebraic approaches can be used to determine minimum fresh resource and waste discharge targets for a resource conservation network (RCN). These minimum targets can be further reduced if the operating conditions of the processes (e.g., flowrate, concentration, and temperature) may be altered. In this chapter, we shall see how process changes may lead to reduced flowrates of fresh resource and waste discharge for an RCN.

5.1 Plus–Minus Principle

The *plus–minus principle* is an established technique for heat exchanger network synthesis (Linnhoff et al., 1982; Smith, 1995, 2005). The same principle applies to RCN equally well. Feng et al. (2009) first extended this principle for water minimization cases where water is used as mass separating agent in mass transfer operation (fixed load problem). Foo (2012) later generalized the principles for a wider range of RCN problems.

As shown in the material recovery pinch diagram (MRPD) in Figure 5.1, the pinch point divides the RCN into higher and lower quality regions. Fresh resource is used in the higher quality region, while waste is discharged in the lower quality region. In principle, any attempt to maximize the use of process sources in the process sinks will lead to a reduction of fresh resource and waste discharge for the RCN.

To reduce fresh resource consumption, and waste discharge for an RCN, the following process change strategies shall be implemented in the higher quality region:

1. Decrease/eliminate the flowrates of process sinks.
2. Reduce the quality requirement of process sinks.
3. Improve the quality of process sources with/without increasing the flowrate of process sources.

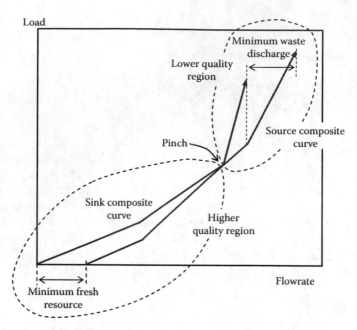

FIGURE 5.1
Higher and lower quality regions.

Note that reduced fresh resource consumption will automatically lead to reduced waste discharge from the RCN too. On the other hand, if the intention was only to reduce waste discharge in the lower quality region, alone (e.g. for cases were fresh resources has been reduced extensively), the following process change strategies can be implemented:

4. Increase the flowrates of process sinks in the lower quality region, with/without decreasing the quality levels of process sinks.

5. Decrease/eliminate the flowrates of process sources in the lower quality region, with/without increasing the quality levels of process sources.

6. Improve the quality of process sources in the higher quality region, with or without increasing the flowrate of process source(s).

Note that points (3) and (6) are essentially the same. In other words, improving the quality levels of process sources in the higher quality region (this include the pinch-causing source) leads to simultaneous reduction in fresh resource and waste discharge of the RCNs (see Figure 5.2). Hence, this point is taken as the highest priority for process change implementation. Note also that strategies 3, 4, and 6 are essentially the plus principles; while strategies 1, 2 and 5 correspond to minus principles.

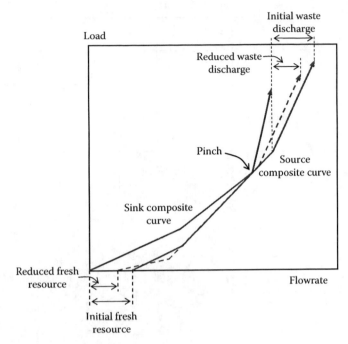

FIGURE 5.2
Decrease in the quality level of process source in the higher quality region leads to simultaneous reduction in fresh resource and waste discharge.

Example 5.1 Water minimization for acrylonitrile (AN) production

The water minimization case study of the AN process in Examples 2.1 and 3.1 is revisited here. By implementing direct water reuse/recycle, the process requires 2.1 kg/s of freshwater and generates 8.2 kg/s of wastewater, as shown in the MRPD in Figure 5.3 (reproduced from Figure 3.4). One option in process changes is to replace the steam-jet ejector with a vacuum pump (El-Halwagi, 1997). Determine the amount of freshwater and wastewater for this case.

Solution

Replacing the steam-jet ejector with a vacuum pump will eliminate the freshwater demand of the boiler (SK1) as well as the condensate from the steam-jet ejector completely (SR4). From Figure 5.3, we notice that SK1 is located in the lower concentration (higher quality) region, while SR4 is found in the higher concentration (lower quality) region. Hence, eliminating SK1 and SR4 corresponds to points (1) and (5) of the process change strategies, which leads to the reduction in freshwater and wastewater flowrates.

The resulting MRPD for this case is shown in Figure 5.4. The freshwater flowrate is reduced to 0.9 kg/s (from 2.1 kg/s), while the wastewater flowrate is reduced to 6.8 kg/s (from 8.2 kg/s). These flowrates correspond to

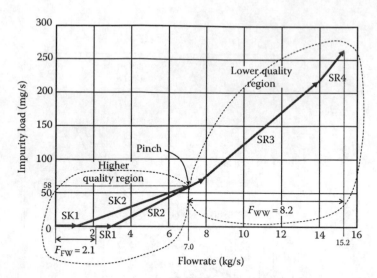

FIGURE 5.3
MRPD for direct water reuse/recycle case.

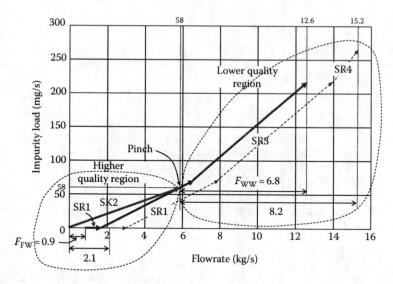

FIGURE 5.4
MRPD for Example 5.1 after process changes.

the reduction of 57% (freshwater) and 17% (wastewater), respectively, as compared with the reuse/recycle case.

Example 5.2 Solvent recovery for metal degreasing process

The metal degreasing process in Examples 2.5 and 3.7 is revisited here. The Reid vapor pressure (RVP) for Condensate I (SR1), i.e., RVP_{SR1}, is a function of the operating temperature (T, given in Kelvin) of the thermal processing unit (Kazantzi and El-Halwagi, 2005), given as

$$RVP_{SR1} = 0.56e^{\left(\frac{T-100}{175}\right)} \tag{5.1}$$

The acceptable operating temperature range is between 430 and 520 K. The thermal processing unit currently operates at 515 K, leading to an RVP of 6.0 atm. This operating condition can be modified to enable better solvent recovery between SR1 and the process sinks.

Determine the fresh solvent and solvent waste that can be further reduced for the RCN of this process.

Solution

The MRPD for the direct solvent reuse/recycle case is shown in Figure 5.5 (reproduced from Figure 3.18). Both the fresh solvent (F_{FS}) and solvent waste (F_{WS}) for the RCN are targeted as 2.4 kg/s each. It is also observed that SR1 is the pinch-causing source that belongs to both the higher and lower purity regions. Hence, process modification for this source will lead to the simultaneous reduction in fresh solvent and solvent waste. This corresponds to points (3) and (6) of the process change strategies.

Based on Equation 5.1, the operating temperature range of the thermal processing unit of 430 and 520 K corresponds to the RVP values of 3.69

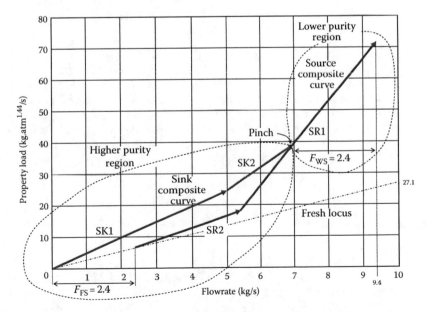

FIGURE 5.5
MRPD for direct solvent reuse/recycle case. (From Kazantzi, V. and El-Halwagi, M.M., *Chem. Eng. Prog.*, 101(8), 28, 2005. Reproduced with permission. Copyright © 2005. American Institute of Chemical Engineers (AIChE)).

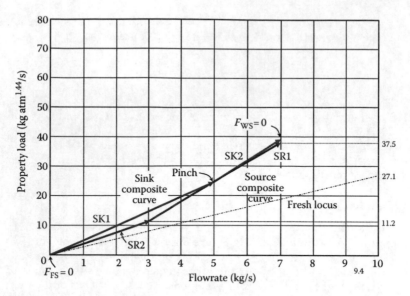

FIGURE 5.6
MRPD for Example 5.2 after process changes.

and 6.17 atm, respectively. Equation 2.25 shows that the property operator (ψ) of the RVP is given as

$$\psi = RVP_i^{1.44} \tag{2.25}$$

Hence, the RVP values of 3.69 and 6.17 atm are converted to the property operator (ψ) values of 6.56 and 13.74 $atm^{1.44}$, respectively. Obviously, for the case with fresh resource of superior operator value (see Section 2.5 for discussion of superior and inferior property operators of fresh resource), sources with a lower operator value are considered a better quality feed. Figure 5.6 shows the MRPD with SR1 of $\psi = 6.56$ $atm^{1.44}$. This leads to the complete removal of both fresh solvent requirement and solvent waste, resembling the special case of *threshold problem*, as discussed in Chapters 3 and 4.

Problems

5.1 Revisit the tire-to-fuel process in Problems 2.1 and 3.1 (see process flowsheet in Figure 5.7), and determine the minimum flowrates of freshwater (0 ppm) and wastewater for each of the following process changes. Determine the corresponding strategies associated with these changes.

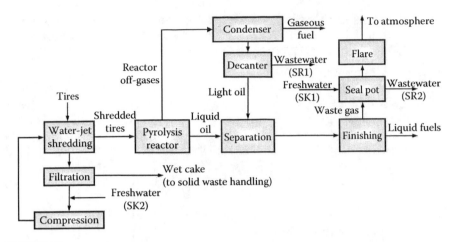

FIGURE 5.7
Process flowsheet for Problem 5.1.

TABLE 5.1

Limiting Data for Problem 3.1

		Base Case		Option 1		Option 2		Option 3	
j	Sinks, SK_j	F_{SKj}	C_{SKj}	F_{SKj}	C_{SKj}	F_{SKj}	C_{SKj}	F_{SKj}	C_{SKj}
1	Seal pot	0.15	200	0.15	200	0.15	200	0.15	200
2	Water jet makeup	0.25	50	0.25	50	0.20	50	0.20	50
i	Sources, SR_i	F_{SRi}	C_{SRi}	F_{SRi}	C_{SRi}	F_{SRi}	C_{SRi}	F_{SRi}	C_{SRi}
1	Decanter	0.27	500	0.27	500	0.24	500	0.20	500
2	Seal pot	0.15	200	0.15	500	0.15	500	0.15	500

Note: All flowrate terms in kg/s; concentration in ppm.

Limiting data for all changes are given in Table 5.1 (Noureldin and El-Halwagi, 1999; Bandyopadhyay, 2006).

a. Option 1—inlet concentration of the seal pot (SK1) is raised from 200 to 500 ppm.

b. Option 2 (to be implemented with Option 1)—reduction of makeup water to the compression station by increasing the compressor outlet pressure to 95 atm (current operating pressure is 70 atm; adjustable operating pressure range is 70–95 atm). The makeup water flowrate ($F_{make\text{-}up}$) is related to compressor outlet pressure (P_{comp}) by the following equation:

$$F_{make\text{-}up} = 0.47e^{-0.009P_{comp}} \tag{5.2}$$

Following this change, the makeup water is reduced to 0.20 kg/s (from 0.25 kg/s in Option 1) and decanter effluent is reduced to 0.24 kg/s (from 0.27 kg/s) in Option 1.

c. Option 3 (to be implemented with Option 2)—reduction of water generation in the reactor by increasing pyrolysis temperature to 710 K in the reactor (present temperature is set at 690 K, which leads to a water generation of 0.12 kg/s) (T_{pyro}, in Kelvin), which leads to the change in water flowrate generation in the reactor (F_{gen}). The two variables are related as follows:

$$F_{gen} = 0.152 + (5.37 - 0.00784 T_{pyro})e^{27.4 - 0.04 T_{pyro}} \tag{5.3}$$

Assuming that the same flowrate is separated in the decanter, the effluent of the latter will be reduced to 0.20 kg/s.

References

Bandyopadhyay, S. 2006. Source composite curve for waste reduction. *Chemical Engineering Journal*, 125, 99–110.

Feng, X., Liu, Y., Huang, L., and Deng, C. 2009. Graphical analysis of process changes for water minimization. *Industrial and Engineering Chemistry Research*, 48, 7145–7151.

Foo, D. C. Y. 2012. A generalised guideline for process changes for resource conservation networks. *Clean Technologies and Environmental Policy*, in press, DOI 10.1007/S10098-012-0475-4.

Kazantzi, V. and El-Halwagi, M. M. 2005. Targeting material reuse via property integration. *Chemical Engineering Progress*, 101(8), 28–37.

Linnhoff, B., Townsend, D. W., Boland, D., Hewitt, G. F., Thomas, B. E. A., Guy, A. R., and Marshall, R. H. 1982, last edition 1994. *A User Guide on Process Integration for the Efficient Use of Energy*. Warwickshire, U.K.: IChemE,.

Noureldin, M. B. and El-Halwagi, M. M. 1999. Interval-based targeting for pollution prevention via mass integration. *Computers and Chemical Engineering*, 23, 1527–1543.

Smith, R. 1995. *Chemical Process Design*. New York: McGraw-Hill.

Smith, R. 2005. *Chemical Process Design and Integration*. New York: John Wiley & Sons.

6

Algebraic Targeting Approach for Material Regeneration Networks

When the potential for flowrate reduction via material reuse/recycle is exhausted, a common mean to further reduce the flowrates of fresh resource and waste discharge for resource conservation network (RCN) is through material *regeneration*. The regenerated source possesses better quality and hence may be further reused/recycled to the process sinks. In this chapter, an algebraic approach developed by Ng et al. (2007, 2008) is introduced to determine the *ultimate flowrate targets* for a material regeneration network, which include the minimum flowrates for regeneration, fresh resource, and waste discharge. The technique is based on the *material cascade analysis* (MCA) technique in Chapter 4, in which better insights are used to assist stream selection for regeneration.

6.1 Types of Interception Units

There are two broad types of interception units, i.e., *single pass* and *partitioning* units (Ng et al., 2009; Tan et al., 2009). Both unit types may be further classified as *fixed outlet quality* or *fixed removal ratio* based on their performance. This chapter focuses on the single pass units of the fixed outlet quality type.[*] A schematic diagram for this unit is shown in Figure 6.1. It is commonly assumed that the unit has uniform inlet and outlet streams without flowrate loss (i.e., $F_{Rin} = F_{Rout}$). In principle, sources of different quality (i.e., inlet concentration, q_{Rin}) may be fed to these units and will be purified to a constant quality, q_{Rout} (e.g., 20 ppm, 5 wt%, etc.), at its outlet stream. Common examples for this kind of units include sand filter, membrane filtration, plate-and-frame filter press, etc.

The remaining of the chapter will focus on concentration-based water minimization problems, where the technique is mainly used. However, the technique is generic in nature and is hence applicable for all kinds of RCNs with different quality indices.

---

[*] See chapter 14 for targeting of other types of interception units.

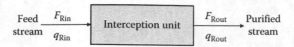

FIGURE 6.1
Single pass interception unit.

6.2 Targeting for Single Pass Interception Unit

The overall procedure for the algebraic targeting technique may be summarized as in Figure 6.2 (Ng et al., 2007, 2008). In principle, the procedure consists of the following six major steps:

1. An appropriate regeneration outlet concentration (C_{Rout}) is selected. In practice, we can evaluate several interception units of different outlet quality to compare the resource conservation potential of these units.

2. *Preliminary allocation* is performed to distribute the sources and sinks to two separate regions based on the chosen regeneration outlet concentration, i.e., the *freshwater region* (FWR) and *regenerated water region* (RWR). All sinks and sources with concentrations lower than the regeneration outlet concentration (i.e., $C_{SKj} < C_{Rout}$ and $C_{SRi} < C_{Rout}$) are sent to the FWR, while the rest to the RWR. In the FWR, both freshwater and regenerated water may be used to supplement the insufficient sources for the sinks. In contrast, only regenerated water should be used in the RWR. In principle, all sources are sent for reuse/recycle to the sinks, with or without regeneration.

3. *Flowrate reallocation*—Since the interception unit has uniform inlet and outlet flowrates, we need to ensure that the sinks and sources in the RWR have equal total flowrates. We next inspect the total sink and source flowrates in the RWR. If the total flowrate of the sinks is higher than that of the sources in the RWR, go to Step 3a or else to Step 3b:

   a. Total or partial flowrate of the lowest concentration sink in the RWR is reallocated to the FWR so that the total flowrate of the sinks is equal to that of the sources ($\Sigma_j F_{SKj} = \Sigma_i F_{SRi}$) in the RWR. Sink of the second lowest concentration is to be used if the first sink is insufficient (and so on). Move to Step 4a directly.

   b. Total or partial flowrate of the highest concentration source in the RWR is reallocated to the FWR so that the total flowrate of the sinks is equal to that of the sources ($\Sigma_j F_{SKj} = \Sigma_i F_{SRi}$) in the RWR. Source of the second highest concentration is to be used

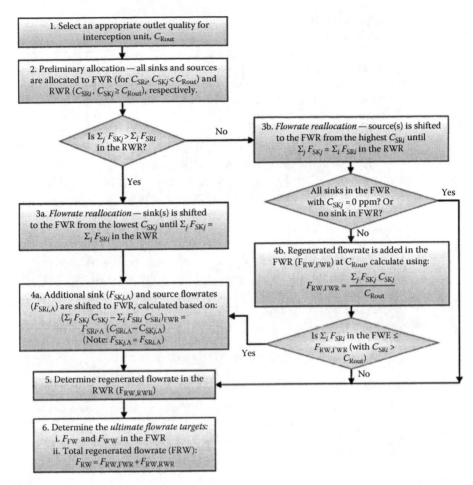

FIGURE 6.2
Algebraic targeting procedure for single pass source interception unit. (From *Chem. Eng. Res. Des.*, 86(10), Ng, D.K.S., Foo, D.C.Y., Tan, R.R., and Tan, Y.L., Extension of targeting procedure for 'Ultimate flowrate targeting with regeneration placement' by Ng et al., *Che. Eng. Res. Des.* 85 (A9): 1253–1267, 1182–1186, Copyright 2008, with permission from Elsevier.)

if the first source is insufficient (and so on). We next check if all sinks in the FWR have concentrations of 0 ppm. If only 0 ppm sinks (or no sink at all) are found in the FWR, then move to Step 5 or else to Step 4b.

4. Steps 4a and 4b are to be used in conjunction with Steps 3a and 3b, respectively:

a. *Additional sink/source shift*—Since the sink that was reallocated to the FWR (in Step 3a) has a concentration higher than C_{Rout}, it may tolerate sources with higher concentrations. Most of these high concentration sources are found in the RWR and hence may be considered for further shifting to the FWR. However, doing this affects the flowrate balance in the RWR. Hence, in order to maintain the flowrate balance, a pair of sink and source of equal flowrates are to be shifted to the FWR simultaneously. Note that both sink and source to be shifted should be among the lowest concentration sinks/sources in the RWR. In addition, the shifted source should originate from a concentration level with net positive flowrate (i.e., column $\Sigma_i F_{SRi} - \Sigma_j F_{SKj}$ in the cascade table) in the RWR and processes concentration higher than the shifted sink. The *additional shifted flowrates* of the source ($F_{SRi,A}$, with concentration $C_{SRi,A}$) and the sink ($F_{SKj,A}$, with concentration $C_{SKj,A}$) to the FWR may be calculated based on the following equation:

$$(\Sigma_j F_{SKj} C_{SKj} - \Sigma_i F_{SRi} C_{SRi})_{FWR} = F_{SRi,A} C_{SRi,A} - F_{SKj,A} C_{SKj,A} \qquad (6.1)$$

Note that the left side of the equation corresponds to the additional load that the sinks can accept in the FWR. Since the additional shifted flowrates of both the sinks and source are essentially equal (i.e., $F_{SRi,A} = F_{SKj,A}$), Equation 6.1 may take a simpler form that follows:

$$(\Sigma_j F_{SKj} C_{SKj} - \Sigma_i F_{SRi} C_{SRi})_{FWR} = F_{SRi,A}(C_{SRi,A} - C_{SKj,A}) \qquad (6.2)$$

Next proceed to Step 5.

b. When there are nonzero concentration sinks in the FRW, regenerated flowrate should be used to fulfill the flowrate requirement of the sinks, instead of using freshwater. The regenerated flowrate in the FWR ($F_{RW,FWR}$) is then calculated using Equation 6.3:

$$F_{RW,FWR} = \frac{\sum_j F_{SKj} C_{SKj}}{C_{Rout}} \qquad (6.3)$$

We next proceed to check if sources in the FWR are sufficient to be sent for regeneration. In other words, the total flowrate of sources in the FWR (with concentration higher than C_{Rout}) should be equal or higher than that determined with Equation 6.3. If the total source flowrate is sufficient, we shall proceed to Step 5. If the total source flowrate is insufficient, the total available source is used as regenerated flowrate. We then proceed to Step 4a to shift the additional sink and source pair from the RWR to the FWR to maximize the reuse/recycle potential of the process sources.

5. We next determine the regeneration flowrate in the RWR ($F_{RW,RWR}$). This can be done using the MCA technique, taking the regeneration outlet concentration as the concentration of the fresh resource in the RWR.

6. Determination of *ultimate flowrate targets*—The freshwater and waste-water flowrates can then be determined for the FWR, similarly with MCA. The total regenerated flowrate is given by the summation of $F_{RW,FWR}$ (Step 4b, if applicable) and $F_{RW,RWR}$ (Step 5).

 Example 6.1 Problem 3.4 revisited

The hypothetical case study from Polley and Polley (2000) is revisited here to illustrate the flowrate targeting procedure. Table 6.1 shows the limiting water data for the case study (reproduced from Table 3.10). If a direct reuse/recycle scheme is implemented, the minimum freshwater (F_{FW}, 0 ppm) and wastewater (F_{WW}) flowrate targets are determined as 70 and 50 t/h, respectively, which may be determined either with graphical (Chapter 3) or MCA techniques (Chapter 4). The result obtained with the latter is shown in Table 6.2.* Assuming that an interception unit with outlet concentration of 10 ppm is used to purify water sources for further reuse/recycle to the sinks, determine the minimum flowrate targets for freshwater, wastewater, and regeneration.

Solution

The six-step procedure is followed to identify the flowrate targets for the case study:

1. *An appropriate regeneration outlet concentration (C_{Rout}) is selected.*
2. *Preliminary allocation is performed to distribute the sources and sinks to two separate regions based on the chosen regeneration outlet concentration, i.e., the FWR and RWR.*

   With the given regeneration outlet concentration (C_{Rout}) of 10 ppm, the FWR and RWR are created. The FWR contains all water sinks and sources that are lower than 10 ppm, while all other sinks and sources with 10 ppm and above are allocated to the RWR. For this case, since all sinks and sources are higher than 10 ppm (lowest is SK1 with 20 ppm), all of them are allocated to the RWR. The results of the preliminary allocation are shown in Table 6.3.

TABLE 6.1

Limiting Water Data for Example 6.1

Sinks, SK_j	F_{SKj} (t/h)	C_{SKj} (ppm)	Sources, SR_i	F_{SRi} (t/h)	C_{SRi} (ppm)
SK1	50	20	SR1	50	50
SK2	100	50	SR2	100	100
SK3	80	100	SR3	70	150
SK4	70	200	SR4	60	250
$\Sigma_j F_{SKj}$	300		$\Sigma_i F_{SRi}$	280	

---

* This is solution for Problem 4.11a.

TABLE 6.2

Flowrate Targeting for Direct Reuse/Recycle Case for Example 6.1

C_k (ppm)	$\Sigma_j F_{SKj}$ (t/h)	$\Sigma_i F_{SRi}$ (t/h)	$\Sigma_i F_{SRi} - \Sigma_j F_{SKj}$ (t/h)	$F_{C,k}$ (t/h)	Δm_k (kg/h)	Cum. Δm_k (kg/h)
				$F_{FW} = 70$		
0						
				70	1.4	
20	50		−50			1.4
				20	0.6	
50	100	50	−50			2.0
				−30	−1.5	
100	80	100	20			0.5
				−10	−0.5	
150		70	70			0.0
				60	3.0	(PINCH)
200	70		−70			3.0
				−10	−0.5	
250		60	60			2.5
				$F_{WW} = 50$	49,987.5	
1,000,000						49,990.0

TABLE 6.3

Results of Preliminary Allocation

(a) FWR	C_k (ppm)	$\Sigma_j F_{SKj}$ (t/h)	$\Sigma_i F_{SRi}$ (t/h)	(b) RWR	C_k (ppm)	$\Sigma_j F_{SKj}$ (t/h)	$\Sigma_i F_{SRi}$ (t/h)
	0				10		
	20				20	50	
	50				50	100	50
	100				100	80	100
	150				150		70
	200				200	70	
	250				250		60
	1,000,000				1,000,000		
	Total					300	280

3. *Flowrate reallocation—To ensure the sinks and sources in the RWR have equal total flowrates. If the total flowrate of the sinks is higher than that of the sources in the RWR, go to Step 3a or else to Step 3b:*

   a. *Total or partial flowrate of the lowest concentration sink in the RWR is reallocated to the FWR so that the total flowrate of the sinks is equal to that of the sources ($\Sigma_j F_{SKj} = \Sigma_i F_{SRi}$). Sink*

TABLE 6.4

Results of Flowrate Reallocation

(a) FWR	C_k (ppm)	$\Sigma_j F_{SKj}$ (t/h)	$\Sigma_i F_{SRi}$ (t/h)	(b) RWR	C_k (ppm)	$\Sigma_j F_{SKj}$ (t/h)	$\Sigma_i F_{SRi}$ (t/h)	$\Sigma_i F_{SRi} - \Sigma_j F_{SKj}$ (t/h)
	0				10			
	20	20			20	30		−30
	50				50	100	50	−50
	100				100	80	100	20
	150				150		70	70
	200				200	70		−70
	250				250		60	60
	1,000,000				1,000,000			
	Total	20	0			280	280	

of the second lowest concentration is to be used if the first sink is insufficient (and so on). Move to Step 4a directly.

From the last row of Table 6.3(a) and (b), it is observed that the total flowrate of the sinks (300 t/h) is higher than that of the sources (280 t/h). Hence, 20 t/h of sinks are to be reallocated to the FWR, in order to achieve the flowrate balance in the RWR. Since the lowest concentration sink is to be moved, this corresponds to SK1 at 20 ppm. Hence, 20 t/h of SK1 is to be moved to the FWR, while the remaining flowrate (30 t/h) will stay in the RWR. The results of the flowrate reallocation are shown in Table 6.4.

4a. *Additional source shift—To maintain the flowrate balance, a pair of sink and source of equal flowrates are to be shifted to the FWR simultaneously. Note that both sink and source to be shifted should be among the lowest concentration sinks/sources in the RWR. In addition, the shifted source should originate from a concentration level with net positive flowrate in the RWR, and posseses higher concentration than the shifted sink.*

In Step 3a, 20 t/h of sink SK1 of 20 ppm has been shifted to the FWR. This sink may tolerate sources of higher concentrations. Since no other sources are found in the FWR, the source flowrate should be shifted from the RWR. To maintain the flowrate balance in the RWR, a pair of sink and source are to be shifted simultaneously. The sink should originate from the 20 ppm level (SK1), since it is the lowest concentration sink in the RWR (which means $C_{SKj,A} = 20$ ppm). On the other hand, the source should originate from 100 ppm (SR2, i.e., $C_{SRi,A} = 100$ ppm), since it is the lowest concentration level with positive net flowrate (observed from the last column of Table 6.4. Based on Equation 6.2, the additional shifted flowrates ($F_{SRi,A}$ and $F_{SKj,A}$) are determined as

$$20 \text{ t/h } (20 \text{ ppm}) = F_{SRi,A}(100 - 20 \text{ ppm})$$

i.e.,

$$F_{SRi,A} = F_{SKj,A} = 5 \text{ t/h}.$$

Hence, flowrates for both SK1 and SR2 in the RWR are reduced by 5 t/h (see Table 6.5). In contrast, their flowrates in the FWR are increased by 5 t/h (see Table 6.6).

5. *We next determine the regeneration flowrate in the RWR ($F_{RW,FWR}$). This can be done using the MCA technique, taking the regeneration outlet concentration as the concentration of the fresh resource in the RWR.*

In the RWR, all sources are sent to satisfy the flowrate requirement of the sinks, with or without water regeneration. Hence, a cascade analysis is performed to determine the minimum regeneration flowrate in the RWR ($F_{RW,FWR}$) as 53.57 t/h, with the result shown in Table 6.5.

A detailed inspection of Table 6.5 indicates that in the lower quality region (i.e., with concentration higher than 150 ppm), 6.43 t/h of source SR4 (250 ppm) is sent for reuse in sink SK4 (200 ppm), while the rest (53.57 t/h) for regeneration. Hence, it can be concluded that the source for regeneration should originate from the 250 ppm source. However, note that this is one of

TABLE 6.5

Regeneration Flowrate Targeting for RWR in Example 6.1

C_k (ppm)	$\Sigma_j F_{SKj}$ (t/h)	$\Sigma_i F_{SRi}$ (t/h)	$\Sigma_i F_{SRi} - \Sigma_j F_{SKj}$ (t/h)	$F_{C,k}$ (t/h)	Δm_k (kg/h)	Cum. Δm_k (kg/h)
				$F_{RW,RWR} = 53.57$		
10						
				53.57	0.54	
20	25		−25			0.54
				28.57	0.86	
50	100	50	−50			1.39
				−21.43	−1.07	
100	80	95	15			0.32
				−6.43	−0.32	
150		70	70			**0.00**
				63.57	3.18	**(PINCH)**
200	70		−70			3.18
				−6.43	−0.32	
250		60	60			2.86
				$F_{RW,RWR} = 53.57$	53,558.04	
1,000,000						53,560.89

TABLE 6.6

Freshwater and Wastewater Flowrate Targeting for FWR in Example 6.1

C_k (ppm)	$\Sigma_j F_{SKj}$ (t/h)	$\Sigma_i F_{SRi}$ (t/h)	$\Sigma_i F_{SRi} - \Sigma_j F_{SKj}$ (t/h)	$F_{C,k}$ (t/h)	Δm_k (kg/h)	Cum. Δm_k (kg/h)
				$F_{FW} = 20$		
0						
				20	0.2	
10						0.2
				20	0.2	
20	25		−25			0.4
				−5	−0.2	
50						0.3
				−5	−0.3	
100		5	5			0.0
				$F_{WW} = 0$	0.0	(PINCH)
1,000,000						0.0

the many options; other possible options for stream selection for regeneration are discussed in Chapter 8.

6. *Determination of ultimate flowrate targets—The freshwater and wastewater flowrates can then be determined for the FWR. The total regenerated flowrate is given by the summation of $F_{RW,FWR}$ (Step 4b, if applicable) and $F_{RW,RWR}$ (Step 5).*

Carrying out the cascade analysis determines that 20t/h of freshwater (F_{FW}) is needed in the FWR, while no wastewater is produced ($F_{WW} = 0$ t/h). The result of cascade analysis is shown in Table 6.6. Since no regenerated water is used in the FWR for this case, (i.e. $F_{RW,FWR} = 0$) the regeneration flowrate (F_{RW}) takes the value of $F_{RW,RWR}$ (53.57 t/h) as determined in Step 5.

For a better representation, we may also merge the targeting results in Tables 6.5 and 6.6 as a combined cascade table in Table 6.7. Note that SR4 at 250ppm has a flowrate of 6.43t/h, as the rest of this source has been purified to 10ppm as regeneration flowrate (F_{RW}).

Example 6.2

For Example 6.1, determine the minimum flowrate targets for freshwater, wastewater, and regenerated water when an interception unit with outlet concentration of 25ppm is used for water regeneration.

Solution

The six-step procedure is followed to identify the flowrate targets for the case study:

1. *An appropriate regeneration outlet concentration (C_{Rout}) is selected.*

TABLE 6.7

Combined Cascade Table for Example 6.1

C_k (ppm)	$\Sigma_j F_{SKj}$ (t/h)	$\Sigma_i F_{SRi}$ (t/h)	$\Sigma_i F_{SRi} - \Sigma_j F_{SKj}$ (t/h)	$F_{C,k}$ (t/h)	Δm_k (kg/h)	Cum. Δm_k (kg/h)
				$F_{FW}=20.00$		
0						
				20.00	0.20	
10		$F_{RW}=53.57$	53.57			0.20
				73.57	0.74	
20	50		−50			0.94
				23.57	0.71	
50	100	50	−50			1.64
				−26.43	−1.32	
100	80	100	20			0.32
				−6.43	−0.32	
150		70	70			**0.00**
				63.57	3.18	**(PINCH)**
200	70		−70			3.18
				−6.43	−0.32	
250		6.43	6.43			2.86
				$F_{WW}=0.00$	0.00	
1,000,000						2.86

2. *Preliminary allocation is performed to distribute the sources and sinks to two separate regions based on the chosen regeneration outlet concentration, i.e., the FWR and the RWR.*

For the given regeneration outlet concentration (C_{Rout}) of 25 ppm, water sink SK1 is allocated to FWR, while all other sinks and sources are left in the RWR, with results shown in Table 6.8.

TABLE 6.8

Results of Preliminary Allocation

(a) FWR	C_k (ppm)	$\Sigma_j F_{SKj}$ (t/h)	$\Sigma_i F_{SRi}$ (t/h)	(b) RWR	C_k (ppm)	$\Sigma_j F_{SKj}$ (t/h)	$\Sigma_i F_{SRi}$ (t/h)
	0						
	20	50			25		
	50				50	100	50
	100				100	80	100
	150				150		70
	200				200	70	
	250				250		60
	1,000,000				1,000,000		
	Total	50	0			250	280

3. *Flowrate reallocation—To ensure the sinks and sources in the RWR have equal total flowrates. If the total flowrate of the sinks is higher than that of the sources in the RWR, go to Step 3a or else to Step 3b:*

 b. *Total or partial flowrate of the highest concentration source in the RWR is reallocated to the FWR so that the total flowrate of the sinks is equal to that of the sources ($\Sigma_j F_{SKj} = \Sigma_i F_{SRi}$). Source of the second highest concentration is to be used if the first source is insufficient (and so on). We next check if all sinks in the FWR have concentrations of 0 ppm. If only 0 ppm sinks (or no sink at all) are found in the FWR, then move to Step 5; or else move to Step 4b.*

From Table 6.8, the total flowrate of the sinks (250 t/h) is found to be lower than that of the sources (280 t/h) in the RWR. Hence, 30 t/h of SR4 at 250 ppm (highest concentration) source is reallocated to the FWR (i.e., 30 t/h of SR4 is left in the RWR) to achieve an equal total flowrate for the sinks and sources. The results of flowrate reallocation are shown in Table 6.9. We then move to Step 4b since the water sink in the FWR (SK1) has a nonzero concentration.

4b. *When there are nonzero concentration sinks in the FRW, regenerated flowrate may be used to fulfill the flowrate requirement of the sinks, instead of using freshwater. The regenerated flowrate in the FWR ($F_{RW,FWR}$) is then calculated using Equation 6.3. We next proceed to check if sources in the FWR are sufficient to be sent for regeneration. If the total source flowrate is insufficient, the total available source is used as regenerated flowrate. We then proceed to Step 4a to shift additional sink and source pair from the RWR to the FWR to maximize reuse/recycle potential of the process sources.*

Since the sink in the FWR is at 20 ppm, it should receive regenerated water instead of freshwater. The minimum

TABLE 6.9

Results of Flowrate Reallocation

(a) FWR	C_k (ppm)	$\Sigma_j F_{SKj}$ (t/h)	$\Sigma_i F_{SRi}$ (t/h)	(b) RWR	C_k (ppm)	$\Sigma_j F_{SKj}$ (t/h)	$\Sigma_i F_{SRi}$ (t/h)	$\Sigma_i F_{SRi} - \Sigma_j F_{SKj}$ (t/h)
	0							
	20	50			25			
	50				50	100	50	−50
	100				100	80	100	20
	150				150		70	70
	200				200	70		−70
	250		30		250		30	30
	1,000,000				1,000,000			
	Total	50	30			250	250	

regeneration flowrate for the sink is then calculated with Equation 6.3 as

$$F_{RW,FWR} = \frac{50\,t/h(20\,ppm)}{25\,ppm} = 40\,t/h$$

However, it is then determined that only 30 t/h source at 250 ppm is available for regeneration in the FWR (reallocated during Step 3b). Hence, the whole source is regenerated, i.e., $F_{RW,RWR} = 30\,t/h$. We hence proceed to Step 4a.

4a. *Additional source shift—To maintain the flowrate balance, a pair of sink and source of equal flowrates are to be shifted to the FWR simultaneously. Note that both sink and source to be shifted should be among the lowest concentration sinks/sources in the RWR. In addition, the shifted source should originate from a concentration level with net positive flowrate in the RWR, and posseses concentration higher than the shifted sink.*

Since 30 t/h of sources are regenerated in the FWR to 25 ppm, it does not maximize the total acceptable load of the sink at 20 ppm. In other words, the 20 ppm sink may accept additional sources with nonzero concentration. However, additional sources are only available in the RWR. Similar to the case in Example 6.1, a pair of water sink and source with equal flowrates are to be shifted from the RWR to the FRW, in order to maintain the flowrate balance in the former. The sink should originate from SK2 of 50 ppm (lowest concentration sink in the RWR, i.e., $C_{SKj,A} = 50\,ppm$). On the other hand, the source should originate from SR2 of 100 ppm (i.e., $C_{SRi,A} = 100\,ppm$), since it is the lowest concentration level with net positive flowrate (see last column of Tables 6.9(a) and (b). The additional shifted flowrates ($F_{SRi,A}$, $F_{SKj,A}$) are determined from Equation 6.2 as

$$50\ t/h\ (20\ ppm) - 30\ t/h\ (25\ ppm) = F_{SRi,A}\ (100 - 50\ ppm)$$

i.e.,

$$F_{SRi,A} = F_{SKj,A} = 5\ t/h$$

Note that the source terms of the left side include the regeneration flowrate at 25 ppm. Hence, Equation 6.3 determines that additional 5 t/h of SK2 and SR2 flowrates in the RWR are shifted to the FWR (see the resulting flowrates in Tables 6.10 and 6.11).

5. *We next determine the regeneration flowrate in the RWR ($F_{RW,FWR}$). This can be done using the MCA technique, taking the regeneration outlet concentration as the concentration of the fresh resource in the RWR.*

TABLE 6.10

Regeneration Flowrate Targeting in RWR for Example 6.2

C_k (ppm)	$\Sigma_j F_{SKj}$ (t/h)	$\Sigma_i F_{SRi}$ (t/h)	$\Sigma_i F_{SRi} - \Sigma_j F_{SKj}$ (t/h)	$F_{C,k}$ (t/h)	Δm_k (kg/h)	Cum. Δm_k (kg/h)
				$F_{RW,RWR} = 30$		
25						
				30	0.75	
50	95	50	−45			0.75
				−15	−0.75	
100	80	95	15			0.00
				0	0.00	
150		70	70			**0.00**
				70	3.50	**(PINCH)**
200	70		−70			3.50
				0	0.00	
250		30	30			3.50
				$F_{RW,RWR} = 30$	29,992.50	
1,000,000						29,996.00

TABLE 6.11

Flowrate Targeting for FWR in Example 6.2

C_k (ppm)	$\Sigma_j F_{SKj}$ (t/h)	$\Sigma_i F_{SRi}$ (t/h)	$\Sigma_i F_{SRi} - \Sigma_j F_{SKj}$ (t/h)	$F_{C,k}$ (t/h)	Δm_k (kg/h)	Cum. Δm_k (kg/h)
				$F_{FW} = 20$		
0						
				20	0.4	
20	50					0.4
				−30	−0.2	
25		$F_{RW,FWR} = 30$				0.3
				0	0.0	
50	$F_{SKj,A} = 5$		−45			0.3
				−5	−0.3	
100		$F_{SRi,A} = 5$	15			0.0
				0	0.0	
150			70			0.0
				0	0.0	
200			−70			0.0
				0	0.0	
250			30			0.0
				$F_{WW} = 0$	0.0	
1,000,000						0.0

Cascade analysis shows that $F_{RW,FWR}$ takes a value of 30 t/h, with the result shown in Table 6.10. Table 6.10 also indicates that $F_{RW,FWR}$ is originated from the source at 250 ppm, i.e., SR4.

6. *Determination of ultimate flowrate targets—The freshwater and wastewater flowrates can then be determined for the FWR. The total regenerated flowrate is given by the summation of $F_{RW,FWR}$ (Step 4b if applicable) and $F_{RW,RWR}$ (Step 5).*

Cascade analysis next determines that the freshwater (F_{FW}) and wastewater (F_{WW}) flowrates in the FWR are 20 and 0 t/h, respectively (see results in Table 6.11). The total regeneration flowrate (F_{RW}) is determined from the summation of $F_{RW,FWR}$ (Step 4b, if applicable) and $F_{RW,RWR}$ (Step 5), i.e.,

$$F_{RW} = 30\ t/h + 30\ t/h = 60\ t/h$$

The result may also be represented in Table 6.12. Note that the 250 ppm source is no longer available as it has been fully regenerated to 25 ppm (F_{RW}).

At this point, a simple comparison may be made for both solutions in Examples 6.1 and 6.2 (where water regeneration is used), as well as that with the base case and direct reuse/recycle scheme. The results are shown in Table 6.13. Before any water recovery scheme is implemented, the minimum flowrates are given as

TABLE 6.12

Combined Cascade Table for Example 6.2

C_k (ppm)	$\Sigma_j F_{SKj}$ (t/h)	$\Sigma_i F_{SRi}$ (t/h)	$\Sigma_i F_{SRi} - \Sigma_j F_{SKj}$ (t/h)	$F_{C,k}$ (t/h)	Δm_k (kg/h)	Cum. Δm_k (kg/h)
				$F_{FW}=20$		
0						
				20	0.40	
20	50		−50			0.40
				−30	−0.15	
25		$F_{RW}=60$	60			0.25
				30	0.75	
50	100	50	−50			1.00
				−20	−1.00	
100	80	100	20			0.00
				0	0.00	
150		70	70			**0.00**
				70	3.50	**(PINCH)**
200	70		−70			3.50
				0	0.00	
250			0.00			3.50
				$F_{WW}=0$	0.00	
1,000,000						3.50

TABLE 6.13

Summary of Comparison for Various Water-Using Schemes

Flowrate Targets	Base Case	Direct Reuse/ Recycle	Example 6.1	Example 6.2
Freshwater, F_{FW} (t/h)	300	70	20	20
Wastewater, F_{WW} (t/h)	280	50	0	0
Regenerated water, F_{RW} (t/h)	—	—	53.57	60

300 and 280 t/h for freshwater and wastewater, respectively. With direct reuse/recycle scheme, these flowrates are reduced to 70 and 50 t/h, respectively. With water regeneration (both Examples 6.1 and 6.2), the freshwater flowrate is further reduced to 20 t/h, while the wastewater flowrate is completely removed.

Since both freshwater and wastewater flowrates remain the same for both Examples 6.1 and 6.2, we next compare the regeneration flowrates for both cases. With C_{Rout} of 10 ppm, regeneration flowrate is determined as 53.57 t/h (Example 6.1). When C_{Rout} is raised to 25 ppm in Example 6.2, the regeneration flowrate increases slightly to 60 t/h. The increase in flowrate indicates that a larger interception unit is to be used. We can then decide the selection of the interception units based on detailed economic analysis (beyond the scope of this chapter).

Finally, note that we may also utilize other equivalent flowrate targeting tools to obtain the same ultimate flowrate targets for an RCN, as long as the same principles outlined here are followed. Note also that the same technique applies to other RCNs other than water networks, so long as a single pass interception unit is utilized.

6.3 Modeling of Mass Exchange Operation as Interception Unit

El-Halwagi (1997, 2006) defined mass exchange operation as a separation process that employs a mass separating agent (MSA, termed as *lean stream*) to selectively remove certain components (e.g., impurities, pollutants, by-products, products) from a component-rich stream (e.g., waste, product, etc.). To perform the separation task, the MSA has to be immiscible (partially or completely) in the rich stream. Common examples for mass exchange operation include absorption, adsorption, stripping, ion exchange, extraction, leaching, etc. These units may be used as interception units in a material regeneration network. In this section, the modeling of a mass exchange operation is presented from the perspective of RCN synthesis. Since the mass exchange operations are mainly concentration-based processes, their quality indices are always given in concentration basis.

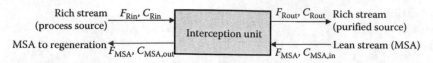

FIGURE 6.3
A mass exchange operation used as interception unit in an RCN. (From *Pollution Prevention through Process Integration: Systematic Design Tools*, Academic Press, San Diego, CA, El-Halwagi, M.M., Copyright 1997, with permission from Elsevier.)

A schematic representation for a mass exchange operation that acts as an interception unit is shown in Figure 6.3, where the rich and lean streams pass through the mass exchange operation in countercurrent flow. In this case, the process source that acts as an impurity-rich stream (with feed quality C_{Rin}) is purified to higher quality (C_{Rout}) and exists at the top outlet stream of the interception unit. Its impurity load (Δm) is transferred to the MSA (lean stream, with flowrate F_{MSA}) and is represented by Equation 6.4. In turn, the impurity load raises the concentration of the MSA from $C_{MSA,in}$ to $C_{MSA,out}$. Note that flowrate losses for both rich and lean streams are negligible for most cases:

$$\Delta m = F_{Rout}\,(C_{Rin} - C_{Rout}) = F_{MSA}\,(C_{MSA,out} - C_{MSA,in}) \qquad (6.4)$$

Figure 6.4 shows a graphical representation of the mass exchange operation in a concentration versus load diagram similar to that in Figure 2.4. Note however that in this case, the process source becomes the rich stream, as its impurity content is to be removed to the MSA that acts as the lean stream.

For a feasible mass exchange operation, a minimum mass transfer driving force is needed to operate the mass exchange unit. As shown in Figure 6.4, the minimum driving force is found at both ends of the mass exchange operation. A few points are worth mentioning here. First, any process source

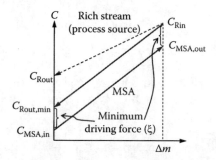

FIGURE 6.4
A mass exchange operation represented in a concentration versus load diagram. (From *Chem. Eng. Sci.*, 49, Wang, Y.P. and Smith, R., Wastewater minimization, 981–1006, Copyright 1994, with permission from Elsevier.)

to be purified by the MSA has to lie in the region that is higher than the minimum driving force from the MSA line. In other words, a process source may be purified to any outlet concentration C_{Rout} that is higher than the minimum threshold value, i.e., $C_{Rout,min}$. Note that the latter is given as the summation of the MSA inlet concentration (to the mass exchange operation, $C_{MSA,in}$) with the minimum driving force (ξ), given as in Equation 6.5:

$$C_{Rout,min} = C_{MSA,in} + \xi \qquad (6.5)$$

However, it is always desired to keep the regeneration flowrate of the process source to the minimum, as this leads to minimum operating cost of the interception unit, as well as reduced capital cost indirectly (which is often a function of regeneration flowrate). Since the source flowrate is given by the inverse slope of the plot in Figure 6.4, it should be purified to the minimum outlet concentration ($C_{Rout,min}$) in order to keep a minimum regenerated source flowrate.

Similarly, to keep the MSA flowrate (and hence its operating cost) to the minimum, the MSA line in Figure 6.4 should have the steepest slope. In other words, the MSA should be operated with minimum driving force difference from the process source, at the rich end of the mass exchange operation. This is conceptually similar to the case of *limiting water profile* (see Section 2.3 for detailed discussion).

In some cases, there are more than one MSAs that may be used for a particular regeneration task. In such cases, it will be good to evaluate the performance of these MSAs from thermodynamic perspective. In equilibrium condition, the concentration of process source SRi (C_{SRi}) may be related to that of MSA l (C_{MSAl}^*) with a linear relationship (El-Halwagi, 1997, 2006), given as in Equation 6.6:

$$C_{SRi} = \alpha_l C_{MSAl}^* + \beta_l \qquad (6.6)$$

where α_l and β_l are specific parameters for a given MSA l. As described earlier, in order for a feasible mass transfer operation to take place, the process source is to be operated with a minimum driving force difference from the MSA. This is depicted in Figure 6.5 (El-Halwagi, 1997, 2006).

As shown in Figure 6.5, the operating point with coordinate (C_{MSAl}, C_{SRi}) is separated from the equilibrium point (C_{MSAl}^*, C_{SRi}^*) by the minimum driving force (ξ_l), given as in Equation 6.7. Note that the latter may have different values for different MSA l:

$$C_{MSAl}^* = C_{MSAl} + \xi_l \qquad (6.7)$$

Combining Equations 6.6 and 6.7 yields the operating line equation for MSA l:

$$C_{SRi} = \alpha_l \left(C_{MSAl} + \xi_l \right) + \beta_l \qquad (6.8)$$

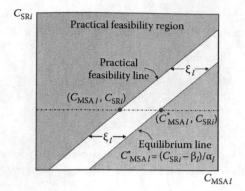

FIGURE 6.5
Equilibrium and practical feasibility lines. (From El-Halwagi, M.M., *Pollution Prevention through Process Integration: Systematic Design Tools*, Academic Press, San Diego, CA, 1997. With permission.)

When multiple MSA candidates are available for use, the MSA that can purify the source to the lowest concentration will normally be chosen, as it will lead to the minimum fresh resource and waste flowrates for the RCN.

Example 6.3 Water minimization for acrylonitrile (AN) production

For the AN process in Examples 3.1, 4.1, and 5.1, regeneration is considered in order to further reduce the water flowrates of the RCN. Limiting data for this case are given in Table 6.14. Note that SK1 (boiler feed water) and SR4 (ejector condensate) in the original case (Examples 3.1 and 4.1) have been eliminated with the removal of the steam-jet ejector in Example 5.1.

Three mass exchange operations are feasible to be used as interception units to purify water sources for further reuse/recycle to the sinks, i.e., air stripping, activated carbon, and resin adsorption units. Due to space constraint, only one mass exchange operation may be installed in the RCN. Data for the MSAs are given in Table 6.15 (El-Halwagi, 1997, 2006). Solve the following tasks:

TABLE 6.14

Limiting Water Data for Example 6.3

Water Sinks, SK_j	Flowrate	NH₃ Concentration	
j	Stream	F_{SKj} (kg/s)	C_{SKj} (ppm)
2	Scrubber inlet	5.8	10
Water Sources, SR_i	Flowrate	NH₃ Concentration	
i	Stream	F_{SRi} (kg/s)	C_{SRi} (ppm)
1	Distillation bottoms	0.8	0
2	Off-gas condensate	5	14
3	Aqueous layer	5.9	25

TABLE 6.15

Data for MSAs in Example 6.3

L	MSA	Inlet Concentration, $C_{MSA,in}$ (ppm)	Outlet Concentration, $C_{MSA,out}$ (ppm)	α_l	β_l	Minimum Driving Force, ξ_l (ppm)
1	Air	0	6	1.4	0	2
2	Activated carbon	10	400	0.02	0	5
3	Resin	3	1100	0.01	0	5

1. Determine which MSA will lead to the minimum freshwater and wastewater flowrates of the RCN. Determine the ultimate flowrate targets for the selected MSA.
2. Derive a water recovery scheme based on the targeting insights.
3. Determine the impurity removal and MSA flowrate in the interception unit.
4. Perform a comparison for various water recovery schemes for this case. See Examples 3.1, 4.1, and 5.1 for other schemes.

Solution

Task 1—MSA selection and ultimate flowrate targeting

With Equation 6.8, we can calculate the lowest outlet concentration of a process source when it is purified with MSA *l*. For instance, when air stripping is utilized, the outlet concentration for the process source is calculated as follows:

$$C_{SR1} = 1.4\,(0 + 2) = 2.8 \text{ ppm}$$

Likewise, we can determine the outlet concentration for the process source when it is purified with activated carbon and resin, i.e., 0.3 ($=0.02\,(10+5)$) and 0.08 ppm ($=0.01\,(3+5)$), respectively. Hence, resin is selected for use as an interception unit, as it purifies the process source to the lowest concentration level, which leads to lowest water flowrates.

We next follow the six-step procedure to identify the ultimate flowrate targets for the RCN:

1. *An appropriate regeneration outlet concentration (C_{Rout}) is selected.*
2. *Preliminary allocation is performed to distribute the sources and sinks to two separate regions based on the chosen regeneration outlet concentration, i.e., the FWR and the RWR.*

    With the calculated regeneration outlet concentration (C_{Rout}) of 0.08 ppm, water source SR1 is allocated to FWR, while sink SK2 and other sources are allocated to the RWR, with results shown in Table 6.16(a) and (b).
3. *Flowrate reallocation—To ensure the sinks and sources in the RWR have equal total flowrates. If the total flowrate of the sinks is*

TABLE 6.16

Results of Preliminary Allocation

(a) FWR	C_k (ppm)	$\Sigma_j F_{SKj}$ (kg/s)	$\Sigma_i F_{SRi}$ (kg/s)	(b) RWR	C_k (ppm)	$\Sigma_j F_{SKj}$ (kg/s)	$\Sigma_i F_{SRi}$ (kg/s)
	0		0.8				
	0.08				0.08		
	10				10	5.8	
	14				14		5
	25				25		5.9
	1,000,000				1,000,000		
	Total	0	0.8			5.8	10.9

higher than that of the sources in the RWR, go to Step 3a or else to Step 3b:

   b. *Total or partial flowrate of the highest concentration source in the RWR is reallocated to the FWR so that the total flowrate of the sinks is equal to that of the sources ($\Sigma_j F_{SKj} = \Sigma_i F_{SRi}$). Source of the second highest concentration is to be used if the first source is insufficient (and so on). We next check if all sinks in the FWR have concentrations of 0 ppm. If only 0 ppm sinks (or no sink at all) are found in the FWR, then move to Step 5; or else move to Step 4b.*

   Table 6.16(a) and (b) show that water sink SK2 has a lower flowrate (5.8 kg/s) as compared with the total source flowrate (10.9 kg/s) in the RWR. Hence, 5.1 kg/s (=10.9 – 5.8 kg/s) of the highest concentration source is reallocated to the FWR. The results of flowrate reallocation are shown in Table 6.17(a) and (b).

   Since no water sink is found in the FWR, we then move to Step 5.

5. *We next determine the regeneration flowrate in the RWR (F_{RW}, F_{WR}). This can be done using the MCA technique, taking the regeneration*

TABLE 6.17

Results of Flowrate Reallocation

(a) FWR	C_k (ppm)	$\Sigma_j F_{SKj}$ (kg/s)	$\Sigma_i F_{SRi}$ (kg/s)	(b) RWR	C_k (ppm)	$\Sigma_j F_{SKj}$ (kg/s)	$\Sigma_i F_{SRi}$ (kg/s)
	0		0.8		0		
	0.08				0.08		
	10				10	5.8	
	14				14		5
	25		5.1		25		0.8
	1,000,000				1,000,000		
	Total	0	5.9			5.8	5.8

outlet concentration as the concentration of the fresh resource in the RWR.

6. *Determination of ultimate flowrate targets*—The freshwater and wastewater flowrates can then be determined for the FWR. The total regenerated flowrate is given by the summation of $F_{RW,FWR}$ (Step 4b if applicable) and $F_{RW,RWR}$ (Step 5).

Targeting is then carried out for both FWR and RWR, with the results shown in Tables 6.18 and 6.19. Alternatively, the targeting results may also be shown in the combined cascade in Table 6.20. As shown, the ultimate flowrate targets for this case correspond to freshwater (F_{FW}), wastewater (F_{WW}), and regeneration (F_{RW}) flowrates of 0, 5.9, and 1.67 kg/s, respectively.

Task 2—Derivation of water recovery scheme

Table 6.18 indicates that the entire SR1 (0.8 kg/s) and a big portion of SR3 (5.1 kg/s) in the FWR will be discharged as wastewater streams (5.9 kg/s). On the other hand, we can observe from Table 6.19 that in the RWR, 4.13 kg/s of SR2 (off gas condensate) at 14 ppm is sent for direct reuse/

TABLE 6.18

Regeneration Flowrate Targeting for FWR in Example 6.3

C_k (ppm)	$\Sigma_j F_{SKj}$ (kg/s)	$\Sigma_i F_{SRi}$ (kg/s)	$\Sigma_i F_{SRi} - \Sigma_j F_{SKj}$ (kg/s)	$F_{C,k}$ (kg/s)	Δm_k (mg/s)	Cum. Δm_k (mg/s)
				$F_{FW} = 0.0$		
0		0.8	0.8			
				0.8	20	
25		5.1	5.1			20
				$F_{WW} = 5.9$	5,899,852.5	
1,000,000						5,899,872.5

TABLE 6.19

Flowrate Targeting for RWR in Example 6.3

C_k (ppm)	$\Sigma_j F_{SKj}$ (kg/s)	$\Sigma_i F_{SRi}$ (kg/s)	$\Sigma_i F_{SRi} - \Sigma_j F_{SKj}$ (kg/s)	$F_{C,k}$ (kg/s)	Δm_k (mg/s)	Cum. Δm_k (mg/s)
				$F_{RW} = 1.67$		
0.08						
				1.67	16.53	
10	5.8		−5.8			16.53
				−4.13	−16.53	
14		5	5			0.00
				0.87	9.53	(PINCH)
25		0.8	0.8			9.53
				$F_{RW} = 1.67$	1,666,625.00	
1,000,000						1,666,634.53

TABLE 6.20

Combined Cascade Table for Example 6.3 ($C_{Rout} = 0.08\,ppm$)

C_k (ppm)	$\Sigma_j F_{SKj}$ (kg/s)	$\Sigma_i F_{SRi}$ (kg/s)	$\Sigma_i F_{SRi} - \Sigma_j F_{SKj}$ (kg/s)	$F_{C,k}$ (kg/s)	Δm_k (mg/s)	Cum. Δm_k (mg/s)
				$F_{FW} = 0.0$		
0		0.8	0.8			
				0.80	0.06	
0.08		$F_{RW} = 1.67$	1.67			0.06
				2.47	24.47	
10	5.8		−5.8			24.53
				−3.33	−13.33	
14		4.14	4.14			11.20
				0.80	8.80	
25		5.1	5.1			20.00
				$F_{WW} = 5.90$	5,899,852.50	
1,000,000						5,899,872.50

recycle to SK2 (observed from $F_{C,k}$ column between interval 10–14 ppm). Besides, flowrates sent for regeneration are originated from sources in the lower quality region, i.e., 0.87 kg/s from 14 ppm (SR2) and 0.8 kg/s from 25 ppm (SR3—decanter aqueous layer).

Based on the insights discussed earlier, the RCN for the AN case may be derived as shown in Figure 6.6. A few remarks are worth mentioning for this RCN. First, it is observed that no freshwater is needed

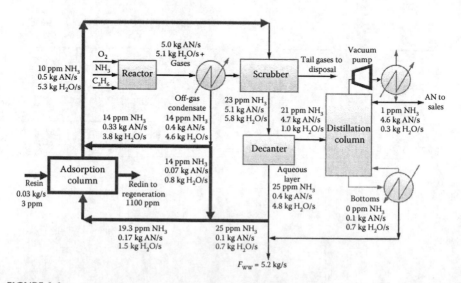

FIGURE 6.6
RCN for AN case in Example 6.3. (From El-Halwagi, M.M., *Pollution Prevention through Process Integration: Systematic Design Tools*, Academic Press, San Diego, CA, 1997. With permission.)

for the process, and it resembles the threshold problem discussed in Chapter 3. Nevertheless, wastewater is discharged from the process, which is mainly due to water generation as a by-product in the reactor. Besides, scrubber is now fed with 5.8 kg/s of regenerated water, which is slightly smaller than its original feed flowrate (6.0 kg/s; see Example 2.1). Note that the purified water from the resin adsorption unit contains a small fraction of AN product (0.5 kg/s), which is now recovered to the process. Hence, upon detailed examination of mass balance on the flowsheet, the AN product from the distillation top stream has increased to 4.6 kg/s (from 3.9 kg/s in Example 2.1). Besides, the terminal wastewater stream has a reduced flowrate of 5.2 kg/s (due to reduced flowrate and water content in the scrubber feed stream).

Task 3—Determination of impurity removal and MSA flowrate in the interception unit

With the identification of process sources for regeneration in Task 2, the impurity load removed from these sources can be determined using Equation 6.4:

$$0.87 \text{ kg/s } (14 - 0.08 \text{ ppm}) + 0.8 \text{ kg/s } (25 - 0.08 \text{ ppm}) = 32 \text{ mg/s}$$

Figure 6.7 shows the removal of impurity load represented on the concentration versus load diagram. Since impurity loads are transferred from two rich streams, the latter may form a *rich composite curve* by combining the individual segment within the same concentration intervals.[*] For instance, within the interval of 0.08–14 ppm, the impurity loads of SR2 (0.87 kg/s (14 − 0.08 ppm) = 12.06 mg/s) and SR3 (0.8 kg/s (14 − 0.08 ppm) = 11.14 mg/s) are added as 23.20 kg/s (Figure 6.7a).

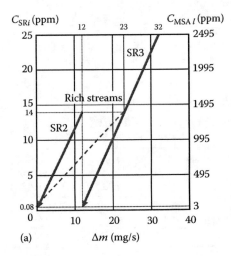

(a)

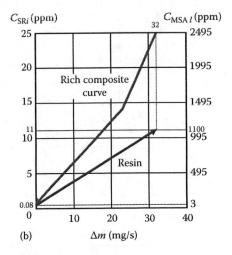

(b)

FIGURE 6.7
Plotting of mass transfer composite curves.

---

[*] See El-Halwagi (1997, 2006) for detailed discussion of the mass transfer composite curves.

The MSA (resin) is also plotted on the same diagram, within its corresponding concentration intervals of 3–1100 ppm (Figure 6.7b). Note that the corresponding concentration of the MSA is determined using Equation 6.8. We may then determine the flowrate of the MSA from the inverse slope of the MSA line or alternatively via Equation 6.4, i.e.,

$$F_{MSA3} = \frac{32 \text{ mg/s}}{(1100 - 3)\text{ppm}} = 0.029 \text{ kg/s}$$

It is also worth noting that there are many other options for the selection of process sources for regeneration. In this example, the selection of sources is guided by targeting insights. Other sources (or different source flowrates) may also be regenerated. Note however that the selection of different sources/flowrates will lead to different impurity load removal.* The final selection of appropriate source candidates/flowrates for regeneration hence lies with detailed economic evaluation of the RCN.

Task 4—Comparison of various water recovery schemes

A comparison of various water recovery schemes for the AN case study is shown in Table 6.21. Before any water recovery and process change takes place, the total freshwater and wastewater flowrates of the entire process are determined as 7.0 and 13.1 kg/s, respectively (see limiting data in Table 3.1). With direct reuse/recycle schemes (Examples 3.1 and 4.1), these flowrates are reduced to 2.1 and 8.2 kg/s, respectively. Next, when the steam ejector is replaced by the vacuum pump (process change option in Example 5.1), the freshwater and wastewater flowrates are further reduced to 0.9 and 6.8 kg/s, respectively. Finally, in this example, when 1.67 kg/s of process source is sent for regeneration in the interception unit for further reuse/recycle, the freshwater flowrate is completely eliminated, while the wastewater flowrate is reduced to the ultimate minimum of 5.9 kg/s (5.2 kg/s upon detailed mass balance check).

Finally, it is worth mentioning that the original objective to examine water recovery schemes for this case is to increase the AN production

TABLE 6.21

Summary of Comparison for Various Water-Using Schemes for AN Production Case

Flowrate Targets	Base Case	Direct Reuse/ Recycle (Example 3.1/4.1)	Process Changes (Example 5.1)	Regeneration (Example 6.3)
Freshwater, F_{FW} (kg/s)	7.0	2.1	0.9	0
Wastewater, F_{WW} (kg/s)	13.1	8.2	6.8	5.2
Regenerated water, F_{RW} (kg/s)	—	—	—	1.67

* Solve Problem 6.5 for regeneration of different water source flowrate.

throughput via debottlenecking its biotreatment facility (see Example 2.1). Hence, with the reduced wastewater flowrate of 5.2 kg/s, a reduction of 40% of current hydraulic capacity has been met. Hence, the AN throughout can now be expanded for 2.5 times of its current capacity (provided that other processing facilities do not encounter process bottlenecks).

6.4 Additional Readings

Apart from the algebraic targeting technique in this chapter, other techniques have also been developed for regeneration flowrate targeting. These include the graphical and algebraic approaches that are based on the *limiting composite curve* (Wang and Smith, 1994; Kuo and Smith, 1998; Agrawal and Shenoy, 2006; Bai et al., 2007; Feng et al., 2007) for the fixed flowrate problem; as well as *source composite curve* (Bandyopadhyay and Cormos, 2008) for the fixed flowrate problem, etc. A review that compares these techniques may be found in Foo (2009).

El-Halwagi (1997, 2006) also has further details on the optimization of mass exchange optimization and the synthesis of mass exchange networks.

Problems

6.1 For Example 6.1, determine the ultimate flowrate targets when a interception unit with $C_{Rout} = 50$ ppm is used.

6.2 For the classical water minimization example in Problem 3.3, determine the following:

a. Minimum freshwater and wastewater flowrates for direct reuse/recycle scheme with MCA (verify the solutions obtained from graphical technique in Problem 3.3).

b. Ultimate flowrate targets when an interception unit with $C_{Rout} = 5$ ppm is used.

c. Identify the source(s) that should be sent for regeneration.

Note: see solution in Examples 8.1 and 8.2.

6.3 For the specialty chemical production process in Problem 4.6 (limiting data given in Table 4.24), perform the following tasks:

a. Determine the ultimate flowrate targets for the RCN when an interception unit with $C_{Rout} = 60$ ppm is used.

TABLE 6.22

Limiting Water Data for Problem 6.4

SK$_j$	F_{SKj} (t/h)	C_{SKj} (ppm)	SR$_i$	F_{SRi} (t/h)	C_{SRi} (ppm)
SK1	120	0	SR1	120	100
SK2	80	50	SR2	80	140
SK3	80	50	SR3	140	180
SK4	140	140	SR4	80	230
SK5	80	170	SR5	195	250
SK6	195	240			

  b. Determine the ultimate flowrate targets for another interception unit with $C_{Rout} = 200$ ppm is used (Deng et al., 2011).

  c. Identify the source(s) that should be sent for regeneration in (a) and (b).

  d. Do a comparison of the flowrate reduction for cases in (a) and (b) with that of direct reuse/recycle scheme (note: minimum flowrate targets for direct reuse/recycle scheme are found by solving Problem 4.6).

6.4 Table 6.22 shows the limiting data for a water minimization example from Sorin and Bédard (1999). Determine the following:

  a. Minimum freshwater and wastewater flowrates for direct reuse/recycle scheme with the algebraic technique in Chapter 4.

  b. Ultimate flowrate targets when an interception unit with $C_{Rout} = 0$ ppm is used.

  c. Identify which source(s) should be sent for regeneration.

  Note: see answers in Problem 7.12.

6.5 Rework the water minimization case for AN production in Example 6.3. In this case, assume that the off-gas condensate (SR2) is the only water source to be purified by the resin adsorption unit to an outlet concentration of 11.6 ppm (El-Halwagi, 1997; Manan et al., 2004). Solve the following tasks as well as the impurity removal of the adsorption unit.

  a. Determine the ultimate flowrate targets for this case.

  b. Determine the impurity removal and the resin flowrate of the adsorption unit.

  c. Derive the water recovery scheme for this case.

6.6 For the Kraft pulping process in Problems 2.2 and 4.1, solve the following tasks (use the limiting data in Table 4.20):

  a. Determine the ultimate flowrate targets for the RCN when an interception unit with $C_{Rout} = 10$ ppm is used.

  b. Determine the ultimate flowrate targets for another interception unit with $C_{Rout} = 15$ ppm.

   c. Determine the ultimate flowrate targets when air stripping is used as an interception unit. The equilibrium relation of the stripping air with the process source is given in Equation 6.9, while the minimum composition difference is given as 4.275 ppm (El-Halwagi, 1997). It may be assumed that the MSA (air) has zero impurity content.

$$C_{SRi} = 0.38 \, C_{MSA} \tag{6.9}$$

   d. Identify which source(s) should be sent for regeneration in tasks (a) through (c).

   e. Do a comparison of the flowrate reduction for cases in (a) through (c), along with the direct reuse/recycle scheme (found by solving Problem 4.1).

Note: answers are found in Problems 7.15 and 8.3.

References

Agrawal, V. and Shenoy U. V. 2006. Unified conceptual approach to targeting and design of water and hydrogen networks. *AIChE Journal*, 52(3), 1071–1081.

Bai, J., Feng, X., and Deng, C. 2007. Graphical based optimization of single-contaminant regeneration reuse water systems. *Chemical Engineering Research and Design*, 85(A8), 1178–1187.

Bandyopadhyay, S. and Cormos, C. C. 2008. Water management in process industries incorporating regeneration and recycle through a single treatment unit. *Industrial and Engineering Chemistry Research*, 47, 1111–1119.

Deng, C., Feng, X., Ng, D. K. S., and Foo, D. C. Y. 2011. Process-based graphical approach for simultaneous targeting and design of water network. *AIChE Journal*, 57(11), 3085–3104.

El-Halwagi, M. M. 1997. *Pollution Prevention through Process Integration: Systematic Design Tools*. San Diego, CA: Academic Press.

El-Halwagi, M. M. 2006. *Process Integration*. San Diego, CA: Elsevier Inc.

Feng, X. Bai, J., and Zheng, X. 2007. On the use of graphical method to determine the targets of single-contaminant regeneration recycling water systems. *Chemical Engineering Science*, 62, 2127–2138.

Foo, D. C. Y. 2009. A state-of-the-art review of pinch analysis techniques for water network synthesis. *Industrial and Engineering Chemistry Research*, 48(11), 5125–5159.

Kuo, W. C. J. and Smith R. 1998. Design of water-using systems involving regeneration. *Process Safety and Environmental Protection*, 76, 94–114.

Manan, Z. A., Tan, Y. L., and Foo, D. C. Y. 2004. Targeting the minimum water flowrate using water cascade analysis technique. *AIChE Journal*, 50(12), 3169–3183.

Ng, D. K. S., Foo, D. C. Y., and Tan, R. R. 2009. Automated targeting technique for single-component resource conservation networks—Part 2: Single pass and partitioning waste interception systems. *Industrial and Engineering Chemistry Research*, 48(16), 7647–7661.

Ng, D. K. S., Foo, D. C. Y., Tan, R. R., and Tan, Y. L. 2007. Ultimate flowrate targeting with regeneration placement. *Chemical Engineering Research and Design*, 85(A9), 1253–1267.

Ng, D. K. S., Foo, D. C. Y., Tan, R. R., and Tan, Y. L. 2008. Extension of targeting procedure for 'Ultimate flowrate targeting with regeneration placement' by Ng et al., *Chemical Engineering Research and Design*, 85 (A9): 1253–1267; *Chemical Engineering Research and Design*, 86(10), 1182–1186.

Polley, G. T. and Polley, H. L. 2000. Design better water networks. *Chemical Engineering Progress*, 96(2), 47–52.

Sorin, M. and Bédard, S. 1999. The global pinch point in water reuse networks. *Process Safety and Environmental Protection*, 77, 305–308.

Tan, R. R., Ng, D. K. S., Foo, D. C. Y., and Aviso, K. B. 2009. A superstructure model for the synthesis of single-contaminant water networks with partitioning regenerators. *Process Safety and Environmental Protection*, 87, 197–205.

Wang, Y. P. and Smith, R. 1994. Wastewater minimisation. *Chemical Engineering Science*, 49, 981–1006.

7

Network Design and Evolution Techniques

To synthesize a resource conservation network (RCN) using pinch analysis technique, a two-step approach that consists of *targeting* and *design* is normally adopted. In the last few chapters, we have seen how minimum flowrate targets may be determined ahead of detailed design, for various direct reuse/recycle and material regeneration networks. For simple cases (e.g., Examples 3.2 and 4.1), we might determine the material recovery scheme based on insights from the targeting stage. However, for more complicated cases, a systematic design procedure is needed to synthesize an RCN that achieves the flowrate targets. In this chapter, a well-established technique called the *nearest neighbor algorithm* (NNA) is introduced for network synthesis. Besides, two evolution techniques are also introduced to simplify the preliminary RCN.

7.1 Procedure for Nearest Neighbor Algorithm

In order to synthesize an RCN that achieves the established flowrate targets, two criteria are to be met by all process sinks, i.e., *flowrate* (F_{SKj}) and *load* (m_{SKj}) requirements. The latter is given by the product of its flowrate and quality index (i.e., $m_{SKj} = F_{SKj} \, q_{SKj}$). To fulfill both of these requirements, one should follow the individual steps of the NNA, which are given as follows (Prakash and Shenoy, 2005a):

1. Arrange all material sinks and the sources in descending order of quality levels (i.e., ascending order of impurity concentration, for concentration-based problems). Note that the sources should include the external fresh resource (FR) and regenerated sources if any, with their respective flowrates obtained in the targeting stage. Start the design from the sink that requires highest quality index (q_{SKj}).

2. Match the selected sink SK_j with source(s) SR_i of the same quality level, if any are found.

3. Mix two source candidates SR_i (with flowrate F_{SRi} and quality q_{SRi}) and SR_{i+1} (with flowrate F_{SRi+1} and quality q_{SRi+1}) to fulfill the flowrate and load requirements of sink SK_j. Note that the source candidates

SR_i and SR_{i+1} are the nearest available "neighbors" to the sink SK_j, with quality levels just lower and just higher than that of the sink, i.e., $q_{SRi} < q_{SKj} < q_{SRi+1}$. The respective flowrate between the source and the sink is calculated via the mass balance Equations 7.1 and 7.2:

$$F_{SRi,SKj} + F_{SRi+1,SKj} = F_{SKj} \tag{7.1}$$

$$F_{SRi,SKj}\, q_{SRi} + F_{SRi+1,SKj}\, q_{SRi+1} = F_{SKj}\, q_{SKj} \tag{7.2}$$

where $F_{SRi,SKj}$ is the allocation flowrate sent from SR_i to SK_j. If SR_i has sufficient flowrate to be allocated to SK_j, i.e., $F_{SRi} \geq F_{SRi,SKj}$, go to Step 5 or else to Step 4.

A simple example is shown in Figure 7.1 for illustration. For Figure 7.1a, sink SK1 can tolerate a concentration of 50 ppm. So the two source candidates correspond to SR1 and SR2, with concentration just lower (0 ppm) and just higher (80 ppm) than that of SK1. On the other hand, SR2 and SR3 are identified as the source candidates for Figure 7.1b, since they fulfill the criteria of being just lower (20 ppm) and just higher (100 ppm) than the concentration of SK1.

4. If the source has insufficient flowrate to be used as the allocation flowrate, i.e., $F_{SRi,SKj} > F_{SRi}$, then whatever is available of that source is used completely. A new pair of neighbor candidates is considered to satisfy the sink. For instance, sink SK1 in Figure 7.1c can tolerate a concentration of 50 ppm. Hence, the source candidates that serve as its neighbor are SR1 and SR2. It is then determined that 50 t/h of each of these sources are needed to fulfill the requirement of SK1. However, SR2 has a flowrate of 20 t/h, which is insufficient to fulfill the sink requirement. Hence, SR2 will fully be recovered to SK1. SR1 and SR3 next become the new pair of neighbor candidates for SK1.

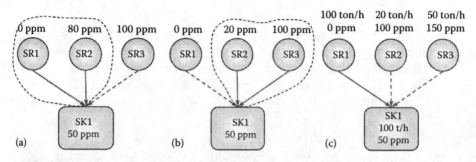

FIGURE 7.1
Identification of source candidates in NNA.

5. Repeat Steps 2–4 for all other sinks. Once all sinks are fulfilled, the unutilized source(s) are discharged as waste.

Note that for threshold problems, Equation 7.2 may be omitted in matching the last process sink in the RCN.*

7.2 Design for Direct Material Reuse/Recycle and the Matching Matrix

For a direct material reuse/recycle scheme, the NNA can be used to synthesize an RCN that achieves the minimum freshwater and wastewater targets. The latter may be identified with the graphical or algebraic approaches introduced in previous chapters.

A convenient representation for the RCN is the *matching matrix* (Prakash and Shenoy, 2005b), which takes the form in Figure 7.2. The sources are arranged as rows while the sinks as columns, both in descending order of quality index. Note that the external fresh resource is located in the first row of the matching matrix, which acts as an external sources along with other process sources of the RCN. In contrast, the waste discharge (WD) is placed in the last column of the matching matrix, which acts as an external sink. Variables in the cells indicate the allocation flowrates ($F_{SRi,SKj}$) of the sink–source matches. Note that the shaded area in the matching matrix indicates that matches between process sinks and sources across higher and

		F_{SKj}	F_{SK1}	F_{SK2}	F_{SK3}	F_{SKj}	F_D
		q_{SKj}	q_{SK1}	q_{SK2}	q_{SK3}	q_{SKj}	
F_{SRi}	q_{SRi}	SK$_j$ / SR$_i$	SK1	SK2	SK3	SK$_j$	WD
F_R	q_R	FR	$F_{FW,SK1}$		$F_{FW,SK3}$		
F_{SR1}	q_{SR1}	SR1		$F_{SR1,SK2}$			
F_{SR2}	q_{SR2}	SR2			$F_{SR2,SK3}$		
F_{SR3}	q_{SR3}	SR3		$F_{SR3,SK2}$		$F_{SR3,SKj}$	
F_{SRi}	q_{SRi}	SR$_i$				$F_{SRi,SKj}$	$F_{SRi,WW}$

FIGURE 7.2
The matching matrix. (From *Chem. Eng. Sci.*, 60(7), Prakash, R. and Shenoy, U.V., Design and evolution of water networks by source shifts, 2089–2093. Copyright 2005b, with permission from Elsevier.)

* See Sections 3.4 and 4.3 for types of threshold problems.

TABLE 7.1

Limiting Water Data for Example 7.1

Sinks, SK$_j$	F_{SKj} (t/h)	C_{SKj} (ppm)	Sources, SR$_i$	F_{SRi} (t/h)	C_{SRi} (ppm)
SK1	50	20	SR1	50	50
SK2	100	50	SR2	100	100
SK3	80	100	SR3	70	150
SK4	70	200	SR4	60	250
			FW	70	0

lower quality regions are forbidden. Note also that the pinch-causing source (SR3 as in Figure 7.2) is an exception, as it belongs to both regions.

Example 7.1 Direct reuse/recycle network (Example 6.1 revisited)

For the hypothetical case study of Polley and Polley (2000) in Example 6.1 (limiting data are summarized in Table 7.1, reproduced from Table 6.1), flowrate targeting indicates that for direct water reuse/recycle schemes, the minimum freshwater (F_{FW}, 0 ppm) and wastewater (F_{WW}) flowrate targets are determined as 70 and 50 t/h, respectively (see Table 6.2). Synthesize the RCN that achieves the flowrate targets and represent the RCN in a matching matrix.

Solution

The NNA procedure is followed to synthesize the RCN, with the following detailed steps:

1. *Arrange all material sinks and the sources in descending order of quality levels. Start the design from the sink that requires the highest quality feed, i.e., q_{SKj} of highest quality.*

   As shown in Figure 7.3, the sinks are arranged horizontally, with the lowest concentration (highest quality) on the leftmost to the highest concentration (lowest quality) on the rightmost side of the diagram. Similarly, all sources are arranged vertically, with the lowest concentration source being on the highest level, while the highest concentration is at the lowest level. We shall start the design with SK1, which requires the most stringent (highest quality) feed.

2. *Match the selected sink SK$_j$ with source(s) SR$_i$ of the same quality level, if any are available.*

3. *Mix two source candidates SR$_i$ and SR$_{i+1}$ to fulfill the flowrate and quality requirements of sink SK$_j$. Note that the source candidates SR$_i$ and SR$_{i+1}$ are the nearest available neighbors to the sink SK$_j$, with quality levels just lower and just higher than that of the sink. The respective flowrate between the source and the sink is calculated via Equations 7.1 and 7.2. If SR$_i$ has sufficient flowrate to be allocated to SK$_j$, i.e., $F_{SRi} \geq F_{SRi,SKj}$, go to Step 5 or else to Step 4.*

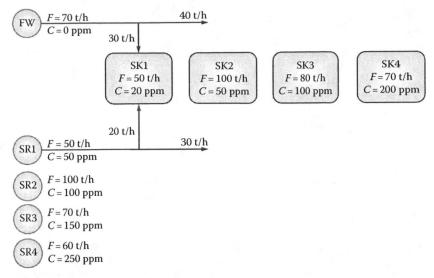

FIGURE 7.3
Results of allocating freshwater and SR1 to SK1. (From *Chem. Eng. Sci.*, 60(1), Prakash, R. and Shenoy, U.V., Targeting and design of water networks for fixed flowrate and fixed contaminant load operations, 255–268, Copyright 2005a, with permission from Elsevier.)

For the selected sink, i.e., SK1 (50 t/h, 20 ppm), there is no source with the same concentration. Hence we proceed to choose two sources, i.e., freshwater (FW) and SR1 that serve as its neighbor candidates, since they have concentrations just lower (FW, 0 ppm) and just higher (SR1, 50 ppm) than that of SK1. Equations 7.1 and 7.2 are then used to determine the allocation flowrates from freshwater ($F_{FW,SK1}$) and SR1 ($F_{SR1,SK1}$) to SK1:

$$F_{FW,SK1} + F_{SR1,SK1} = 50 \text{ t/h}$$

$$F_{FW,SK1} \ (0 \text{ ppm}) + F_{SR1,SK1} \ (50 \text{ ppm}) = 50 \text{ t/h} \ (20 \text{ ppm})$$

Hence,

$$F_{FW,SK1} = 30 \text{ t/h; and } F_{SR1,SK1} = 20 \text{ t/h}$$

This means that the leftover flowrates of FW and SR1 are 40 t/h (=70 – 30 t/h) and 30 t/h (=50 – 20 t/h), respectively. Figure 7.3 shows the result of this step. Since both freshwater and SR1 have sufficient flowrate to be allocated to SK1, we then move to Step 5.

5. *Repeat Steps 2–4 for all other sinks.*
   2. *Match the selected sink SK$_j$ with source(s) SR$_i$ of the same quality level, if any are available.*

      We next look at the second sink, i.e., SK2 (100 t/h, 50 ppm). Since SR1 has the same concentration as SK2,

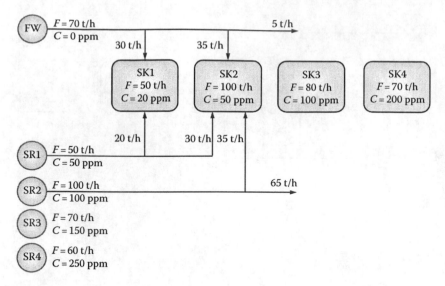

FIGURE 7.4
Results of allocating SR1, freshwater, and SR2 to SK2. (From *Chem. Eng. Sci.*, 60(1), Prakash, R. and Shenoy, U.V., Targeting and design of water networks for fixed flowrate and fixed contaminant load operations, 255–268, Copyright 2005a, with permission from Elsevier.)

its leftover flowrate (30 t/h) is sent to the latter. Hence, SK2 will only require a flowrate of 70 t/h (=100 − 30 t/h, Figure 7.4).

3. *Mix two source candidates SR$_i$ and SR$_{i+1}$ to fulfill the flowrate and quality requirements of sink SK$_j$. Note that the source candidates SR$_i$ and SR$_{i+1}$ are the nearest available "neighbors" to the sink SK$_j$, with quality levels just lower and just higher than that of the sink. The respective flowrate between the source and the sink is calculated via Equations 7.1 and 7.2. If SR$_i$ has sufficient flowrate to be allocated to SK$_j$, i.e., $F_{SRi} \geq F_{SRi,SKj}$, go to Step 5 or else to Step 4.*

After SR1 is exhausted, the new neighbor candidates for SK2 are freshwater (with leftover flowrate of 40 t/h) and SR2 (100 t/h, 100 ppm). Equations 7.1 and 7.2 are used again to determine the allocation flowrates from freshwater ($F_{FW,SK2}$) and SR2 ($F_{SR2,SK2}$) to SK2:

$$F_{FW,SK2} + F_{SR2,SK2} = 70 \text{ t/h}$$

$$F_{FW,SK2}(0 \text{ ppm}) + F_{SR2,SK2}(100 \text{ ppm}) = 70 \text{ t/h } (50 \text{ ppm})$$

Hence,

$$F_{FW,SK2} = F_{SR2,SK2} = 35 \text{ t/h}$$

Both freshwater and SR2 have sufficient flowrate to be allocated to SK2. However, the leftover flowrates now of freshwater and SR1 are 5 t/h (=40 − 35 t/h) and 65 t/h (=100 − 35 t/h), respectively. Figure 7.4 shows the result of this step. We then move to Step 5 again.

5. *Repeat Steps 2–4 for all other sinks.*

   2. *Match the selected sink SK_j with source(s) SR_i of the same quality level, if any are available.*

We next proceed to sink SK3 (80 t/h, 100 ppm). The remaining flowrate from SR2 (65 t/h) is sent to SK3 since they have the same concentrations. This means that the flowrate requirement of SK2 is reduced to 15 t/h (= 80 − 65 t/h).

   3. *Mix two source candidates SR_i and SR_{i+1} to fulfill the flowrate and quality requirements of sink SK_j. Note that the source candidates SR_i and SR_{i+1} are the nearest available neighbors to the sink SK_j, with quality levels just lower and just higher than that of the sink. The respective flowrate between the source and the sink is calculated via Equations 7.1 and 7.2. If SR_i has sufficient flowrate to be allocated to SK_j, i.e., $F_{SRi} \geq F_{SRi,SKj}$, go to Step 5 or else to Step 4.*

Since SR2 has been fully utilized, the new neighbor candidates for SK3 are freshwater (with leftover flowrate of 5 t/h) and SR3 (70 t/h, 150 ppm). Equations 7.1 and 7.2 next determine the allocation flowrates from freshwater ($F_{FW,SK3}$) and SR3 ($F_{SR3,SK3}$) to SK3:

$$F_{FW,SK3} + F_{SR3,SK3} = 15\ t / h$$

$$F_{FW,SK3}\ (0\ ppm) + F_{SR3,SK3}\ (150\ ppm) = 15\ t/h\ (100\ ppm)$$

Hence,

$$F_{FW,SK3} = 5\ t/h; \quad and \quad F_{SR3,SK3} = 10\ t/h$$

The leftover freshwater (5 t/h) is now sent to SK3 completely. On the other hand, the leftover flowrate of SR3 is determined as 60 t/h (=70 − 10 t/h). Figure 7.5 shows the result of this step. We then move to Step 5.

5. *Repeat Steps 2–4 for all other sinks.*

   2. *Match the selected sink SK_j with source(s) SR_i of the same quality level, if any are available.*

   3. *Mix two source candidates SR_i and SR_{i+1} to fulfill the flowrate and quality requirements of sink SK_j. Note that the source candidates SR_i and SR_{i+1} are the nearest available neighbors to the sink SK_j, with quality levels just lower and just higher than that of the sink. The respective flowrate between the source and the sink is calculated via Equations 7.1 and 7.2. If SR_i has sufficient flowrate to be allocated to SK_j, i.e., $F_{SRi} \geq F_{SRi,SKj}$, go to Step 5 or else to Step 4.*

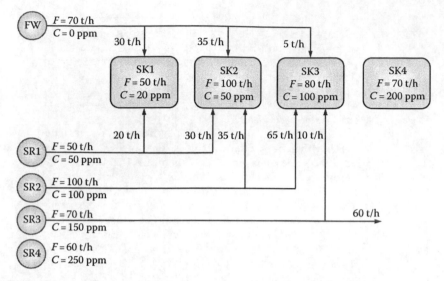

FIGURE 7.5
Results of allocating SR2, freshwater, and SR3 to SK3. (From *Chem. Eng. Sci.*, 60(1), Prakash, R. and Shenoy, U.V., Targeting and design of water networks for fixed flowrate and fixed contaminant load operations, 255–268, Copyright 2005a, with permission from Elsevier.)

We next proceed to analyze the last sink, i.e., SK4 (70 t/h, 200 ppm). There is no source with the same concentration as this sink; hence, we proceed to identify the neighbor candidates for SK4. These correspond to SR3 (with leftover flowrate of 60 t/h, 150 ppm) and SR4 (60 t/h, 250 ppm). Equations 7.1 and 7.2 determine the allocation flowrates from SR3 ($F_{SR3,SK4}$) and SR4 ($F_{SR4,SK4}$) to SK4 as

$$F_{SR3,SK4} + F_{SR4,SK4} = 70 \text{ t/h}$$

$$F_{SR3,SK4} (150 \text{ ppm}) + F_{SR3,SK4} (250 \text{ ppm}) = 70 \text{ t/h} (200 \text{ ppm})$$

Hence,

$$F_{SR3,SK4} = F_{SR4,SK4} = 35 \text{ t/h}$$

Both SR3 and SR4 have sufficient flowrates for SK4. The flowrates of these sources are both reduced to 25 t/h (=60 − 35 t/h). We then move to Step 5.

5. *Repeat Steps 2–4 for all other sinks. Once all sinks are fulfilled, the unutilized source(s) are discharged as waste.*

Since all sinks are fulfilled, the leftover flowrates in sources SR3 and SR4 are discarded as wastewater. This corresponds to a flowrate of 25 t/h for respective sources, or a total wastewater

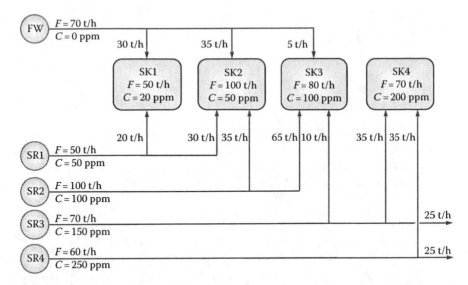

FIGURE 7.6
Final resulting RCN for Example 7.1. (From Prakash, R. and Shenoy, U.V., *Chem. Eng. Sci.*, 60(1), 255, 2005a.)

F_{SRi} (t/h)	C_{SRi} (ppm)	F_{SKj} (t/h)	50	100	80	70	50
		C_{SKj} (ppm)	20	50	100	200	
		SK_j / SR_i	SK1	SK2	SK3	SK4	WW
70	0	FW	30	35	5		
50	50	SR1	20	30			
100	100	SR2		35	65		
70	150	SR3			10	35	25
60	250	SR4				35	25

FIGURE 7.7
Matching matrix for RCN in Example 7.1. (From *Chem. Eng. Sci.*, 60(7), Prakash, R. and Shenoy, U.V., Design and evolution of water networks by source shifts, 2089–2093. Copyright 2005b, with permission from Elsevier.)

flowrate of 50 t/h matches the identified target. Figure 7.6 shows the resulting RCN for the example, while its representation in the matching matrix is given in Figure 7.7.

Example 7.2 Property reuse/recycle network (Example 3.6 revisited)

The metal degreasing process in Examples 2.5 and 3.7 is revisited here. The main property of concern for a reuse/recycle scheme is the Reid

TABLE 7.2

Limiting Data for Example 7.2

Sink, SK$_j$	F_{SKj} (kg/s)	ψ_{SKj} (atm$^{1.44}$)	Source, SR$_i$	F_{SRi} (kg/s)	ψ_{SRi} (atm$^{1.44}$)
Degreaser (SK1)	5.0	4.87	Condensate I (SR1)	4.0	13.20
Absorber (SK2)	2.0	7.36	Condensate II (SR2)	3.0	3.74
			Fresh solvent	2.4	2.713

vapor pressure (RVP), which is converted to the corresponding operator values using Equation 2.25. The limiting data are summarized in Table 7.2 (reproduced from Table 2.3). Flowrate targeting indicates that for direct reuse/recycle schemes, the minimum fresh solvent (F_{FS}) and waste solvent (F_{WS}) flowrate targets are determined as 2.4 kg/s, respectively (see Figure 3.18). Synthesize an RCN that achieves the flowrate targets, and represent the RCN in a conventional process flow diagram (PFD).

Solution

The NNA procedure is followed to synthesize the RCN for both cases, given as follows:

1. *Arrange all material sinks and the sources in descending order of quality levels respectively. Start the design from the sink that requires the highest quality feed, i.e., q_{SKj} of highest quality.*
2. *Match the selected sink SK$_j$ with source(s) SR$_i$ of the same quality level, if any are available.*
3. *Mix two source candidates SR$_i$ and SR$_{i+1}$ to fulfill the flowrate and quality requirements of sink SK$_j$. Note that the source candidates SR$_i$ and SR$_{i+1}$ are the nearest available neighbors to the sink SK$_j$, with quality levels just lower and just higher than that of the sink. The respective flowrate between the source and the sink is calculated via Equations 7.1 and 7.2. If SR$_i$ has sufficient flowrate to be allocated to SK$_j$, i.e., $F_{SRi} \geq F_{SRi,SKj}$, go to Step 5 or else to Step 4.*
4. *If the source has insufficient flowrate to be used as the allocation flowrate, i.e., $F_{SRi,SKj} > F_{SRi}$, then whatever is available of that source is used completely. A new pair of neighbor candidates is considered to satisfy the sink.*

As shown in Figure 7.8, SK1 with the lowest property operator value (i.e., most stringent quality requirement) is placed on the left, while SK2 with a higher property operator value is placed on the right. Similarly, all sources are arranged vertically, with the highest quality external fresh solvent source (FS) being on the highest level while the lowest quality SR1 at the lowest level.

Since no source and sink pairs have the same property operator values, we proceed to identify two sources, i.e., SR1 and SR2, with operators being just higher and lower than SK1, to serve as the neighbor candidates for the latter. Note that FS is not the

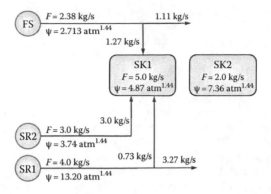

FIGURE 7.8
Results of allocating SR2, fresh solvent, and SR1 to SK1. (From Foo, D.C.Y. et al., *Chem. Eng. Sci.*, 61(8), 2626, 2006b.)

neighbor candidate at this stage. Equations 7.1 and 7.2 are then used to determine the allocation flowrates from SR1 ($F_{SR1,SK1}$) and SR2 ($F_{SR2,SK1}$) to SK1:

$$F_{SR1,SK1} + F_{SR2,SK1} = 5 \text{ kg/s}$$

$$F_{SR1,SK1}(13.2 \text{ atm}^{1.44}) + F_{SR2,SK1}(3.74 \text{ atm}^{1.44})$$
$$= 5 \text{ kg/s} (4.87 \text{ atm}^{1.44}).$$

Hence,

$$F_{SR1,SK1} = 0.6 \text{ kg/s}; \text{ and } F_{SR2,SK1} = 4.4 \text{ kg/s}.$$

However, SR2 has a smaller flowrate of 3.0 kg/s. Hence, its entire flowrate is sent to SK1. This means that the leftover flowrate requirement of SK1 is now reduced to 2 kg/s (=5 − 3 kg/s, see Figure 7.8). We then identify the next pair of source candidates to fulfill the requirement of SK1. These correspond to FS and SR1. Equations 7.1 and 7.2 are used again to determine the allocation flowrates from FS ($F_{FS,SK1}$) and SR1 ($F_{SR1,SK1}$) to SK1:

$$F_{FS,SK1} + F_{SR1,SK1} = 2 \text{ kg/s}$$

$$F_{FS,SK1}(2.713 \text{ atm}^{1.44}) + F_{SR1,SK1}(13.2 \text{ atm}^{1.44})$$
$$= 5 \text{ kg/s} (4.87 \text{ atm}^{1.44}) - 3 \text{ kg/s} (3.74 \text{ atm}^{1.44}).$$

Hence,

$$F_{FS,SK1} = 1.27 \text{ kg/s}; \text{ and } F_{SR1,SK1} = 0.73 \text{ kg/s}.$$

Since both sources have sufficient flowrates, the requirement of SK1 is fulfilled. The leftover flowrates of FS and SR1 are determined as 1.11 kg/s (=2.38 − 1.27 kg/s) and 3.27 kg/s (=4 − 0.73 kg/s), respectively. Figure 7.8 shows the result of this step.

5. *Repeat Steps 2–4 for all other sinks.*

   2. *Match the selected sink SK_j with source(s) SR_i of the same quality level, if any are available.*

   3. *Mix two source candidates SR_i and SR_{i+1} to fulfill the flowrate and quality requirements of sink SK_j. Note that the source candidates SR_i and SR_{i+1} are the nearest available neighbors to the sink SK_j, with quality levels just lower and just higher than that of the sink. The respective flowrate between the source and the sink is calculated via Equations 7.1 and 7.2. If SR_i has sufficient flowrate to be allocated to SK_j, i.e., $F_{SRi} \geq F_{SRi,SKj}$, go to Step 5 or else to Step 4.*

5. Repeat Steps 2–4 for all other sinks. Once all sinks are fulfilled, the unutilized source(s) are discharged as waste.

We next proceed to sink SK2. Similar to the case of SK1, there is no source with the same property operator value. Hence, we identify sources FS and SK1 as the neighbor candidates for this sink. Equations 7.1 and 7.2 determine the allocation flowrates from FS ($F_{FS,SK2}$) and SR1 ($F_{SR1,SK2}$) to SK2 as

$$F_{FS,SK2} + F_{SR1,SK2} = 2 \text{ kg/s}$$

$$F_{FS,SK2}(2.713 \text{ atm}^{1.44}) + F_{SR1,SK2}(13.2 \text{ atm}^{1.44}) = 2 \text{ kg/s} \ (7.36 \text{ atm}^{1.44}).$$

Hence,

$$F_{FS,SK2} = 1.11 \text{ kg/s; and } F_{SR1,SK2} = 0.89 \text{ kg/s}.$$

Since both sources have sufficient flowrates, the requirement of SK2 is fulfilled. Note that the fresh solvent source is used completely. The leftover flowrate of SR1, i.e., 2.38 kg/s (=3.27 − 0.89 kg/s), is discharged as waste solvent, matching the flowrate target identified earlier. Figure 7.9 shows the resulting RCN for the case, while Figure 7.10 shows the solvent recovery scheme in a conventional PFD.

7.3 Design for Material Regeneration Network

The NNA may be used to synthesize an RCN with an interception unit (in short *material regeneration network*) that achieves the minimum freshwater,

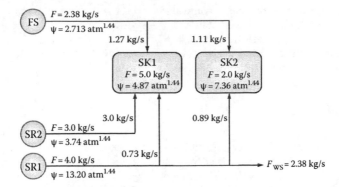

FIGURE 7.9
Final resulting RCN for Example 7.2. (From Foo, D.C.Y. et al., *Chem. Eng. Sci.*, 61(8), 2626, 2006b.)

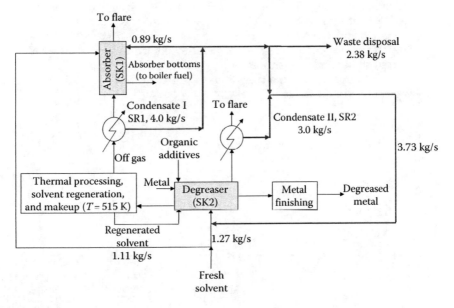

FIGURE 7.10
Direct solvent reuse/recycle scheme for metal degreasing problem (Example 7.2). (From Foo, D.C.Y. et al., *Chem. Eng. Sci.*, 61(8), 2626, 2006b.)

wastewater, and regeneration flowrate targets. However, other insights gained from the targeting stage, particularly the identified sources to be sent for regeneration, should be incorporated to ease the network design stage.

Example 7.3 Water regeneration network (Example 6.1 revisited)

For the hypothetical case study of Polley and Polley (2000) in Examples 6.1 and 7.1 (limiting data in Table 7.1), the targeting stage identifies that the minimum freshwater, wastewater, and regenerated water (F_{RW}) have the flowrates of 20, 0, and 53.57 t/h, respectively, when an interception unit with an outlet concentration (C_{Rout}) of 10 ppm is used for water regeneration (see Example 6.1 and Table 6.9). Synthesize the RCN that achieves the flowrate targets. Make good use of the insights gained from the targeting stage (Example 6.1).

Solution

As discussed in Example 6.1, several insights are identified from the targeting stage. In Table 6.7, 53.57 t/h of regeneration flowrate is to be purified to 10 ppm for further reuse/recycle for the sinks in the regenerated water region (RWR). This regenerated flowrate is to be originated from source SR4 (250 ppm). Besides, in Table 6.8, it is identified that 20 t/h of freshwater (FW) is needed in the freshwater region (FWR). This freshwater source is to be mixed with 5 t/h of SR2 to partially fulfill the requirement of SK1 (5 t/h) in the FWR. From these insights, the associated flowrates should first be allocated between the respective sinks and sources, before the NNA is applied. This is shown in Figure 7.11. Note that after the allocation of these flowrates, the freshwater source has ceased, while SR2 has a reduced flowrate of 95 t/h.

The NNA procedure can then be used to match the remaining sinks and sources in the RCN.

1. *Start the design from the sink that requires the highest quality feed, i.e., q_{SKj} of highest quality.*
2. *Match the selected sink SK_j with source(s) SR_i of the same quality level, if any are available.*

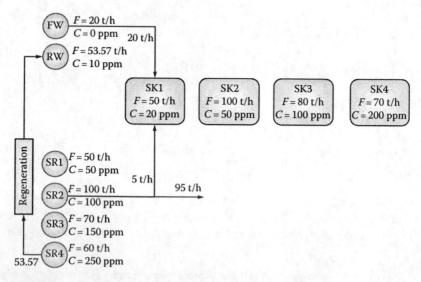

FIGURE 7.11
Insight gain from the targeting stage.

3. *Mix two source candidates SR$_i$ and SR$_{i+1}$ to fulfill the flowrate and quality requirements of sink SK$_j$. Note that the source candidates SR$_i$ and SR$_{i+1}$ are the nearest available neighbors to the sink SK$_j$, with quality levels just lower and just higher than that of the sink. The respective flowrate between the source and the sink is calculated via Equations 7.1 and 7.2. If SR$_i$ has sufficient flowrate to be allocated to SK$_j$, i.e., $F_{SRi} \geq F_{SRi,SKj}$, go to Step 5 or else to Step 4.*

We first look at SK1 whose flowrate requirement has not been completely fulfilled. After the flowrate allocation from freshwater and SR2, the flowrate requirement of SK1 is reduced to 25 t/h. Since no source and sink pairs have the same concentration we then identify RW (10 ppm) and SR1 (50 ppm) as the neighbor candidates for SK1. Equations 7.1 and 7.2 then determine the allocation flowrates as 18.75 and 6.25 t/h, respectively (see result in Figure 7.12). The flowrates of RW and SR1 are reduced to 34.82 and 43.75 t/h, respectively. Since both sources have sufficient flowrates, we then proceed to the next sink, i.e., SK2.

5. *Repeat Steps 2–4 for all other sinks.*

   2. *Match the selected sink SK$_j$ with source(s) SR$_i$ of the same quality level, if any are available.*

   3. *Mix two source candidates SR$_i$ and SR$_{i+1}$ to fulfill the flowrate and quality requirements of sink SK$_j$. Note that the source candidates SR$_i$ and SR$_{i+1}$ are the nearest available neighbors to the sink SK$_j$, with quality levels just lower and just higher than that of the sink. The respective flowrate between the source and*

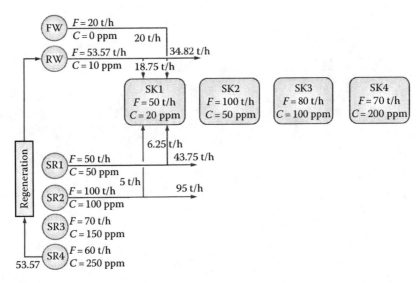

FIGURE 7.12
Results of allocating FW, SR2, SR1, and RW to SK1.

the sink is calculated via Equations 7.1 and 7.2. If SR_i has sufficient flowrate to be allocated to SK_j, i.e., $F_{SRi} \geq F_{SRi,SKj}$, go to Step 5 or else to Step 4.

For SK2 (100 t/h, 50 ppm), the leftover flowrate from SR1 (43.75 t/h) of the same concentration is fully utilized. We then identify RW and SR2 as the new neighbor candidates for SK2. With Equations 7.1 and 7.2, the allocated flowrates from these neighbors are determined as 31.25 (RW) and 25 t/h (SR2), respectively. Both sources have sufficient flowrate for the allocation; hence, we then proceed to SK3. Note that the leftover flowrates of RW and SR2 are now reduced to 3.57 and 70 t/h, respectively (see Figure 7.13).

5. *Repeat Steps 2–4 for all other sinks.*

2. *Match the selected sink SK_j with source(s) SR_i of the same quality level, if any are available.*

3. *Mix two source candidates SR_i and SR_{i+1} to fulfill the flowrate and quality requirements of sink SK_j. Note that the source candidates SR_i and SR_{i+1} are the nearest available neighbors to the sink SK_j, with quality levels just lower and just higher than that of the sink. The respective flowrate between the source and the sink is calculated via Equations 7.1 and 7.2. If SR_i has sufficient flowrate to be allocated to SK_j, i.e., $F_{SRi} \geq F_{SRi,SKj}$, go to Step 5 or else to Step 4.*

5. *Repeat Steps 2–4 for all other sinks. Once all sinks are fulfilled, the unutilized source(s) are discharged as waste.*

Since SR2 has the same concentration as SK3, its leftover flowrate (70 t/h) is fully allocated to SK3. The new neighbor candidates hence correspond to RW and SR3 (70 t/h, 150 ppm).

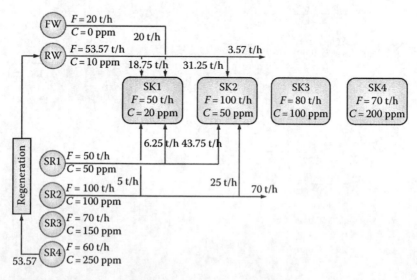

FIGURE 7.13
Results of allocating SR1, RW, and SR2 to SK2.

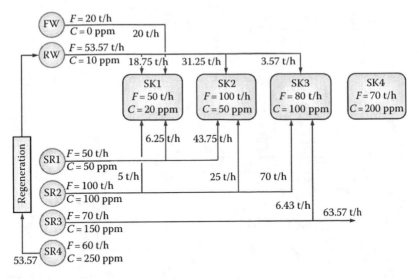

FIGURE 7.14
Results of allocating SR2, RW, and SR3 to SK3.

Equations 7.1 and 7.2 then determine that the allocated flow-rates from these sources are 3.57 (RW) and 6.43 t/h (SR3), respectively. Hence, RW is completely utilized, while SR3 has a reduced flowrate of 63.57 t/h (=70 − 6.43 t/h; see Figure 7.14). We then proceed to SK4.

Since SK4 is the last sink to be fulfilled by the source, there is no flexibility in allocating the flowrate from the sources to this sink. Hence, leftover flowrates from SR3 and SR4 are sent to this sink. The summation of these leftover flowrates is determined as 70 t/h, sufficient to meet the flowrate requirement of SK4. Besides, it is worth noting that since SR3 contributes a larger flowrate than that of SR4, the resulting mixture has a concentration of 159 ppm, lower than the limiting concentration of SK4 (200 ppm). The final RCN is shown in Figure 7.15, while its matrix representation is given in Figure 7.16.

Finally, note that for a material regeneration network, insights from the targeting stage serve as guidance in the network design stage. The NNA can equally be used even without insights from the targeting stage. For such a case, a *degenerate* solution exists for the problem (i.e., different network structure that achieves the same network targets).[*] However, it is a good practice to incorporate those insights, as they will certainly be useful when a more complicated RCN is to be synthesized.[†]

---

[*] Solve Problem 7.13a for a degenerate solution for this case.
[†] See Problems in Chapter 9 and Section 10.3 for extended applications of NNA in various RCN synthesis problem.

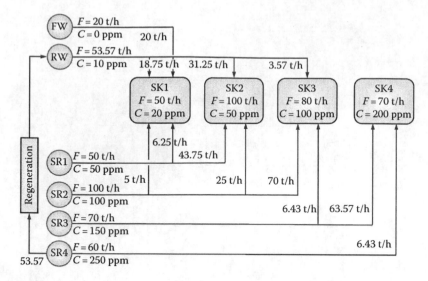

FIGURE 7.15
Final resulting RCN for Example 7.3.

F_{SRi} (t/h)	C_{SRi} (ppm)	F_{SKj} (t/h) / SK$_j$ SR$_i$	50	100	80	70	53.57
		C_{SKj} (ppm)	20	50	100	200	250
			SK1	SK2	SK3	SK4	RW
20	0	FW	20				
53.57	10	RW	18.75	31.25	3.57		
50	50	SR1	6.25	43.75			
100	120	SR2	5	25	70		
70	150	SR3			6.43	63.57	
60	250	SR4				6.43	53.57

FIGURE 7.16
Matching matrix of the RCN in Example 7.3.

7.4 Network Evolution Techniques

In this section, two well established network evolution techniques are introduced. These techniques are used to simplify a preliminary RCN that is synthesized using the NNA.

7.4.1 Source Shift Algorithm

In the first technique, i.e., the *source shift algorithm* (SSA), some matches between sinks and sources (e.g., established using the NNA) are removed, aiming to reduce their number of interconnections.

To evolve an RCN with SSA, the following steps are to be followed (Prakash and Shenoy, 2005b; Ng and Foo, 2006):

1. Identify a sink–source pair that fulfills both the following criteria:
   a. The chosen sink–source matches should have the same quality value, i.e., $q_{SKj} = q_{SRi}$.
   b. The source flowrate should be larger than or equal to that of the sink, i.e., $F_{SRi} \geq F_{SKj}$.
2. Feed the selected sink fully with the source.
3. Shift the excess flowrate experienced by the sink to the sink (where the source was matched to originally) that experiences a flowrate deficit.

A simple illustration is shown in Figure 7.17, where two process sinks are matched with an external fresh resource (FR) as well as two process sources. As observed, a total of six interconnections were established between the sinks and sources (Figure 7.17a). Note that both sink–source pairs SK1–SR1 and SK2–SR1 fulfill both the aforementioned criteria (in step 1). We shall first illustrate for the SK1–SR1 pair.

Following Step 2, SK1 has to be fully allocated by SR1 flowrate. Hence, SR1 that was earlier allocated to SK2 is shifted to SK1 (Figure 7.17b). However, this means that SK1 now has an excess flowrate of 50 t/h, while SK2 experiences a flowrate deficit of 50 t/h. To restore the flowrate deficit of the latter,

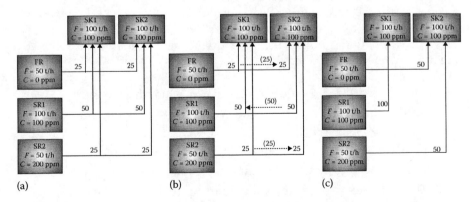

(a)   (b)   (c)

FIGURE 7.17
Application of SSA for SK1–SR1 pair. (From Ng, D.K.S. and Foo, D.C.Y., Evolution of water network with improved source shift algorithm and water path analysis, *Ind. Eng. Chem. Res.*, 45(24), 8095. Copyright 2006 American Chemical Society. With permission.)

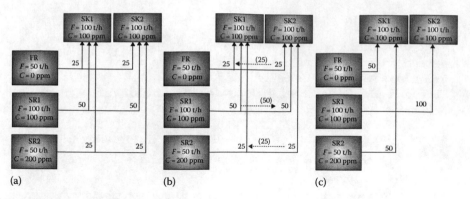

FIGURE 7.18
Application of SSA for SR2–SR1 pair. (From Ng, D.K.S. and Foo, D.C.Y., Evolution of water network with improved source shift algorithm and water path analysis, *Ind. Eng. Chem. Res.*, 45(24), 8095. Copyright 2006 American Chemical Society. With permission.)

50 t/h of excess source flowrate in SK1 is shifted to SK2, using the matches of SK2–FR and SK2–SR2, as shown in Figure 7.17b. A net result of these source shifts is the reduced number of interconnections. As shown in Figure 7.17c, the numbers of interconnections in the RCN are reduced by half, i.e., from six to three matches.

Similar steps may also be applied for the SK2–SR1 pair in Figure 7.18. Following Step 2, flowrate from SR1 is fully allocated to SK2. This causes a 50 t/h flowrate deficit in SK1, while excess flowrate is experienced in SK2. Hence, to restore the flowrate balances, 50 t/h flowrate is shifted from SK2 to SK1 using the SK1–FR and SK1–SR2 matches (Figure 7.18b). The net result is an RCN with three interconnections (Figure 7.18c). Note that the latter has a different configuration as observed in Figure 7.17c.

Apart from the earlier-described procedure, SSA may also be applied to interconnections that involve waste streams in the *lower quality region* of an RCN. To carry out the SSA for the lower quality region, the revised SSA steps are implemented:

1. Identify a set of sink–source pairs that fulfill both criteria:
   a. The sink–source pairs should consist of two sets of sink–source and source–waste matches.
   b. All matches should be served by the same source pairs.
2. Feed the sink fully with the source of higher quality.
3. Shift the excess flowrate experienced by the sink to the waste that was originally fed by the source.

A simple illustration is given in Figure 7.19, showing the lower quality region of an RCN as designed by NNA. A total of six interconnections are observed

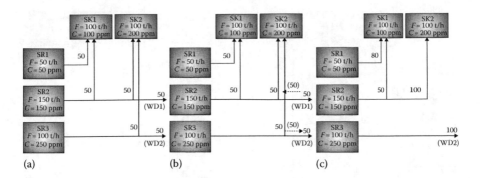

FIGURE 7.19
Application of SSA for lower quality region.

between three sources and two sinks, as well as the waste discharge (WD)
streams (Figure 7.19a). Following Step 1 of the revised procedure, the following sink–source matches are identified, i.e., SK2–SR2, SK2–SR3, WD1–SR2,
and WD2–SR3. SK2 is then fully fed with SR2 of higher quality (Step 2). The
excess flowrate of 50 t/h is then discharged as waste using the WD2–SR3
match (Step 3; see Figure 7.19b). The net result of the shift is the reduced
interconnection with four matches (Figure 7.19c). This may also be viewed as
degenerate solutions for the problem.

Note that the revised SSA also works well for threshold problems with
zero freshresource.

7.4.2 Material Path Analysis

Material Path Analysis (MPA) aims to evolve an RCN by having additional
fresh resource usage (termed as *"penalty"*). In essence, it explores the trade-
off between the minimum external fresh resource usage and network complexity. A material path is defined as *a continuous route that starts from an
external fresh resource, linked with the sink–source connections, and ends at the
waste discharge* (Ng and Foo, 2006). When a material path is observed, optimization may be performed to remove some interconnections by having fresh
resource penalty. In order to minimize the penalty, a useful heuristic is to
remove the smallest sink–source match(es) along the material path (Ng and
Foo, 2006).

A simple example is given in Figure 7.20. A material path is identified in
Figure 7.20a that connects FR, SK1–SR1 match, and WD1. By sending an
additional 50 t/h of fresh resource to SK1, the original allocated flowrate
between SR1 and SK1 is no longer necessary. Hence, the excess flowrate of
SR1 is discharged as waste through WD1 (see Figure 7.20b). In other words,
the removal of the SK1–SR1 match (i.e., the RCN with one reduced interconnection) is with an expense of 50 t/h fresh resource penalty.

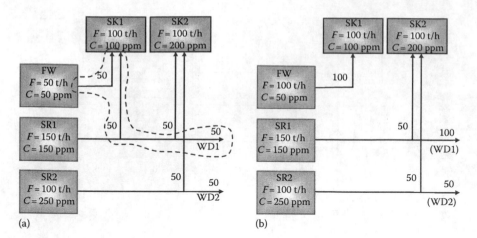

FIGURE 7.20
Example of MPA.

Note that SSA is normally applied before MPA, as the latter normally involves fresh resource penalty, which is not the case for SSA. Note also that both SSA and MPA are easily implemented through the matching matrix (Prakash and Shenoy, 2005b; Ng and Foo, 2006).

Example 7.4 Network evolution with SSA and MPA

Revisit the preliminary water network synthesized using NNA in Example 7.1. Reduce the number of interconnections in the RCN with the following techniques:

1. SSA—reduce interconnections in both higher and lower quality regions.
2. MPA—identify the water penalty incurred with the removal of interconnections.

Solution

Case 1—Interconnection reduction using SSA
  The matching matrix in Figure 7.7 is reproduced in Figure 7.21, which shows that the RCN has a total of 12 sink–source matches (indicated by cells with figures), including those that connect with the freshwater (FW) and wastewater (WW) streams. To evolve the RCN using the SSA, the three-step procedure is followed:

1. *Identify a sink–source pair that fulfills both the following criteria:*
   a. *The chosen sink–source matches should have the same quality value, i.e., $q_{SKj} = q_{SRi}$.*
   b. *The source flowrate should be larger than or equal to that of the sink, i.e., $F_{SRi} \geq F_{SKj}$.*

F_{SRi} (t/h)	C_{SRi} (ppm)	F_{SKj} (t/h) C_{Skj} (ppm) SK_j SR_i	SK1	SK2	SK3	SK4	WW
		50 / 20					
		100 / 50					
		80 / 100					
		70 / 200					
70	0	FW	30	35	5		
50	50	SR1	20	30			
100	100	SR2		35	65		
70	150	SR3			10	35	25
60	250	SR4				35	25

FIGURE 7.21
Identification of sink–source pairs for SSA.

From Figure 7.21, it is observed that both SK2–SR1 and SK3–SR2 pairs fulfill the first criteria, i.e., concentrations of both sink and source are the same (indicated by the dashed boxes). However, only the SK3–SR2 pair fulfills the second criteria, i.e., $F_{SR3} \geq F_{SK2}$, and hence is selected to undergo Steps 2 and 3.

2. *Feed the selected sink fully with the source.*
3. *Shift the excess flowrate experienced by the sink to the sink (where the source was matched to originally) that experiences a flowrate deficit.*

Following Step 2, the flowrate requirement of SK3 should be fulfilled by SR2. However, at present, only 65 t/h of the SR2 flowrate is allocated to SK3, while the remaining (35 t/h) to SK2. To fully satisfy the SK3 flowrate with SR2, 15 t/h of flowrate is to be shifted from the current SK2–SR2 match. This causes a 15 t/h flowrate deficit for SK2. In contrast, SK3 experiences a flowrate surplus of 15 t/h due to the shift. To restore flowrate balance for both SK2 and SK3, 15 t/h of flowrate is to be shifted to SK2 from other existing sink–source matches associated with SK3. These correspond to the matches of SK3–FW and SK3–SR3, with allocated flowrates of 5 and 10 t/h, respectively (Figure 7.22).

The net result of the source shifts is shown in Figure 7.23 (newly allocated flowrates are highlighted in bold). Note that the SK2–FW match row has higher flowrate; while a new SK2–SP3 match emerges in the RCN. The total number of interconnections has now been reduced to 11 matches.

Besides, the revised SSA is also applied for the lower quality (higher concentration) region:

1. *Identify a set of sink–source pairs that fulfill both criteria:*
   a. *The sink–source pairs should consist of two sets of sink–source and source–waste matches.*
   b. *All matches should be served by the same source pairs.*

Process Integration for Resource Conservation

F_{SRi} (t/h)	C_{SRi} (ppm)	F_{SKj} (t/h)	50	100	80	70	50
		C_{Skj} (ppm)	20	50	100	200	
		SK$_j$ / SR$_i$	SK1	SK2	SK3	SK4	WW
70	0	FW	30	35 ◁ 5	5		
50	50	SR1	20	30			
100	100	SR2		35 [15] ▷	65		
70	150	SR3		◁ 10	10	35	25
60	250	SR4				35	25

FIGURE 7.22
Source shifts for Example 7.4. (From Ng, D.K.S. and Foo, D.C.Y., Evolution of water network with improved source shift algorithm and water path analysis, *Ind. Eng. Chem. Res.*, 45(24), 8095. Copyright 2006 American Chemical Society. With permission.)

F_{SRi} (t/h)	C_{SRi} (ppm)	F_{SKj} (t/h)	50	100	80	70	50
		C_{SKj} (ppm)	20	50	100	200	
		SK$_j$ / SR$_i$	SK1	SK2	SK3	SK4	WW
70	0	FW	30	**40**			
50	50	SR1	20	30			
100	100	SR2		**20**	**80**		
70	150	SR3		**10**		35	25
60	250	SR4				35	25

FIGURE 7.23
Results of source shift between SK2 and SK3 matches. (From Ng, D.K.S. and Foo, D.C.Y., Evolution of water network with improved source shift algorithm and water path analysis, *Ind. Eng. Chem. Res.*, 45(24), 8095. Copyright 2006 American Chemical Society. With permission.)

2. Feed the sink fully with the source of higher quality.
3. Shift the excess flowrate experienced by the sink to the waste that was originally fed by the source.

Four matches that may undergo SSA in the higher concentration region are identified, i.e., SK4–SR3, SK4–SR4, WW–SR3, and WW–SR4. SK4 is then fully satisfied by SR3 (higher quality as compared with SR4), by shifting its flowrate that is originally allocated to WW (25 t/h). The excess flowrate in SK4 is then shifted to WW (Figure 7.24). Doing this leads to the removal of the WW–SR3 match. As shown in Figure 7.25, the resulting RCN has a total of 10 interconnections.

F_{SRi} (t/h)	C_{SRi} (ppm)	F_{SKj} (t/h)	50	100	80	70	50
		C_{Skj} (ppm)	20	50	100	200	
		SK_j / SR_i	SK1	SK2	SK3	SK4	WW
70	0	FW	30	40			
50	50	SR1	20	30			
100	100	SR2		20	80		
70	150	SR3		10		35 ◁25	25
60	250	SR4				35 25▷	25

FIGURE 7.24
Source shift for lower quality region. (From Prakash, R. and Shenoy, U.V. *Chemical Engineering Science*, 60(7), 2089, 2006.)

F_{SRi} (t/h)	C_{SRi} (ppm)	F_{SKj} (t/h)	50	100	80	70	50
		C_{SKj} (ppm)	20	50	100	200	
		SK_j / SR_i	SK1	SK2	SK3	SK4	WW
70	0	FW	30	40			
50	50	SR1	20	30			
100	100	SR2		20	80		
70	150	SR3		10		60	
60	250	SR4				10	50

FIGURE 7.25
Results of SSA for lower quality region. (From Ng, D.K.S. and Foo, D.C.Y., *Ind. Eng. Chem. Res.*, 45(24), 8095, 2006.)

Case 2—Interconnection reduction using MPA

We next approach the problem with MPA. One of the *water paths* for this case is shown in Figure 7.26, which connects the matches of SK2–FW, SK2–SR3, SK4–SR3, SK4–SR4, and WW–SR4. It is then observed that this path consists of two matches (SK2–SR3 and SK4–SR4) with the smallest flowrates, i.e., 10t/h. These matches may be removed from the path by adding 10t/h of freshwater flowrates to the RCN.

As shown in Figure 7.27, 10t/h of water is moved from the SK2–SR3 match along the water path to the SK4–SR3 match. Having fed with 10t/h of water, SK4 now experiences an excess of water. Hence, the SK4–SR4 match that supplies water to SK4 (10t/h) is no longer needed. The excess water flowrate of SK4–SR4 match can then be moved along the path to the WW–SR4 match (Figure 7.27). This results in a larger waste-water flowrate of 60t/h (from 50t/h in Figure 7.26).

		F_{SKj} (t/h)	50	100	80	70	50
		C_{SKj} (ppm)	20	50	100	200	
F_{SRi} (t/h)	C_{SRi} (ppm)	SK$_j$ / SR$_i$	SK1	SK2	SK3	SK4	WW
70	0	FW	30	40			
50	50	SR1	20	30			
100	100	SR2		20	80		
70	150	SR3		10		60	
60	250	SR4				10	50

FIGURE 7.26
Identification of material path. (From Ng, D.K.S. and Foo, D.C.Y., *Ind. Eng. Chem. Res.*, 45(24), 8095, 2006.)

		F_{SKj} (t/h)	50	100	80	70	50
		C_{SKj} (ppm)	20	50	100	200	
F_{SRi} (t/h)	C_{SRi} (ppm)	SK$_j$ / SR$_i$	SK1	SK2	SK3	SK4	WW
70	0	FW	30	40			
50	50	SR1	20	30			
100	100	SR2		20	80		
70	150	SR3		10	10	60	
60	250	SR4				10	10	50

FIGURE 7.27
Shifting of source flowrate along material path. (From Ng, D.K.S. and Foo, D.C.Y., *Ind. Eng. Chem. Res.*, 45(24), 8095, 2006.)

Next, we also notice that SK2 has a flowrate deficit of 10 t/h. The water deficit is supplemented by freshwater, which means that the freshwater flowrate is also increased by 10 t/h, i.e., to 80 t/h (from 70 t/h in Figure 7.26). In other words, two interconnections, i.e., SK2–SR3 and SK4–SR4, are removed from the network with the expense of 10 t/h of freshwater and wastewater penalties. The result of the MPA is shown in Figure 7.28, i.e., an RCN with eight interconnections.

Note that a new material path may also be defined even though some of its matches may not exist in the network yet. One such water path is shown for the problem in Figure 7.29. The path connects the matches of SK1–FW, SK1–SR1, SK2–SR1, SK2–SR2, and WW–SR2. For this path, the WW–SR2 match does not exist in the network yet, but may be easily added as the discharge flowrate for SR2.

F_{SRi} (t/h)	C_{SRi} (ppm)	F_{SKj} (t/h)	50	100	80	70	60
		C_{SKj} (ppm)	20	50	100	200	
		SK$_j$ / SR$_i$	SK1	SK2	SK3	SK4	WW
80	0	FW	30	50			
50	50	SR1	20	30			
100	100	SR2		20	80		
70	150	SR3				70	
60	250	SR4					60

FIGURE 7.28
Removal of SK2–SR3 and SK4–SR4 matches with 10 t/h freshwater penalty. (From Ng, D.K.S. and Foo, D.C.Y., *Ind. Eng. Chem. Res.*, 45(24), 8095, 2006.)

F_{SRi} (t/h)	C_{SRi} (ppm)	F_{SKj} (t/h)	50	100	80	70	60
		C_{SKj} (ppm)	20	50	100	200	
		SK$_j$ / SR$_i$	SK1	SK2	SK3	SK4	WW
80	0	FW	30	50			
50	50	SR1	20	30			
100	100	SR2		20	80		
70	150	SR3				70	
60	250	SR4					60

FIGURE 7.29
Path involving a new wastewater stream. (From Ng, D.K.S. and Foo, D.C.Y., *Ind. Eng. Chem. Res.*, 45(24), 8095, 2006.)

As observed from Figure 7.29, the matches with smallest flowrate in this path correspond to SK1–SR1 and SK2–SR2. By adding an additional 20 t/h freshwater flowrate to SK1 (via SK1–FW match), the SK1–SR1 and SK2–SR2 matches are removed, coupled with the emergence of the new wastewater match, i.e., WW–SR2 of 20 t/h. Even though a new match is created for this case, it is offset by the removal of two other matches (SK1–SR1 and SK2–SR2), i.e., a net result of one match removal. Note that with the addition of 20 t/h freshwater, the total freshwater and wastewater flowrates are increased to 100 and 80 t/h, respectively (from 80 and 60 t/h, respectively, Figure 7.28). In other words, the removal of the sink–source match is at the expense of 20 t/h of water penalty. The result for this path analysis is a network with seven matches, as shown in Figure 7.30.

F_{SKj} (t/h)			50	100	80	70	80
		C_{SKj} (ppm)	20	50	100	200	
F_{SRi} (t/h)	C_{SRi} (ppm)	SK$_j$ / SR$_i$	SK1	SK2	SK3	SK4	WW
100	0	FW	50	50			
50	50	SR1		30			
100	100	SR2			80		20
70	150	SR3				70	
60	250	SR4					60

FIGURE 7.30
Removal of SK1–SR1 and SK2–SR2 matches with additional 20 t/h freshwater penalty. (From Ng, D.K.S. and Foo, D.C.Y., *Ind. Eng. Chem. Res.*, 45(24), 8095, 2006.)

TABLE 7.3

Comparison of Various Network Evolution Schemes

Schemes	Number of Matches	Freshwater Flowrate (t/h)	Wastewater Flowrate (t/h)	Water Penalty (t/h)
NNA	12			
SSA	11	70	50	0
	10			
MPA	8	80	60	10
	7	100	80	30

A simple comparison among all schemes is shown in Table 7.3. Network evolution with SSA does not lead to any flowrate penalty but can only reduce the number of interconnections to 10 matches. Hence, RCN synthesized by NNA and evolved by SSA are degenerate solutions with the same water flowrate targets. On the other hand, evolution via MPA reduces the number of interconnection further to seven matches but with additional water flowrate penalty. These options serve as alternatives in trading off water recovery and network complexity.

7.5 Additional Readings

Apart from the NNA, there exist many other network design techniques for RCN synthesis, e.g., *water grid diagram* (Wang and Smith, 1994), water main method (Kuo and Smith, 1998; Feng and Seider, 2001), source sink mapping diagram (El-Halwagi, 1997), water source diagram (Gomes et al., 2006), etc. A review that compares these techniques is found in the review paper by Foo (2009). Besides, the concepts of SSA and MPA are also extended into a mathematical optimization model (Das et al., 2009).

Problems

Water Minimization Problems

7.1 For Example 7.1, synthesize the water reuse/recycle networks for the following cases:

a. The RCN has an impure freshwater feed of 10 ppm impurity. The minimum freshwater flowrate is determined as 75 t/h.

b. Two freshwater feeds are available for the RCN, i.e., pure (0 ppm) and impure freshwater feeds (80 ppm), with targeted flowrates of 56.26 and 43.75 t/h, respectively.

c. Three freshwater feeds are available for the RCN, i.e., one pure (0 ppm) and two impure freshwater sources (20 and 50 ppm), with targeted flowrates of 0, 50, and 50 t/h, respectively.

7.2 For the acrylonitrile case study in Examples 3.1, 5.1, and 6.3, synthesize the RCN for the following cases. Show the RCN in a conventional PFD. For all cases, assume that pure freshwater feed is used.

a. A reuse/recycle network where recovery between process source(s) and sink(s) is given higher priority, before freshwater is used. The minimum freshwater and wastewater flowrates are targeted as 2.1 and 8.2, respectively. Use the limiting data in Table 3.1.

b. Redo (a) by considering an additional process constraint, i.e., no water recovery between any process sources to boiler feed water sink (SK1). Note that the water flowrate targets remain the same as in (a) (see the resulting flowsheet in Figure 3.10).

c. Reconstruct the RCN in (b) when the steam-jet ejector is replaced by a vacuum pump (see details in Example 5.1). Use the limiting data in Table 6.18.

d. An interception unit is added to the RCN in (c), with an outlet concentration of 11.6 ppm. Synthesize a water regeneration network that meets the wastewater flowrate target of 5.9 kg/s. The targeting stage indicates that no freshwater is needed for the RCN (see Example 6.3 for details; the resulting flowsheet is given in Figure 6.3).

7.3 Synthesize a water reuse/recycle network for the tire-to-fuel process in Problems 2.1 and 3.1, assuming that pure freshwater is used. The minimum freshwater flowrate is determined as 0.135 kg/s. Use the limiting data in Table 3.8. Present the RCN in a conventional PFD.

7.4 Synthesize a water reuse/recycle network for the palm oil mill in Problem 4.8, assuming that pure freshwater is used. The minimum freshwater flowrate is determined as 1.63 t/h.

7.5 Synthesize a water reuse/recycle network for the steel plant in Problem 4.9, assuming that pure freshwater is used. The minimum freshwater flowrate is determined as 2234.21 t/h.

7.6 Synthesize a water reuse/recycle network for the bulk chemical production case in Problem 3.5. Use the limiting data in Table 3.11. Assume that pure freshwater is used. The minimum freshwater flowrate is determined as 11.1 kg/s. Evolve the preliminary network to remove interconnections of small flowrates (<1 t/d) but do not increase the flowrate targets. Present the RCN in a conventional PFD.

7.7 Synthesize the following water networks for the specialty chemical production process in Problem 4.6. Limiting data for water minimization are summarized in Table 4.25. Assume that pure freshwater is used.

  a. A direct reuse/recycle network with minimum freshwater flowrate of 90.64 t/h. Can you synthesize another alternative reuse/recycle scheme to avoid the reuse of effluent from the cooling system to the reactor?

  b. A water regeneration network with an interception unit (C_{Rout} = 200 ppm). The ultimate flowrate targeting procedure determines that the freshwater, regenerated water and wastewater have a flowrate of 42.25, 67.75 and 2.25 t/h respectively. Ignore the process constraint in (a), i.e., the reuse of effluent from cooling system to reactor is allowed. Make use of the targeting results in Problem 6.3 (b) and (c).

7.8 Synthesize the water reuse/recycle network in Examples 3.6 and 4.4 (limiting data are given in Table 4.13). The targeting step (see Figure 3.17 and Table 4.16) indicates that it is a zero discharge network with a 60 t/h freshwater flowrate (0 ppm impurity content). (hint: Equation 7.2 may be omitted when matching the last water sink). Is the network structure similar to that developed by insights from targeting insights (solve Problem 4.7 to compare answers)?

7.9 For the water minimization case in Examples 3.5 and 4.3 (limiting data are given in Table 4.11), complete the following tasks:

  a. Synthesize the water network. The targeting step indicates that it is a zero fresh resource network with a wastewater discharge of 20 t/h (Figure 3.15 and Table 4.12; hint: Equation 7.2 may be omitted when matching the last water sink). Is the network structure similar to

TABLE 7.4

Limiting Water Data for Problem 7.10

SK$_j$	F_{SKj} (t/h)	C_{SKj} (ppm)	SR$_i$	F_{SRi} (t/h)	C_{SRi} (ppm)
SK1	1200	120	SR1	500	100
SK2	800	105	SR2	2000	110
SK3	500	80	SR3	400	110
			SR4	300	60

that developed by insights from targeting insights (solve Problem 4.7 to compare answers)?

b. Evolve the preliminary network with SSA to reduce one wastewater stream but do not increase its water flowrates (hint: treat this as a lower quality region).

7.10 Table 7.4 shows the limiting data for a water minimization case study (Jacob et al., 2002). Complete the following tasks:

a. Synthesize the water reuse/recycle network for this case. The targeting stage indicates that this is a zero fresh resource network, with a minimum wastewater flowrate of 700 t/h. (hint: Equation 7.2 may be omitted when matching the last water sink).

b. Evolve the preliminary network with SSA to reduce network connection but do not increase its water flowrates (hint: treat this as a lower quality region).

c. If we were to remove one additional interconnection, what would be the minimum freshwater flowrate penalty to be incurred? Which sink–source match should be removed to achieve this target?

7.11 Synthesize the water reuse/recycle network for the organic chemical production case in Problem 3.6 (limiting data are found in Figure 3.22). The targeting step indicates that this is a network with zero freshwater and zero wastewater flowrates. Verify this with the water network (hint: Equation 7.2 may be omitted when matching the last water sink) (answer in JEMA paper). Present the RCN in a conventional PFD.

7.12 For the water minimization case in Problem 6.4 (Sorin and Bédard, 1999), perform the following tasks:

a. Synthesize a water reuse/recycle network, assuming that pure freshwater feed is used. The minimum freshwater flowrate is reported as 200 t/h. Represent the RCN in matching matrix form.

b. Synthesize a water regeneration network. The minimum freshwater and regenerated flowrates are reported as 120 (0 ppm) and 88.89 t/h ($C_{Rout} = 10$ ppm), respectively.

    c. Evolve the RCN obtained in (a) using MPA. How many interconnections are removed having 30 t/h of freshwater penalty?

    d. If the resulting RCN in (c) is to be evolved further, what is the minimum freshwater penalty that is to be incurred? How many interconnections are being removed in this case?

7.13 For Example 7.3, synthesize the water regeneration network for the following case:

    a. By ignoring the insights gained from the targeting stage.

    b. When the interception unit has an outlet concentration (C_{Rout}) of 25 ppm, make use of the insights gained from the targeting stage (Example 6.2).

7.14 Obtain a different network configuration for Example 7.4 for the following case:

    a. Identify another water path different from that given in Figure 7.26. Evolve the water network by adding 10 t/h of freshwater penalty.

    b. Identify another water path different from that given in Figure 7.29. Evolve the water network by adding 20 t/h of freshwater penalty. Why does the number of interconnections remain the same?

7.15 For the Kraft paper process in Problems 2.2, 4.1, and 6.6 synthesize the water network for the following water schemes (assume that pure freshwater is used):

    a. Direct water reuse/recycle—the minimum freshwater flowrate is targeted as 89.9 t/h.

    b. Water regeneration network with interception unit of $C_{Rout} = 10$ ppm—the ultimate flowrate targets are given as 0 t/h (freshwater), 101.96 t/h (wastewater), and 134.85 t/h (regenerated flowrate), respectively.

    c. Water regeneration network with interception unit of $C_{Rout} = 15$ ppm—the ultimate flowrate targets are given as 2.73 t/h (freshwater), 104.69 t/h (wastewater), and 174.34 t/h (regenerated flowrate), respectively (answer for Problem 6.6).

    d. Water regeneration network with air stripping being used as interception unit with $C_{Rout} = 1.625$ ppm. The ultimate flowrate targets are given as 0 t/h (freshwater), 101.96 t/h (wastewater), and 95.05 t/h (regenerated water), respectively.

       Note: For (b)–(d), sources to be sent for regeneration may be obtained from insights gained from the targeting stage (i.e., by solving Problem 6.6(d)).

7.16 For the case study in Problem 6.2, synthesize the water network for the following water recovery schemes, assuming that pure freshwater is used. Limiting data are given in Table 3.9.

    a. Direct water reuse/recycle—the minimum freshwater flowrate is targeted as 90 t/h.

    b. Water regeneration network with interception unit of $C_{Rout} = 5$ ppm. The ultimate flowrate targets are given as 20 t/h for both freshwater (0 ppm) and wastewater as well as 73.68 t/h for regenerated flowrate, respectively (see solution in Example 8.2).

7.17 For the water minimization case in Problem 4.4 (Savelski and Bagajewicz, 2001), perform the following tasks:

    a. Synthesize a water reuse/recycle network, assuming that a pure freshwater feed is used. The minimum freshwater flowrate is reported as 166.266 t/h. Represent the RCN in matching matrix form.

    b. Evolve the RCN obtained in (a) using SSA. How many interconnections are removed as a result of SSA?

Utility Gas Recovery Problems

7.18 For the magnetic tape manufacturing process in Problem 2.5, design the RCN and present the nitrogen recovery scheme in a conventional PFD. The minimum pure fresh nitrogen needed for the network is targeted as 5.6 kg/s. Use the limiting data in Table 3.12.

7.19 For Problem 3.8, design the RCN and present the oxygen recovery scheme in a conventional PFD. The targeting stage identifies that the minimum fresh oxygen (with 5% impurity) needed for the network is 30 kg/s (*hint:* Equation 7.2 may be omitted when matching the last two sinks).

7.20 Synthesize a reuse/recycle network for the refinery hydrogen network in Example 2.4. The limiting data are given in Table 2.2. For a fresh hydrogen feed with 1% impurity, the targeting stage identifies that its minimum flowrate is 182.86 MMscfd (see Figure 3.12). Present the RCN in a conventional PFD.

7.21 Synthesize a reuse/recycle network for the refinery hydrogen network in Problems 2.6 and 4.10. The fresh hydrogen feed has an impurity content of 5%. The minimum flowrate for the hydrogen feed is identified as 268.82 mol/s.

7.22 Synthesize a reuse/recycle network for the hydrogen recovery in Problem 4.16. For the fresh hydrogen feed of 0.1% impurity content, the minimum flowrate for hydrogen feed is identified as 125.2 mol/s.

Property Integration Problems

7.23 Revisit the metal degreasing process in Example 5.2. Synthesize the RCN for the case where the thermal processing unit is operated at a lower temperature, i.e., 430 K, which in turn leads to lower RVP value as well as lower operator value (see discussion in Example 5.2). Make use of the limiting data in Table 7.5. Represent the RCN in a conventional PFD (see answer in Figure 1.4).

TABLE 7.5

Limiting Data for Problem 7.22

Sink, SK$_j$	F_{SKj} (kg/s)	Ψ_{SKj} (atm$^{1.44}$)	Source, SR$_i$	F_{SRi} (kg/s)	Ψ_{SRi} (atm$^{1.44}$)
Degreaser (SK1)	5.0	4.87	Condensate I (SR1)	4.0	6.56
Absorber (SK2)	2.0	7.36	Condensate II (SR2)	3.0	3.74
			Fresh solvent	0	2.713

7.24 Synthesize an RCN for fiber recovery in the paper making process in Examples 2.6 and 4.5. The limiting data are given in Table 4.17. The targeting stage identifies that the minimum fresh fiber needed for the RCN is 14.98 t/h (see Example 4.5). Present the RCN in a conventional PFD. Hint: be careful with the inferior property operator value of the fresh fiber feed.

7.25 For the microelectronics manufacturing facility in Problem 3.11, synthesize the RCN for the recovery of its ultrapure water. The ultrapure water (fresh resource) has a resistivity value of 18,000 kΩ/cm. The targeting stage indicates that the minimum flowrate for the ultrapure water is 972 gal/min. Present the RCN in a conventional PFD.

7.26 For the wafer fabrication process in Problem 4.19, synthesize the reuse/recycle network for the following section of the plant:

a. Fabrication section (FAB) only—flowrate targeting indicates that the minimum ultrapure water (UPW, 18 MΩ m) requirement is 1516.47 t/h (Problem 4.19a). Present the RCN in a conventional PFD (refer to Figure 2.19).

b. Entire plant—this includes cleaning, cooling water makeup, and scrubber operations. Note that apart from ultrapure water, reject streams from ultrafiltration (UF) and reverse osmosis (RO) units may also be used. The targeting stage indicates that the minimum flowrates for these sources are 1516.47, 928.45, and 649.92 t/h (Problem 4.19b). Develop two alternative designs for this case, with and without the recovery of UF and RO reject streams, respectively. Present the RCN in a conventional PFD (with reference to Figure 4.3).

References

Das, A. K., Shenoy, U. V., and Bandyopadhyay, S. 2009. Evolution of resource allocation networks. *Industrial and Engineering Chemistry Research*, 48, 7152–7167.

El-Halwagi, M. M. 1997. *Pollution Prevention through Process Integration: Systematic Design Tools*. San Diego, CA: Academic Press.

Feng, X. and Seider, W. D. 2001. New structure and design method for water networks. *Industrial and Engineering Chemistry Research*, 40, 6140–6146.

Foo, D. C. Y. 2009. A state-of-the-art review of pinch analysis techniques for water network synthesis. *Industrial and Engineering Chemistry Research*, 48(11), 5125–5159.

Gomes, J. F. S., Queiroz, E. M., and Pessoa, F. L. P. 2006. Design procedure for water/wastewater minimization: Single contaminant. *Journal of Cleaner Production*, 15, 474–485.

Jacob, J., Kaipe, H., Couderc, F., and Paris, J. 2002. Water network analysis in pulp and paper processes by pinch and linear programming techniques. *Chemical Engineering Communications*, 189(2), 184–206.

Kuo, W. C. J. and Smith R. 1998. Designing for the interactions between water-use and effluent treatment. *Chemical Engineering Research and Design*, 76, 287–301.

Ng, D. K. S. and Foo, D. C. Y. 2006. Evolution of water network with improved source shift algorithm and water path analysis. *Industrial and Engineering Chemistry Research*, 45(24), 8095–8104.

Polley, G. T. and Polley, H. L. 2000. Design better water networks. *Chemical Engineering Progress*, 96(2), 47–52.

Prakash, R. and Shenoy, U. V. 2005a. Targeting and design of water networks for fixed flowrate and fixed contaminant load operations. *Chemical Engineering Science*, 60(1), 255–268.

Prakash, R. and Shenoy, U. V. 2005b. Design and evolution of water networks by source shifts. *Chemical Engineering Science*, 60(7), 2089–2093.

Savelski, M. J. and Bagajewicz, M. J. 2001. Algorithmic procedure to design water utilization systems featuring a single contaminant in process plants. *Chemical Engineering Science*, 56, 1897–1911.

Sorin, M. and Bédard, S. 1999. The global pinch point in water reuse networks. *Process Safety and Environmental Protection*, 77, 305–308.

8

Targeting for Waste Treatment and Total Material Networks

In previous chapters, we learnt how a resource conservation network (RCN) that achieves the minimum resource target is synthesized. This includes the direct reuse/recycle as well as the material regeneration networks. In this chapter, we learn how to identify waste streams from an RCN from a targeting view point, using graphical and algebraic techniques. We then learn how to target the minimum treatment flowrate needed for these waste streams. We then combine the techniques to synthesize the RCN in an overall framework known as the *total material network* (TMN).

8.1 Total Material Network

The concept of TMN was first introduced by Kuo and Smith (1998) for water minimization of fixed load problems (i.e. water using processes that perform as mass transfer operations). As shown in the modified framework in Figure 8.1, the TMN consists of three individual elements, i.e., direct reuse/recycle, regeneration, and waste treatment. Since there are interactions among these individual elements, they are analyzed as an overall framework. After material recovery potential is maximized with a direct reuse/recycle scheme, the fresh resource and waste flowrates can be further reduced by employing an interception unit for partial purification of process sources. Waste treatment is finally employed to treat the effluent prior to environmental discharge.

The concept of TMN applies for all types of RCNs, e.g., water, utility gas, and property networks. One of the established procedures to identify the various targets for a TMN is given in Figure 8.2, adopted from the targeting procedure for *total water network* (Ng et al., 2007a,b).

From Figure 8.2, it is observed that the procedure requires the targeting task for various elements in the TMN to be performed in sequence:

1. Direct reuse/recycle
2. Identification of waste streams
3. Targeting for regeneration
4. Targeting for waste treatment

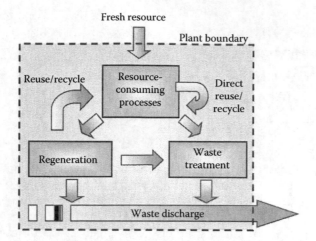

FIGURE 8.1
Concept of a TMN. (Adapted from *Chem. Eng. Res. Des.*, 76, Kuo, W.C.J. and Smith, R., Designing for the interactions between water-use and effluent treatment, 287–301, Copyright 1998, with permission from Elsevier.)

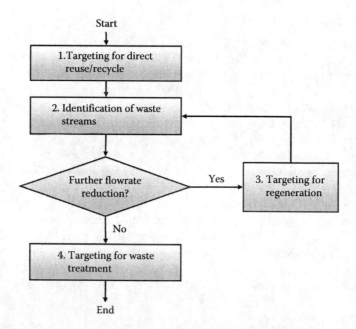

FIGURE 8.2
Targeting procedure for a TMN. (Adapted from Foo, D.C.Y., *Ind. Eng. Chem. Res.*, 48(11), 5125, 2009. With permission.)

Targeting techniques for Tasks 1 and 3 have been discussed in Chapters 3, 4, and 6, respectively. Techniques for Tasks 2 and 4 are discussed in the following sections of this chapter.

8.2 Generic Procedure for Waste Stream Identification

As discussed in Section 3.2, an RCN may be divided into two separate regions with respect to the pinch quality, i.e., *higher* (HQR) and *lower quality regions* (LQR). In the LQR, the total flowrate of the process sources is always more than that of the process sinks. In this case, the source that is unutilized (i.e., not reused/recycled to the sinks) will emit as waste streams from the RCN. Hence, to identify the individual waste streams, we should focus on the LQR.

Both graphical and algebraic techniques may be used in identifying waste streams from the RCN. The overall procedure for targeting is given as follows (Ng et al., 2007a):

1. Carry out the targeting procedure to identify the minimum flowrates for the RCN (e.g., fresh resource, waste, and/or regenerated flowrates). Identify also the pinch quality and the pinch-causing source, as well as the allocated flowrates of the pinch-causing source to the LQR (F_{LQR}).

2. Target the minimum flowrate of the pinch-causing source needed for the LQR (in short, the *minimum pinch flowrate, F_{PN}*). Note that the pinch-causing source is excluded from the source but will act as the "fresh resource" for this region during targeting.

3. Identify the highest quality waste from the pinch-causing source. This is given by the difference between the minimum pinch flowrate and the allocated flowrates of the pinch-causing source to the LQR, i.e.,

$$F_{W1} = F_{LQR} - F_{PN} \tag{8.1}$$

4. Identify other waste streams from the RCN. In most cases, one of the waste streams will always emit from the lowest quality source of the RCN. Note that there may be other unutilized sources that emit as waste streams for the RCN too.

We will now demonstrate two examples that make use of graphical and algebraic targeting procedures, respectively, for both direct reuse/recycle and regeneration networks.

Example 8.1 Graphical approach for waste identification for direct reuse/recycle network

Consider the classical water minimization example from Wang and Smith (1994), with the limiting data given in Table 8.1 (reproduced from Table 3.10). Use the *material recovery pinch diagram* (MRPD) introduced in Chapter 3 to identify the individual wastewater streams emitted from direct reuse/recycle network of this problem. Pure freshwater (0 ppm) is used in the RCN.

Solution

The following steps are used to identify the individual waste streams from the RCN:

1. *Carry out the targeting procedure to identify the minimum flow-rates for the RCN (e.g., fresh resource, waste, and/or regenerated flowrates). Identify also the pinch quality and the pinch-causing source, as well as the allocated flowrates of the pinch-causing source to the LQR.*

   Figure 8.3 shows the MRPD for flowrate targeting for direct water reuse/recycle schemes (see Chapter 3 for detailed procedure). As shown, the minimum freshwater (F_{FW}, 0 ppm) and wastewater (F_{WW}) flowrates are determined as 90 t/h, respectively. The pinch concentration is identified as 100 ppm, with SR1 and SR2 being the pinch-causing sources (both with the same concentration of 100 ppm). The allocated flowrates of this pinch-causing source to the HQR and LQR are identified as 70 (F_{HQR}) and 50 t/h (F_{LQR}), respectively.

2. *Target the minimum flowrate of the pinch-causing source needed for the LQR (in short, the minimum pinch flowrate, F_{PN}). Note that the pinch-causing source is excluded from the source but will act as the "fresh resource" for this region during targeting.*

   From Figure 8.3, we observe that only one water sink is found in the LQR, i.e., SK4. Besides, sources SR3 and SR4 and a small portion of the pinch-causing sources (SR1 and SR2) are also found in this region. The MRPD is replotted for this region by excluding the pinch-causing sources, as shown in Figure 8.4.

TABLE 8.1

Limiting Water Data for Example 8.1

Sinks, SK$_j$	F_{SKj} (t/h)	C_{SKj} (ppm)	Sources, SR$_i$	F_{SRi} (t/h)	C_{SRi} (ppm)
SK1	20	0	SR1	20	100
SK2	100	50	SR2	100	100
SK3	40	50	SR3	40	800
SK4	10	400	SR4	10	800
$\Sigma_i F_{SKj}$	170		$\Sigma_i F_{SRi}$	170	

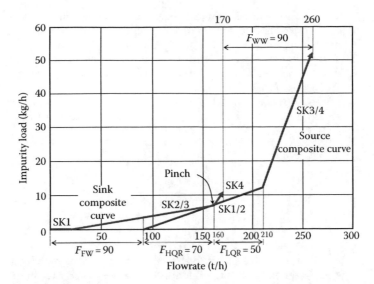

FIGURE 8.3
MRPD for flowrate targeting for direct water reuse/recycle schemes. (From Ng, D.K.S., Foo, D.C.Y., and Tan, R.R., Targeting for total water network—Part 1: Waste stream identification, *Ind. Eng. Chem. Res.*, 46(26), 9107. Copyright 2007a American Chemical Society. With permission.)

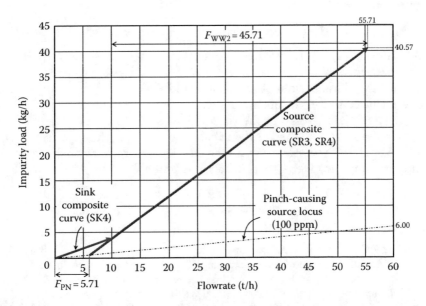

FIGURE 8.4
MRPD for flowrate targeting for LQR. (From Ng, D.K.S., Foo, D.C.Y., and Tan, R.R., Targeting for total water network—Part 1: Waste stream identification, *Ind. Eng. Chem. Res.*, 46(26), 9107. Copyright 2007a American Chemical Society. With permission.)

Note that the source composite curve is slid on a locus, with its slope corresponding to the pinch concentration of 100 ppm. The minimum pinch flowrate (F_{PN}) is identified as 5.71 t/h.

3. *Identify the highest quality waste from the pinch-causing source.*
4. *Identify other waste streams from the RCN. This is given by the waste discharge in Step 2 of the targeting procedure.*

Based on Equation 8.1, the flowrate of the first wastewater stream (F_{WW1}) that is originated from the pinch-causing source is identified as

$$F_{WW1} = 50 - 5.71 \text{ t/h} = 44.29 \text{ t/h}.$$

From Figure 8.4, it is then observed that 45.71 t/h of wastewater (F_{WW2}) is emitted from the unutilized sources, i.e., SR3 and SR4.

Finally, we should also do a verification to ensure that the total flowrates of the individual wastewater streams match the wastewater flowrate target, identified in Step 1 of the procedure. For this case, the total flowrate of the wastewater streams sums up to 90 t/h (=44.29 + 45.71 t/h), identical to the earlier identified target.

8.3 Waste Identification for Material Regeneration Network

In Figure 8.2, waste stream identification (Task 2) is again needed after targeting for a material regeneration network (Task 3) has been carried out. For this case, the identification of the waste streams enables the selection of the process source to be sent to the interception unit (for reuse/recycle in the RCN). After the recovery of the regenerated source(s), new waste streams (with less flowrate) can then be identified from the material regeneration network, to be sent for waste treatment. This is demonstrated with the following example, using the algebraic approach in Chapter 4.

Example 8.2 Algebraic approach for waste identification for regeneration network

The water minimization case in Example 8.1 is revisited here. An interception unit with outlet concentration (C_{Rout}) of 5 ppm is used to purify water for further water recovery. Use the *material cascade analysis* (MCA) in Chapter 4 to identify the individual wastewater streams emitted from this RCN. Pure freshwater (0 ppm) is used for this RCN (note to readers: remember to solve Problem 6.2 before you look at the solutions here).

Solution

Following the procedure of the algebraic targeting technique in Chapter 6, the minimum freshwater (F_{FW}) and wastewater (F_{WW}) flowrates are both identified as 20 t/h (see flowrate targeting result for *freshwater region*-FWR, Table 8.2), while the minimum regeneration flowrate (F_{RW}) is targeted as 73.68 t/h (see flowrate targeting result for *regenerated water region*-RWR, Table 8.3).

In Chapter 6, we discussed how to identify the process sources to be sent for regeneration (see Example 6.1). As observed in Table 8.3, only a single water sink (SK4 400 ppm) exists in the LQR (concentration higher than 100 ppm). Besides, SR4 (30 t/h, 800 ppm) and 53.68 t/h of the pinch-causing source (identified from interval between 100 and 400 ppm) are available for recovery to SK4. One option to satisfy the requirement of SK4 is to make use of the pinch-causing source, since

TABLE 8.2

Flowrate Targeting for FWR for Example 8.2

C_k (ppm)	$\Sigma_j F_{SKj}$ (t/h)	$\Sigma_i F_{SRi}$ (t/h)	$\Sigma_i F_{SRi} - \Sigma_j F_{SKj}$ (t/h)	$F_{C,k}$ (t/h)	Δm_k (kg/h)	Cum. Δm_k (kg/h)
				$F_{FW}=20$		
0	20		20			
				0	0	
800		20	20			0
				$F_{WW}=20$	19,984	(PINCH)
1,000,000						19,984

TABLE 8.3

Flowrate Targeting for RWR for Example 8.2

C_k (ppm)	$\Sigma_j F_{SKj}$ (t/h)	$\Sigma_i F_{SRi}$ (t/h)	$\Sigma_i F_{SRi} - \Sigma_j F_{SKj}$ (t/h)	$F_{C,k}$ (t/h)	Δm_k (kg/h)	Cum. Δm_k (kg/h)
				$F_{RW}=73.68$		
$C_{Rout}=5$						
				73.68	3.32	
50	140		−140			3.32
				−66.32	−3.32	
100		120	120			0.00
				53.68	16.11	(PINCH)
400	10		−10			16.11
				43.68	17.47	
800		30	30			33.58
				$F_{RW}=73.68$	73,625.26	
1,000,000						73,658.84

the latter has sufficient flowrate. This leaves the leftover flowrate of the pinch-causing source (43.68 t/h) and SR4 (30 t/h) to be sent for regeneration, as indicated by the values in the fifth column of Table 8.3.

Note, however, that the earlier-described flowrate allocation scheme does not make full use of the limiting load of SK4. This is due to the fact that SK4 may accept sources up to 400 ppm. However, it was only fed with the pinch-causing source of much lower concentration (100 ppm) with the current allocation scheme. Besides, fulfilling SK4 with the pinch-causing source also means that the interception unit is receiving sources of lower concentration (i.e., higher flowrate from SR4). In other words, the interception unit is removing higher impurity load from the RCN in order to achieve the flowrate targets. A better strategy to overcome this pitfall is to identify the appropriate sources for regeneration by making use of the individual waste streams identified earlier. This is described next.

In Example 8.1, it is found that two wastewater streams are emitted from the reuse/recycle network of this case study. They are originated from the pinch-causing source (44.29 t/h), as well as from SR3 and SR4 (45.71 t/h), respectively. Since a targeted regeneration flowrate of 73.68 t/h is needed for the RCN, we shall first make full use of the lowest concentration wastewater (i.e., 44.29 t/h from the pinch-causing source), in order to keep the regeneration load to the lowest. We then supplement the remaining flowrate requirement of the interception unit with the wastewater stream from SR3 and SR4, i.e., 29.39 t/h (=73.68 − 44.29 t/h).

It is always useful to crosscheck if this newly proposed allocation scheme will lead to a feasible RCN. This can be done by carrying out another cascade analysis for the RWR, by removing the waste stream flowrates that are sent for regeneration. As shown in Table 8.4, both sources at 100

TABLE 8.4

Flowrate Targeting for RWR for Example 8.2 (After Removal of Sources for Regeneration)

C_k (ppm)	$\Sigma_j F_{SKj}$ (t/h)	$\Sigma_i F_{SRi}$ (t/h)	$\Sigma_i F_{SRi} - \Sigma_j F_{SKj}$ (t/h)	$F_{C,k}$ (t/h)	Δm_k (kg/h)	Cum. Δm_k (kg/h)
				$F_{RW} = 73.68$		
$C_{Rout} = 5$						
				73.68	3.32	
50	140		−140			3.32
				−66.32	−3.32	
100		75.71	75.7			0.00
				9.40	2.82	(PINCH)
400	10		−10			2.82
				−0.60	−0.24	
800		0.60	0.62			2.58
				0.00	0.00	
1,000,000						2.58

TABLE 8.5

Combined Cascade Representation for Water Regeneration Network for Example 8.2

C_k (ppm)	$\Sigma_j F_{SKj}$ (t/h)	$\Sigma_i F_{SRi}$ (t/h)	$\Sigma_i F_{SRi} - \Sigma_j F_{SKj}$ (t/h)	$F_{C,k}$ (t/h)	Δm_k (kg/h)	Cum. Δm_k (kg/h)
				$F_{FW} = 20.00$		
0	20		−20.00			
				0.00	0.00	
$C_{Rout} = 5$		$F_{RW} = 73.68$	73.68			0.00
				73.68	3.32	(PINCH)
50	140		−140.00			3.32
				−66.32	−3.32	
100		75.71	75.71			0.00
				$F_{LQR} = 9.40$	2.82	(PINCH)
400	10		−10.00			2.82
				−0.60	−0.24	
800		20.60	20.60			2.58
				$F_{WW} = 20.00$	19,984.00	
1,000,000			0.00			19,986.58

and 800 ppm are now reduced to 75.71 t/h (=120 − 44.29 t/h) and 0.60 t/h (=30 − 29.39 t/h), respectively. The network is still feasible as indicated by the positive (or zero) values in the flowrate and impurity load cascades.

We can now combine Tables 8.2 and 8.4 into a single cascade representation for the water regeneration network, as given in Table 8.5.

Next, we proceed to identify the waste streams originated from the water regeneration network. This can be done by following the earlier-described steps on waste stream identification:

1. *Carry out the targeting procedure to identify the minimum flowrates for an RCN (e.g., fresh resource, waste, and/or regenerated flowrates). Identify also the pinch quality and the pinch-causing source, as well as the allocated flowrates of the pinch-causing source to the LQR.*

   Flowrate targeting for the regeneration schemes in Table 8.5 shows that two pinch concentrations are identified at 5 and 100 ppm, respectively. Since wastewater will only be generated in the LQR, i.e., with concentration higher than 100 ppm, this region shall be analyzed in detail. From Table 8.1, we can identify that SR1 and SR2 are the pinch-causing sources at 100 ppm pinch. From column 5 of Table 8.5, we can observe that 9.4 t/h of these pinch-causing sources is allocated to the LQR (F_{LQR}).

2. *Target the minimum flowrate of the pinch-causing source needed for the LQR (in short, the minimum pinch flowrate, F_{PN}). Note that the pinch-causing source is excluded from the source but will act as the fresh resource for this region during targeting.*

Table 8.6

Flowrate Targeting for LQR for Example 8.2

C_k (ppm)	$\Sigma_j F_{SKj}$ (t/h)	$\Sigma_i F_{SRi}$ (t/h)	$\Sigma_i F_{SRi} - \Sigma_j F_{SKj}$ (t/h)	$F_{C,k}$ (t/h)	Δm_k (kg/h)	Cum. Δm_k (kg/h)
				$F_{PN} = 5.71$		
100			0			
				5.71	1.71	
400	10		−10			1.71
				−4.29	−1.71	
800		20.6	20.6			0.00
				$F_{WW2} = 16.31$	16,302.74	(PINCH)
1,000,000			0			16,302.74

Flowrate targeting is then performed for the LQR. Table 8.6 indicates that the minimum pinch flowrate (F_{PN}) is targeted at 5.71 t/h. Note that the pinch-causing source acts as a fresh resource for this case, and hence, it is supplied at the concentration of 100 ppm.

3. *Identify the highest quality waste from the pinch-causing source.*
4. *Identify other waste streams from the RCN. This is given by the waste discharge in Step 2 of the targeting procedure.*

Based on Equation 8.1, the first wastewater stream that is originated from the pinch-causing source has a flowrate of

$$F_{WW1} = 9.40 - 5.71 \, t/h = 3.69 \, t/h.$$

Another wastewater stream is then observed to emit from the unutilized sources at 800 ppm, with a flowrate (F_{WW2}) of 16.31 t/h. Finally, a verification confirms that the total flowrate of the individual wastewater streams does match the target identified in Step 1 of the procedure (see Table 8.2 or 8.5), i.e., 20 t/h (=3.69 + 16.31 t/h). One may also use the network design technique in Chapter 7 to synthesize the RCN, with the result shown in Figure 8.5.*

8.4 Targeting for Minimum Waste Treatment Flowrate

The procedure in Figure 8.2 indicates that the last step in the synthesis of TMN is the targeting for waste treatment. To minimize the waste treatment costs, waste load is discharged to the environment at its maximum allowable limit (q_D). Hence, the treatment unit should have an outlet quality (q_T) that is equal or lower than the maximum allowable discharge

---

* Solve Problem 7.16 to verify the answer.

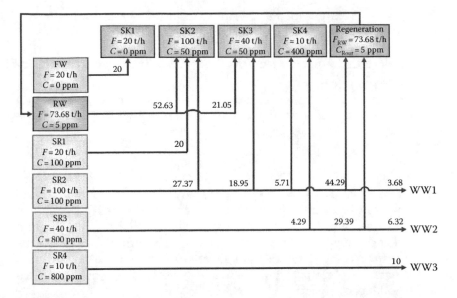

FIGURE 8.5
RCN for Example 8.2. (From Ng, D.K.S., Foo, D.C.Y., and Tan, R.R., Targeting for total water network—Part 2: Waste treatment targeting and interactions with water system elements, *Ind. Eng. Chem. Res.*, 46(26), 9114. Copyright 2007b American Chemical Society. With permission.)

quality (i.e., $q_T \geq q_D$). Similar to the source interception units discussed in Chapter 6, the waste treatment units may be classified as *fixed outlet quality* or *fixed removal ratio* (RR) types based on their performance. The RR value of a treatment unit is given as the ratio of the impurity load removed by the treatment unit (Δm_R) to the total impurity load feed to the treatment unit (m_F) (Wang and Smith, 1994):

$$RR = \frac{\Delta m_R}{m_F} = \frac{F_T \left(C_{Rout} - C_{Rin}\right)}{F_T C_{Rin}} \tag{8.2}$$

where F_T and C_{Rin} are the treatment/regenerated flowrate and the inlet concentration of the interception unit, respectively. For interception units without water loss, Equation 8.2 may be simplified as

$$RR = \frac{C_{Rout} - C_{Rin}}{C_{Rin}} \tag{8.3}$$

To determine the minimum treatment requirement for the waste streams, we can make use of the *waste treatment pinch diagram* (WTPD), which is applicable for both fixed outlet quality or fixed RR type treatment units. The procedure to construct the WTPD is similar to that of the MRPD (see Chapter 3).

Detailed steps to target the minimum treatment flowrate using the WTPD are given as follows (Ng et al., 2007b):

1. Arrange the waste streams in descending order of their quality levels.

2. Calculate the load possessed by the waste streams (m_{Wy}), which are given as the product of the waste flowrate (F_{Wy}) with its corresponding quality level (q_{Wy}), as given in Equation 8.4:

$$m_{Wy} = F_{Wy} \, q_{Wy} \tag{8.4}$$

3. All waste streams ($W1$, $W2$, ..., Wy) are plotted on a load-versus-flowrate diagram in descending order of their quality levels. The individual waste segments are connected by linking the tail of an earlier segment to the arrowhead of a latter to form the *waste composite curve* (WCC). The horizontal distance of the composite curve represents the total waste flowrate, while its vertical distance represents its total load (Figure 8.6).

4. A discharge locus is added in the WTPD, with its slope equal to the *maximum allowable discharge quality* (q_D). Note that the discharge locus starts at the origin of the WTPD and ends at the *discharge point* on the right (see Figure 8.6). The vertical distance of this locus sets the *maximum allowable waste discharge load* to the environment. Move to Step 5a for a waste treatment unit of fixed outlet quality or to Step 5b for a unit of fixed RR type.

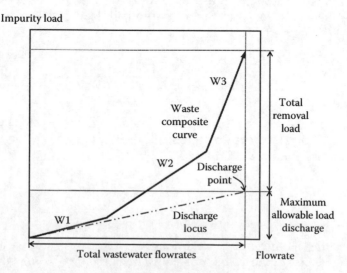

FIGURE 8.6
Waste treatment pinch diagram. (From Ng, D.K.S. et al., *Ind. Eng. Chem. Res.*, 46(26), 9114, 2007b.)

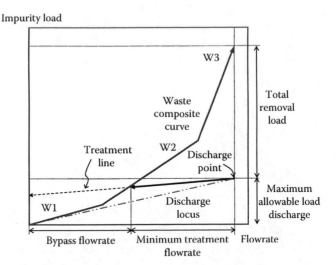

FIGURE 8.7
Targeting for waste treatment unit of fixed outlet quality type. (From Ng, D.K.S. et al., *Ind. Eng. Chem. Res.*, 46(26), 9114, 2007b.)

5a. Targeting for a waste treatment unit of fixed outlet quality type

A *treatment line* is placed in the WTPD, with its slope corresponding to the outlet quality of the treatment unit (q_T). The treatment line is connected to the discharge point and intersects with the WCC (Figure 8.7). The minimum treatment and bypass flowrates for the waste streams are given by the horizontal distance of the treatment line that lies below and above the WCC, respectively.

5b. Targeting for a waste treatment unit of fixed RR type

The total impurity load fed to the treatment unit (m_F, given by Equation 8.2) is first labeled on the WTPD, starting from the end point of the WCC (Figure 8.8). The horizontal distance of the WCC portion that contributes to the treatment feed load indicates the minimum treatment flowrate required for the waste streams. The horizontal distance of the remaining WCC portion indicates the bypass waste flowrate.

Example 8.3 Targeting for minimum waste treatment

The water minimization case in Example 8.2 is revisited here. Target the minimum treatment flowrate needed for the wastewater streams generated from the water regeneration network for the following cases:

1. Waste treatment unit with fixed outlet concentration (C_T) of 10 ppm, for an environmental discharge limit (C_D) of 20 ppm
2. Waste treatment unit with fixed RR of 0.9, for an environmental discharge limit (C_D) of 100 ppm

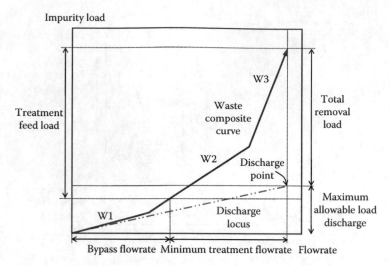

FIGURE 8.8
Targeting for waste treatment unit of fixed RR type. (From Ng, D.K.S. et al., *Ind. Eng. Chem. Res.*, 46(26), 9114, 2007b.)

Solution

1. Case 1—Waste treatment unit with $C_T = 10\,$ppm, for $C_D = 20\,$ppm
   Example 8.2 determines that two wastewater streams are emitted from the water regeneration network, i.e., 3.69 t/h at 100 ppm (WW1) and 16.31 t/h at 800 ppm (WW2). Following the detailed procedure for plotting the WTPD,

   1. *Arrange the waste streams in descending order of their quality levels.*
   2. *Calculate the load possessed by the waste streams (m_{W_y}), which are given as the product of the waste flowrate (F_{W_y}) with its corresponding quality level (q_{W_y}).*

      The individual wastewater streams are arranged in ascending order of their impurity concentration. Their impurity loads are then calculated, given as in column 4 of Table 8.7. Besides, the cumulative flowrate (Cum. F_{WW_y}) and cummulative load (Cum. m_{WW_y}) values are also summarized in the final two columns of the table.

TABLE 8.7

Impurity Load Possessed by the Waste Streams in Example 8.3

Wastewater, WWy	F_{WW_y} (t/h)	C_{WW_y} (ppm)	m_{WW_y} (kg/h)	Cum. F_{WW_y} (t/h)	Cum. m_{WW_y} (kg/h)
WW1	3.69	100	0.37	3.69	0.37
WW2, WW3	16.31	800	13.05	20.00	13.42

3. *All waste streams (W1, W2, ..., Wy) are plotted on a load-versus-flowrate diagram with a descending order of their quality levels. The individual waste segments are connected by linking the tail of an earlier segment to the arrowhead of a latter to form the WCC.*

4. *A discharge locus is added in the WTPD, with its slope equal to the maximum allowable discharge quality (q_D).*

    All wastewater streams are plotted to form the WCC on the WTPD, given as in Figure 8.9. As shown, the wastewater streams have a total flowrate of 20 t/h and impurity load of 13.42 kg/h. For Case 1, where C_D is set at 20 ppm, the maximum allowable discharge load is calculated as 0.4 kg/h (= 20 t/h × 20 ppm). The discharge point is also located in the WTPD (see Figure 8.9). In other words, a total of 13.02 kg/h (= 13.42 – 0.4 kg/h) of impurity load (Δm_R) is to be removed from the wastewater streams.

5a. *Targeting for a waste treatment unit of fixed outlet quality type*

    A treatment line is placed in the WTPD, with its slope corresponding to the outlet quality of the treatment unit (q_T). The minimum treatment and bypass flowrates for the waste streams are given by the horizontal distance of the treatment line that lies below and above the WCC, respectively.

    A treatment line with a slope corresponding to = C_T 10 ppm is added to the WTPD in Figure 8.9. The minimum treatment

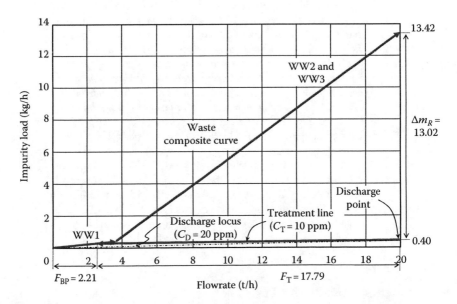

FIGURE 8.9
Treatment flowrate targeting for Example 8.3 Case 1. (From Ng, D.K.S. et al., *Ind. Eng. Chem. Res.*, 46(26), 9114, 2007b.)

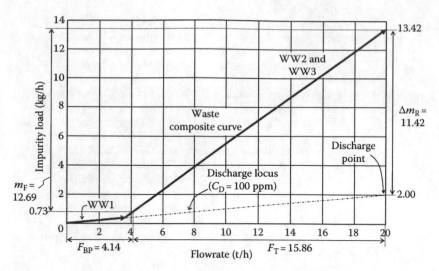

FIGURE 8.10
Treatment flowrate targeting for Example 8.3 Case 2. (From Ng, D.K.S. et al., *Ind. Eng. Chem. Res.*, 46(26), 9114, 2007b.)

(F_T) and bypass (F_{BP}) flowrates are then determined as 17.79 and 2.21 t/h, respectively.

2. Case 2—Waste treatment unit with fixed RR=0.9, and C_D = 100 ppm

For Case 2, where C_D is set at 100 ppm, the maximum allowable discharge load is calculated as 2 kg/h (=20 t/h × 100 ppm). In other words, a total of 11.42 kg/h (=13.42 − 2 kg/h; see Figure 8.10) impurity load (Δm_R) is to be removed from the wastewater streams.

5b. *Targeting for a waste treatment unit of fixed RR type*

The total impurity load fed to the treatment unit (m_F, given by Equation 8.2) is first labeled on the WTPD, starting from the end point of the WCC. The horizontal distance of the WCC portion that contributes to the treatment feed load indicates the minimum treatment flowrate required for the waste streams. The horizontal distance of the remaining WCC portion indicates the bypass flowrates of the waste streams.

For a waste treatment unit with RR=0.9, Equation 8.2 determines that the total treatment feed load (m_F) is 12.69 kg/h (=11.42/0.9 kg/h). The WTPD in Figure 8.10 then determines that the minimum treatment (F_T) and bypass (F_{BP}) flowrates are 15.86 and 4.14 kg/h, respectively.

8.5 Insights from the WTPD

Similar to the case in MRPD, insights from the WTPD may be used to determine the structure of the waste treatment network. For instance, the individual

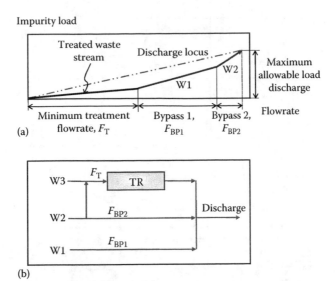

FIGURE 8.11
(a) Revised WCC and (b) waste treatment network for Figure 8.7. (From Ng, D.K.S. et al., *Ind. Eng. Chem. Res.*, 46(26), 9114, 2007b.)

segments in Figure 8.7, i.e., the *treated waste stream* (the treatment line below the WCC) and the bypass streams may be combined to form the revised WCC in Figure 8.11. Note that in Figure 8.11a, the individual segments have been rearranged with descending quality levels. The treatment and bypass flowrates of the individual wastewater streams can be identified from the segments of the WCC. As shown in Figure 8.11, the entire waste W1 and a small portion of W2 bypass the treatment unit (with flowrates of F_{BP1} and F_{BP2} respectively); while the remaining W2 and entire W3 are sent for treatment (with flowrate F_T).

Example 8.4 Synthesis of a wastewater treatment network

Determine the structure of the wastewater treatment network for both cases in Example 8.3. Combine the wastewater treatment network with the network structure in Figure 8.5 to form the *total water network*.

Solution

1. Case 1—Waste treatment unit with $C_T = 10\,ppm$ and $C_D = 20\,ppm$

    Figure 8.9 determines that the minimum treatment (F_T) and bypass (F_{BP}) flowrates for this case are 17.79 and 2.21 t/h, respectively. The entire wastewater streams of WW2 and WW3 and 1.47 t/h (=3.68 − 2.21 t/h) of WW1 are sent for treatment before environmental discharge. The wastewater treatment network hence takes the form shown in Figure 8.12.

    Combining the wastewater treatment network with the rest of the network in Figure 8.5 yields the total water network in Figure 8.13.

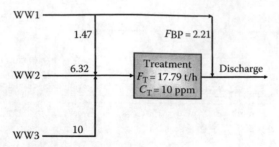

FIGURE 8.12
Structure of wastewater treatment network for Example 8.4. (From Ng, D.K.S. et al., *Ind. Eng. Chem. Res.*, 46(26), 9114, 2007b.)

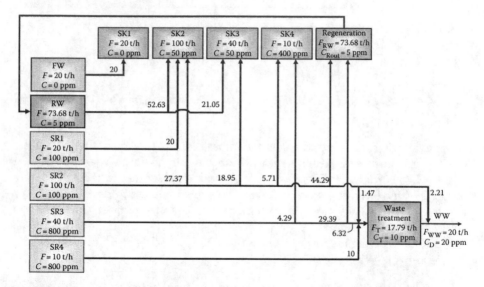

FIGURE 8.13
Total water network for Case 1. (From Ng, D.K.S. et al., *Ind. Eng. Chem. Res.*, 46(26), 9114, 2007b.)

2. Case 2—Waste treatment unit with RR = 0.9 and C_D = 100 ppm

For this case, Figure 8.10 determines that the minimum treatment (F_T) and bypass (F_{BP}) flowrates for the wastewater streams are 15.86 and 4.14 kg/h, respectively. Hence, the entire wastewater stream WW1 (3.69 t/h) will bypass the treatment unit, along with 0.45 t/h (=4.14 − 3.69 t/h) of WW2. On the other hand, the remaining flowrate of WW2 (5.86 t/h = 6.31 − 0.45 t/h) along with the entire WW3 stream is sent for treatment before environmental discharge. The structure of the wastewater treatment network hence takes the form as in Figure 8.14, while its associated total water network is given in Figure 8.15.

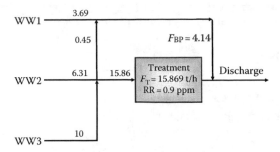

FIGURE 8.14
Structure of wastewater treatment network for Case 2. (From Ng, D.K.S. et al., *Ind. Eng. Chem. Res.*, 46(26), 9114, 2007b.)

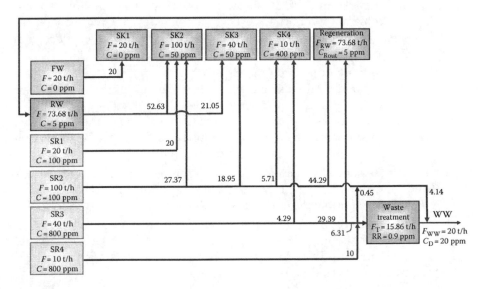

FIGURE 8.15
Total water network for Case 2. (From Ng, D.K.S. et al., *Ind. Eng. Chem. Res.*, 46(26), 9114, 2007b.)

8.6 Additional Readings

We have thus far discussed the targeting for a single waste treatment unit in this chapter. In some cases, multiple treatment units are needed before the waste may be discharged to the environment. Targeting for multiple treatment units is discussed in Ng et al. (2007b) and Sao et al. (2010).

Problems

Water Minimization Problems

8.1 For the water minimization case study in Problem 3.4 (Polley and Polley, 2000), solve the following cases:

a. Use the algebraic targeting technique to identify the individual wastewater streams that emit from direct reuse/recycle network with single pure freshwater feed—Does it coincide with the wastewater streams found in the detailed network design stage (Figure 7.6)? Note: you may make use of the flowrate targeting result in Table 6.2 (see answers in Example 10.1).

b. Use the algebraic targeting technique to identify the individual wastewater streams that emit from direct reuse/recycle network with single impure freshwater feed (10 ppm impurity content) in Problems 3.4b and 4.11—Does it coincide with the wastewater streams found in the detailed network design stage? (Solve Problem 7.1a to verify answer.)

c. For network in (a) and (b), target the minimum treatment flowrate for the identified wastewater streams. The treatment unit has an outlet concentration of 25 ppm, while the environmental discharge limit is given as 30 ppm. Determine the structure of the RCN. Note that network structure for (a) is available in Example 7.1; while that for (b) is available by solving Problem 7.1(a).

8.2 Revisit the specialty chemical production process in Problems 4.6, 6.3, and 7.7, and perform the following tasks: Limiting data for water minimization is summarized in Table 4.25. Assume pure freshwater is used.

a. Make use of the targeting results in Problem 4.6 (for direct reuse/recycle network) to identify the individual wastewater streams that emit from this RCN. Does it match the wastewater streams obtained in the network design stage (Problem 7.7)?

b. For a water regeneration network with an interception unit ($C_{Rout} = 200$ ppm), identify the new wastewater streams that will emit from the water regeneration network. Make use of the targeting results in Problem 6.3b.

c. A wastewater treatment unit with a fixed outlet concentration (C_T) of 250 ppm is used for treating the wastewater streams for RCN in (b) before environmental discharge (Deng et al., 2011). The environmental limit is given as 300 ppm. Target the minimum treatment flowrate and synthesize the total water network for this case. Make use of the RCN structure in problem 7.6b.

8.3 For the Kraft pulping process in Problems 4.1, 6.6, and 7.15 do the following tasks:

a. Make use of the targeting results in Problem 4.1 (for direct reuse/recycle scheme) to identify the individual wastewater streams that emit from this RCN. Does it match the wastewater streams obtained in the network design stage (Problem 7.15a)?

b. When an interception unit ($C_{Rout} = 10\,ppm$) is added in the RCN, the ultimate flowrate targeting procedure determines that the freshwater, regenerated water, and wastewater have a flowrate of 0, 134.85, and 101.96 t/h, respectively. Select the wastewater streams identified in part (a) to be sent for regeneration, and reidentify the new wastewater streams that will emit from the water regeneration network.

c. A biotreatment facility with $C_T = 15\,ppm$ (El-Halwagi, 1997) is used to remove methanol prior to final discharge. The maximum discharge limit of methanol to the river is given as 15 ppm. Identify the structure of the total water network. Make use of the solution of Problem 7.15(b).

d. Repeat the parts (b) and (c) when a single treatment facility with $C_{Rout} = 15\,ppm$ is used for both water regeneration and wastewater treatment. The ultimate flowrate targeting procedure determines that the freshwater, regenerated flowrate, and wastewater have a flowrate of 2.73, 174.34, and 104.69 t/h, respectively. Make use of the solution of Problem 7.15(c).

e. Repeat the parts (b) and (c) when air stripping is used as an interception unit for the RCN, while the biotreatment facility (with $C_T = 15\,ppm$, El-Halwagi, 1997) is used for wastewater treatment. The data for the interception unit are found in Problem 6.6c. The ultimate flowrate targeting procedure determines that the freshwater, regenerated flowrate, and wastewater have a flowrate of 0, 95.05, and 101.96 t/h, respectively. Make use of the solution of Problem 7.15d.

8.4 For the water network case given in Problems 6.4 and 7.12 (Sorin and Bédard, 1999), perform the following tasks:

a. Identify the individual wastewater streams that emit from the direct reuse/recycle RCN. Make use of the targeting results in Problem 6.4a. Verify the answers with the solution in Problem 7.12(a).

b. For the RCN with an interception unit of $C_{Rout} = 10\,ppm$, it is estimated that the minimum regeneration flowrate needed is 88.89 t/h. Determine the flowrates of the individual wastewater streams that should be sent for regeneration.

c. Determine the flowrates of the individual wastewater streams that emit from the water regeneration network in part (b). Make use of the targeting results in Problem 6.4b.

d. For a wastewater treatment unit of RR = 0.95, target the minimum treatment flowrate for the wastewater streams. The environmental limit is given as 20 ppm. Synthesize the total water network for this case.

Utility Gas Recovery Problems

8.5 For the magnetic tape manufacturing process in Problems 4.14 and 4.18, determine the flowrates of the individual water gas streams that are sent for disposal. Are these streams identical to those found during the design stage in Problem 7.18?

Property Integration Problems

8.6 The microelectronics manufacturing facility in Problems 3.11 and 7.25 is revisited. Determine the flowrates of the individual wastewater streams that are sent to the waste treatment section. Does the answer match that in Problem 7.25?

8.7 For the wafer fabrication process in Problems 4.19 and 7.26, several wastewater streams may be identified after network design stage. The environmental discharge limit (C_D) for these wastewater streams is given as 2 ppm for their heavy metal content. For a wastewater treatment facility with an outlet concentration (C_T) of 0.5 ppm, determine

TABLE 8.8

Limiting Data for Wastewater Streams in Problem 8.7

y	WWy	F_{WWy} (t/h)	C_{WWy} (ppm)
(a) UF and RO reject streams are recovered for the use			
1	Wet I	250	5.0
2	Wet II	180.16	4.5
3	CMP I	50	10.0
4	CMP II	86.31	4.5
5	Etc.	280	5.0
6	Cleaning	180	15.0
7	Scrubber	250	10.0
8	UF reject	278.45	1.5
9	RO reject	649.92	10.0
(b) UF and RO reject streams are not recovered for use			
1	Wet I	207.79	5
2	Wet II	180.16	4.5
3	CMP II	86.31	4.5
4	Cleaning	152.21	15
5	UF reject	928.45	1.5
6	RO reject	649.92	10

the minimum wastewater treatment flowrate needed to fulfill the discharge limit for the following cases.

a. UF and RO reject streams are recovered for the use of various water sinks. The limiting data for the individual wastewater streams are given in Table 8.8a.

b. The reject streams of UF and RO are not recovered for use. The limiting data for the individual wastewater streams are given in Table 8.8b.

c. Construct the total network for (a) and (b). Make use of the RCN design in Problem 7.26.

References

Deng, C., Feng, X., Ng, D. K. S., and Foo, D. C. Y. 2011. Process-based graphical approach for simultaneous targeting and design of water network. *AIChE Journal*, 57(11), 3085–3104.

Foo, D. C. Y. 2009. A state-of-the-art review of pinch analysis techniques for water network synthesis. *Industrial and Engineering Chemistry Research*, 48(11), 5125–5159.

Kuo, W. C. J. and Smith, R. 1998. Designing for the interactions between water-use and effluent treatment. *Chemical Engineering Research and Design*, 76, 287–301.

Ng, D. K. S., Foo, D. C. Y., and Tan, R. R. 2007a. Targeting for total water network— Part 1: Waste stream identification. *Industrial and Engineering Chemistry Research*, 46(26), 9107–9113.

Ng, D. K. S., Foo, D. C. Y., and Tan, R. R. 2007b. Targeting for total water network— Part 2: Waste treatment targeting and interactions with water system elements. *Industrial and Engineering Chemistry Research*, 46(26), 9114–9125.

Polley, G. T. and Polley, H. L. 2000. Design better water networks. *Chemical Engineering Progress*, 96(2), 47–52.

Sorin, M. and Bédard, S. 1999. The global pinch point in water reuse networks. *Process Safety and Environmental Protection*, 77, 305–308.

Soo, S. S. T., Toh, E. H., Yap, K. K. K., Ng, D. K. S., and Foo, D. C. Y. 2010. Targeting for minimum flowrate for multiple treatment processes. 1st International Conference on Process Engineering and Advanced Materials (ICPEAM 2010), 24th Symposium of Malaysian Chemical Engineers (SOMCHE 2010), Kuala lampur, June 15–17, 2010.

Wang, Y. P. and Smith, R. 1994. Wastewater minimisation. *Chemical Engineering Science*, 49, 981–1006.

9

Synthesis of Pretreatment Network

In the previous chapters, it has been assumed that fresh resources can always fulfill the needs for various sinks in a resource conservation network (RCN). However, there are cases where this assumption does not hold. For instance, water pretreatment units are always used in the process plant to produce dematerialized water (e.g., for boiler feed water, specialty chemical production) or ultrapure water (e.g., semiconductor manufacturing). In most cases, the pretreatment units are normally used for purifying fresh resources before they are sent to the process sinks, which require stringent quality. In this chapter, the two-step insight-based pinch analysis technique is used to synthesize a pretreatment network. In the first step, the *material recovery pinch diagram* introduced in Chapter 3 is extended as the *pretreatment pinch diagram* (PPD), which is used for targeting the minimum treatment flowrate for a *pretreatment network*. In step two, the design of a pretreatment network with nearest neighbor algorithm introduced in Chapter 7 is also discussed.

9.1 Basic Modeling of a Partitioning Interception Unit

As mentioned in Chapter 6, there exist two broad types of interception units for use in an RCN, i.e., *single pass* and *partitioning* units. The former has been discussed in Chapter 6. Figure 9.1, on the other hand, shows a partitioning type interception unit that is commonly used in pretreatment networks. The interception unit consists of two outlet streams, i.e., a higher quality *purified* stream, and a lower quality *reject* stream. The overall flowrate and material balances of the unit are given in Equations 9.1 and 9.2, respectively.

$$F_{Rin} = F_{RP} + F_{RJ} \tag{9.1}$$

$$F_{Rin}q_{Rin} = F_{RP}q_{RP} + F_{RJ}q_{RJ} \tag{9.2}$$

where
  F_{Rin}, F_{RP}, and F_{RJ} are the flowrates of inlet, purified, and reject streams of the unit
  q_{Rin}, q_{RP}, and q_{RJ} are the quality indices (e.g., impurity concentration, operator, etc.)

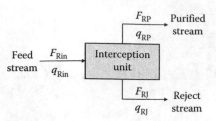

FIGURE 9.1
A partitioning interception unit.

Since the interception unit possesses two outlet streams, an overall *product recovery factor* (*RC*) is used to define the recovery of the desired product from the inlet stream. For a dilute system (which is commonly found in pretreatment network), the equation is approximated as follows:

$$RC = \frac{F_{RP}}{F_{Rin}} \tag{9.3}$$

Note that with Equations 9.1 through 9.3, we can determine the quality value of either streams for a partitioning interception unit, when the quality values of two other streams are given. Similar to single pass units, partitioning interception units are also characterized as *fixed outlet quality* or *removal ratio* types. For the former type, the interception unit purifies a feed stream (with quality q_{Rin}) to a purified stream with uniform quality (q_{RP}), e.g., 20 ppm, 5 wt%, etc. On the other hand, interception units of the removal ratio type purify a given feed stream by removing a portion of its impurity/property load. The *removal ratio index* (*RR*) for these units may be described in both generic quality index and concentration forms, given as in Equations 9.4 and 9.5, respectively:

$$RR = \frac{F_{RJ}\left(q^{max} - q_{RJ}\right)}{F_{Rin}\left(q^{max} - q_{Rin}\right)} \tag{9.4}$$

$$RR = \frac{F_{RJ}C_{RJ}}{F_{Rin}C_{Rin}} \tag{9.5}$$

where q^{max} is the highest possible value for a given purity index, e.g., 100%, etc., for a pure fresh resource (e.g., freshwater, gas feed, etc.).

9.2 Pretreatment Pinch Diagram

Steps to construct the PPD are given as follows (Tan et al., 2010), resembling those for plotting the MRPD in Chapter 3:

1. Arrange the process sinks in descending order of quality levels.
2. Calculate the maximum load acceptable by each sink (m_{SKj}), which is given as the product of the sink flowrate (F_{SKj}) with its corresponding quality level (q_{SKj}), i.e.,

$$m_{SKj} = F_{SKj}q_{SKj} \tag{9.6}$$

3. All process sinks are plotted from the origin on a load-versus-flowrate diagram, with descending order of their quality levels to form the *sink composite curve*. The individual sink segments are connected by linking the tail of the latter to the arrowhead of an earlier segment (Figure 9.2). The vertical distance of the sink composite curve hence represents the total load acceptable by all sinks in Equation 9.6.
4. A *fresh resource line* is plotted from the origin on the load-versus-flowrate diagram (i.e., the PPD), with its slope corresponding to its quality level (q_R). The length of the fresh resource line will be decided in the following step.
5. The *reject stream line* is plotted on the right and just touches the sink composite curve on the PPD, with its slope corresponding to its quality level (q_{RJ}). The latter may be given as a parameter of the interception unit or calculated from the model as described earlier. The point where the reject stream line intersects the fresh resource line

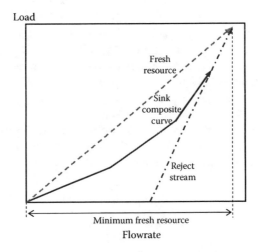

FIGURE 9.2
Plotting of sink composite curve, fresh resource, and reject stream lines on the PPD. (From *Chem. Eng. Res. Des.*, 88(4), Tan, R.R., Ng, D.K.S., and Foo, D.C.Y., Graphical approach to minimum flowrate targeting for partitioning water pretreatment unit, 393–402, Copyright 2010, with permission from Elsevier.).

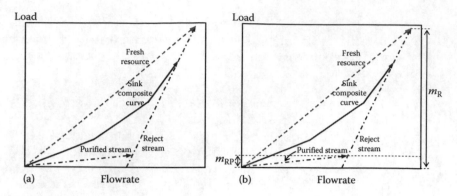

FIGURE 9.3
The PPD for interception unit of (a) fixed outlet quality and (b) removal ratio types. (From *Chem. Eng. Res. Des.*, 88(4), Tan, R.R., Ng, D.K.S., and Foo, D.C.Y., Graphical approach to minimum flowrate targeting for partitioning water pretreatment unit, 393–402, Copyright 2010, with permission from Elsevier.)

determines the length of both stream lines. The minimum flowrate of fresh resource (F_R) is then determined based on the horizontal distance of its stream line (Figure 9.2).

6. The targeting procedure for an interception unit of the fixed outlet quality type is outlined in Step 6a, while that for the fixed removal ratio type unit is given in Step 6b:

   a. The *purified stream line* is plotted from the origin, with its slope corresponding to its quality level (q_{RP}), and touches the reject stream line (Figure 9.3a).

   b. Impurity/property load of the fresh resource (m_R) is calculated as the product of its flowrate and quality level (Equation 9.7). Load of the purified stream (m_{RP}) is then calculated based on Equation 9.8. A horizontal line is then plotted on the PPD, with a vertical distance from the x-axis that represents the load of the purified stream. The *purified stream line* can then be plotted from the origin to the intersect point of the horizontal and reject stream lines (Figure 9.3b). The quality level (q_{RP}) of the purified stream is then determined from its slope:

$$m_R = F_R q_R \tag{9.7}$$

$$m_{RP} = F_R q_R (1 - RR) \tag{9.8}$$

7. A *bypass stream line* is drawn parallel to the fresh resource line (i.e., both streams have equal quality levels) such that it is tangent to the sink composite curve (see Figure 9.4a). The intersect points of

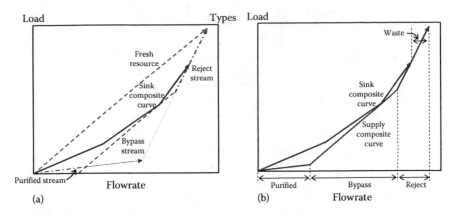

FIGURE 9.4
(a) Plotting of bypass stream and (b) a complete PPD. (From *Chem. Eng. Res. Des.*, 88(4), Tan, R.R., Ng, D.K.S., and Foo, D.C.Y., Graphical approach to minimum flowrate targeting for partitioning water pretreatment unit, 393–402, Copyright 2010, from Elsevier.)

the bypass stream line with the purified and reject stream lines define the three segments on the *supply composite curve*, i.e., the purified, bypass, and reject streams. The horizontal distances of these segments indicate their respective flowrates. The overhang of the supply composite curve from the sink composite curve defines the waste discharge from the network, with its flowrate determined by the horizontal distance, as shown in the complete PPD in Figure 9.4b.

Example 9.1 Concentration-based water pretreatment network

Table 9.1 shows the limiting data for four water sinks. A freshwater feed is available for service, however, with an impurity concentration (C_{FW}) of 25 ppm. A partitioning interception unit is used to purify the freshwater feed to fulfill the requirement of the water sinks. Determine the minimum flowrate targets for freshwater feed as well as the purified, bypass, and reject streams for the following cases:

TABLE 9.1

Limiting Data for Example 9.1

Sinks, SK$_j$	F_{SKj} (t/h)	C_{SKj} (ppm)
SK1	4	10
SR2	1	20
SR3	2	30
SR4	1	40

1. The interception unit has a water recovery factor (*RC*) of 0.75 and purifies water to a fixed concentration of 5 ppm at the purified stream (C_{RP}).
2. The interception unit has an RR value of 0.9 and has a reject stream concentration (C_{RJ}) of 100 ppm. Determine also the concentration of the purified stream as well as the water recovery factor (*RC*) of the interception unit.

Solution

Case 1—Interception unit with $RC = 0.75$ and $C_{RP} = 5$ ppm

Equations 9.1 through 9.3 are first used to determine the missing concentration of the reject stream (C_{RJ}). For this case, the freshwater feed of 25 ppm is purified to 5 ppm (C_{RP}), with an interception unit of $RC = 0.75$. The equations determine the reject stream concentration (C_{RJ}) as 85 ppm. Next, the seven-step procedure is followed to plot the PPD:

1. *Arrange the process sinks in descending order of their quality levels.*
2. *Calculate the maximum load acceptable by each sink (m_{SKj}).*

   From Table 9.1, it is observed that all water sinks have been arranged in descending order of their quality levels (i.e., in ascending order of impurity concentrations). The maximum load acceptable by each sink is then calculated and shown in the fourth column of Table 9.2.
3. *All process sinks are plotted from the origin on a load-versus-flowrate diagram, with descending order of their quality levels to form the sink composite curve. The individual sink segments are connected by linking the tail of the latter to the arrowhead of an earlier segment.*
4. *A fresh resource line is plotted from the origin on the load-versus-flowrate diagram (i.e., the PPD), with its slope corresponding to its quality level (q_R). The length of the fresh resource line will be decided in the following step.*

   Since the flowrate and load values of the individual segments of the sink composite curve are cumulative in nature, cumulative flowrates (Cum. F_{SKj}) and cumulative loads (Cum. m_{SKj}) for all segments are also calculated, as given in the last two columns of Table 9.2. These values are then plotted to form the sink composite curve on the PPD, which takes the form of a load-versus-flowrate diagram in Figure 9.5. A freshwater line

TABLE 9.2

Data for Plotting PPD in Example 9.1

Sinks, SK_j	F_{SKj} (t/h)	C_{SKj} (ppm)	m_{SKj} (g/h)	Cum. F_{SKj} (t/h)	Cum. m_{SKj} (g/h)
SK1	4	10	40	4	40
SR2	1	20	20	5	60
SR3	2	30	60	7	120
SR4	1	40	40	8	160

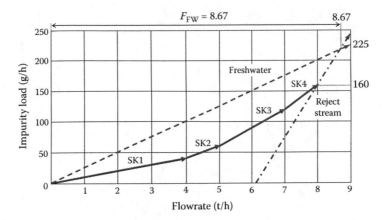

FIGURE 9.5

Plotting of sink composite curve, freshwater, and reject stream lines for Example 9.1 (case 1). (From *Chem. Eng. Res. Des.*, 88(4), Tan, R.R., Ng, D.K.S., and Foo, D.C.Y., Graphical approach to minimum flowrate targeting for partitioning water pretreatment unit, 393–402, Copyright 2010, with permission from Elsevier.)

is plotted from the origin and takes the maximum span of the PPD. Note that the slope of the individual segments of the sink composite curve, as well as that of the freshwater line, correspond to their concentration levels.

5. *The reject stream line is plotted on the right and just touches the sink composite curve on the PPD, with its slope corresponding to its quality level (q_{RJ}). The point where the reject stream line intersects with the fresh resource line determines the length of both stream lines. The minimum flowrate of fresh resource (F_R) is then determined based on the horizontal distance of its stream line.*

   A reject stream line with its slope corresponding to 85 ppm is then plotted on the PPD on the right and just touches the sink composite curve. Note that the reject stream line also intersects the freshwater line. The intersection hence determines the exact length of both lines. We can then determine the minimum flowrate of freshwater (F_{FW}) from the horizontal distance of its stream line, i.e., 8.67 t/h (Figure 9.5).

6a. *The purified stream line is plotted from the origin, with its slope corresponding to its quality level (q_{RP}), and touches the reject stream line.*

7. *A bypass stream line is drawn parallel to the fresh resource line (i.e., both streams have equal quality levels) such that it is tangent to the sink composite curve. The intersect points of the bypass stream line with the purified and reject stream lines define the three segments on the supply composite curve, i.e., the purified, bypass, and reject streams. The horizontal distances of these segments indicate their respective flowrates. The overhang of the supply composite curve from the sink composite curve defines the waste discharge from the network, with its flowrate determined by the horizontal distance.*

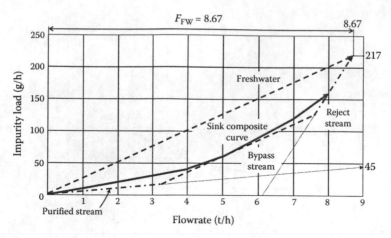

FIGURE 9.6
Plotting of purified and bypass stream lines for Example 9.1. (From *Chem. Eng. Res. Des.*, 88(4), Tan, R.R., Ng, D.K.S., and Foo, D.C.Y., Graphical approach to minimum flowrate targeting for partitioning water pretreatment unit, 393–402, Copyright 2010, with permission from Elsevier.)

The purified stream with its slope corresponding to 5 ppm is then plotted on the PPD and touches the reject stream line. A bypass stream line is then plotted in parallel with the freshwater line (with a concentration of 25 ppm) and forms a tangent with the sink composite curve (Figure 9.6). A supply composite curve is next formed by combining the three individual segments of purified, bypass, and reject streams. The minimum purified (F_{RP}), bypass (F_{BP}), and reject (F_{RJ}) flowrates are indicated by the horizontal distance of their respective segments, i.e., 3.25, 4.33, and 1.09 t/h (Figure 9.7). Finally, the wastewater flowrate (F_{WW}) is also identified as 0.67 t/h from Figure 9.7.

Case 2—Interception unit with $RR = 0.9$ and $C_{RJ} = 100$ ppm
Steps 1–5 in this case are similar to those in Case 1. The resulting PPD is shown in Figure 9.8. Note that the slope of the reject stream corresponds to the concentration of 100 ppm for this case. The minimum flowrate of freshwater (F_{FW}) is determined as 8.53 t/h, as shown in Figure 9.8. Steps 6b and 7 of the procedure are discussed next.

6b. Impurity/property load of the fresh resource (m_R) is calculated as the product of its flowrate and quality level (Equation 9.7). Load of the purified stream (m_{RP}) is then calculated based on Equation 9.8. A horizontal line is then plotted on the PPD, with a vertical distance from the x-axis that represents the load of the purified stream. The purified stream line can then be plotted from the origin to the intersect point of the horizontal and reject stream lines. The quality level (q_{RP}) of the purified stream is then determined from its slope.

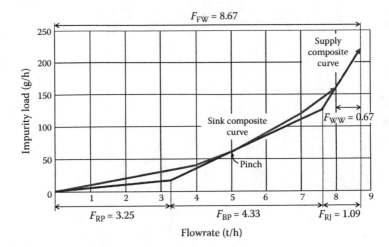

FIGURE 9.7
A complete PPD for Example 9.1 (Case 1). (From *Chem. Eng. Res. Des.*, 88(4), Tan, R.R., Ng, D.K.S., and Foo, D.C.Y., Graphical approach to minimum flowrate targeting for partitioning water pretreatment unit, 393–402, Copyright 2010, with permission from Elsevier.)

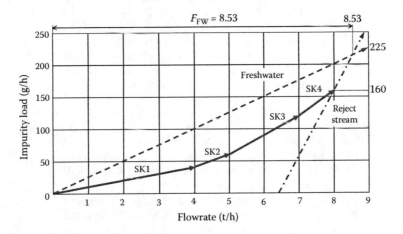

FIGURE 9.8
Plotting of sink composite curve, freshwater, and reject stream lines for Example 9.1 (Case 2). (From *Chem. Eng. Res. Des.*, 88(4), Tan, R.R., Ng, D.K.S., and Foo, D.C.Y., Graphical approach to minimum flowrate targeting for partitioning water pretreatment unit, 393–402, Copyright 2010, with permission from Elsevier.)

Following Equations 9.7 and 9.8, the impurity loads of freshwater (m_{FW}) and purified stream (m_{RP}) are calculated as 213 g/h (=8.53 t/h × 25 ppm) and 21.3 g/h (=213 g/h × (1−0.9)), respectively. A horizontal line is then plotted with a vertical distance of 21.3 g/h from the x-axis and intersects the reject stream line (Figure 9.9). The purified stream line is then plotted

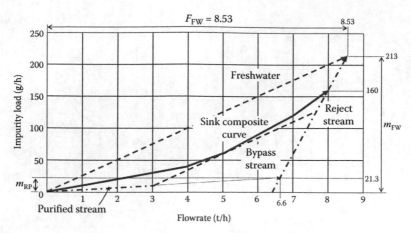

FIGURE 9.9
Plotting of purified and bypass stream lines for Example 9.1 (Case 2). (From *Chem. Eng. Res. Des.*, 88(4), Tan, R.R., Ng, D.K.S., and Foo, D.C.Y., Graphical approach to minimum flowrate targeting for partitioning water pretreatment unit, 393–402, Copyright 2010, with permission from Elsevier.)

between the origin and the intersection point, corresponding to a horizontal distance of 6.6 t/h. The slope of this line corresponds to the concentration of the purified stream (C_{RP}), which is then determined as 3.2 ppm.

7. A bypass stream line is drawn parallel to the fresh resource line (i.e., both streams have equal quality levels) such that it is tangent to the sink composite curve (see Figure 9.4a). The intersect points of the bypass stream line with the purified and reject stream lines define the three segments on the supply composite curve, i.e., the purified, bypass, and reject streams. The horizontal distances of these segments indicate their respective flowrates. The overhang of the supply composite curve from the sink composite curve defines the waste discharge from the network, with its flowrate determined by the horizontal distance.

Step 7 for this case is also identical to that in Case 1. The bypass stream that is parallel with the freshwater line defines the purified, bypass, and reject segments of the supply composite curve. The minimum flowrates of these segments are determined by their horizontal distances, i.e., 2.99, 4.68, and 0.86 t/h for purified (F_{RP}), bypass (F_{BP}), and reject (F_{RJ}) streams, respectively (Figure 9.10). The wastewater flowrate (F_{WW}) can also be identified from Figure 9.7 as 0.53 t/h.

With Equation 9.1, the inlet flowrate of the interception unit (F_{Rin}) is determined as 3.85 t/h (= 2.99 + 0.86). Note that in some cases, the inlet flowrate of the interception unit is taken as the treatment capacity of the interception unit. In other cases (e.g., reverse osmosis network for ultrapure water production), one may also report the treatment capacity based on the purified stream flowrate (F_{RP}). Finally, the *RC* value of the interception unit is determined using Equation 9.3 as 0.78 (= 2.99/(3.85)).

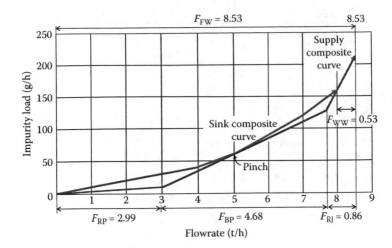

FIGURE 9.10
A complete PPD for Example 9.1 (Case 2). (From *Chem. Eng. Res. Des.*, 88(4), Tan, R.R., Ng, D.K.S., and Foo, D.C.Y., Graphical approach to minimum flowrate targeting for partitioning water pretreatment unit, 393–402, Copyright 2010, with permission from Elsevier.)

9.3 Insights on Design Principles from PPD

Similar to the case of MRPD in Chapter 3, design principles may be derived based on the insights of PPD. Figure 9.11 shows that the purified stream lies completely in the *higher quality region* (HQR, below the pinch) of the PPD, whereas the reject stream is found in the *lower quality region* (LQR, above the pinch). On the other hand, the bypass stream is found to exist in both regions and can be viewed as the *pinch-causing source* as that in the MRPD (Section 3.2). Similar to the case in MRPD, the allocated flowrates of the bypass stream for each respective region are given by the horizontal distance of its segment in these regions, respectively (see Figure 9.11).

The two most important "golden rules" can then be derived as follow (Tan et al., 2010):

1. In the HQR, flowrate requirements of the sinks are to be fulfilled by the purified stream and the allocation flowrate of the bypass stream. In other words, reject stream should not be utilized in this region, as it belongs to the LQR.

2. In the LQR, flowrate requirements of the sinks are to be fulfilled by the reject stream, as well as the allocation flowrate of the bypass stream. Similarly, the purified stream should not be utilized in this region, as it belongs to the HQR.

Following these rules during the design stage will ensure that the various flowrate targets are fulfilled for a pretreatment network.

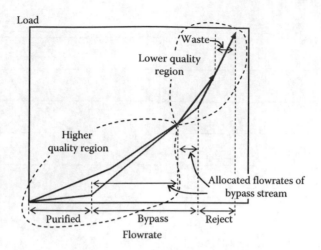

FIGURE 9.11
Higher and lower quality regions in the PPD.

9.4 Pretreatment Network Design with Nearest Neighbor Algorithm

It has been shown in Chapter 7 that the *nearest neighbor algorithm (NNA)* and *matching matrix* can be used for designing an RCN that fulfils the flowrate targets. In this section, the two tools are used for designing a pretreatment network. A generic form of the matching matrix for pretreatment network design is shown in Figure 9.12. Similar to the case of RCN in Chapter 7, all available sources, i.e., purified (RP), bypass (BP), and reject (RJ) streams, are arranged as rows, while sinks (including waste discharge—WD) as columns, both in descending order of quality. Note that fresh resource stream has been excluded, as it becomes the bypass stream of the interception unit. Parameters in the cells indicate the allocation flowrates ($F_{SRi,SKj}$) for each sink–source match. Note that the shaded area in the matching matrix indicates forbidden matches among sinks and sources. This is due to the insights

		F_{SKj}	F_{SK1}	F_{SK2}	F_{SKj}	F_D
		q_{SKj}	q_{SK1}	q_{SK2}	q_{SKj}	q_D
F_{SRi}	q_{SRi}		SK1	SK2	SK$_j$	WD
F_{RP}	q_{RP}	RP	$F_{RP,SK1}$	$F_{RP,SK2}$		
F_{BP}	q_{BP}	BP	$F_{BP,SK1}$	$F_{BP,SK2}$	$F_{BP,SKj}$	
F_{RJ}	q_{RJ}	RJ			$F_{RJ,SKj}$	$F_{RJ,WD}$

FIGURE 9.12
Matching matrix for a pretreatment network.

gained from the PPD as discussed earlier. For instance, the reject stream (RJ) that is located at the LQR should only be fed to the sinks SK_j and waste discharge (WD) in the same region but not to SK1 and SK2 in the HQR.

Example 9.2 Water pretreatment network design

Example 9.1 (Case 1) is revisited. Design a pretreatment network that achieves the flowrate targets identified by the PPD.

Solution

The matching matrix for this case is shown in Figure 9.13. Note that the flowrates for the purified, bypass, and reject streams of the interception unit as well as that of the wastewater stream are identified during the targeting phase (see Figure 9.7). Note also that since SK1 and SK2 belong to the HQR (lower concentration than the pinch-25 ppm), they can only be fed by the purified and bypass streams. On the other hand, SK3 and SK4 that are located on the LQR (higher concentration than pinch) are to be fed bypass and reject streams.

The NNA is then used to determine the flowrate of each sink–source match (readers may refer to Section 7.1 for the detailed procedure). For SK1 of 10 ppm, the neighbor candidates are purified (RP) and bypass (BP) streams. Utilizing Equations 7.1 and 7.2, we can determine the respective flowrates for matches between SKI with RP ($F_{RP,SK1}$) and BP ($F_{BP,SK1}$) as follows:

$$F_{RP,SK1} + F_{BP,SK1} = 4 \text{ t/h}$$

$$F_{RP,SK1}(5 \text{ ppm}) + F_{BP,SK1}(25 \text{ ppm}) = 50 \text{ t/h} (10 \text{ ppm})$$

Hence,

$$F_{RP,SK1} = 3 \text{ t/h; and } F_{BP,SK1} = 1 \text{ t/h}$$

For SK2, the same pair of neighbor candidates are used, since their flowrates are still in excess. Applying the same equation pair leads to the flowrate matches of 0.25 ($F_{RP,SK2}$) and 0.75 t/h ($F_{BP,SK2}$), respectively. The RP flowrate is now exhausted.

The neighbor candidates for SK3 are next identified, i.e., BP and RJ. Applying the NNA equations leads to the flowrates of 1.83 and 0.17 t/h

		F_{SKj}	4	1	2	1	0.67
		C_{SKj}	10	20	30	40	85
F_{SRi}	C_{SRi}		SK1	SK2	SK3	SK4	WW
3.25	5	RP	$F_{RP,SK1}$	$F_{RP,SK2}$			
4.33	25	BP	$F_{BP,SK1}$	$F_{BP,SK2}$	$F_{BP,SK3}$	$F_{BP,SK4}$	
1.09	85	RJ			$F_{RJ,SK3}$	$F_{RJ,SK4}$	$F_{RJ,WW}$

FIGURE 9.13
Matching matrix for a pretreatment network (all flowrate terms in t/h; concentration in ppm).

	F_{SKj}	4	1	2	1	0.67
	C_{SKj}	10	20	30	40	85
F_{SRi}	C_{SRi}		SK1	SK2	SK3	SK4	WW
3.25	5	RP	3	0.25			
4.33	25	BP	1	0.75	1.83	0.75	
1.09	85	RJ			0.17	0.25	0.67

FIGURE 9.14
Pretreatment network design for Example 9.1 (Case 1; all flowrate terms in t/h; concentration in ppm).

for $F_{BP,SK3}$ and $F_{RJ,SK3}$, respectively. The same pair of neighbor candidates are also used for SK4. The NNA equations then determine that the flowrates for matches $F_{BP,SK4}$ and $F_{RJ,SK4}$ are 0.17 and 0.25 t/h, respectively. The BP flowrate is then exhausted, while the remaining flowrate of RJ is sent for wastewater discharge ($F_{RJ,WW}$). A complete matching matrix for the network is found in Figure 9.14.

Problems

Water Minimization Problems

9.1   Rework Case 2 in Example 9.1. In this case, assume that the concentration for the reject stream is lowered to 65 ppm, while all other parameters remain unchanged. Determine the flowrate targets for the pretreatment network. Design the network that achieves the flowrate targets.

Property Integration Problems

9.2   A pretreatment network is used to produce dematerialized water for a manufacturing process (Tan et al., 2010). Water quality is characterized by electrical resistivity, R (MΩ m), which is an index that reflects the total ionic content in the aqueous streams. The mixing rule of this property is given by Equation 2.64. The freshwater supply has a resistivity value of 0.1 MΩ m. The pretreatment network uses multistage membrane filtration units as interception unit, which produces dematerialized water at a fixed resistivity of 10 MΩ m and possesses a recovery factor (RC) of 0.8. Limiting data for the water sinks are given in Table 9.3. Solve the following tasks:

   a.   Determine the resistivity operator of the reject stream (hint: make use of Equation 2.64).
   b.   Determine the flowrate targets of the pretreatment network.
   c.   Design the pretreatment network that achieves the flowrate targets.

TABLE 9.3

Limiting Data for Problem 9.2

Sinks, SK_j	F_{SKj} (t/h)	R_{SKj} (MΩ m)	ψ_{SKj} (MΩ m)$^{-1}$
SK1	20	5	0.2
SR2	10	3.33	0.3
SR3	10	0.05	20
SR4	2	0.017	60

Reference

Tan, R. R., Ng, D. K. S., and Foo, D. C. Y. 2010. Graphical approach to minimum flow-rate targeting for partitioning water pretreatment unit. *Chemical Engineering Research and Design*, 88(4), 393–402.

10

Synthesis of Inter-Plant Resource Conservation Networks

In previous chapters, various flowrate targeting and synthesis techniques have been presented. However, the focus of previous chapters is mainly on a single resource conservation network (RCN). In this chapter, material recovery across different RCNs is analyzed. In practice, these individual RCNs may be different process plants located in different sites or a huge process plant that may be physically segregated into different sections. We shall collectively term this as *inter-plant resource conservation networks* (IPRCNs). In this chapter the two-step pinch analysis technique is used to synthesize an IPRCN. In step one, a generic targeting procedure to determine the overall minimum fresh resource, waste discharge, and cross-plant stream flowrates for an IPRCN is illustrated, based on the work of Chew et al. (2010a,b). Graphical and algebraic techniques presented in earlier chapters are used to carry out the targeting tasks. Network design of IPRCN is then carried using the nearest neighbor algorithm introduced in Chapter 7."

10.1 Types of IPRCN Problems

In general, there exist two types of IPRCN problems, with reference to the direction of the cross-plant streams. For an *unassisted integration* scheme, the cross-plant streams are only found within the *pinch region* of an IPRCN and are exported from RCN(s) of higher quality pinch to those with lower quality pinch (Figure 10.1). In most cases, these cross-plant streams are individual waste streams emitted from the RCNs (e.g., wastewater, purged gas, and waste solvent), after their resource conservation potential is exhausted. The pinch region is defined by the highest and lowest quality pinches among all RCNs. For the case in Figure 10.1, the pinch region is defined by the pinches of Network A (higher quality) and Network B (lower quality). Since the waste streams are emitted from the lower quality region (LQR) of Network A (W1, W2, and W3), they can serve as the potential cross-plant streams for Network B. However, since the latter only experiences resource deficit in its higher quality region (HQR), only waste streams that have quality higher than the pinch of Network B shall be used as cross-plant streams, i.e., W1 and W2.

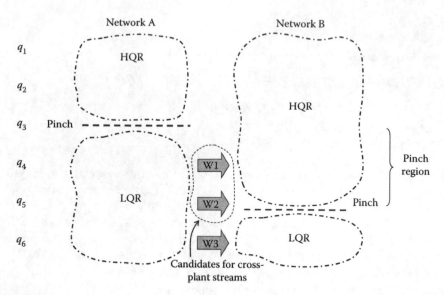

FIGURE 10.1
Principle of unassisted integration scheme. (From Chew, I.M.L. et al., *Ind. Eng. Chem. Res.*, 49(14), 6439, 2010a. With permission.)

On the other hand, in *assisted integration* scheme, some cross-plant streams are also exported from RCN(s) with a lower quality pinch to RCN(s) with a higher quality pinch, apart from those that flow in the reverse direction (as in the unassisted integration scheme). We term these cross-plant streams as *assisted streams* (Chew et al., 2010b), as they enhance the material recovery potential for the IPRCN. Note that the assisted streams do not necessarily exist within the pinch region. In other words, some of the assisted streams may exist outside the pinch region. An example is shown in Figure 10.2. In this case, waste W1 and W4 are the assisted streams that are exported from Network B (with lower quality pinch) to Network A (with higher quality pinch). Besides, waste W2 and W3 serve as cross-plant streams that flow in the reverse direction, such as those in the unassisted integration scheme (see Figure 10.1). However, unlike the case of unassisted integration scheme, the assisted stream may not necessarily be the waste stream emitting from the RCN. Hence, identifying the assisted stream is part of the targeting procedure of the assisted integration scheme.

10.2 Generic Targeting Procedure for IPRCN

The overall procedure for targeting IPRCN may be summarized as in Figure 10.3. In principle, the procedure consists of the following four major steps:

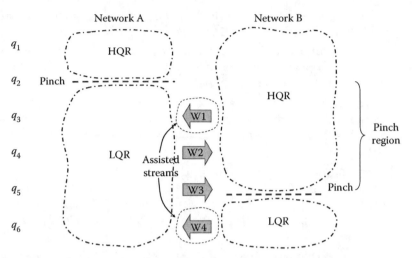

FIGURE 10.2
Principle of assisted integration scheme. (From Chew, I.M.L. et al., *Ind. Eng. Chem. Res.*, 49(14), 6456, 2010b. With permission.)

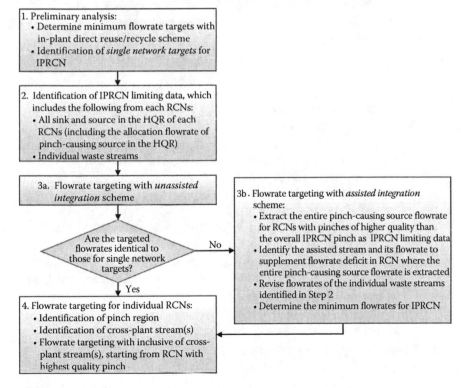

FIGURE 10.3
Targeting procedure for IPRCN.

1. *Preliminary analysis* is performed for all individual RCNs that are involved in the IPRCN. This involves the following detailed steps:

   a. The minimum fresh resource and waste flowrates are determined for an in-plant direct reuse/recycle scheme. This sets the network targets for each RCN before the analysis of IPRCN.

   b. The minimum fresh resource and waste flowrates for the overall IPRCN are then determined by assuming all material sinks and sources to belong to the same network. This is termed as the *single network targets*.

2. Identification of IPRCN limiting data—All sinks and sources in the HQR of each RCNs are extracted. This includes the allocation flowrate of the pinch-causing source to the HQR. Besides, the individual waste streams emitting from the LQR of each RCNs are also extracted.

3. Overall flowrate targeting for IPRCN, with the following steps:

   a. Flowrate targeting with *unassisted integration* scheme—targeting is performed using the data identified in Step 2. If the overall minimum flowrates are identical to those of the single network targets (Step 1b), the targeting procedure proceeds to Step 4, else to Step 3b.

   b. Flowrate targeting with *assisted integration* scheme that involves the following detailed steps:

       i.   For RCN with pinch of higher quality than the overall IPRCN pinch (identified in Steps 1b and 3a), the entire flowrate of its pinch-causing source will be extracted as waste stream in the IPRCN limiting data.

       ii. Identify the assisted stream and its flowrate needed to supplement flowrate deficit in RCN, where the entire pinch-causing source flowrate is extracted. The assisted stream should originate from the pinch-causing source of the overall IPRCN, which has been identified in Step 2.

       iii. Revise the individual waste stream flowrates identified in Step 2, for RCNs in which their entire pinch-causing source flowrate is extracted as waste stream in the IPRCN (substeps (i)), and those that involve the use of assisted stream (substeps (ii)).

       iv. Determine the minimum flowrate targets for IPRCN with the revised limiting data.

4. Flowrate targeting for individual RCNs, with the following detailed steps:

   a. Identify the *pinch region*, which is defined by the highest and lowest quality pinches among all RCNs.

b. Individual waste stream that exists within this pinch region is identified as the potential cross-plant stream.

c. Flowrate targeting for individual RCNs, inclusive of cross-plant stream(s). Perform the targeting step from RCNs with highest to lowest quality pinches.

Note again that flowrate targeting in each of the steps given earlier may be carried out using any graphical or algebraic techniques described in earlier chapters of this book.

Example 10.1 Inter-plant water network with unassisted integration scheme

The water minimization case studies in Examples 8.1 (Network A) and 6.1 (Network B) are revisited here (Bandyopadhyay et al., 2010). In this case, apart from in-plant direct reuse/recycle, water recovery may also be carried out between both networks. The limiting data for these networks are given in Table 10.1 (reproduced from Tables 8.1 and 6.1, respectively). The freshwater feed is available at 0 ppm. Use the graphical targeting technique of *material recovery pinch diagram* (MRPD, Chapter 3*) to identify the overall minimum freshwater and wastewater flowrates needed for the inter-plant water network, water flowrates for individual networks, as well as that of the cross-plant stream(s).

Solution

The targeting procedure is followed to identify the various targets for the inter-plant water network:

TABLE 10.1

Limiting Water Data for Individual RCNs in Example 10.1

Sinks, SK_j	F_{SKj} (t/h)	C_{SKj} (ppm)	Sources, SR_i	F_{SRi} (t/h)	C_{SRi} (ppm)
Network A					
SK1	20	0	SR1	20	100
SK2	100	50	SR2	100	100
SK3	40	50	SR3	40	800
SK4	10	400	SR4	10	800
Network B					
SK5	50	20	SR5	50	50
SK6	100	50	SR6	100	100
SK7	80	100	SR7	70	150
SK8	70	200	SR8	60	250
$\Sigma_i F_{SKj}$	470		$\Sigma_i F_{SRi}$	450	

---

* See Section 3.1 for the detailed step of this targeting technique.

1. *Preliminary analysis is performed for all individual RCNs that are involved in the IPRCN. This involves the following detailed steps:*
   a. *The minimum fresh resource and waste flowrates are determined for an in-plant direct reuse/recycle scheme.*
   b. *The minimum fresh resource and waste flowrates for the IPRCN are then determined by assuming all material sinks and sources to belong to the same network. This is termed as the single network targets.*

   For Network A, flowrate targeting for a direct reuse/recycle scheme has been presented in Example 8.1. MRPD in Figure 10.4a (reproduced from Figure 8.3) indicates that the freshwater (F_{FW}) and wastewater (F_{WW}) flowrates for this RCN are both determined as 90 t/h.

   On the other hand, MRPD for Network B is given in Figure 10.4b, indicating that the RCN requires 70 t/h of freshwater and 50 t/h of wastewater, respectively. Note that the pinch concentrations for these RCNs correspond to 100 and 150 ppm, respectively.

   If we consider all water sinks and sources as if they belong to the same network, the single network targets correspond to 155 t/h of freshwater and 135 t/h of wastewater, respectively, with an overall IPRCN pinch concentration found at 100 ppm. The MRPD for the single network is given in Figure 10.5.

2. *Identification of IPRCN limiting data—All sinks and sources in the HQR of each RCNs are extracted. This includes the allocation flowrate of the pinch-causing source to the HQR. Besides, the individual waste streams emitting from the LQR of each RCNs are also extracted.*

   We first analyze water sinks in both Networks A and B. For Network A, since the pinch concentration is identified at 100 ppm, sinks SK1, SK2, and SK3 that have lower concentrations (hence belong to the HQR) are extracted as limiting water data for IPRCN. Similarly, sinks SK5, SK6, and SK7 that have lower concentrations than 150 ppm (pinch concentration) in Network B are also extracted (see Table 10.2a).

   For water sources, the pinch-causing sources for both networks are first identified. These correspond to SR1 and SR2 for Network A, as well as SR7 for Network B. All sources with lower concentration than that these pinch-causing sources are extracted as limiting data for IPRCN, i.e., SR5 and SR6. For the pinch-causing sources, only their allocated flowrate to the HQR (F_{HQR}) will be extracted. These correspond to 70 t/h of SR1 and SR2 in Network A, as well as 10 t/h of SR7 for Network B, as shown in Figure 10.4. The source data is summarized in Table 10.2b.

   We next identify the individual wastewater streams emitting from the RCNs. For Network A, detailed steps in identifying individual wastewater streams have been presented in Example 8.1. As shown, two wastewater streams are identified, i.e., WW1 of 44.29 t/h (100 ppm) and WW2 of 45.71 t/h (800 ppm).

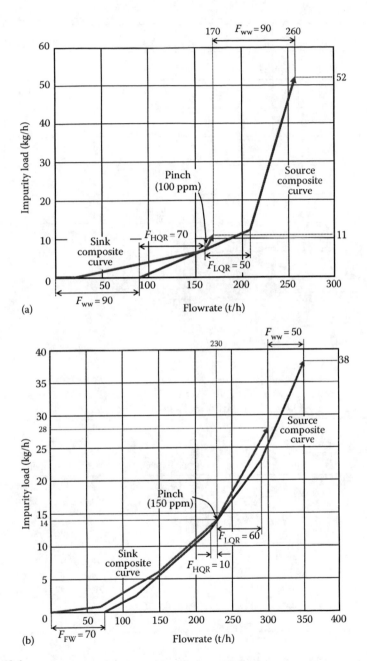

FIGURE 10.4
MRPD for (a) Network A and (b) Network B. (From *Chem. Eng. Sci.*, 60(1), Prakash, R. and Shenoy, U.V., Targeting and design of water networks for fixed flowrate and fixed contaminant load operations, 255–268. Copyright 2005a, with permission from Elsevier.)

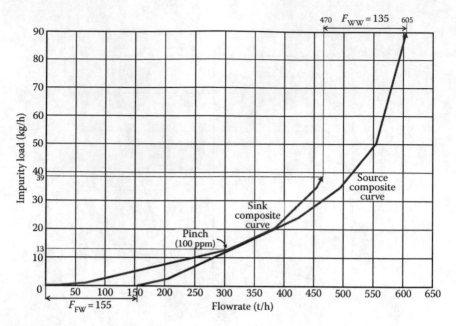

FIGURE 10.5
MRPD for single network targeting.

TABLE 10.2

Limiting Water Data for IPRCN Flowrate Targeting: (a) Sinks in the HQR,
(b) Sources in HQR, and (c) Individual Wastewater Streams

	(a)			(b)			(c)	
Sinks, SK$_j$	F_{SKj} (t/h)	C_{SKj} (ppm)	Sources, SR$_i$	F_{SRi} (t/h)	C_{SRi} (ppm)	Wastewater, WWy	F_{WWy} (t/h)	C_{WWy} (ppm)
Network A								
SK1	20	0	SR1	20	100	WW1	44.29	100
SK2	100	50	SR2	50	100	WW2	45.71	800
SK3	40	50						
Network B								
SK5	50	20	SR5	50	50	WW3	25.00	150
SK6	100	50	SR6	100	100	WW4	25.00	250
SK7	80	100	SR7	10	150			
$\Sigma_j F_{SKj}$	390		$\Sigma_i F_{SRi}$	230		$\Sigma_y F_{WWy}$	140	

For Network B, an MRPD is plotted for the LQR in Figure 10.6,
by excluding the pinch-causing sources (i.e., only SK8 and SR8
are included). Note that the source composite curve is slid on
the pinch-causing source locus, and the slope of the latter corresponds to the pinch concentration of 150 ppm. Figure 10.6

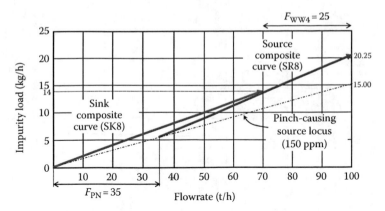

FIGURE 10.6
Wastewater identification for Network B.

shows that the *minimum pinch flowrate* (F_{PN}) needed for the LQR is 35 t/h. Since the allocated flowrate of the pinch-causing source to the LQR is 60 t/h (F_{LQR}, Figure 10.4), this means that the unutilized flowrate of this source will emit as wastewater stream. Hence, wastewater stream WW3 has a flowrate of 35 t/h (=60 – 25 t/h) and possesses the pinch concentration of 150 ppm. Besides, another wastewater stream is observed to emit from the unutilized source SR8 in the LQR. From Figure 10.6, this wastewater stream (WW4) is observed to have a flowrate of 25 t/h, with the same concentration as SR8, i.e., 250 ppm. The individual wastewater stream data are summarized in Table 10.2c.

3a. *Flowrate targeting with an unassisted integration scheme*

We next carry out flowrate targeting with the IPRCN limiting data identified in Table 10.2. As shown in the MRPD in Figure 10.7, the overall freshwater and wastewater flowrates for the IPRCN are identified as 155 and 135 t/h, respectively, which are identical to those of the single network targets identified in Step 1b (Figure 10.5). Hence, this problem is classified as an unassisted integration case. We can then proceed to Step 4.

4. *Flowrate targeting for individual RCNs with the following detailed steps:*

   a. *Identify the pinch region, which is defined by the highest and lowest quality pinches among all RCNs.*

   b. *Individual waste stream that exists within this pinch region is identified as the potential cross-plant stream.*

   c. *Flowrate targeting for individual RCNs, inclusive of cross-plant stream(s). Perform the targeting step from RCNs with highest to lowest quality pinches.*

In Step 1, we have identified that the pinch concentrations for Networks A and B are 100 and 150 ppm, respectively. These pinch concentrations define the pinch region of the IPRCN,

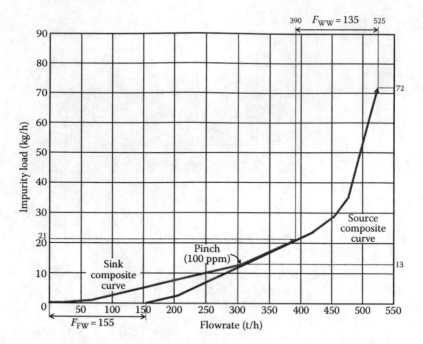

FIGURE 10.7
MRPD for IPRCN.

where a cross-plant stream is found. Among the four waste-water streams identified in Step 1 (see Table 10.2) only WW1 and WW3 exist within the pinch region. Since Network A has a higher quality pinch (100 ppm) as compared with Network B, the cross-plant stream should be exported from the former to the latter. In other words, only wastewater WW1 that emits from Network A will become the cross-plant stream of the IPRCN.

We next perform flowrate targeting for both individual RCNs, by taking into consideration the cross-plant stream. Network A will first be analyzed, as it has higher quality pinch at 100 ppm. Since this RCN does not receive any cross-plant stream, its freshwater flowrate remains unchanged as before, i.e., 90 t/h. However, this RCN might experience reduced wastewater flowrate as part of its wastewater may be used as a cross-plant stream. The exact flowrate of wastewater of Network A will be determined once the exact flowrate of the cross-plant stream is known.

For Network B, since a cross-plant stream is available (from Network A), it should be used prior to the freshwater feed. Since both of these sources have different concentrations, this resembles the multiple resource problems discussed in Chapter 3 (see Section 3.3 for detailed procedure—assume the cross-plant stream as an impure fresh resource that is virtually free). The revised MRPD for Network B is shown in Figure 10.8. However, note that the MRPD is plotted using the data in Table 10.1, instead of those in Table 10.2. We first include a locus for the cross-plant stream, with its slope corresponding to the concentration of 100 ppm. Since this concentration coincides with that of source SR6, the locus is added at the end of the SR6 segment (Figure 10.8a), without having to connect the segments of SR7 and SR8. The segments of SK5 and SK6 are then moved along with the cross-plant stream locus to the right until it touches the sink composite curve and lie completely below and to the right of the latter. This sets the minimum flowrate for the freshwater requirement for Network B, i.e. $F_{FW} = 65$ t/h. The remaining segments of the source composite curve with concentration higher than 100 ppm, i.e., SR7 and SR8, are then slid along the cross-plant stream locus, until they touch the sink composite curve. We can then determine the minimum flowrates of cross-plant stream (F_{CP}) and wastewater (F_{WW}) from the revised MRPD in Figure 10.8b, i.e., 15 and 60 t/h respectively.

After determining the minimum cross-plant flowrate (F_{CP}), we then return to Network A. After sending 15 t/h as the cross-plant stream, the wastewater stream WW1 has a reduced flowrate of 29.29 t/h (=44.29 − 15 t/h). From Table 10.1, we observe that this wastewater stream (100 ppm) is originated from sources SR1 and SR2. In other words, we may assume SR1 to have a reduced flowrate of 5 t/h, after exporting 15 t/h of its flowrate as a cross-plant stream. The revised MRPD for Network A is then constructed with the limiting data in Table 10.1, however with the reduced flowrate of SR1 (5 t/h). The revised MRPD is given in Figure 10.9. We now determine that the wastewater flowrate (F_{WW}) of this RCN has been reduced to 75 t/h. However, note that the freshwater flowrate (F_{FW}) remains identical as before, i.e., 90 t/h.

Finally, note that when the minimum freshwater and wastewater flowrates of both individual RCNs are added, they correspond to the total flowrate of 155 t/h (=90 + 65 t/h) for freshwater and 135 t/h (=75 + 60 t/h) of wastewater, which match the flowrate targets obtained in Steps 1 and 2.

Example 10.2 Inter-plant water network with assisted integration scheme

Table 10.3 shows the limiting data for an inter-plant water network that involves three water networks (Olesen and Polley, 1996). Use the algebraic targeting technique of *material cascade analysis* (MCA; Chapter 4*) to identify the overall minimum freshwater and wastewater flowrates needed for the inter-plant water network, water flowrates for individual networks, as well as that of the cross-plant stream(s). The freshwater feed is available at 0 ppm.

---

* See Section 4.1 for the detailed step of this targeting technique.

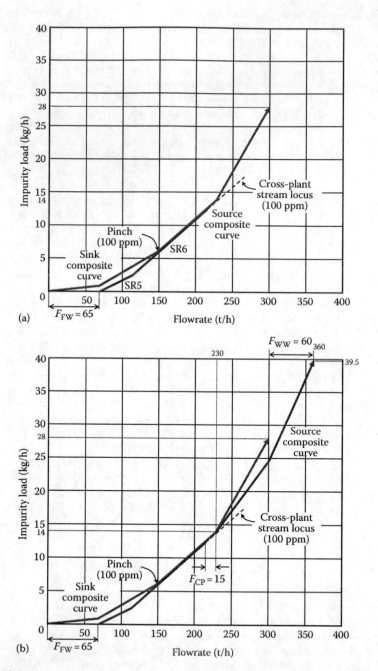

FIGURE 10.8
Revised MRPD for Network B: (a) targeting for fresh water, and (b) targeting for cross-plant stream and wastewater.

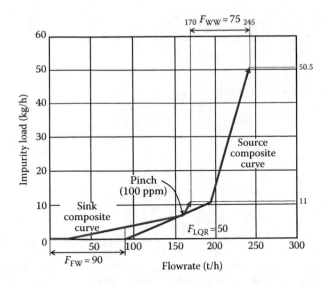

FIGURE 10.9
Revised MRPD for Network A.

TABLE 10.3

Limiting Water Data for Individual RCNs in Example 10.2

Sinks, SK_j	F_{SKj} (t/h)	C_{SKj} (ppm)	Sources, SR_i	F_{SRi} (t/h)	C_{SRi} (ppm)
Network A					
SK1	20.00	0	SR1	20.00	100
SK2	66.67	50	SR2	66.67	80
SK3	100.00	50	SR3	100.00	100
SK4	41.67	80	SR4	41.67	800
SK5	10.00	400	SR5	10.00	800
Network B					
SK6	20.00	0	SR6	20.00	100
SK7	66.67	50	SR7	66.67	80
SK8	15.63	80	SR8	15.63	400
SK9	42.86	100	SR9	42.86	800
SK10	6.67	400	SR10	6.67	1000
Network C					
SK11	20.00	0	SR11	20.00	100
SK12	80.00	25	SR12	80.00	50
SK13	50.00	25	SR13	50.00	125
SK14	40.00	50	SR14	40.00	800
SK15	300.00	100	SR15	300.00	150
$\Sigma_j F_{SKj}$	880.17		$\Sigma_i F_{SRi}$	880.17	

Solution

The targeting procedure is followed to identify the various targets for the inter-plant water network:

1. *Preliminary analysis is performed for all individual RCNs that are involved in the IPRCN. This involves the following detailed steps:*
   a. *The minimum fresh resource and waste flowrates are determined for an in-plant direct reuse/recycle scheme.*
   b. *The minimum fresh resource and waste flowrates for the IPRCN are then determined by assuming all material sinks and sources to belong to the same network. This is termed as the single network targets.*

   To perform flowrate targeting for a direct reuse/recycle scheme in each RCN, a variant of MCA, i.e., *water cascade analysis* (WCA), is utilized. Tables 10.4 through 10.6 show the targeting results of the individual RCNs. As shown, the freshwater flowrate (F_{FW}) for Networks A, B, and C are determined as 98.33, 54.64, and 186.67 t/h, respectively. Note that the flowrates for wastewater (F_{WW}) are identical to those of fresh water for each plant. The pinch concentrations for these RCNs correspond to 100, 400, and 150 ppm, respectively. On the other hand, the single network targets (assuming all water sinks and sources as if they belong to the same network) correspond to 314.36 t/h for both freshwater and wastewater flowrates, with pinch concentration at 150 ppm (Table 10.7).

2. *Identification of IPRCN limiting data—All sinks and sources in the HQR of each RCNs are extracted. This includes the allocation flowrate*

TABLE 10.4

Cascade Table for Network A

C_k (ppm)	$\Sigma_j F_{SKj}$ (t/h)	$\Sigma_i F_{SRi}$ (t/h)	$\Sigma_i F_{SRi} - \Sigma_j F_{SKj}$ (t/h)	$F_{C,k}$ (t/h)	Δm_k (kg/h)	Cum. Δm_k (kg/h)
				$F_{FW} = 98.33$		
0	20.00		−20.00			
				78.33	3.92	
50	166.67		−166.67			3.92
				−88.33	−2.65	
80	41.67	66.67	25.00			1.27
				$F_{HQR} = -63.33$	−1.27	
100		120.00	120.00			0.00
				$F_{LQR} = 56.67$	17.00	(PINCH)
400	10.00		−10.00			17.00
				46.67	18.67	
800		51.67	51.67			35.67
				$F_{WW} = 98.33$	98,254.67	
1,000,000						98,290.33

TABLE 10.5

Cascade Table for Network B

C_k (ppm)	$\Sigma_j F_{SKj}$ (t/h)	$\Sigma_i F_{SRi}$ (t/h)	$\Sigma_i F_{SRi} - \Sigma_j F_{SKj}$ (t/h)	$F_{C,k}$ (t/h)	Δm_k (kg/h)	Cum. Δm_k (kg/h)
				$F_{FW} = 54.64$		
0	20.00		−20.00			
				34.64	1.73	
50	66.67		−66.67			1.73
				−32.02	−0.96	
80	15.63	66.67	51.04			0.77
				19.02	0.38	
100	42.86	20.00	−22.86			1.15
				$F_{HQR} = -3.84$	−1.15	
400	6.67	15.63	8.96			0.00
				$F_{LQR} = 5.12$	2.05	(PINCH)
800		42.86	42.86			2.05
				47.98	9.60	
1,000		6.67	6.67			11.64
				$F_{WW} = 54.64$	54,588.21	
1,000,000						54,599.86

TABLE 10.6

Cascade Table for Network C

C_k (ppm)	$\Sigma_j F_{SKj}$ (t/h)	$\Sigma_i F_{SRi}$ (t/h)	$\Sigma_i F_{SRi} - \Sigma_j F_{SKj}$ (t/h)	$F_{C,k}$ (t/h)	Δm_k (kg/h)	Cum. Δm_k (kg/h)
				$F_{FW} = 186.67$		
0	20.00		−20.00			
				166.67	4.17	
25	130.00		−130.00			4.17
				36.67	0.92	
50	40.00	80.00	40.00			5.08
				76.67	3.83	
100	300.00	20.00	−280.00			8.92
				−203.33	−5.08	
125		50.00	50.00			3.83
				$F_{HQR} = -153.33$	−3.83	
150		300.00	300.00			0.00
				$F_{LQR} = 146.67$	95.33	(PINCH)
800		40.00	40.00			95.33
				$F_{WW} = 186.67$	186,517.33	
1,000,000						186,612.67

TABLE 10.7

Cascade Table for Single Network Targeting

C_k (ppm)	$\Sigma_j F_{SKj}$ (t/h)	$\Sigma_i F_{SRi}$ (t/h)	$\Sigma_i F_{SRi} - \Sigma_j F_{SKj}$ (t/h)	$F_{C,k}$ (t/h)	Δm_k (kg/h)	Cum. Δm_k (kg/h)
				$F_{FW} = 314.36$		
0	60.00		−60.00			
				254.36	6.36	
25	130.00		−130.00			6.36
				124.36	3.11	
50	273.33	80.00	−193.33			9.47
				−68.98	−2.07	
80	57.29	133.33	76.04			7.40
				7.06	0.14	
100	342.86	160.00	−182.86			7.54
				−175.79	−4.39	
125		50.00	50.00			3.14
				$F_{HQR} = -125.79$	−3.14	
150		300.00	300.00			0.00
				$F_{LQR} = 174.21$	43.55	**(PINCH)**
400	16.67	15.63	−1.04			43.55
				173.16	69.27	
800		134.52	134.52			112.82
				307.69	61.54	
1,000		6.67	6.67			174.36
				$F_{WW} = 314.36$	314,040.80	
1,000,000						314,215.16

of the pinch-causing source to the HQR. Besides, the individual waste streams emitting from the LQR of each RCNs are also extracted.

In Step 1, the pinch concentrations for the individual RCNs have been determined as 100, 400, and 150 ppm, respectively. Hence, all water sinks with lower concentrations in the HQR of each network are extracted as the limiting data for IPRCN. These correspond to SK1–SK4 in Network A, SK6–SK9 in Network B, and SK11–SK15 in Network C (see Table 10.9a).

For water sources, the pinch-causing sources for all networks are first identified, i.e., SR1 and SR3 for Network A, SR8 for Network B, as well as SR15 of Network C. The allocation flowrates of these pinch-causing sources to the HQR (F_{HQR}) and LQR (F_{LQR}) of each RCN are also given in Tables 10.4 through 10.6. However, only allocation flowrates of these pinch-causing sources to the HQR of each RCN will be extracted, i.e., 63.33 t/h of SR3 (Network A), 3.84 t/h of SR8 (Network B), and 153.33 t/h of SR15 (Network C). Besides, all sources with concentration lower than the pinch concentrations of each RCN (SR2, SR6, SR7, SR11–SR13) are also extracted (see Table 10.9b).

TABLE 10.8

Wastewater Identification for Network A

C_k (ppm)	$\Sigma_j F_{SKj}$ (t/h)	$\Sigma_i F_{SRi}$ (t/h)	$\Sigma_i F_{SRi} - \Sigma_j F_{SKj}$ (t/h)	$F_{C,k}$ (t/h)	Δm_k (kg/h)	Cum. Δm_k (kg/h)
				$F_{PN} = 5.71$		
100			0.00			
				5.71	1,714.29	
400	10.00		−10.00			1,714.29
				−4.29	−1,714.29	
800		51.67	51.67			0.00
				$F_{WW2} = 47.38$	47,343,047.62	
1,000,000						47,343,047.62

We next move to identification of the individual wastewater streams emitting from the RCNs. For Network A, flowrate targeting is conducted for the LQR. Note that the pinch-causing sources (SR1 and SR3) have been excluded from the source data but serve as a "fresh resource" for the LQR.* As shown in Table 10.8, flowrate targeting indicates that the minimum pinch flowrate (F_{PN}) for the LQR is determined as 5.71 t/h. Since the allocated flowrate of the pinch-causing source to the LQR is 56.67 t/h (F_{LQR}, Table 10.4), this means that the unutilized flowrate of this source will emit as wastewater stream (WW1), with a flowrate of 50.95 t/h (=56.67−5.71 t/h) and concentration of 100 ppm. Besides, another wastewater stream WW2 will emit from the unutilized sources SR4 and SR5, with flowrate of 47.38 t/h (= 56.67 − 5.71 t/h) and concentration of 800 ppm.

For Networks B and C, since no water sinks are found at the LQR of these RCNs, the allocation flowrates of their pinch-causing sources to the LQR, along with the unutilized sources in the LQR, will emit as wastewater streams. These are summarized in Table 10.9c.

3a. *Flowrate targeting with an unassisted integration scheme*

We next carry out flowrate targeting with the IPRCN limiting data identified in Table 10.9. The cascade result is shown in Table 10.10. However, it is observed that the overall freshwater (F_{FW}) and wastewater (F_{WW}) flowrates for the IPRCN are identified as 316.26, respectively, which are higher than those of the single network targets identified in Step 1 (Table 10.7). In other words, the assisted integration scheme has failed to locate the correct flowrate targets for the IPRCN. Hence, we proceed to Step 3b, to perform the targeting step using the assisted integration schem.

3b. *Flowrate targeting with assisted integration scheme, that involves the following detailed steps:*

---

* See Example 8.2 for detailed steps.

TABLE 10.9

Limiting Water Data for Unassisted Integration Scheme: (a) Sinks in the HQR, (b) Sources in HQR, and (c) Individual Wastewater Streams

	(a)			(b)			(c)	
Sinks, SK_j	F_{SKj} (t/h)	C_{SKj} (ppm)	Sources, SR_i	F_{SRi} (t/h)	C_{SRi} (ppm)	Wastewater, WWy	F_{WWy} (t/h)	C_{WWy} (ppm)
Network A								
SK1	20.00	0	SR2	66.67	80	WW1	50.95	100
SK2	66.67	50	SR3	63.33	100	WW2	47.38	800
SK3	100.00	50						
SK4	41.67	80						
Network B								
SK6	20.00	0	SR6	20.00	100	WW3	5.12	400
SK7	66.67	50	SR7	66.67	80	WW4	42.86	800
SK8	15.63	80	SR8	3.84	400	WW5	6.67	1000
SK9	42.86	100						
Network C								
SK11	20.00	0	SR11	20.00	100	WW6	146.67	150
SK12	80.00	25	SR12	80.00	50	WW7	40.00	800
SK13	50.00	25	SR13	50.00	125			
SK14	40.00	50	SR15	153.33	150		339.65	
SK15	300.00	100		523.84				
$\Sigma_j F_{SKj}$	863.50		$\Sigma_i F_{SRi}$			$\Sigma_y F_{WWy}$		

    i. *For an RCN with a pinch of higher quality than the overall IPRCN pinch (identified in Steps 1b and 3a), the entire flowrate of its pinch-causing source will be extracted as a waste stream in the IPRCN limiting data.*

    ii. *Identify the assisted stream and its flowrate needed to supplement the flowrate deficit in the RCN where the entire pinch-causing source flowrate is extracted. The assisted stream should originate from the pinch-causing source of the overall IPRCN, which has been identified in Step 2.*

    iii. *Revise individual wastewater stream flowrates identified in Step 2, for RCNs in which their entire pinch-causing source flowrate is extracted as a waste stream in the IPRCN (substeps (i)) and those that involve the use of an assisted stream (substeps (ii)).*

    iv. *Determine the minimum flowrate targets for IPRCN with the revised limiting data.*

      Both Steps 1b and 3a identify that the overall IPRCN pinch concentration falls at 150 ppm (see Tables 10.7 and 10.10). From Step 1, it is observed that Network A is the only RCN with higher quality pinch i.e., 100 ppm. Hence, the pinch-causing sources of this RCN, i.e., SR1 and SR3, are fully extracted as

TABLE 10.10

Targeting for IPRCN (Unassisted Integration Scheme)

C_k (ppm)	$\Sigma_j F_{SKj}$ (t/h)	$\Sigma_i F_{SRi}$ (t/h)	$\Sigma_i F_{SRi} - \Sigma_j F_{SKj}$ (t/h)	$F_{C,k}$ (t/h)	Δm_k (kg/h)	Cum. Δm_k (kg/h)
				$F_{FW} = 316.26$		
0	60.00		−60.00			
				256.26	6.41	
25	130.00		−130.00			6.41
				126.26	3.16	
50	273.34	80.00	−193.34			9.56
				−67.07	−2.01	
80	57.30	133.33	76.04			7.55
				8.97	0.18	
100	342.86	154.28	−188.58			7.73
				−179.61	−4.49	
125		50.00	50.00			3.24
				$F_{HQR} = -129.61$	−3.24	
150		300.00	300.00			**0.00**
				$F_{LQR} = 170.39$	42.60	**(PINCH)**
400		8.96	8.96			42.60
				179.35	71.74	
800		130.24	130.24			114.34
				309.59	61.92	
1,000		6.67	6.67			176.26
				$F_{WW} = 316.26$	315,939.86	
1,000,000						316,116.11

limiting data for IPRCN. However, in Table 10.8, it is shown that 5.71 t/h of this pinch-causing source is used as the minimum pinch flowrate for the LQR of Network A (while 50.95 t/h are emitted as wastewater stream WW1). We now extract the minimum pinch flowrate as part of wastewater stream WW1, which increases the flowrate of the latter (F_{WW1}) to 56.67 t/h (=5.71 + 50.95 t/h). However, doing this leads to flowrate deficit in the LQR of Network A, as the minimum pinch flowrate for the LRQ of Network A is no longer available. Hence, an assisted stream will now be identified to supplement the flowrate requirement of LQR in Network A. Note that the assisted stream should originate from the pinch-causing source of the overall IPRCN and should have a flowrate that is lower than the allocated flowrate to the LQR of the overall IPRCN.

Since the overall IPRCN pinch concentration falls at 150 ppm, the source candidate that can be used as the assisted stream is identified as SR15 from Network C. In other words, SR15 will replace the original pinch-causing source in Network A (SR1 and SR3) in order to fulfill the flowrate requirement of LQR in

TABLE 10.11

Targeting for Minimum Flowrate for Assisted Stream

C_k (ppm)	$\Sigma_j F_{SKj}$ (t/h)	$\Sigma_i F_{SRi}$ (t/h)	$\Sigma_i F_{SRi} - \Sigma_j F_{SKj}$ (t/h)	$F_{C,k}$ (t/h)	Δm_k (kg/h)	Cum. Δm_k (kg/h)
				$F_{AS} = 6.15$		
150			0.00			
				6.15	1,538.46	
400	10.00		−10.00			15,38.46
				−3.85	−1,538.46	
800		51.67	51.67			0.00
				$F_{WW2} = 47.82$	47,782,256.41	(PINCH)
1,000,000						47,782,256.41

Network A. Since the assisted stream flowrate should be lower than the allocated flowrate of SR15 to the LQR of the overall IPRCN, this means that it has a maximum bound of 174.21 t/h (F_{LQR}; see Table 10.10).

We next perform the targeting step to identify the minimum flowrate of the assisted stream, as well as the newly emitted wastewater streams in the LQR of Network A (due to the use of assisted stream of different quality). The cascade table for this step is shown in Table 10.11, resembling that for wastewater identification step (Table 10.8). As shown, the minimum flowrate requirement for the assisted stream (F_{AS}) is identified as 6.15 t/h, while the wastewater stream (F_{WW2}) from this region has a revised flowrate of 47.82 t/h (note that the original flowrate was 47.38 t/h, when 100 ppm pinch-causing source is available; see Table 10.9). The revised wastewater flowrates of WW1 and WW2 are given in Table 10.12.

On the other hand, since 6.15 t/h of SR15 has been utilized as an assisted stream in Network A, its availability in Network C has been reduced. We may assume that this flowrate is originated from wastewater stream WW6 (which is part of source SR15 originally). Hence, the flowrate of wastewater stream WW6 is reduced to 140.52 t/h (=146.67 − 6.15 t/h). Note that the flowrate of wastewater stream WW7 at 800 ppm remains identical to that in Table 10.9.

The revised limiting data for IPRCN are now summarized in Table 10.12. The data that have been revised are highlighted in bold case, to differentiate them from those given in Table 10.9. Next, flowrate targeting is conducted for IPRCN, with the revised limiting data. As shown in Table 10.13, the minimum freshwater (F_{FW}) and wastewater (F_{WW}) flowrates are determined as 314.36 t/h, which are identical to those of the single network targets in Step 1 (Table 10.7). Note that the pinch concentration is also identical to the earlier case, i.e., 150 ppm.

TABLE 10.12

Limiting Water Data for Assisted Integration Scheme: (a) Sinks in the HQR, (b) Sources in HQR, and (c) Individual Wastewater Streams

(a)			(b)			(c)		
Sinks, SK$_j$	F_{SKj} (t/h)	C_{SKj} (ppm)	Sources, SR$_i$	F_{SRi} (t/h)	C_{SRi} (ppm)	Wastewater, WWy	F_{WWy} (t/h)	C_{WWy} (ppm)
Network A								
SK1	20.00	0	SR2	66.67	80	**WW1**	**56.67**	**100**
SK2	66.67	50	SR3	63.33	100	WW2	47.82	800
SK3	100.00	50						
SK4	41.67	80						
Network B								
SK6	20.00	0	SR6	20.00	100	WW3	5.12	400
SK7	66.67	50	SR7	66.67	80	WW4	42.86	800
SK8	15.63	80	SR8	3.84	400	WW5	6.67	1000
SK9	42.86	100						
Network C								
SK11	20.00	0	SR11	20.00	100	**WW6**	**140.52**	**150**
SK12	80.00	25	SR12	80.00	50	WW7	40.00	800
SK13	50.00	25	SR13	50.00	125			
SK14	40.00	50	SR15	153.33	150			
SK15	300.00	100		523.84				
$\Sigma_j F_{SKj}$	863.50		$\Sigma_i F_{SRi}$			$\Sigma_y F_{WWy}$	339.65	

4. *Flowrate targeting for individual RCNs, with the following detailed steps:*

   a. *Identify the pinch region, which is defined by the highest and lowest quality pinches among all RCNs.*

   b. *Individual waste stream that exists within this pinch region is identified as the potential cross-plant stream.*

   c. *Flowrate targeting for individual RCNs, inclusive of cross-plant stream(s). Perform the targeting step from RCNs with highest to lowest quality pinches.*

The pinch region for the IPRCN is defined by the pinch concentrations of Networks A and B, i.e., 100 and 400 ppm. From Table 10.12, we observe that only wastewater streams WW1 (100 ppm), WW3 (150 ppm), and WW6 (150 ppm) are found within the pinch region, which can be used as cross-plant streams.

We next perform flowrate targeting for each individual RCNs, starting from Network A with highest quality pinch (100 ppm), before moving to Network C (150 ppm) and Network B (400 ppm) with higher pinch concentration (lower quality). The cascade table for Network A is shown in Table 10.14. As shown,

TABLE 10.13

Targeting for IPRCN (Assisted Integration Scheme)

C_k (ppm)	$\Sigma_j F_{SKj}$ (t/h)	$\Sigma_i F_{SRi}$ (t/h)	$\Sigma_i F_{SRi} - \Sigma_j F_{SKj}$ (t/h)	$F_{C,k}$ (t/h)	Δm_k (kg/h)	Cum. Δm_k (kg/h)
				$F_{FW} = 314.36$		
0	60.00		−60.00			
				254.36	6.36	
25	130.00		−130.00			6.36
				124.36	3.11	
50	273.34	80.00	−193.34			9.47
				−68.98	−2.07	
80	57.30	133.34	76.04			7.40
				7.06	0.14	
100	342.86	160.00	−182.86			7.54
				−175.80	−4.39	
125		50.00	50.00			3.14
				−125.80	−3.14	
150		293.85	293.85			0.00
				168.05	42.01	(PINCH)
400		8.96	8.96			42.01
				177.01	70.80	
800		130.68	130.68			112.82
				307.69	61.54	
1,000		6.67	6.67			174.36
				$F_{WW} = 314.36$	314,046.97	
1,000,000						314,221.33

this RCN requires freshwater flowrate (F_{FW}) of 98.33 t/h. Step 3b earlier indicates that this RCN receives an assisted stream from Network C ($F_{AS} = 6.15$) and generates two wastewater streams at 100 ppm (56.67 t/h) and 800 ppm (47.82 t/h). However, only the 100 ppm wastewater can be used as a cross-plant stream, as it exists within the pinch region. Note also that these wastewater streams will be determined at a later stage when the exact flowrate of the cross-plant stream is known. We next move to Network C where this cross-plant stream is used.

In Network C, the cross-plant stream from Network A should be utilized to reduce the freshwater flowrate. The targeting step resembles that for RCN with multiple fresh resources;* hence, only a brief description is given here. Table 10.15 shows the targeting results for Network C. Note that the flowrate of the 150 ppm source has been reduced to 293.85 t/h (from 300 t/h), as 6.15 t/h of this source has been utilized as assisted stream for Network A.

---

* See Section 4.2 for detailed steps and example.

TABLE 10.14

Targeting for Network A: Flowrate Targeting with Assisted Stream from Network C (Wastewater Flowrate is Yet to Finalize)

C_k (ppm)	$\Sigma_j F_{SKj}$ (t/h)	$\Sigma_i F_{SRi}$ (t/h)	$\Sigma_i F_{SRi} - \Sigma_j F_{SKj}$ (t/h)	$F_{C,k}$ (t/h)	Δm_k (kg/h)	Cum. Δm_k (kg/h)
				$F_{FW} = 98.33$		
0	20.00		−20.00			
				78.33	3.92	
50	166.67		−166.67			3.92
				−88.33	−2.65	
80	41.67	66.67	25.00			1.27
				−63.33	−1.27	
100		120.00	120.00			0.00
				56.67	2.83	(PINCH)
150		$F_{AS} = 6.15$	6.15			2.83
				62.82	15.70	
400	10.00		−10.00			18.54
				52.82	21.13	
800		51.67	51.67			39.66
				$F_{WW} = 104.48$	104,399.75	
1,000,000						104,439.41

TABLE 10.15

Targeting for Network C: Flowrate Targeting with Cross-Plant Stream from Network A (Wastewater Flowrate is Yet to Finalize)

C_k (ppm)	$\Sigma_j F_{SKj}$ (t/h)	$\Sigma_i F_{SRi}$ (t/h)	$\Sigma_i F_{SRi} - \Sigma_j F_{SKj}$ (t/h)	$F_{C,k}$ (t/h)	Δm_k (kg/h)	Cum. Δm_k (kg/h)
				$F_{FW} = 167.78$		
0	20.00		−20.00			
				147.78	3.69	
25	130.00		−130.00			3.69
				17.78	0.44	
50	40.00	80.00	40.00			4.14
				57.78	2.89	
100	300.00	20.00 + **56.67** (F_{CP})	−223.34			7.03
				−165.56	−4.14	
125		50.00	50.00			2.89
				−115.56	−2.89	
150		293.85 (= 300 − 6.15)	293.85			0.00
				178.29	115.89	(PINCH)
800		40.00	40.00			115.89
				$F_{WW} = 218.29$	218,117.43	
1,000,000						218,233.32

TABLE 10.16

Targeting for Network A: Revised Cascade with Cross-Plant Stream Sent to Network C (To Be Carried Out after Cross-Plant Stream Flowrate to Network C Is Known)

C_k (ppm)	$\Sigma_j F_{SKj}$ (t/h)	$\Sigma_i F_{SRi}$ (t/h)	$\Sigma_i F_{SRi} - \Sigma_j F_{SKj}$ (t/h)	$F_{C,k}$ (t/h)	Δm_k (kg/h)	Cum. Δm_k (kg/h)
				$F_{FW} = 98.33$		
0	20.00		−20.00			
				78.33	3.92	
50	166.67		−166.67			3.92
				−88.33	−2.65	
80	41.67	66.67	25.00			1.27
				−63.33	−1.27	
100		63.34	63.34			**0.00**
		(=120 − 56.67)		0.00	0.00	**(PINCH)**
150		$F_{AS} = 6.15$	6.15			0.00
				6.15	1.54	
400	10.00		−10.00			1.54
				−3.85	−1.54	
800		51.67	51.67			0.00
				$F_{WW} = 47.82$	47,781.98	
1,000,000						47,781.98

It is determined that the cross-plant stream (56.67 t/h) from Network A can be fully utilized in Network C; hence, the freshwater flowrate (F_{FW}) in Network C is reduced to 167.78 t/h (from 186.67 t/h prior to IPRCN; see Table 10.6). An inspection of its LQR indicates that two wastewater streams are emitted from this RCN in the absence of water sink, i.e., 184.44 t/h at 150 ppm and 40 t/h at 800 ppm levels. The former can be used as a cross-plant stream for Network B as it exists within the pinch region. Similar to the case of Network A, the wastewater stream flowrates of Network C will be revised once the exact cross-plant stream flowrate from this network (to Network B) is known.

Note that since the cross-plant stream from Network A is fully consumed in network C, we should also revise the wastewater stream for the former. Since the cross-plant stream is originated from the source of 100 ppm; hence, the source flowrates at this level (i.e., SR1 and SR3) are reduced to 63.33 t/h (=120 − 56.67 t/h). The targeting result is shown in the revised cascade in Table 10.16, which indicates a reduced wastewater flowrate of 47.82 t/h (from 104.48 t/h; see Table 10.14).

Finally, in Network B, we first determine that only 10.24 t/h of cross-plant stream (F_{CP}) from Network C is needed. This

TABLE 10.17

Targeting for Network B

C_k (ppm)	$\Sigma_j F_{SKj}$ (t/h)	$\Sigma_i F_{SRi}$ (t/h)	$\Sigma_i F_{SRi} - \Sigma_j F_{SKj}$ (t/h)	$F_{C,k}$ (t/h)	Δm_k (kg/h)	Cum. Δm_k (kg/h)
				$F_{FW} = 48.24$		
0	20.00		−20.00			
				28.24	1.41	
50	66.67		−66.67			1.41
				−38.42	−1.15	
80	15.63	66.67	51.04			0.26
				12.62	0.25	
100	42.86	20.00	−22.86			0.51
				−10.24	−0.51	
150		$F_{CP} = 10.24$	10.24			0.00
				0.00	0.00	
400	6.67	15.63	8.96			**0.00**
				8.96	3.58	**(PINCH)**
800		42.86	42.86			3.58
				51.82	10.36	
1,000		6.67	6.67			13.95
				$F_{WW} = 58.48$	58,423.66	
1,000,000			0.00			58,437.61

reduces the freshwater and wastewater flowrates in Network B to 48.24 and 58.48 t/h, respectively (Table 10.17). On the other hand, the source flowrate at 150 ppm in Network C (where the cross-plant stream is originated) is also revised. This leads to reduced wastewater flowrate, as reflected in Table 10.18.

Figure 10.10 shows a simplified cascade diagram for all networks in the IPRCN (only water cascades are shown; while load cascades are omitted). As shown, all cross-plant and assisted streams are found within the pinch region (100–400 ppm). Two cross-plant streams are observed to flow from networks with higher quality pinch to those with lower quality pinch (refer to RCNs before IPRCN analysis; see Tables 10.4 through 10.6), i.e., Network A to Network C and Network C to Network B. In contrast, the assisted stream flows in a reverse direction, i.e., from Network A to Network C. Note that since the assisted stream is exported from the LQR of Network C, it is termed as the *low quality assisted stream*, following the work of Chew et al. (2010b). We can easily determine that the overall flowrates of freshwater (98.33 + 167.78 + 48.24 = 314.36 t/h) and wastewater (47.82 + 208.05 255 + 58.48 = 314.36 t/h) across three individual Networks A–C match the targets identified earlier in both Steps 1 and 3b.

TABLE 10.18

Targeting for Network C: Revised Cascade with Cross-Plant Stream Sent to Network B (To Be Carried Out after Cross-Plant Stream Flowrate to Network C Is Known)

C_k (ppm)	$\Sigma_j F_{SKj}$ (t/h)	$\Sigma_i F_{SRi}$ (t/h)	$\Sigma_i F_{SRi} - \Sigma_j F_{SKj}$ (t/h)	$F_{C,k}$ (t/h)	Δm_k (kg/h)	Cum. Δm_k (kg/h)
				$F_{FW} = 167.78$		
0	20.00		−20.00			
				147.78	3.69	
25	130.00		−130.00			3.69
				17.78	0.44	
50	40.00	80.00	40.00			4.14
				57.78	2.89	
100	300.00	20.00 + 56.67	−223.34			7.03
		(F_{CP})		−165.56	−4.14	
125		50.00	50.00			2.89
				−115.56	−2.89	
150		283.61	283.61			0.00
		(=293.85 − 10.24)		168.05	109.24	**(PINCH)**
800		40.00	40.00			109.24
				$F_{WW} = 208.05$	207,887.53	
1,000,000						207,996.76

10.3 Design of IPRCN

Apart from determining the various network targets, the IPRCN that achieves those targets is to be designed. The *nearest neighbor algorithm* (NNA) and *matching matrix* introduced in Chapter 7 (RCNs for continuous mode) are good tools for use in designing the IPRCN. A generic form of the matching matrix for IPRCN is shown in Figure 10.11. As shown, all sources (including fresh resource—FR) are arranged as rows, while sinks (including waste discharge—WD) are arranged as columns and in descending order of quality index for each respective RCN. Note that a cross-plant stream (CP) will act as a sink in an RCN where it is to be exported and as source where it is imported. Values in the cells indicate the allocation flowrates ($F_{SRi,SKj,p}$) of sink–source matches in RCN p, while those in the parenthesis indicate the cross-plant stream flowrates between different RCNs. Note however that no matches should be formed among fresh resource, cross-plant stream, and with waste discharge. Besides, the shaded area in the matching matrix indicates that sink–source matches in different RCNs are forbidden. In other words, the recovery among sinks and sources of different RCNs can only be carried out via cross-plant streams.

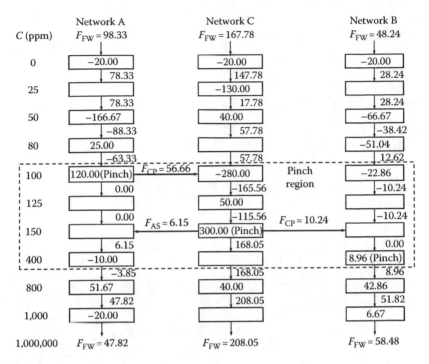

FIGURE 10.10
Simplified cascade diagrams for individual RCNs within the IPRCN (all flowrate terms in t/h).

Example 10.3 Design of inter-plant water network

Design an inter-plant water network for the water minimization problem in Example 10.1. Make use of the various network targets obtained through targeting stage.

Solution

The first step in the design tasks is to locate the various identified IPRCN targets for use in conjunction with the NNA and matching matrix. The latter have been described in detail in Chapter 7 (see Section 7.1 and Example 7.1 for detailed illustration); hence, a brief illustration is given here. Based on the results in Example 10.1, the targeted flowrates for freshwater, wastewater, and cross-plant stream are first located in the matching matrix (see Figure 10.12). For instance, Figures 10.8 and 10.9 indicate that the freshwater flowrates for Networks A and B are 90 and 65 t/h, while their wastewater flowrates are identified as 75 and 60 t/h, respectively. These flowrate targets are embedded in the matching matrix in Figure 10.12. Note however that at this stage, only the wastewater flowrates are known, while their concentration levels remain as unknown. Besides, the targeting stage also identifies that 15 t/h of cross-plant stream is exported from Network A. Since this stream has a concentration of 100 ppm, it may be originated from SR1 or SR2 (given as

RCNp	F_{SRip}	q_{SRip}	RCNp F_{SKip} q_{SKip}	$p=A$ $F_{SK1,A}$ $q_{SK1,A}$ SK1	$p=A$ $F_{CP,A}$ $q_{CP,A}$ CP	$p=A$ $F_{D,A}$ $q_{D,A}$ WD	$p=B$ $F_{SK2,B}$ $q_{SK2,B}$ SK2	$p=B$ $F_{D,B}$ $q_{D,B}$ WD	$p=C$ $F_{SK3,C}$ $q_{SK3,C}$ SK3	$p=C$ $F_{D,C}$ $q_{D,C}$ WD
$p=A$	$F_{R,A}$ $q_{R,A}$		FR	$F_{R,SK1A}$						
	$F_{SR1,A}$ $q_{SR1,A}$		SR1	$F_{SR1,SK1A}$	$(F_{SR1,CPA})$	$F_{SR1,WDA}$				
	$F_{SR2,A}$ $q_{SR2,A}$		SR2	$F_{SR2,SK1A}$	$(F_{SR2,CPA})$	$F_{SR2,WDA}$				
$p=B$	$F_{R,B}$ $q_{R,B}$		FR				$F_{R,SK2,B}$			
	F_{CPB} q_{CPB}		CP				$(F_{CPSK2,B})$			
	$F_{SR3,B}$ $q_{SR3,B}$		SR3				$F_{SR3,SK2,B}$	$F_{SR3,WD,B}$		
$p=C$	$F_{R,C}$ $q_{R,C}$		FR						$F_{R,SK3,C}$	
	F_{CPC} q_{CPC}		CP						$(F_{CPSK3,C})$	
	$F_{SR4,C}$ $q_{SR4,C}$		SR4						$F_{SR4,SK3,C}$	$F_{SR4,WD,C}$

FIGURE 10.11
Matching matrix for an IPRCN.

			Network A						Network B				
	$F_{SKj,p}$		20	100	40	15	10	75	50	100	80	70	60
	$C_{SKj,p}$		0	50	50	100	400	WW	20	50	100	200	WW
			SK1	SK2	SK3	CP	SK4	WW	SK5	SK6	SK7	SK8	WW
RCNp	$F_{SRi,p}$	$C_{SRi,p}$											
Network A	90	0	FW										
	20	100	SR1			(15)							
	100	100	SR2										
	40	800	SR3										
	10	800	SR4										
Network B	65	0	FW										
	50	50	SR5										
	15	100	CP										
	100	100	SR6										
	70	150	SR7										
	60	250	SR8										

FIGURE 10.12
Locating flowrate targets for each RCN (flowrate terms in t/h; concentrations in ppm).

value in parenthesis in Figure 10.12). However, the sink(s) in Network B to be matched with the cross-plant stream will remain as unknown at this stage.

We first determine the flowrate allocation for in-plant water recovery for Network A. For SK1, since it has a limiting concentration of 0 ppm, it is fed with freshwater. Next, both SK2 and SK3 have limiting concentrations of 50 ppm, and they are fed with freshwater, SR1 and SR2 (both with 100 ppm). Equations 7.1 and 7.2 are used to determine the allocated flowrates and are summarized in Figure 10.13 (detailed calculations are omitted). Note that 15 t/h of SR1 has been utilized as a cross-plant stream earlier; hence, its flowrate for in-plant recovery is reduced to 5 t/h. Next, Equations 7.1 and 7.2 are also used to determine the allocated flowrates between SR2, SR3, and SK4. The remaining sources are then discharged as wastewater streams.

Similar steps are followed to determine the allocated flowrates for all sink–source matches in Network B. Apart from in-plant water recovery, 15 t/h of cross-plant stream is also available for use (sent from Network A). For sinks SK5, SK6, and SK8, their allocated flowrates from the sources are essentially the same as that with in-plant recovery with a direct reuse/ recycle scheme (see Example 7.1 for detailed illustration). On the other hand, SK7 is fed with SR6 and a cross-plant stream, which possess equal concentration levels. Finally, the concentrations of the wastewater streams are determined from the mean concentration of the contributed water sources. A graphical representation of the IPRCN is given in Figure 10.14. Since sinks and sources of different RCNs are integrated directly (via cross-plant stream), this is termed as *direct integration* scheme.

10.4 IPRCN with Material Regeneration and Waste Treatment

Targeting procedures for material regeneration and waste treatment for an RCN have been discussed in Chapters 6 and 8, respectively. They are now extended for IPRCN synthesis. Hence, only a brief discussion is given here; readers may refer to previous chapters for detailed steps. Note that in carrying out the targeting steps, we should make use of the limiting data identified for IPRCN flowrate targeting, rather than the original limiting data of the individual RCN (this includes the limiting data for the assisted integration scheme).

For a direct integration case, the interception and waste treatment units are located in individual RCNs. Hence, targeting and synthesis steps are carried out for individual RCNs (with limiting data identified for IPRCN flowrate targeting). Apart from the direct integration scheme, there are also IPRCNs with an *indirect integration* scheme, where sinks and sources of different RCNs are integrated via a *centralized utility facility* (CUF). The latter are commonly used in an *eco-industrial park* where resources of different process plants are shared. In these cases, direct connections are only found for in-plant recovery, i.e.,

RCNp	F_{SRip}	C_{SRip}	F_{SKjp} / C_{SKjp}	Network A						Network B				
			F_{SKjp}	20	100	40	15	10	75	50	100	80	70	60
			C_{SKjp}	0	50	50	100	400	526.63	20	50	100	200	250
				SK1	SK2	SK3	CP	SK4	WW	SK5	SK6	SK7	SK8	WW
Network A	90	0	FW	20	50	20								
	20	100	SR1		5		(15)							
	100	100	SR2		45	20								
	40	800	SR3					5.71	29.29					
	10	800	SR4					4.29	35.71					
									10					
Network B	65	0	FW							30	35			
	50	50	SR5							20	30			
	100	100	SR6								35	65		
	15	100	CP									(15)		
	70	150	SR7										35	35
	60	250	SR8										35	25

FIGURE 10.13

Matching matrix for IPRCN in Example 10.3 (flowrates in t/h; concentrations in ppm). (From Chew, I.M.L. et al., *Ind. Eng. Chem. Res.*, 49(14), 6439, 2010a. With permission.)

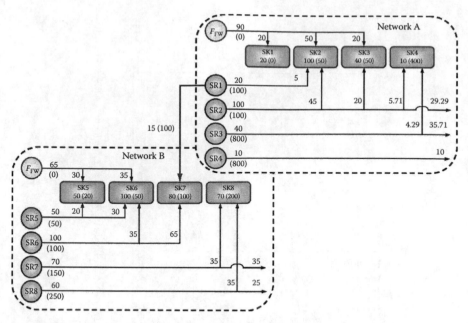

FIGURE 10.14
IPRCN in Example 10.3 (flowrate in t/h: concentration in ppm given in parenthesis).

among sinks and sources within the same RCN but not among those of different RCNs. In most cases, the CUF supplies fresh resources for all RCNs and also provides services to purify material sources for further recovery and to treat waste streams prior to environmental discharge.

 Example 10.4 Targeting for source regeneration and waste treatment for an inter-plant water network

Revisit the inter-plant water network problem in Example 10.1. A CUF is available for use for the IPRCN. The CUF supplies freshwater to both networks and consists of an interception unit for water regeneration; as well as a treatment system that purifies wastewater streams before they are being discharged to the environment.* Make use of the IPRCN limiting data in Table 10.2. Determine the following network targets in sequence:

1. Determine the *ultimate flowrate targets* (i.e., minimum freshwater, wastewater, and regenerated water flowrates) of the IPRCN, assuming that the interception unit has an outlet concentration (C_{Rout}) of 30 ppm. Identify also the individual wastewater streams that will emit from the IPRCN.
2. Determine the minimum treatment flowrate needed for the wastewater streams. The wastewater treatment system has a

---

* This is conceptually similar to the *total material network* as discussed in Chapter 8 for single RCN.

removal ratio (RR) of 0.95. The environmental discharge limit is given as 100 ppm.

Solution

1. Ultimate flowrate targets with interception unit of $C_{Rout} = 30$ ppm

Step 2 of the algebraic targeting procedure (see Section 6.2 for details) indicates that the overall IPRCN is divided into *freshwater region* (FWR) and *regenerated water region* (RWR), with concentrations lower and higher than the outlet concentration of the regeneration unit (30 ppm), respectively. All water sinks and sources are then allocated to these regions according to their respective concentrations, i.e., SK1 and SK5 to FWR, while all other sinks and sources are allocated to RWR. Next, it is then determined that the total flowrate of the water sources is higher than that of the sinks in the RWR. Hence, Step 3b is followed to reallocate 50 l/h of water sources at concentration levels of 250 ppm (4.29 t/h) and 800 ppm (45.71 t/h) from RWR to the FWR, as shown in Table 10.19(a) and (b) respectively.

Since a water sink exists in the FWR at 20 ppm, it may be partially fed with the regenerated water of 30 ppm. Flowrate of the latter is then determined by Equation 6.3 as 33.33 t/h ($= 50$ t/h $\times 20$ ppm/30 ppm) (Step 4b). This regenerated flowrate may be fed by the 800 ppm wastewater source (WW2, as indicated by Table 10.2 (c)) found in the FWR. We then proceed to determine the flowrate targets in both FRW and RWR, respectively (Steps 5 and 6), with the results shown in Tables 10.20 and 10.21. The ultimate flow targets for the entire IPRCN are determined as 36.67 t/h of freshwater (F_{FW}), 16.67 t/h of wastewater (F_{WW}), and 169.05 t/h regenerated water ($F_{RW} = F_{RW,FWR} + F_{RW,RWR} = 33.33 + 135.71$ t/h). A combined cascade for the targeting results is shown in Table 10.22.

From Tables 10.20 through 10.22, we can identify that the total wastewater flowrate of the IPRCN ($F_{WW} = 16.67$ t/h) is

TABLE 10.19

Results of Flowrate Reallocation

(a) FWR	C_k (ppm)	$\Sigma_j F_{SKj}$ (t/h)	$\Sigma_i F_{SRi}$ (t/h)	(b) RWR	C_k (ppm)	$\Sigma_j F_{SKj}$ (t/h)	$\Sigma_i F_{SRi}$ (t/h)
	0	20			0		
	20	50			20		
	50				50	240	50
	100				100	80	214.29
	150				150		35
	250		4.29		250		20.71
	800		45.71		800		
	1,000,000				1,000,000		
	Total	70	50		Total	320	320

TABLE 10.20

Targeting for Regeneration: FWR

C_k (ppm)	$\Sigma_j F_{SKj}$ (t/h)	$\Sigma_i F_{SRi}$ (t/h)	$\Sigma_i F_{SRi} - \Sigma_j F_{SKj}$ (t/h)	$F_{C,k}$ (t/h)	Δm_k (kg/h)	Cum. Δm_k (kg/h)
				$F_{FW} = 36.67$		
0	20		−20			
				16.67	0.33	
20	50		−50			0.33
				−33.33	−0.33	
$C_{Rout} =$ 30		$F_{RW,FWR} = 33.33$	33.33			0.00
				0.00	0.00	
250		4.29	4.29			0.00
				4.29	2.36	(PINCH)
800		12.38	12.38			2.36
				$F_{WW} = 16.67$	16,653.33	
1,000,000						16,655.69

TABLE 10.21

Targeting for Regeneration: RWR

C_k (ppm)	$\Sigma_j F_{SKj}$ (t/h)	$\Sigma_i F_{SRi}$ (t/h)	$\Sigma_i F_{SRi} - \Sigma_j F_{SKj}$ (t/h)	$F_{C,k}$ (t/h)	Δm_k (kg/h)	Cum. Δm_k (kg/h)
				$F_{RW,RWR} = 135.71$		
$C_{Rout} = 30$						
				135.71	2.71	
50	240	50	−190			2.71
				−54.29	−2.71	
100	80	214.29	134.29			0.00
				80.00	4.00	(PINCH)
150		35	35			4.00
				115.00	11.50	
250		20.71	20.71			15.50
				$F_{RW,RWR} = 135.71$	135,680.36	
1,000,000						135,695.86

contributed by two individual wastewater streams, emitted from the lower quality region below the pinch (secondary pinch for Table 10.22), i.e., 4.29 t/h (WW1, 250 ppm) and 12.38 t/h (WW2, 800 ppm), respectively. An inspection on Table 10.1 indicates that these wastewater streams are orginated from SR8 (250 ppm), as well as SR3 and/or SR4 (250 ppm).

2. Minimum treatment flowrate with a treatment system of RR = 0.95

In an earlier case, we identified that two individual wastewater streams emit from the IPRCN. We next utilize the *waste treatment pinch diagram* (WTPD) to determine the minimum

TABLE 10.22

Combined Cascade for Regeneration Targeting

C_k (ppm)	$\Sigma_j F_{SKj}$ (t/h)	$\Sigma_i F_{SRi}$ (t/h)	$\Sigma_i F_{SRi} - \Sigma_j F_{SKj}$ (t/h)	$F_{C,k}$ (t/h)	Δm_k (kg/h)	Cum. Δm_k (kg/h)
				$F_{FW} = 36.67$		
0	20		−20			
				16.67	0.33	
20	50		−50			0.33
				−33.33	−0.33	
$C_{Rout} =$ 30		$F_{RW} = 169.05$	169.05			0.00
				135.71	2.71	(Limiting pinch)
50	240	50	−190			2.71
				−54.29	−2.71	
100	80	134.29	54.29			0.00
				0.00	0.00	
250		4.29	4.29			0.00
				4.29	2.36	(Secondary pinch)
800		12.38	12.38			2.36
				$F_{WW} = 16.67$	16,653.33	
1,000,000						16,655.69

wastewater treatment flowrate (see Section 8.4 for detailed procedure). To construct the WTPD, the individual wastewater streams are arranged in ascending order of impurity concentration (descending quality level). Table 10.23 shows their cumulative flowrate and impurity load values for plotting the WTPD. As shown, a total of 16.67 t/h of wastewater is to be treated, with a total load of 10.98 kg/h.

The WTPD is next plotted on a load-versus-flowrate diagram, as shown in Figure 10.15. For an environmental discharge limit (C_D) of 100 ppm, the maximum allowable discharge load is calculated as 1.67 kg/h (=16.67 t/h × 100 ppm). Hence, a total impurity load of 9.31 kg/h ($\Delta m_R = 10.98 - 1.67$ kg/h) is to be removed from the wastewater streams. For a waste treatment unit with RR=0.95, the total treatment feed load (m_F) is determined as 9.80 kg/h (=9.31/0.95 kg/h, Equation 8.2). Next, the minimum

TABLE 10.23

Cumulative Flowrate and Load for Plotting the WTPD

Wastewater, WWy	F_{WWy} (t/h)	C_{WWy} (ppm)	m_{WWy} (kg/h)	Cum. F_{WWy} (t/h)	Cum. m_{WWy} (kg/h)
WW1	4.29	250	1.07	4.29	1.07
WW2	12.38	800	9.90	16.67	10.98

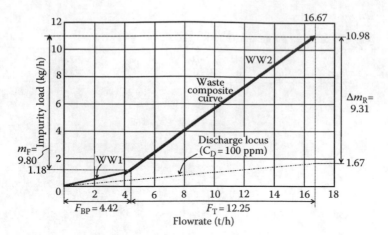

FIGURE 10.15
Targeting for minimum waste treatment flowrate for IPRCN in Example 10.4. (From Chew, I.M.L. et al., *Ind. Eng. Chem. Res.*, 49(14), 6439, 2010a. With permission.)

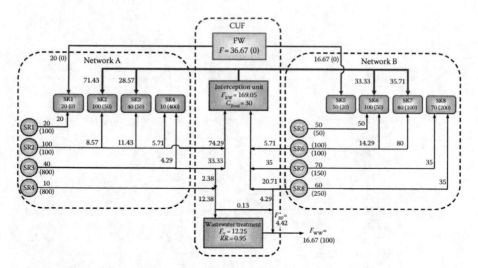

FIGURE 10.16
IPRCN with material regeneration and waste treatment (in Example 10.4) (flowrate in t/h; concentration in ppm given in parenthesis).

treatment (F_T) and bypass (F_{BP}) flowrates are determined as 12.25 and 4.42 t/h, respectively, as shown in the WTPD in Figure 10.15.

An IPRCN that achieves the identified network targets is designed using the NNA (detailed steps are omitted) and is shown in Figure 10.16. From Figure 10.16, we observe that apart from in-plant water recovery, water sources are also recovered via the CUF (after purification). Hence, this is classified as an indirect integration scheme as discussed earlier.

10.5 Additional Readings

In this chapter, we have only discussed a type of assisted integration scheme, i.e., with low quality assisted stream. There are also cases where *higher quality assisted stream(s)* are needed. The latter normally involves RCNs with threshold problems. Besides, there are also cases where new pinch quality emerges with the use of cross-plant stream(s). Hence, the targeting techniques discussed earlier are to be revised to cater for both cases. Readers may refer to the work of Chew et al. (2010a,b) for the revised targeting procedure for both cases.

Problems

Water Minimization Problems

10.1 Water minimization study is carried out in a catalyst plant (Feng et al., 2006). The plant consists of two separate sections, i.e., synthesis and preparation departments. In the former, a large amount of sodium ion is generated. These ions are removed by a washing processes in the preparation department. Limiting water data for these sections is given in Tables 10.24 and 10.25. Note that sodium ion (Na^+) content is identified as the main impurity for water recovery (Feng et al., 2006). Freshwater supply for the plant is free of sodium ion. Perform the following tasks:

    a. Use MRPD or MCA to identify the minimum flowrate targets for freshwater and wastewater of the overall IRRCN, individual networks, and cross-plant stream(s) for a direct reuse/recycle scheme (direct integration).

    b. A regeneration unit (with $C_{Rout} = 30\,ppm$) is used to purify water sources for further reuse/recycle for both sections. Determine the ultimate flowrate targets (i.e., minimum freshwater, wastewater,

TABLE 10.24

Limiting Data for Problem 10.1: Synthesis Department

j	Sinks, SK_j	F_{SKj} (m³/d)	C_{SKj} (ppm)	i	Sources, SR_i	F_{SRi} (m³/d)	C_{SRi} (ppm)
1	Paste formation	34.6	1300	1	Gel recovery	28.3	4000
2	Ion exchange and filtration	48.2	1300	2	Washing and drying	2022.8	1300
3	Washing and drying	3114.5	800	3	Pump cooling	24.4	62
4	Pump cooling	24.4	60				

TABLE 10.25

Limiting Data for Problem 10.1: Preparation Department

j	Sinks, SK$_j$	F_{SKj} (m³/d)	C_{SKj} (ppm)	i	Sources, SR$_i$	F_{Sri} (m³/d)	C_{Sri} (ppm)
5	Paste formation	119.3	400	4	Pump cooling	525.4	62
6	Paste formation and filtration	271.7	400	5	Sedimentation	1380	500
7	Washing, filtration, and drying	1851	80				
8	Pump cooling	525.4	60				

and regenerated water) for the IPRCN. Is the interception unit be used in the CUF or the individual RCN?

c. Design an inter-plant water network that achieves the various identified network targets in tasks (a) and (b).

10.2 Revisit Example 10.2, and solve the following cases:

a. Perform the targeting steps for a direct reuse/recycle scheme using the graphical technique of MRPD (all flowrate targets are exactly the same as those in Example 10.2). Design an inter-plant water network that achieves the various identified network targets.

b. An interception unit with outlet concentration (C_{Rout}) of 10 ppm is used for water regeneration in the CUF. Identify the ultimate flowrate targets (i.e., minimum freshwater, wastewater, and regeneration flowrates) for the problem (note: make use of the limiting water data for assisted integration scheme in Table 10.12.

c. Apart from purifying water sources for recovery, the same interception unit in (b) is also being used as wastewater treatment unit (i.e., a total water network with single treatment unit), to ensure the wastewater streams in complying with environmental limit prior to discharge. The environmental discharge limit is given as 30 ppm. Determine the minimum wastewater treatment and bypass flowrates for this case, and design an IPRCN that achieves the network targets.

Utility Gas Recovery Problems

10.3 Assume that the two refinery hydrogen networks in Problems 4.16 (Network A) and 4.15 (Network B) form an IPRCN with direct integration scheme (with direct reuse/recycle of hydrogen sources). Use MCA to identify the minimum fresh hydrogen feed and purge gas flowrates for the overall IPRCN, individual networks, as well as the cross-plant stream flowrate(s). The limiting data for these networks are given in Tables 4.30 and 4.31. Fresh hydrogen feeds for the individual networks are available at 5% (Network A) and 0.01% (Network B) respectively.

Property Integration Problems

10.4 Inter-plant water recovery is to be carried out between two wafer fabrication plants (originated from Problems 2.7 and 3.11) in an eco-industrial park. To evaluate water recovery potential, the main property in concern is resistivity (R), which reflects the total ionic content in the aqueous streams. Limiting water data of both wafer fabrication plants are summarized in Tables 10.26 and 10.27. Ultrapure water (UPW; resistivity of $18\,M\Omega$ cm) is used to supplement process sources when they are insufficient for use in the sinks. The mixing rule for resistivity is given as follows (reproduced from Equation 2.64):

$$\frac{1}{R_M} = \sum_i \frac{x_i}{R_i} \qquad (10.1)$$

Solve the following cases:

a. Assuming that the CUF only supplies UPW for the wafer fabrication plants, use MRPD to identify the minimum flowrate targets for UPW and wastewater of the overall IRRCN, individual networks, as well as that of the cross-plant stream(s) for a direct reuse/recycle scheme (direct integration).

b. For case (a), two interception units are installed in the individual networks for water regeneration. The interception units possess different outlet resistivity values, i.e., $8\,M\Omega$ cm for that of Network

TABLE 10.26

Limiting Data for Problem 10.4: Network A

j	Sinks, SK_j	F_{SKj} (t/h)	Ψ_{SKj} ([MΩ m]$^{-1}$)	i	Sources, SR_i	F_{SRi} (t/h)	Ψ_{SRi} ([MΩ m]$^{-1}$)
1	Wet	500	0.143	1	Wet I	250	1
2	Lithography	450	0.125	2	Wet II	200	0.5
3	CMP	700	0.100	3	Lithography	350	0.333
4	Etc	350	0.200	4	CMP I	300	10
				5	CMP II	200	0.5
				6	Etc	280	2

TABLE 10.27

Limiting Data for Problem 10.4: Network B

j	Sinks, SK_j	F_{SKj} (t/h)	Ψ_{SKj} ([MΩ m]$^{-1}$)	i	Sources, SR_i	F_{SRi} (t/h)	Ψ_{SRi} ([MΩ m]$^{-1}$)
5	Wafer Fab	182	0.0625	7	50% Spent rinse	227.12	0.125
6	CMP	159	0.1	8	100% Spent rinse	227.12	0.5

A and $11\,M\Omega$ cm for that of Network B. Determine the ultimate flowrate targets (i.e., minimum freshwater, wastewater, and regenerated water) for the individual RCNs.

c. Instead of individual interception units (case (b)), a centralized interception unit with outlet resistivity value of $9\,M\Omega$ cm is installed in the CUF to purify water sources for further reuse/recycle. Determine the ultimate flowrate targets for the IPRCN.

d. Design inter-plant water networks that achieve the various identified network targets in tasks (a–c).

References

Bandyopadhyay, S., Sahu, G. C., Foo, D. C. Y., and Tan, R. R. 2010. Segregated targeting for multiple resources using decomposition principle. *AIChE Journal*, 56(5), 1235–1248.

Chew, I. M. L., Foo, D. C. Y., Ng, D. K. S., and Tan, R. R. 2010a. A new flowrate targeting algorithm for inter-plant resource conservation network. Part 1—Unassisted integration scheme. *Industrial and Engineering Chemistry Research*, 49(14), 6439–6455.

Chew, I. M. L., Foo, D. C. Y., and Tan, R. R. 2010b. A new flowrate targeting algorithm for inter-plant resource conservation network. Part 2—Assisted integration scheme. *Industrial and Engineering Chemistry Research*, 49(14), 6456–6468.

Feng, X., Wang, N., and Chen, E. 2006. Water system integration in a catalyst plant. *Transactions of the Institute of Chemical Engineers, Part A*. 84(A8), 645–651.

Olesen, S. G. and Polley, G. T. 1996. Dealing with plant geography and piping constraints in water network design. *Process Safety and Environmental Protection*, 74, 273–276.

11

Synthesis of Batch Material Networks

In previous chapters, we learnt how resource conservation is carried out for continuous processes. This chapter focuses on resource conservation networks (RCNs) for batch processes or in short *batch material networks* (BMNs). For a BMN, apart from quality constraint, time dimension is another constraint that needs to be considered in synthesizing a BMN. Hence, a mass storage system is commonly employed to recover material across time intervals. Various targeting and design techniques for continuous processes in the previous chapters are extended here for a BMN.

11.1 Types of Batch Resource Consumption Units

Before discussing the use of targeting and design techniques for a BMN, it is best to understand the characteristics of various batch resource consumption units. Figure 11.1 shows two typical types of batch operation mode. For Process 1 in Figure 11.1a, its actual operation takes place between the duration of t_2 and t_3. However, before the operation commences, fresh resource (e.g., water, gas) is charged to the operation within the duration of $t_1 - t_2$ (as sink). On the other hand, waste is discharged from the process (as source) at a later duration ($t_3 - t_4$), upon the completion of its operation. This type of operation is classified as the *truly batch* process (Gouws et al., 2010). A batch polymerization reaction is a typical example for this type of operation. In most cases, water (resource) is used as a carrier for the polymerization reaction. Freshwater is charged to the reactor before the reaction commences; and wastewater is discharged at the end of the polymerization reaction.

Another type of batch process mode is *semicontinuous* operation, as shown in Figure 11.1b. For this case, the charging of fresh resource to the process (sink) and its discharge of waste (source) are both carried out during the cause of operation, i.e., between t_5 and t_6. Vessel washing operation is a typical example of this type. During the washing operation, freshwater is charged to the vessel, while wastewater is withdrawn; both are carried out simultaneously. Note that some operations (e.g., washing) may either be operated in truly batch or semicontinuous operation mode, depending on the process requirement.

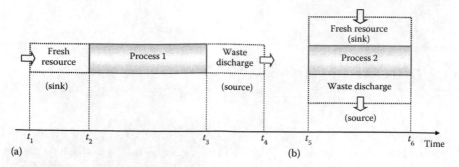

FIGURE 11.1
Types of batch resource consumption units: (a) truly batch and (b) semicontinuous operations. (Adapted with permission from Gouws, J., Majozi, T., Foo, D.C.Y., Chen, C.L., and Lee, J.-Y., Water minimisation techniques for batch processes, *Ind. Eng. Chem. Res.*, 49(19), 8877. Copyright 2010 American Chemical Society.)

Besides, one should also note that both processes in Figure 11.1 serve as process sink and source at the same time. There are also many cases where only process sink or source is found in the operations. Typical example for the former is a batch reactor where a particular resource (e.g., water, gas, etc.) is used as reactant. Being a limiting reactant, the resource is consumed completely upon the completion of the reaction. In other words, the reactor is treated as a process sink for the resource in concern. On the other hand, a chemical reactor may also serve as a process source that produces useful material. Typical example for this operation is fermentation processes, where a significant amount of water is produced. The graphical representation for these type of processes is shown in Figure 11.2 for truly batch or semicontinuous processes.

11.2 Targeting Procedure for Direct Reuse/Recycle in a BMN without Mass Storage System

Procedure for flow targeting in a BMN without mass storage system is given as follows (Foo et al., 2005):

1. Define the time intervals for the BMN based on start and end times of the process sinks and sources.
2. Locate the process sinks and sources in their respective time intervals and determine their *interval flows*, based on the ratio of the duration of the time interval to that of the sink/source.

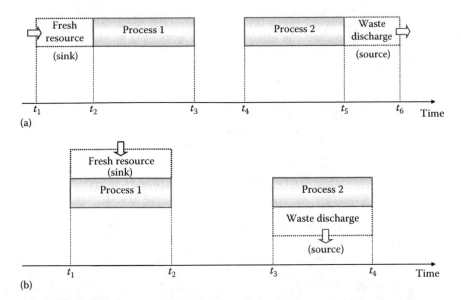

FIGURE 11.2

Batch resource consumption units with either process sink or source that operate in (a) truly batch and (b) semicontinuous mode. (Adapted with permission from Gouws, J., Majozi, T., Foo, D.C.Y., Chen, C.L., and Lee, J.-Y., Water minimisation techniques for batch processes, *Ind. Eng. Chem. Res.*, 49(19), 8877. Copyright 2010 American Chemical Society.)

3. Carry out flow targeting for each time interval to identify the minimum fresh resource and waste discharge, with either graphical or algebraic targeting technique. The summation of flows across all time intervals gives the overall flows for the entire BMN.

Example 11.1 Graphical targeting for a batch water network without a water storage tank

A classical *batch water network* example (Wang and Smith, 1995) is visited here. The network consists of three water-using processes, each consisting of a pair of water sink and source. The limiting water data of the processes are given in Figure 11.3.

Complete the following tasks:

1. Classify the three water-using processes. Are they truly batch or semicontinuous operations?
2. Identify the limiting data for water minimization study for the batch water network.
3. Use the *material recovery pinch diagram* (MRPD) in Chapter 3 to identify the overall minimum freshwater and wastewater flows for this network, assuming no water storage system is in use. Freshwater supply is assumed to be free of impurity (i.e., 0 ppm).

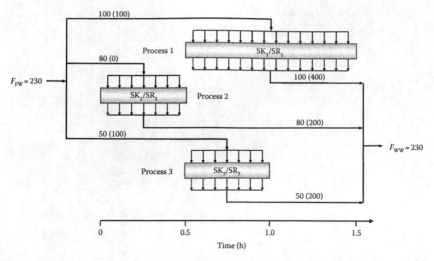

FIGURE 11.3
Batch water network in Example 11.1 (all flowrate terms in t/h; concentration in ppm, given in parenthesis). (From Foo, D.C.Y. et al., *J. Clean. Prod.*, 13(15), 1381, 2005. With permission.)

Solution

1. Classification of water-using processes

   From Figure 11.3, it is observed that the water sinks and sources are present in the same duration as that of the water-using processes. Hence, they are classified as the semicontinuous operations. Note also that the flowrate terms have a unit of t/h.

2. Identification of limiting data

   To carry out water minimization study, the flowrates of the water sinks and sources in Table 11.1 are converted into flow terms. The latter are obtained by the product of sink/source flowrates with their corresponding duration. For instance, SK2 and SR2 have start (T^{STT}) and end time (T^{END}) of 0 and 0.5 h, respectively. Hence, for a duration of 0.5 h, the limiting flows for

TABLE 11.1

Limiting Water Data for Example 11.1

SK_j	F_{SKj} (ton)	C_{SKj} (ppm)	T^{STT} (h)	T^{END} (h)
SK1	100	100	0.5	1.5
SK2	40	0	0	0.5
SK3	25	100	0.5	1.0
SR_i	F_{SRi} (ton)	C_{SRi} (ppm)	T^{STT} (h)	T^{END} (h)
SR1	100	400	0.5	1.5
SR2	40	200	0	0.5
SR3	25	200	0.5	1.0

these sink and source are given as 40 ton ($=80\,t/h \times 0.5\,h$). The limiting data for all water sinks and sources and their start and end time are summarized in Table 11.1.

3. Flow targeting with MRPD

The three-step procedure is followed to identify the minimum freshwater and wastewater flows of the batch water network:

1. *Define the time intervals for the BMN based on start and end times of the process sinks and sources.*
2. *Locate the process sinks and sources in their respective time intervals, and determine their interval flows, based on the ratio of the duration of the time interval to that of the sink/source.*

   Based on the start and end time of water sinks and sources, three time intervals may be defined accordingly, i.e., 0–0.5 h (time interval $t = A$), 0.5–1.0 h ($t = B$), and 1.0–1.5 h ($t = C$). Following Step 2 of the procedure, water sinks and sources are located in their respective time intervals. The interval flows of the water sinks and sources are next determined and are given in Table 11.2.

   For instance, sink SK1 that exists between 0.5–1.5 h is located in time intervals B and C. Since both time intervals B and C have a duration of 0.5 h, the interval flows of SK1 in these intervals are determined as half of its total flow, i.e.,

$$F_{SK1,B} = F_{SK1,C} = 100\,\text{ton}\left(\frac{0.5\,h}{1.0\,h}\right) = 50\,\text{ton}$$

   Interval flows for other sinks and sources are determined with similar steps.

3. *Carry out flow targeting for each time interval to identify the minimum fresh resource and waste discharge, with either graphical*

TABLE 11.2

Time Intervals and Water Flows for Process Sinks and Sources in Their Respective Time Intervals (All Flow Terms in Ton)

	Time Interval, t		
SK_j/SR_i	$t = A$ (0–0.5 h)	$t = B$ (0.5–1.0 h)	$t = C$ (1.0–1.5 h)
SK1		$F_{SK1,B} = 50$	$F_{SK1,C} = 50$
SK2	$F_{SK2,A} = 40$		
SK3		$F_{SK3,B} = 25$	
SR1		$F_{SR1,B} = 50$	$F_{SR1,C} = 50$
SR2	$F_{SR2,A} = 40$		
SR3		$F_{SR3,B} = 25$	

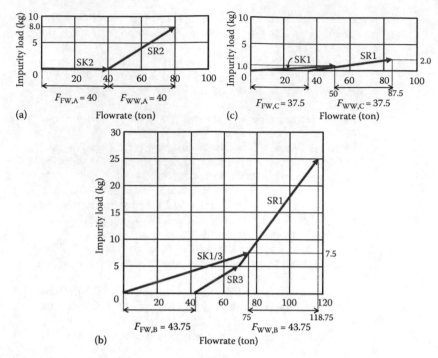

FIGURE 11.4
Flow targeting with MRPD for (a) 0–0.5 h, (b) 0.5–1.0 h, and (c) 1.0–1.5 h.

or algebraic targeting techniques. The summation of flows across all time intervals gives the overall flows for the entire BMN.

The MRPD is constructed in each time interval to determine their respective flow targets, given as in Figure 11.4. Note that the flowrate term used for continuous processes (Chapter 3) is replaced with flow terms here. The targeting result indicates the minimum freshwater and wastewater flows are identified as 40 ton (0–0.5 h), 43.75 ton (0.5–1.0 h), and 37.5 ton (1.0–1.5 h), respectively. Hence, the summation of the individual flows across all time intervals gives the overall flow targets for the batch water network, i.e., 121.25 ton (= 40 + 43.75 + 37.5 ton) for both freshwater and wastewater flows, respectively.

11.3 Targeting Procedure for Direct Reuse/Recycle in a BMN with Mass Storage System

Procedure for flow targeting in a BMN with a mass storage system is given as follows. Note that algebraic targeting technique is more convenient for use in BMN with a mass storage system. Note also that Steps 1 and 2 are essentially the same as the targeting procedure for BMN without a mass storage system.

1. Define the time intervals for the BMN based on start and end times of the process sinks and sources.
2. Locate the process sinks and sources in their respective time intervals, and determine their interval flows, based on the ratio of the duration of the time interval to that of the sink/source.
3. Carry out flow targeting for the entire BMN, assuming that the network is operated continuously. In other words, all process sinks and sources are assumed to be present in the same time interval. Note that this is only feasible with the presence of a mass storage system, where process sources may be stored for recovery to sinks across different time intervals.
4. Analyze the material flow cascades across the quality levels to determine the allocation of fresh resource flows in each time interval.*
5. For time intervals that experience material deficit (after allocation of fresh resource), explore the opportunity to recover process sources from lower quality level to fulfill the material requirement.*
6. Identify excess material flow that need to be recovered *directly* (within the same time interval) or *indirectly* (via mass storage from other time intervals). For the latter, it is assumed that mass storage is available at each quality level (where a source is present), and no mixing of different quality sources is allowed in the mass storage.*
7. Excess material flow that is not recovered (in Step 6) will leave as waste from the respective time interval.*
8. Explore opportunity to remove excessive storages. This may be done by storing material of different quality in the same storage tank, providing that their storing durations are not overlapping.

Example 11.2 Algebraic targeting for a batch water network

Example 11.1 is revisited here. For freshwater supply that is free of impurity (i.e., 0 ppm), use the *material cascade analysis* (MCA) technique in Chapter 4 to identify the overall minimum freshwater and wastewater flows for the batch water network for the following cases:

1. Network without water storage tank
2. Network with water system. Determine also the minimum size of the water storage tank(s).

---

* These steps are only applicable when there is only single sink with non-zero quality index in the higher quality region of the BMN.

Solution

1. Case 1—Network without water storage tank

The three-step procedure outlined in Section 11.2 is followed to identify the minimum flow targets of the batch water network:

1. *Define the time intervals for the BMN based on start and end times of the process sinks and sources.*
2. *Locate the process sinks and sources in their respective time intervals, and determine their interval flows, based on the ratio of the duration of the time interval to that of the sink/source.*
3. *Carry out flow targeting for each time interval to identify the minimum fresh resource and waste discharge, with either graphical or algebraic targeting technique. The summation of flows across all time intervals gives the overall flows for the entire BMN.*

Steps 1 and 2 have been previously described in Example 11.1, and are omitted here. Step 3 is next followed to carry out flow targeting with the variant of MCA, i.e. *water cascade analysis* (WCA) technique, by making use of the interval flows in Table 11.2. Note however that the flowrate terms in the previous case (RCN in continuous mode) is replaced with flow terms here for the WCA technique.

Tables 11.3 through 11.5 show the results of flow targeting in each time interval for the BMN without a water storage tank. Note that the infeasible cascades are not shown for brevity (see Chapter 4 for detailed targeting procedure of MCA). As shown, the minimum freshwater ($F_{FW,t}$) and wastewater ($F_{WW,t}$) flow targets for the time intervals are identified as 40 ton (0–0.5 h), 43.75 ton (0.5–1.0 h), and 37.5 ton (1.0–1.5 h), respectively. This leads to overall freshwater ($\Sigma_t F_{FW,t}$) and wastewater flows ($\Sigma_t F_{WW,t}$) of 121.25 ton for the batch water network, identical to that obtained using the graphical technique (Example 11.1).

2. Case 2—Network with water storage system.

The eight-step procedure is followed to identify the flow targets for the network with water storage tank(s).

TABLE 11.3

Flow Targeting with MCA for 0–0.5 h

C_k (ppm)	$\Sigma_j F_{SKj,t}$ (ton)	$\Sigma_i F_{SRi,t}$ (ton)	$\Sigma_i F_{SRi,t} - \Sigma_j F_{SKj,t}$ (ton)	$F_{C,k,t}$ (ton)	$\Delta m_{k,t}$ (kg)	Cum. $\Delta m_{k,t}$ (kg)
				$F_{FW,A} = 40$		
0	40		−40			
				0	0.00	
200		40	40			0.00
				$F_{WW,A} = 40$	39,992.00	**(PINCH)**
1,000,000						39,992.00

TABLE 11.4

Flow Targeting with MCA for 0.5–1.0 h

C_k (ppm)	$\sum_j F_{SKj,t}$ (ton)	$\sum_i F_{SRi,t}$ (ton)	$\sum_i F_{SRi,t} - \sum_j F_{SKj,t}$ (ton)	$F_{C,k,t}$ (ton)	$\Delta m_{k,t}$ (kg)	Cum. $\Delta m_{k,t}$ (kg)
				$F_{FW,B} = 43.75$		
0						
				43.75	4.38	
100	75		−75			4.38
				−31.25	−3.13	
200		25	25			1.25
				−6.25	−1.25	
400		50	50			0.00
				$F_{WW,B} = 43.75$	43,732.50	(PINCH)
1,000,000						43,732.50

TABLE 11.5

Flow Targeting with MCA for 1.0–1.5 h

C_k (ppm)	$\sum_j F_{SKj,t}$ (ton)	$\sum_i F_{SRi,t}$ (ton)	$\sum_i F_{SRi,t} - \sum_j F_{SKj,t}$ (ton)	$F_{C,k,t}$ (ton)	$\Delta m_{k,t}$ (kg)	Cum. $\Delta m_{k,t}$ (kg)
				$F_{FW,C} = 37.5$		
0						
				37.5	3.75	
100	50		−50			3.75
				−12.5	−3.75	
400		50	50			0.00
				$F_{WW,C} = 37.5$	37,485.00	(PINCH)
1,000,000						37,485.00

1. *Define the time intervals for the BMN based on start and end times of the process sinks and sources.*
2. *Locate the process sinks and sources in their respective time intervals, and determine their interval flows, based on the ratio of the duration of the time interval to that of the sink/source.*
3. *Carry out flow targeting for the entire BMN, assuming that the network is operated continuously. In other words, all process sinks and sources are assumed to be present in the same time interval. Note that this is only feasible with the presence of a mass storage system, where process source(s) may be stored for recovery to sinks across different time intervals.*

The flow targeting is carried out by assuming that all water sinks and sources are present in the same time interval. As shown in Table 11.6, the minimum freshwater (F_{FW}) and wastewater (F_{WW}) flow targets for the overall

TABLE 11.6

Flow Targeting with MCA for Batch Water Network with Storage System

C_k (ppm)	$\Sigma_t \Sigma_j$ $F_{SKj,t}$ (ton)	$\Sigma_t \Sigma_i$ $F_{SRi,t}$ (ton)	$\Sigma_t \Sigma_i F_{SRi,t}$ $- \Sigma_t \Sigma_j F_{SKj,t}$ (ton)	$\Sigma_t F_{C,k,t}$ (ton)	$\Sigma_t \Delta m_{k,t}$ (kg)	Cum. $\Sigma_t \Delta m_{k,t}$ (kg)
				$F_{FW} = 102.5$		
0	40		−40			
				62.5	6.25	
100	75 + 50		−125			6.25
				−62.5 ($F_{A,HPR}$)	−6.25	
200		25 + 40	65			**0.00**
				2.5 ($F_{A,LPR}$)	0.50	**(PINCH)**
400		50 + 50	100			0.50
				$F_{WW} = 102.5$	102,459.00	
1,000,000						102,459.50

network are determined as 102.5 ton. The pinch concentration for the overall network is identified as 200 ppm. The corresponding pinch-causing source at this concentration has a total water flow of 65 ton. As indicated in Chapter 4, 62.5 ton of the pinch-causing source flow is to be allocated to the higher quality region (HQR, F_{HQR}), while the remaining (2.5 ton) is allocated to the lower quality region (LQR, F_{LQR}).

4. *Analyze the material flow cascades across the quality levels to determine the allocation of fresh resource flows in each time interval.*

We next analyze the allocation of freshwater for each time interval based on the flow target identified in Step 3. In Table 11.7, the *net flow column* ($\Sigma_t \Sigma_i F_{SRi,t} - \Sigma_t \Sigma_j F_{SKj,t}$) is further segregated into individual time intervals (t = A, B, and C), while the last two columns of Table 11.6 are now omitted. Note that a negative entry indicates flow deficit is encountered, while positive entry indicates flowrate surplus. The flow cascade (column 5) indicates that 40 ton of freshwater is consumed at 0 ppm level, leaving the remaining flow (62.5 ton) for lower level (100 ppm). Since only time interval A (0–0.5 h) at 0 ppm encounters water deficit, 40 ton freshwater is sent to this time interval.

At 100 ppm level, both time intervals B and C (0.5–1.5 h) encounter flow deficits. Hence, the leftover freshwater flow of 62.5 ton will be allocated here. The freshwater flows allocated to these intervals are determined based on the ratio of their flow requirement, i.e., 75:50, which are determined as 37.5 ton (= 62.5 × 75/125) and 25 ton (= 62.5 × 50/125), respectively.

The freshwater allocation to all time intervals are given in parenthesis in Table 11.7.

TABLE 11.7

Freshwater Allocation for Each Time Interval
(Values Given in Parenthesis)

C_k (ppm)	$\Sigma_i F_{SRi,t} - \Sigma_j F_{SKj,t}$ (ton)			$\Sigma_t F_{C,k,t}$ (ton)
	$t = A$	$t = B$	$t = C$	
	(40)			$F_{FW} = 102.5$
0	−40			
		(37.5)	(25)	62.5
100		−75	−50	
				−62.5
200	40	25		(PINCH)
				2.5
400		50	50	
				$F_{WW} = 102.5$
1,000,000				

5. *For time intervals that experience material deficit (after allocation of fresh resource), explore the opportunity to recover process sources from lower quality level to fulfill the material requirement.*
6. *Identify excess material flow that needs to be recovered directly (within the same time interval) or indirectly (via mass storage from other time intervals). For the latter, it is assumed that mass storage is available at each quality level (where a source is present), and no mixing of different quality sources is allowed in the mass storage.*

Since the time intervals B and C encounter water deficit of 75 and 50 ton, (at 100 ppm level) respectively, the allocated freshwater flow is insufficient. These time intervals still encounter water deficits of 37.5 ton (=75 − 37.5 ton) and 25 ton (=50 − 25 ton) of water flow, respectively. On the other hand, the flow cascade in column 5 (see Table 11.8) indicates that these flow requirements, i.e., 62.5 ton (=37.5 + 25 ton), may be fulfilled by water flow from a lower concentration level. We next proceed to inspect the 200 ppm level. It is observed that time intervals A and B have surplus of water flows (40 and 25 ton, respectively) after fulfiling the flow requirement of time intervals B and C. In order to minimize storage size, the excess flow at 200 ppm (25 ton) is fed directly to fulfill the requirement at 100 ppm within time interval B (indicated by the upward arrow in Table 11.8). This leaves a flow deficit of 37.5 ton, i.e., 12.5 (=37.5 − 25 ton in internal B) ton for interval B and 25 ton for interval C. These flow requirements can be fulfilled by performing indirect recovery, i.e., by storing excess water flow in interval A (200 ppm).

TABLE 11.8

Direct and Indirect Recovery of Water Sources

C_k (ppm)	$\Sigma_i F_{SRi,t} - \Sigma_j F_{SKj,t}$ (ton)			$\Sigma_t F_{C,k,t}$ (ton)
	$t = A$	$t = B$	$t = C$	
	(40)			$F_{FW} = 102.5$
0	−40			
		(37.5)	(25)	62.5
100		−75	−50	
		⇧		−62.5
200	40	25		(PINCH)
Cum. F_{ST}	37.5	25 (=37.5−12.5)	0 (=25−25)	2.5
400		50	50	
				$F_{WW} = 102.5$
1,000,000				

The values (in box) within the concentration interval of 200 and 400 ppm in Table 11.8 indicate the cumulative flow of the water storage tank (Cum. F_{ST}). As shown, 37.5 ton of water is first stored at interval A. In interval B, 12.5 ton of the stored water is first consumed, while the remaining (25 ton) is left for interval C. The largest value among the cumulative flows determines the size of the water storage tank, i.e., 37.5 ton. Note that since only water of 200 ppm is to be recovered via storage, a single water storage tank is sufficient for use in this case.

7. *Excess material flow(s) that is not recovered (in Step 6) will leave as waste from the respective time interval.*

8. *Explore the opportunity to remove excessive storages. This may be done by storing water of different quality in the same storage tank, providing that their storing durations are not overlapping.*

Since 37.5 ton of excess water flow in time interval A has been recovered (via water storage to intervals B and C), the remaining flow of 2.5 ton of water flow surplus at 200 ppm (=40−37.5 ton) will be emitted as wastewater from this interval. Besides, 50 ton of wastewater flows are also emitted from time intervals B and C, respectively (400 ppm). The individual wastewaters are indicated by the downward arrows in Table 11.9, with their respective flows located in the last row of the table. Summing the individual values across all time intervals results in the total wastewater flow of 102.5 ton ($\Sigma_t F_{WW,t} = F_{WW}$), as identified in Step 3 of the targeting procedure.

Since the batch water network in this case requires only a single water storage (200 ppm), Step 8 of the targeting procedure to remove the excessive storage tanks is not applicable here.

TABLE 11.9

Identification of Wastewater Streams in Each Time Interval

C_k (ppm)	$\Sigma_i F_{SRi,t} - \Sigma_j F_{SKj,t}$ (ton)			$\Sigma_t F_{C,k,t}$ (ton)
	$t=A$	$t=B$	$t=C$	
	(40)			$F_{FW}=102.5$
0	−40			
		(37.5)	(25)	62.5
100		−75	−50	
		⇧		−62.5
200	40	25		(PINCH)
Cum. F_{ST}	37.5	25	0	2.5
400	⇩	50	50	
	⇩	⇩	⇩	$F_{WW}=102.5$
1,000,000	⇩	⇩	⇩	
$\Sigma_t F_{WW,t}$	2.5	50	50	

One should also take note that the stored material (i.e., water in this case) may also be utilized for a new batch of production, apart from being recovered to the current batch. For such cases, material may be stored in the last time interval (of previous batch) and be used in the first or second time intervals (of current batch). In other words, the storage tank helps to overcome the time limitation of the batch process.

Besides, it is worth noting that the allocation flow targets of the pinch-causing source to the HQR and LQR are conserved. In Table 11.6, it is observed that the pinch-causing source (200 ppm) has a total flow of 65 ton, with allocated flows of 62.5 ton (F_{HQR}) and 2.5 ton (F_{LQR}) for HQR and LQR, respectively. The former is contributed by time intervals A and B via direct (37.5 ton) and indirect integration (25 ton), respectively.

In contrast, the allocated flow to the LQR is contributed by interval A, where wastewater is withdrawn for discharge (see column 2 of Table 11.9). Note also that the allocated flow targets will not be conserved if one were to use the excess water flow at 400 ppm for flow deficits at 100 ppm, even though both of them are present within intervals B and C. Doing that will also lead to higher freshwater and wastewater flows for the overall network.

11.4 Targeting for Batch Regeneration Network

To further reduce the flow targets of fresh resource and waste discharge of a BMN, regeneration should be considered. The algebraic targeting technique in Chapter 6 may be used to set the flow targets for a BMN with interception unit of the single pass and fixed outlet quality type (with flowrate term replaced with flow term). The overall targeting procedure for this *batch regeneration network* is the same as that of BMN with mass storage system (Section 11.3), with the regeneration targeting being incorporated in Step 3.

 Example 11.3 Targeting for regeneration in batch water network

The water minimization case in Examples 11.1 and 11.2 is revisited here. An interception unit with outlet concentration (C_{Rout}) of 20 ppm is used to regenerate water sources. Use the algebraic targeting approach in Chapter 4 to identify the *ultimate flow targets* (i.e., minimum fresh and regenerated water, as well as wastewater) for this BMN, assuming pure freshwater supply (0 ppm) is used.

Solution

The targeting procedure in Section 11.3 is followed to identify the ultimate flow targets for the batch water network. Note that Steps 1 and 2 are omitted here, as they are the same as in earlier examples.

3. *Carry out flow targeting for the entire BMN, assuming that the network is operated continuously. In other words, all process sinks and sources are assumed to be present in the same time interval. Note that this is only feasible with the presence of a mass storage system, where process sources may be stored for recovery to sinks across different time intervals.*

   The algebraic targeting approach in Chapter 6 is used to determine the ultimate flow targets for a batch regeneration network. However, only a brief summary of the important steps are given here. The readers should refer to Chapter 6 for the detailed targeting procedure (see Figure 6.2 for summary of the procedure).

   Following Step 2 of the algebraic targeting procedure the overall network is divided into *freshwater region* (FWR) and *regenerated water region* (RWR), with concentration lower and higher than $C_{Rout}=200$ ppm. Water sinks and sources are then allocated to these regions based on their respective concentrations.

   Next, it is then determined that the total water flow of the sinks is less than that of the sources in the RWR. Hence, Step 3b (of the algebraic targeting procedure) is followed to reallocate 40 ton of sources at the highest concentration (400 ppm) from the RWR to the FWR. Since the water sink in the FWR can only be fed with pure freshwater (i.e., $C_{SKj}=0$ ppm), we may proceed to steps 5 and 6 to determine the flow targets in both FRW and RWR, respectively. The targeting result is shown in Tables 11.10 and 11.11.

   From Tables 11.10 and 11.11, the ultimate flow targets for the batch water network correspond to 40 ton freshwater (F_{FW}),

TABLE 11.10

Targeting for Batch Regeneration Network: FWR

C_k (ppm)	$\Sigma_t \Sigma_j F_{SKj,t}$ (ton)	$\Sigma_t \Sigma_i F_{SRi,t}$ (ton)	$\Sigma_t \Sigma_i F_{SRi,t} - \Sigma_t \Sigma_j F_{SKj,t}$ (ton)	$\Sigma_t F_{C,k,t}$ (ton)	$\Sigma_t \Delta m_{k,t}$ (kg)	Cum. $\Sigma_t \Delta m_{k,t}$ (kg)
				$F_{FW}=40$		
0	40		−40			
					0 0	
400		40	40			0
				$F_{WW}=40$	39,984	(PINCH)
1,000,000						39,984

TABLE 11.11

Targeting for Batch Regeneration Network: RWR

C_k (ppm)	$\Sigma_t \Sigma_j F_{SKj,t}$ (ton)	$\Sigma_t \Sigma_i F_{SRi,t}$ (ton)	$\Sigma_t \Sigma_i F_{SRi,t} - \Sigma_t \Sigma_j F_{SKj,t}$ (ton)	$\Sigma_t F_{C,k,t}$ (ton)	$\Sigma_t \Delta m_{k,t}$ (kg)	Cum. $\Sigma_t \Delta m_{k,t}$ (kg)
				$F_{RW}=69.44$		
0					69.44	5.56
100	75 + 50		−125			5.56
				−55.56	−5.56	
200		25 + 40	65			0.00
				9.44	1.89	(PINCH)
400		50 + 10	60			1.89
				$F_{RW}=69.44$	69,416.67	
1,000,000						69,418.56

40 ton wastewater (F_{WW}), and 69.44 ton regenerated water (F_{RW}), respectively. A combined cascade for the targeting results is shown in Table 11.12.

4. *Analyze the material flow cascades across the quality levels to determine the allocation of fresh resource flows in each time interval.*

Step 4 is then followed to determine the allocation of fresh and regenerated water (both are considered as fresh resources) to each time interval. The net flow column ($\Sigma_t \Sigma_i F_{SRi,t} - \Sigma_t \Sigma_j F_{SKj,t}$) is segregated into individual time intervals ($t=$A, B, and C) in Table 11.13. Note that the last two columns of Table 11.12 are excluded here. From Table 11.13, it can be easily observed that 40 ton freshwater is sent entirely to remove the flow deficit (0 ppm) in time interval A (indicated by the values in parenthesis).

At lower concentration level, i.e., 20 ppm, the regenerated water source is available at 69.44 ton. This source will be allocated to 100 ppm level, where flow deficit is encountered (in time

TABLE 11.12

Combined Cascade for Batch Regeneration Network

C_k (ppm)	$\Sigma_t \Sigma_j F_{SKj,t}$ (ton)	$\Sigma_t \Sigma_i F_{SRi,t}$ (ton)	$\Sigma_t \Sigma_i F_{SRi,t} - \Sigma_t \Sigma_j F_{SKj,t}$ (ton)	$\Sigma_t F_{C,k,t}$ (ton)	$\Sigma_t \Delta m_{k,t}$ (kg)	Cum. $\Sigma_t \Delta m_{k,t}$ (kg)
				$F_{FW}=40$		
0	40		−40			
				0	0.00	
$C_{Rout}=20$		$F_{RW}=69.44$	69.44			0.00
				69.44	5.56	
100	125		−125			5.56
				−55.56	−5.56	
200		55.56	55.56			0.00
				0	0.00	
400		40	40			0.00
				$F_{WW}=40$	39,984.00	(PINCH)
1,000,000						39,984.00

TABLE 11.13

Fresh and Regenerated Water Allocation for Each Time Interval

C_k (ppm)	$\Sigma_i F_{SRi,t} - \Sigma_j F_{SKj,t}$ (ton) $t=A$	$t=B$	$t=C$	$\Sigma_t \Sigma_j F_{SKj,t}$ (ton)	$\Sigma_t \Sigma_i F_{SRi,t}$ (ton)	$\Sigma_t F_{C,k,t}$ (ton)
						$F_{FW}=40$
0	(40)			40		
	−40					0
$C_{Rout}=20$					$F_{RW}=69.44$	
		($F_{RW,A}=41.67$)	($F_{RW,B}=27.78$)			69.44
100		−75	−50	125		
						−55.56
200	40	25			65	(PINCH)
						9.44
400		8.33	22.22		30.56	
		(=50 − 41.67)	(=50 − 27.78)			
						$F_{WW}=40$
1,000,000						

intervals B and C). The flows allocated to these intervals are determined based on the ratio of their flow requirement, i.e., 75:50, which are determined as 41.67 ton ($F_{RW,A}=69.44 \times 75/125$) and 27.78 ton ($F_{RW,B}=69.44 \times 50/125$), respectively. Besides, it is good to determine the original source of these regenerated flows. Since there are sufficient flows at the water sources at

400 ppm, they are sent for regeneration. Hence, these sources have reduced flows (see columns 3 and 4 in Table 11.13).

5. *For time intervals that experience material deficit (after allocation of fresh resource), explore the opportunity to recover process sources from lower quality level to fulfill the material requirement.*

6. *Identify excess material flow that need to be recovered directly (within the same time interval) or indirectly (via mass storage from other time intervals). For the latter, it is assumed that mass storage is available at each quality level (where a source is present), and no mixing of different quality sources is allowed in the mass storage.*

Since the time intervals B and C encounter water deficit of 75 and 50 ton (100 ppm), respectively, the allocated flow from regenerated water is insufficient. These time intervals still encounter water flow deficits of 33.33 ton (= 75 − 41.67 ton) and 22.22 ton (= 50 − 27.78 ton), respectively. These flow requirements can be supplemented by sources at a higher concentration level. However, the last column of Table 11.13 indicates that only 55.56 ton of water can be recovered from the pinch-causing source as allocated flow to the HQR, and they are available in intervals A and B. To minimize storage requirement, 25 ton of water source in interval B can be recovered directly within the same interval to remove its water deficit (indicated by the upward arrow in Table 11.14). The remaining source flow (55.56 − 25 = 30.56 ton) is to be recovered indirectly by the storage tank. A flow of 8.33 ton (= 75 − 41.67 − 25 ton) of this stored water will be recovered to interval B, while the remaining (30.56 − 8.33 = 22.22 ton) is recovered to interval C. The cumulative flow of the water storage tank (Cum. F_{ST}) is indicated by the values (in box) within the concentration interval of 200 and 400 ppm in Table 11.14. The largest value among the cumulative flows determines the size of the water storage tank, i.e., 30.56 ton.

7. *Excess material flow that is not recovered (in Step 6) will leave as waste from the respective time interval.*

8. *Explore opportunity to remove excessive storages. This may be done by storing water of different quality in the same storage tank, providing that their storing durations are not overlapping.*

Upon the recovery of water flow, the excess water flows in all time intervals are discharged as wastewater streams from the intervals. These correspond to sources at 200 ppm for interval A and 400 ppm for intervals B and C. These individual wastewaters are indicated by the downward arrows in Table 11.15, with their respective flows located in the last row of the table. Summing the individual values across all time intervals results in the total wastewater flow of 40 ton ($\Sigma_t F_{WW,t} = F_{WW}$), as identified in the last column of Table 11.15.

Step 8 of the targeting procedure is not applicable since the batch water network only requires single water storage (at 200 ppm).

TABLE 11.14

Direct and Indirect Recovery of Water Sources

C_k (ppm)	$\Sigma_i F_{SRi,t} - \Sigma_j F_{SKj,t}$ (ton)			$\Sigma_t \Sigma_j F_{SKj,t}$ (ton)	$\Sigma_t \Sigma_i F_{SRi,t}$ (ton)	$\Sigma_t F_{C,k,t}$ (ton)
	$t = A$	$t = B$	$t = C$			
						$F_{FW} = 40$
0	(40)			40		
	−40					0
$C_{Rout} = 20$					$F_{RW} = 69.44$	
		$(F_{RW,A} = 41.67)$	$(F_{RW,B} = 27.78)$			69.44
100		−75	−50	125		
		⇧				−55.56
200	40	25			65	(PINCH)
Cum. F_{ST}	30.56	22.22 (=30.56 − 8.33)	0 (=22.22 − 22.22)			9.44
400		8.33	22.22		30.56	
		(=50 − 41.67)	(=50 − 27.78)			
						$F_{WW} = 40$
1,000,000						

TABLE 11.15

Identification of Wastewater Streams in Each Time Interval

C_k (ppm)	$\Sigma_i F_{SRi,t} - \Sigma_j F_{SKj,t}$ (ton)			$\Sigma_t \Sigma_j F_{SKj,t}$ (ton)	$\Sigma_t \Sigma_i F_{SRi,t}$ (ton)	$\Sigma_t F_{C,k,t}$ (ton)
	$t = A$	$t = B$	$t = C$			
						$F_{FW} = 40$
0	(40)			40		
	−40					0
$C_{Rout} = 20$					$F_{RW} = 69.44$	
		$(F_{RW,A} = 41.67)$	$(F_{RW,B} = 27.78)$			69.44
100		−75	−50	125		
		⇧				−55.56
200	40	25			65	(PINCH)
Cum. F_{ST}	30.56	22.22 (=30.56 − 8.33)	0 (=22.22 − 22.22)			9.44
400	⇩	8.33	22.22		30.56	
		(=50 − 41.67)	(=50 − 27.78)			
	⇩	⇩	⇩			$F_{WW} = 40$
1,000,000	⇩	⇩	⇩			
$\Sigma_t F_{WW,t}$	9.44	8.33	22.22			

11.5 Design of a BMN

The *nearest neighbor algorithm* (NNA) and *matching matrix* introduced in Chapter 7 (RCNs for continuous mode) are ready for use in designing a BMN (all flowrate terms are replaced by flow terms). A generic form of the matching matrix is shown in Figure 11.5 (similar to that for inter-plant resource conservation network in Figure 10.11). Similar to earlier cases, sources (including regenerated source and fresh resource—FR) are arranged as rows, while sinks (including waste discharge—WD) are arranged as columns. However, note that only the eligible sinks and sources are located within each time interval and are arranged in a descending order of quality index within each time interval. Note also that mass storage (ST) may act as sink and source, depending on their role in the intervals. When source is to be stored for recovery at a later time interval, the mass storage takes the role as a sink. In contrast, when stored source is to be recovered to the sinks, the mass storage acts as a source in the time interval. Values in the cells indicate the allocation flows ($F_{SRi,SKj,t}$) of the sink–source matches within time interval t. Note that the shaded area in the matching matrix indicates that the matching of sinks and sources in different time intervals (e.g., between $t = A$ and $t = B$) is forbidden.

Example 11.4 Design of a BMN

The water minimization case in Example 11.3 is revisited here. Through targeting, the *ultimate water flows* for this BMN are identified as 40 ton of freshwater and wastewater, respectively, as well as 69.44 ton of regenerated water. Use the matching matrix to design the batch regeneration network.

Solution

Based on the targeting results in Table 11.15, the targeted flows for Freshwater, wastewater, and regenerated water are first located in each time interval, as shown in Figure 11.6. Besides, the targeted flows for water

				Time interval, t				
				$t = A$			$t = B$	
			$F_{SKj,t}$	$F_{SK1,A}$	$F_{ST,A}$	$F_{D,A}$	$F_{SK2,B}$	$F_{D,B}$
			$q_{SKj,t}$	$q_{SK1,A}$	$q_{ST,A}$	$q_{D,A}$	$q_{SK2,B}$	$q_{D,B}$
Time interval, t	$F_{SRi,t}$	$q_{SRi,t}$		SK1	ST	WD	SK2	WD
$t = A$	$F_{R,A}$	$q_{R,A}$	FR	$F_{R,SK1,A}$				
	$F_{SR1,A}$	$q_{SR1,A}$	SR1	$F_{SR1,SK1,A}$	$F_{SR1,ST,A}$	$F_{SR1,WD,A}$		
$t = B$	$F_{R,B}$	$q_{R,B}$	FR				$F_{R,SK2,B}$	
	$F_{ST,B}$	$q_{ST,B}$	ST				$F_{ST,SK2,B}$	
	$F_{SR2,B}$	$q_{SR2,B}$	SR2				$F_{SR2,SK2,B}$	$F_{SR2,WD,B}$

FIGURE 11.5
Matching matrix for a BMN.

Time interval, t	$F_{SR,t}$	$C_{SR,t}$	
t = A (0–0.5 h)	40	0	FW
	40	200	SR2
t = B (0.5–1.0 h)	41.67	20	RW
	25	200	SR3
	8.33	200	ST1
	8.33	400	SR1
t = C (1.0–1.5 h)	27.78	20	RW
	22.22	200	ST1

Time interval, t

	t = A (0–0.5 h)			t = B (0.5–1.0 h)			t = C (1.0–1.5 h)	
$F_{SK,t}$	40	30.56	9.44	50	25	8.33	50	22.22
$C_{SK,t}$	0	200		100	100		100	
	SK2	ST1	WW	SK1	SK3	WW	SK1	WW

FIGURE 11.6
Locating flow targets for each time intervals for Example 11.4 (all flow terms in ton; concentration in ppm).

storage (ST1) are also located as sink in time interval A (where 30.56 ton water is stored) and sources in time intervals B and C (where stored water is to be recovered to sinks, with flows of 8.33 and 22.22 ton respectively).

The nearest neighbor algorithm (see Section 7.1 for detailed procedure) is then used to determine the water flow of each sink–source match in each time interval. For time interval A (0–0.5 h), no direct recovery is found, as sink SK2 requires freshwater feed. Hence, a big portion of water from source SR2 is stored for indirect recovery (30.56 ton, value given in parenthesis; see Figure 11.7), while the remaining is discharged as waste-water (9.44 ton).

In time interval B (0.5–1.0 h), apart from the regenerated water (RW—originated from SR1; see Table 11.13) and SR3 that are recovered directly to sinks SK1 and SK3, 8.33 ton of stored water is also recovered indirectly (value given in parenthesis). The unutilized SR1 is then discharged as wastewater. Note that the recovery flows among the sink–source matches are determined via Equations 7.1 and 7.2 (detailed steps are omitted here).

In the final time interval (1.0–1.5 h), regenerated water (originated from SR1; see Table 11.13) and stored water are recovered directly and indi-rectly (value in parenthesis), respectively, while the unutilized SR1 is dis-charged as wastewater.

Finally the concentrations for wastewater streams are determined from the contributed water sources.

11.6 Waste Treatment and Batch Total Network

The targeting procedures for waste treatment and *total material network* in Chapter 8 are readily extended here for a BMN (with flowrate term replaced with flow term). However, due to the time dimension of the BMN, mass stor-age is needed so that waste streams may be stored and mixed at a later time (if necessary) to achieve the desired quality for waste treatment and for envi-ronmental discharge. This is shown in the following example.

Example 11.5 Wastewater treatment for a BMN

The water minimization case in Example 11.4 is revisited here. From Figure 11.7, the wastewater streams for the batch water network are extracted and summarized in Table 11.16. Use the *waste treatment pinch diagram* (WTPD) in Chapter 8 to determine the minimum treatment flow needed for these wastewater streams. The outlet concentration of the wastewater treatment system is assumed as 20 ppm, while the dis-charge limit is given as 50 ppm. Determine also when water storage tank is needed to store the wastewater streams for mixing purpose.

Solution

To construct the WTPD, the individual wastewater streams are first arranged in ascending order of impurity concentration (i.e. descending

Time interval, t	$F_{SR/t}$	$C_{SR/t}$	$F_{SK/t}$ / $C_{SK/t}$	Time interval, t t = A (0–0.5h) SK2 40 / 0	ST1 30.56 / 200	WW 9.44 / 200	t = B (0.5–1.0h) SK1 50 / 100	SK3 25 / 100	WW 8.33 / 400	t = C (1.0–1.5h) SK1 50 / 100	WW 22.22 / 400
t = A (0–0.5h)	40	0	FW	40							
	40	200	SR2		(30.56)	9.44					
t = B (0.5–1.0h)	41.67	20	RW				27.78	13.89			
	25	200	SR3				22.22	2.78			
	8.33	200	ST1					(8.33)			
	8.33	400	SR1						8.33		
t = C (1.0–1.5h)	27.78	20	RW							27.78	
	22.22	200	ST1							(22.22)	
	22.22	400	SR1								22.22

FIGURE 11.7
Matching matrix for Example 11.4 (all flow terms in ton; concentration in ppm).

TABLE 11.16

Limiting Data for Wastewater Streams in Example 11.5

y	WWy	F_{WW_y} (ton)	C_{WW_y} (ppm)	T^{STT} (h)	T^{END} (h)
1	WW1	9.44	200	0.5	1.5
2	WW2	8.33	400	0	0.5
3	WW3	22.22	400	0.5	1.0

TABLE 11.17

Cumulative Flow and Load for Plotting the WTPD

Wastewater, WWy	F_{WW_y} (ton)	C_{WW_y} (ppm)	m_{WW_y} (kg)	Cum. F_{WW_y} (ton)	Cum. m_{WW_y} (kg)
WW1	9.44	200	1.89	9.44	1.89
WW2, WW3	30.56	400	12.22	40.00	14.11

order of quality). Their cumulative flow and impurity load are next calculated in Table 11.17. As shown, the wastewater streams consist of a
total flow of 40 ton, with an impurity load of 14.11 kg. At this point, the
start and end times of these wastewater streams are ignored.

The WTPD is next plotted on a load-versus-flow diagram, as shown in
Figure 11.8. For an environmental discharge limit (C_D) of 50 ppm, the maximum allowable discharge load is calculated as 2 kg (= 40 ton × 50 ppm).

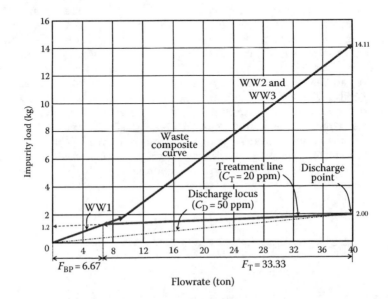

FIGURE 11.8

Targeting for minimum waste treatment flow. (From Foo, D.C.Y. et al., *Clean Technol. Environ. Policy*, in review.)

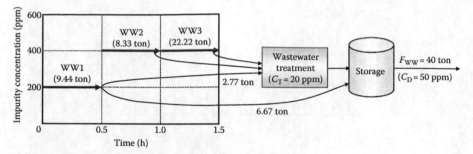

FIGURE 11.9
Wastewater treatment scheme for Example 11.5. (From Foo, D.C.Y. et al., *Clean Technol. Environ. Policy*, in review.)

Hence, a total impurity load of 12.11 kg (= 14.11 − 2.00 kg) is to be removed from the wastewater streams.

A treatment line with slope that corresponds to the outlet concentration (C_T) of 20 ppm is added to the WTPD, starting at the discharge point. The minimum treatment (F_T) and bypass (F_{BP}) flows are then determined as 33.33 and 6.67 ton, respectively.

However, we should note that the individual wastewater streams are actually present at different time intervals. Hence, they are not treated at the same time. In order to meet the discharge concentration limit (50 ppm), the wastewater streams treated earlier are temporarily stored in the storage tank. Once all wastewater streams are treated and mixed in the storage to the desired concentration, they are then discharged to the environment.

For this example, wastewater WW1 (9.44 ton) is found in the first time interval (0–0.5 h). A big portion of this wastewater stream (6.67 ton) will bypass the treatment unit (WWT) and hence is sent to the storage tank directly. The remaining flow of the wastewater stream (2.77 ton) is treated before being sent to the storage tank. At later time intervals (0.5–1.5 h), WW2 and WW3 will then be treated and stored. Once they are mixed well in the storage tank (to 50 ppm), the terminal wastewater will be discharged to the environment. This is shown in Figure 11.9.

The total water network for this example is shown in Figure 11.10. Note again that since all water-using processes are operated in a semicontinuous mode, their water sinks and sources exist when the processes are in operation.

11.7 Additional Readings

We have only discussed one targeting technique for BMN in this chapter. There are also other targeting and design techniques that may be used for BMN synthesis, e.g., those proposed by Wang and Smith (1995), Majozi et al.

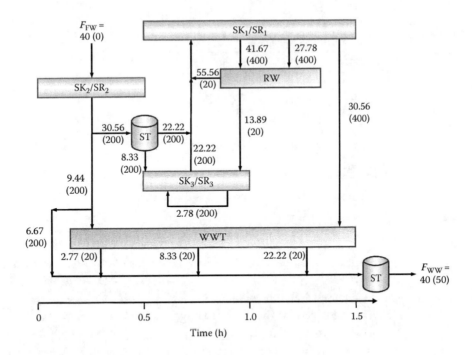

FIGURE 11.10
Total water network for Example 11.5 (all flow terms in ton; concentration in ppm, given in parenthesis). (From Foo, D.C.Y. et al., *Clean Technol. Environ. Policy*, in review.)

(2006), Liu et al. (2007), and Chen and Lee (2008, 2010). A comparison of these techniques is found in the review by Gouws et al. (2010).

Problems

Water Minimization Problems

11.1 Use the NNA and matching matrix to design the batch water network in Examples 11.1 and 11.2. The targeted water flows are given in their respective examples (solution for Example 11.2 is found in Figure 17.6).

11.2 A batch water network that consists of four semicontinuous water-using processes (Kim and Smith, 2004) is analyzed for its water recovery potential. Limiting water data for this case study are given in Table 11.18. freshwater feed is assumed to be free of impurity. Complete the following tasks:

a. Target the minimum freshwater and wastewater flows for direct reuse/recycle scheme for the network when no water storage tank

TABLE 11.18

Limiting Water Data for Problem 11.2

SK$_j$	F_{SKj} (t/h)	C_{SKj} (ppm)	T^{STT} (h)	T^{END} (h)
SK1	20	0	0	1
SK2	100	50	1	3.5
SK3	40	50	3	5
SK4	10	400	1	3
SR$_i$	F_{SRi} (t/h)	C_{SRi} (ppm)	T^{STT} (h)	T^{END} (h)
SR1	20	100	0	1
SR2	100	100	1	3.5
SR3	40	800	3	5
SR4	10	800	1	3

is present (hint: beware of the water flowrates of the semicontinuous processes).

b. Repeat (a) for repeated batch processes with water storage tank. Determine the size of the water storage.

c. An interception unit with outlet concentration (C_{Rout}) of 10 ppm is used to regenerate water sources for further recovery. Determine the *ultimate flowrate targets* (minimum freshwater, wastewater, and regenerated water) for the batch regeneration network.

d. A wastewater treatment facility with outlet concentration of 20 ppm is added to the batch regeneration network in (c). Determine the minimum treatment flow needed for the wastewater streams of this total water network when the discharge limit is 50 ppm. Also determine when a water storage tank is needed to store the wastewater streams for mixing purpose.

e. Design water networks that achieve the flow targets in (a) through (d).

11.3 Table 11.19 shows the limiting water data for a batch water network that consists of four truly batch water-using processes (Chen and Lee, 2008). Freshwater feed is assumed to be free of impurity. Complete the following tasks:

a. Target the minimum freshwater and wastewater flows for repeated batch processes with a water storage tank.

b. An interception unit with outlet concentration (C_{Rout}) of 0.1 kg/kg is used to regeneration water sources. Determine the *ultimate flowrate targets* (minimum freshwater, wastewater, and regenerated water) for the water regeneration network. Synthesize the RCN for this case.

TABLE 11.19

Limiting Water Data for Problem 11.3

SK_j	F_{SK_j} (ton)	C_{SK_j} (kg/kg)	T^{STT} (h)	T^{END} (h)
SK1	20	0	0	1.0
SK2	16	0.25	0	0.5
SK3	15	0.1	5.0	6.5
SK4	24	0.25	2.0	2.5
SK5	15	0.1	7.0	8.5
SR_i	F_{SR_i} (ton)	C_{SR_i} (kg/kg)	T^{STT} (h)	T^{END} (h)
SR1	20	0.2	4.0	5.0
SR2	16	0.5	4.5	5.0
SR3	15	0.1	5.0	6.5
SR4	24	0.4	6.5	7.0
SR5	15	0.12	7.0	8.5

11.4 Figure 11.11 shows five water-using processes for an agrochemical manufacturing plant, where water is used extensively as washing
agent and reaction solvent to produce three types of products, i.e., A,
B, and C (Majozi et al., 2006). To reduce the freshwater and wastewater
flows, a water recovery study is to be conducted. However, during the

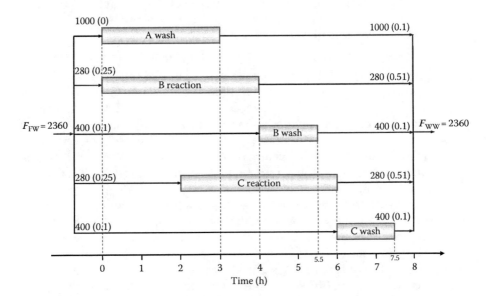

FIGURE 11.11
Water-using processes for an agrochemical plant in Problem 11.4 (all flow terms in kg; values
in parenthesis represent limiting concentration in kg salt/kg water).

manufacturing process, sodium chloride (NaCl) is formed as a by-product, which is the main impurity of concern for the water recovery study.

Product A was produced in an organic solvent that is immiscible in water. Water is hence used for washing the NaCl content in product A. For products B and C, water is used for both reaction solvent as well as the washing agent. However, since the reaction solvent water has removed most of the NaCl content, the wash water for products B and C do not remove any NaCl content. The feed water and wastewater flows for the five water-using processes are given in Figure 11.11.

Besides, the water-using processes have the following characteristics. The feed water of the processes can only be charged prior to the start of their actual operation, while wastewater is discharged upon the completion of the operation. In other words, no water flow is charged to and discharged from the processes during the operation. Note that the exact duration of water charge and wastewater discharge is insignificant as compared with that of the actual operation.

Perform the following tasks:

a. Should the water-using processes be classified as truly batch or semicontinuous operations? Identify the limiting data (which should include the info on time dimension) for a water minimization study.

b. Use the MCA technique to target the minimum freshwater and wastewater flows for direct reuse/recycle scheme for repeated batch processes with water storage tank. Determine also the size of the water storage tank.

c. Design the batch water network that achieves the flow targets in (b).

11.5 The process flow diagram for a batch polyvinyl chloride (PVC) resin manufacturing plant is shown in Figure 11.12 (Chan et al., 2008). Fresh vinyl chloride monomer (VCM) is mixed with recycled VCM to be fed to a series of batch polymerization reactors. A huge amount of water is used as reaction carrier in the reactor. Besides, toward the end of the polymerization process, water is added to maintain the reactor temperature. Hence, upon the completion of the polymerization reaction, the reactor content is in a slurry form that contains PVC resin and water. The reactor effluent is sent to a blowdown vessel, which serves as a buffer for the downstream purification processes that are operated in continuous mode. A steam stripper is used to recover the unconverted VCM from the reactor effluent. The stripped VCM, along with the VCM that emits from the reactors and blowdown vessel, are sent to the VCM recovery system. Effluent from the stripper is sent to a decanter where the PVC is separated from its water content. PVC product from the decanter is then forwarded to the dryer to reduce its moisture content before sending to the product storage tank.

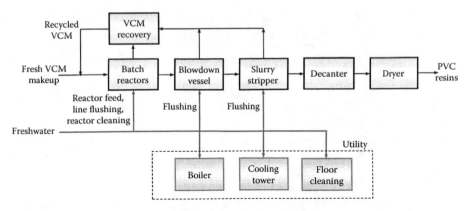

FIGURE 11.12
Process flow diagram for PVC manufacturing. (With kind permission from Springer Science+Business Media: *Clean Technol. Environ. Policy*, An integrated approach for water minimisation in a PVC manufacturing process, 10(1), 2008, 67, Chan, J.H., Foo, D.C.Y., Kumaresan, S., Aziz, R.A., and Hassan, M.A.A.)

Apart from the process water being used in the reactor, a significant amount of utility water is consumed in the plant. Utility water is used to flush the reactors, blowdown vessel, and the stripper as well as for reactor and general plant cleaning. Besides, boiler and cooling tower makeup consume a huge amount of water too. Water recovery potential is explored to reduce the overall water flows of the process. Suspended solid content is the main impurity of concern for water recovery.

Due to the polymerization reactors that are operated in batch mode, water sinks associated with these units are also available in batch mode, as given in Table 11.20 (Chan et al., 2008). Note that since the new batch of the process starts prior to the completion of the previous batch, there are three batches of operation ($N − 1$, N, and $N + 1$) that exist within the time interval of analysis (0–14.78 h). In contrast, the single water source of the process is originated from the decanter that is operated in continuous mode, with its limiting flowrate given in Table 11.21.

Complete the following tasks:

a. Identify the interval flows for the water sinks and sources for the process (hint: the water source can be modeled as a semi-continuous stream).

b. Target the minimum freshwater and wastewater flows for direct reuse/recycle scheme for the network with/without a water storage tank. Freshwater feed is free from suspended solid content. Determine also the size of the water storage.

TABLE 11.20

Limiting Data for Water Sinks in Problem 11.5

Batch	SK$_j$	Operation	F_{SKj} (kg)	C_{SKj} (ppm)	T^{STT} (h)	T^{END} (h)
	2a	Line flushing	12,500	10	0	4
N − 1	5a	Reactor cleaning	5,400	20	0	4.5
	6a	Cooling tower	28,350	50	0	4.5
	7a	Flushing	8,000	200	0	4
	1b	Reactor feed	105,000	10	0	3.33
	2b	Line flushing	15,625	10	9.28	14.28
	3b	Boiler makeup 1	3,300	10	0	0.25
	4b	Boiler makeup 2	3,300	10	8.00	8.25
N	5b	Reactor cleaning	6,000	20	9.78	14.78
	6b	Cooling tower	93,114	50	0	14.78
	7b	Flushing	10,000	200	9.28	14.28
	8b	Floor cleaning 1	1,500	500	0	0.25
	9b	Floor cleaning 2	1,500	500	8.00	8.25
	1c	Reactor feed	105,000	10	10.28	13.61
N + 1	3c	Boiler makeup 1	3,300	10	10.28	10.53
	6c	Cooling tower	28,350	50	10.28	14.78
	8c	Floor cleaning 1	1,500	500	10.28	10.53

TABLE 11.21

Limiting Data for Water Source in Problem 11.5

SR$_i$	Operation	Flowrate (kg/h)	C_{SRi} (ppm)
1	Decanter	9244	50

Property Integration Problems

11.6 Table 11.22 shows the limiting data for a batch property integration problem (Foo, 2010), where two water sources are to be recovered to two water sinks. Both water sources and sinks are operated in semicontinuous mode. In considering water recovery, pH is the main stream quality of concern. To supplement the water sources, two external fresh resources are eligible, i.e., acid (pH = 5) and freshwater (pH = 7). The mixing rule for pH is given as

$$10^{-pH} = \sum_i x_i 10^{-pH_i} \qquad (11.1)$$

Assuming that the process is to be operated repeatedly with water storage tanks, complete the following tasks:

TABLE 11.22

Limiting Water Data for Problem 11.6

SK_j	F_{SKj} (kg/s)	pH (SK_j)	T^{STT} (h)	T^{END} (h)
SK1	8	7.2	5	6
SK2	3	7.35	7	8
SR_i	F_{SRi} (kg/s)	pH (SR_i)	T^{STT} (h)	T^{END} (h)
SR1	6	9	0	1
SR2	2	7.7	3	4

a. Assuming only freshwater is used as external fresh resource (i.e., acid is not considered), determine the minimum freshwater and wastewater flow for the batch water network. What is the size of the storage tanks needed to achieve the flowrate targets?

b. Due to technical reasons, it is determined that the total flow of freshwater to be added to both sinks should not be more than 2.97 kg/s. Determine the amount of acid needed for the RCN for this case.* Determine also the size of the storage tanks needed for this case.

c. Design the batch water networks that achieve the flowrate targets in (a) and (b).

References

Chan, J. H., Foo, D. C. Y., Kumaresan, S., Aziz, R. A., and Hassan, M. A. A. 2008. An integrated approach for water minimisation in a PVC manufacturing process. *Clean Technologies and Environmental Policy*, 10(1), 67–79.

Chen, C.-L. and Lee, J.-Y. 2008. A graphical technique for the design of water-using networks in batch processes. *Chemical Engineering Science*, 63, 3740–3754.

Chen, C.-L. and Lee, J.-Y. 2010. On the use of graphical analysis for the design of batch water networks. *Clean Technologies and Environmental Policy*, 12, 117–123.

Foo, D. C. Y. 2010. Automated targeting technique for batch process integration. *Industrial and Engineering Chemistry Research*, 49(20), 9899–9916.

Foo, D. C. Y., Lee, J.-Y., Ng, D. K. S., and Chen, C.-L. Targeting and design for batch regeneration and total networks. *Clean Technologies and Environmental Policy* (in review).

Foo, D. C. Y., Manan, Z. A., and Tan, Y. L. 2005. Synthesis of maximum water recovery network for batch process systems. *Journal of Cleaner Production*, 13(15), 1381–1394.

---

* Hint: Refer to Sections 4.2 and 4.3 for the targeting procedure for multiple fresh resources and threshold problems.

Gouws, J., Majozi, T., Foo, D. C. Y., Chen, C. L., and Lee, J.-Y. 2010. Water minimisation techniques for batch Processes. *Industrial and Engineering Chemistry Research*, 49(19), 8877–8893.

Kim, J.-K. and Smith, R. 2004. Automated design of discontinuous water systems. *Process Safety and Environmental Protection*, 82(B3), 238–248.

Liu, Y. J., Yuan, X. G., and Luo, Y. Q. 2007. Synthesis of water utilisation system using concentration interval analysis method (II) discontinuous process. *Chinese Journal of Chemical Engineering*, 15, 369–375.

Majozi, T., Brouckaert, C. J., and Buckley, C. A. 2006. A graphical technique for wastewater minimization in batch processes. *Journal of Environmental Management*, 78, 317–329.

Wang, Y. P. and Smith, R. 1995. Time pinch analysis. *Chemical Engineering Research and Design*, 73, 905–914.

Part II

Mathematical Optimization Techniques

In Part II of this book, the readers will learn how to use mathematical techniques for the synthesis of a resource conservation network (RCN). Two main types of mathematical optimization techniques are introduced, i.e., superstructural approach (Chapter 12) and automated targeting techniques (Chapters 13 through 17). The latter has been given greater importance, as it is based on the targeting concept introduced in Part I of this book. Both techniques can be learnt individually, since there is no connection between them. Note however that Chapters 13 through 15 should be read sequentially, as they cover basic techniques for direct material reuse/recycle, regeneration, and waste treatment. The readers can find advanced topics such as inter-plant RCN and batch material networks in Chapters 16 and 17, respectively, which are based on the techniques discussed in the earlier chapters. Chapters 16 and 17 stand alone, with no interconnection among them.

The author has laid emphasis on linear program (LP) models, since they can be solved using a wide range of software, even with spreadsheet software (e.g., Microsoft Excel). In all the chapters in Part II, various optimization models are solved using commercial optimization software known as LINGO. A demo version of this software is made available on publisher's website (www.crcpress.com/product/isbn/9781439860489). Updated version of the software can be downloaded from the website of LINDO Systems Inc. (www.lindo.com). To facilitate self-learning, LINGO files of selected examples are also made available to readers on publishers website (please look for the icon 🖴). However, it is advisable that the readers redo the coding of these example to have a better understanding of the models.

12

Synthesis of Resource Conservation Networks: A Superstructural Approach

In this chapter, the most commonly used mathematical optimization technique, i.e., *superstructural approach*, is used to synthesize resource conservation networks (RCNs). These include direct reuse/recycle and material regeneration network. The model may also be extended to include process constraints such as forbidden/compulsory matches, reduced piping connection, and cost consideration. A simple extension is also shown for the synthesis of inter-plant resource conservation networks (IPRCNs).

12.1 Superstructural Model for Direct Reuse/Recycle Network

To synthesize an RCN, a superstructural representation is first constructed. For a direct reuse/recycle network, a common representation is shown in Figure 12.1 (reproduced from Figure 1.17). As shown, all sources are matched to the sinks via connections. Note that a fresh resource is also considered as a source, while waste discharge is considered as a sink. Each connection is represented by a flowrate variable ($F_{SRi,SKj}$), where its value is to be determined as the result of the optimization model.

Another representation for ease of model formulation takes the form of a *matching matrix* as introduced in Chapter 7, with the revised form shown in Figure 12.2. In the matching matrix, all sources including fresh resource (FR) are arranged as rows, while the sinks including waste discharge (WD) as columns, both in a descending order of quality index. Figures in the cells indicate the allocation flowrates ($F_{SRi,SKj}$) between sources and sinks in the RCN. The unutilized source flowrate will be sent for waste discharge ($F_{SRi,D}$). Note however that no match should be built between fresh resource and waste discharge (i.e., $F_{FR,WD} = 0$). As mentioned earlier, the values of these allocation flowrates are the results of the optimization model.

Next, the optimization model is formulated. The optimization objective can be set to determine the minimum total flowrate of fresh resource (F_R), given by Equation 12.1:

$$\text{Minimize } F_R \tag{12.1}$$

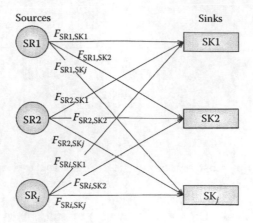

FIGURE 12.1
Superstructural model for direct reuse/recycle.

		F_{SKj}	F_{SK1}	F_{SK2}	F_{SK3}	F_{SKj}	F_D
		q_{SKj}	q_{SK1}	q_{SK2}	q_{SK3}	q_{SKj}	q_D
F_{SRi}	q_{SRi}	SK_j SR_i	SK1	SK2	SK3	SK_j	WD
F_R	q_R	FR	$F_{R,SK1}$	$F_{R,SK2}$	$F_{R,SK3}$	$F_{R,SKj}$	
F_{SR1}	q_{SR1}	SR1	$F_{SR1,SK1}$	$F_{SR1,SK2}$	$F_{SR1,SK3}$	$F_{SR1,SKj}$	$F_{SR1,D}$
F_{SR2}	q_{SR2}	SR2	$F_{SR2,SK1}$	$F_{SR2,SK2}$	$F_{SR2,SK3}$	$F_{SR2,SKj}$	$F_{SR2,D}$
F_{SR3}	q_{SR3}	SR3	$F_{SR3,SK1}$	$F_{SR3,SK2}$	$F_{SR3,SK3}$	$F_{SR3,SKj}$	$F_{SR3,D}$
F_{SRi}	q_{SRi}	SR_i	$F_{SRi,SK1}$	$F_{SRi,SK2}$	$F_{SRi,SK3}$	$F_{SRi,SKj}$	$F_{SRi,D}$

FIGURE 12.2
Superstructural model for direct reuse/recycle in matching matrix form.

where the total fresh resource flowrate is given by the sum of the individual fresh resource flowrate terms in the fourth row, i.e.,

$$F_R = \Sigma_j F_{R,SKj} \qquad (12.2)$$

Alternatively, the optimization objective can be set to minimize the cost of the RCN (Equation 12.3), which may take the form of *minimum operating cost* (MOC), *annual operating cost* (AOC), or the *total annualized cost* (TAC):

$$\text{Minimize cost} \qquad (12.3)$$

The previous optimization objective is subject to the following constraints:

$$\Sigma_i F_{SRi,SKj} + F_{R,SKj} = F_{SKj} \quad \forall j \qquad (12.4)$$

$$\sum_i F_{SRi,SKj} q_{SRi} + F_{R,SKj} q_R \geq F_{SKj} q_{SKj} \quad \forall j \tag{12.5}$$

$$\sum_j F_{SRi,SKj} + F_{SRi,D} = F_{SRi} \quad \forall i \tag{12.6}$$

$$F_{SRi,SKj}, F_{R,SKj}, F_{SRi,D} \geq 0 \quad \forall i,j \tag{12.7}$$

The constraint in Equation 12.4 describes that the flowrate requirement of the sink is fulfilled by various process sources ($F_{SRi,SKj}$) as well as the fresh resource ($F_{R,SKj}$). Equation 12.5 describes that the process sources and fresh resource feed must fulfill the minimum quality requirement of the process sinks. The flow-rate balance in Equation 12.6 indicates that the flowrate of each process source may be allocated to sinks, while the unutilized source will be sent for waste discharge. Equation 12.7 indicates that all variables of the superstructural model should take non-negative values. Since the flowrates and quality of the process sink and source as well as the fresh resource feed are parameters, Equations 12.4 through 12.7 are linear. In other words, the superstructural model is a linear program (LP) for which any solution found is globally optimal.

 Example 12.1 Direct water reuse/recycle network

The Acrylonitrile (AN) production case study in Examples 3.1 and 4.1 (El-Halwagi, 1997) is revisited. Synthesize a direct reuse/recycle network with minimum freshwater consumption. The limiting water data for the case study are shown in Table 12.1 (reproduced from Table 4.2).

Solution

A superstructure for this case is first constructed, as shown in the matching matrix form in Figure 12.3. All sources including freshwater (FW)

TABLE 12.1

Limiting Water Data for AN Production

Water Sinks, SKj		Flowrate	NH₃ Concentration
j	Stream	F_{SKj} (kg/s)	C_{SKj} (ppm)
1	Boiler feed water	1.2	0
2	Scrubber inlet	5.8	10
Water Sources, SR$_i$		Flowrate	NH₃ Concentration
i	Stream	F_{SRi} (kg/s)	C_{SRi} (ppm)
1	Distillation bottoms	0.8	0
2	Off-gas condensate	5	14
3	Aqueous layer	5.9	25
4	Ejector condensate	1.4	34

F_{SRi}	C_{SRi}	F_{SKj}	1.2	5.8	F_{WW}
		C_{SKj}	0	10	
F_{SRi}	C_{SRi}	SK_j / SR_i	SK1	SK2	WW
F_{FW}	0	FW	$F_{FW,SK1}$	$F_{FW,SK2}$	
0.8	0	SR1	$F_{SR1,SK1}$	$F_{SR1,SK2}$	$F_{SR1,WW}$
5	14	SR2	$F_{SR2,SK1}$	$F_{SR2,SK2}$	$F_{SR2,WW}$
5.9	25	SR3	$F_{SR3,SK1}$	$F_{SR3,SK2}$	$F_{SR3,WW}$
1.4	34	SR4	$F_{SR4,SK1}$	$F_{SR4,SK2}$	$F_{SR4,WW}$

FIGURE 12.3
Matching matrix for direct reuse/recycle scheme in Example 12.1.

are arranged as rows, while sinks including wastewater discharge (WW) are arranged as columns.

To formulate an optimization model to minimize the total freshwater flowrate (F_{FW}) of the RCN, the objective takes the revised form of Equation 12.1:

$$\text{Minimize } F_{FW} \tag{12.8}$$

subject to the following constraints:

Equation 12.2 relates the total freshwater flowrate with the individual freshwater flowrate requirement of the sinks ($F_{FW,SKj}$). One may easily derive this flowrate balance from the fourth row of the matching matrix in Figure 12.3:

$$F_{FW} = F_{FW,SK1} + F_{FW,SK2}$$

Equation 12.4 describes the flowrate balances for process sinks. Similarly, one can derive these flowrate balances by observing columns 4 and 5 of Figure 12.3:

$$F_{SR1,SK1} + F_{SR2,SK1} + F_{SR3,SK1} + F_{SR4,SK1} + F_{FW,SK1} = 1.2$$
$$F_{SR1,SK2} + F_{SR2,SK2} + F_{SR3,SK2} + F_{SR4,SK2} + F_{FW,SK2} = 5.8$$

The quality requirement of the sinks is described by Equation 12.5. Note however that the inequality sign takes a reverse form as that in Equation 12.5, as the quality index is described by impurity concentration. In other words, impurity sent from the process sources (and those from freshwater source[s], if any) must not be more than what can be tolerated in the process sink:

$$F_{SR1,SK1}(0) + F_{SR2,SK1}(14) + F_{SR3,SK1}(25) + F_{SR4,SK1}(34) + F_{FW,SK1}(0) \le 1.2\,(0)$$
$$F_{SR1,SK2}(0) + F_{SR2,SK2}(14) + F_{SR3,SK2}(25) + F_{SR4,SK2}(34) + F_{FW,SK2}(0) \le 5.8\,(10)$$

Equation 12.6 describes the flowrate balance of the process sources. As shown, each process source may be allocated to any process sinks, while the unutilized source will be sent for wastewater discharge ($F_{SRi,WW}$):

$$F_{SR1,SK1} + F_{SR1,SK2} + F_{SR1,WW} = 0.8$$

$$F_{SR2,SK1} + F_{SR2,SK2} + F_{SR2,WW} = 5$$

$$F_{SR3,SK1} + F_{SR3,SK2} + F_{SR3,WW} = 5.9$$

$$F_{SR4,SK1} + F_{SR4,SK2} + F_{SR4,WW} = 1.4$$

It is also convenient to determine the total flowrate of the wastewater stream (F_{WW}), given by the last column of the matching matrix in Figure 12.3:

$$F_{WW} = F_{SR1,WW} + F_{SR2,WW} + F_{SR3,WW} + F_{SR4,WW}$$

Equation 12.7 indicates that all flowrate variables take non-negative values. These include the flowrate terms for the matches of sink–source (e.g., $F_{SR1,SK1}$, $F_{SR2,SK1}$, etc.), freshwater–sink ($F_{FW,SK1}$, $F_{FW,SK2}$), and source–wastewater pairs (e.g., $F_{SR1,WW}$, $F_{SR2,WW}$, etc.):

$$F_{SR1,SK1}, F_{SR2,SK1}, F_{SR3,SK1}, F_{SR4,SK1} \geq 0$$

$$F_{SR1,SK2}, F_{SR2,SK2}, F_{SR3,SK2}, F_{SR4,SK2} \geq 0$$

$$F_{FW,SK1}, F_{FW,SK2} \geq 0$$

$$F_{SR1,WW}, F_{SR2,WW}, F_{SR3,WW}, F_{SR4,WW} \geq 0$$

In terms of LINGO formulation, the previous model can be written as follows*:

Model:
```
!Optimisation objective : to minimise freshwater (FW)
flowrate ;
min = FW;
FW = FW1 + FW2;

!Flowrate balance for process sinks;
F11 + F21 + F31 + F41 + FW1 = 1.2;
F12 + F22 + F32 + F42 + FW2 = 5.8;

!Impurity constraints for process sinks;
F11*0 + F21*14 + F31*25 + F41*34 + FW1*0 <= 1.2*0;
F12*0 + F22*14 + F32*25 + F42*34 + FW2*0 <= 5.8*10;
```

---

* The exclaimation mark in LINGO formulation enables the software to ignore the remainder of the line until next semi-colon (;). Hence, it is used to include remarks throughout chapters 12 through 17.

```
!Flowrate balance for process sources;
F11 + F12 + F1W=0.8;
F21 + F22 + F2W=5;
F31 + F32 + F3W=5.9;
F41 + F42 + F4W=1.4;

!Total wastewater flowrate (WW);
WW=F1W + F2W + F3W + F4W;

!All variables should take non-negative values;
F11 >= 0; F21 >= 0; F31 >= 0; F41 >= 0;
F12 >= 0; F22 >= 0; F32 >= 0; F42 >= 0;
FW1 >= 0; FW2 >= 0;
F1W >= 0; F2W >= 0; F3W >= 0; F4W >= 0;
END
```

Note that the last set of constraints (all variables to take non-negative values) may be omitted, as LINGO assumes all variables as non-negative values by default. However, they are needed if other optimization software is used.

The following is the solution report generated by LINGO:

```
Global optimal solution found.
Objective value:                    2.057143
Total solver iterations:                   3

                Variable         Value
                FW               2.057143
                FW1              0.4000000
                FW2              1.657143
                F11              0.8000000
                F21              0.000000
                F31              0.000000
                F41              0.000000
                F12              0.000000
                F22              4.142857
                F32              0.000000
                F42              0.000000
                F1W              0.000000
                F2W              0.8571429
                F3W              5.900000
                F4W              1.400000
                WW               8.157143
```

As shown, the minimum freshwater flowrate of the RCN is determined as 2.06 kg/s, while the wastewater flowrate is determined as 8.15 kg/s. This matches the answer in Examples 3.1 and 4.1 (see Figure 3.4 or Table 4.5). A revised matching matrix with optimized variables is shown in Figure 12.4.

Finally, note that the concentration of the wastewater discharge is not determined as part of the optimization model. This is to avoid bilinear term (product of unknown flowrate and concentration of wastewater stream) that would be present in the optimization model, which in turn leads to nonlinear program (NLP). Once the optimization model

F_{SRi}	C_{SRi}	F_{SKj}	1.2	5.8	8.16
		C_{SKj}	0	10	25.4
		SK_j / SR_i	SK1	SK2	WW
2.06	0	FW	0.4	1.66	
0.8	0	SR1	0.8		
5	14	SR2		4.14	0.86
5.9	25	SR3			5.9
1.4	34	SR4			1.4

FIGURE 12.4
Network structure for RCN in Example 12.1 (all flowrate terms in kg/s; concentration in ppm).

is solved, we can now determine the concentration of the wastewater discharge via mass balance. For this case, the concentration of the wastewater stream (C_{WW}) is calculated as follows:

$$C_{WW} = \frac{0.86(14) + 5.9(25) + 1.4(34)}{0.86 + 5.9 + 1.4} = 25.4\,\text{ppm}$$

12.2 Incorporation of Process Constraints

The advantage of the superstructural approach is its flexibility to incorporate various process constraints, such as forbidden or compulsory matches between process sources and sinks, etc. The reasons for imposing forbidden matches may include impurity buildup (due to local recycling), geographical factors (process sink and source that are too far apart), contamination issues (e.g., for certain processes, the source contains other contaminant that may poison the catalyst in the reactor), etc. In contrast, one may impose compulsory matches to connect the process source that consists of a given species (e.g., solvent, catalyst, etc.) to a sink where the species is needed (e.g., solvent extraction unit, reactor, etc.).

Example 12.2 Incorporation of process constraints

Example 12.1 is revisited here. Figure 12.4 shows that the superstructural model in Example 12.1 determines that water source SR1 (distillation bottoms) is matched to sink SK1 (boiler feed water). In most cases, this is not a favorable solution, as boiler feed water would normally prefer freshwater feed, rather than reuse/recycle water. Besides, it is decided that SR1 is to be fully recovered to SK2 (scrubber inlet). Incorporate these process constraints into the superstructural model to synthesize an alternative network structure for the case study.

Solution

The superstructural model for this case remains similar to that in Example 12.1. Two additional constraints are added to impose the forbidden (Equation 12.9) and compulsory (Equation 12.10) match criteria:

$$F_{SR1,SK1} = 0 \qquad\qquad (12.9)$$

$$F_{SR1,SK2} = 0.8 \qquad\qquad (12.10)$$

LINGO formulation for this case can be written as follows:

Model:

```
! Optimisation objective : to minimise freshwater (FW)
flowrate ;
min = FW;
FW = FW1 + FW2;

!Flowrate balance for process sinks;
F11 + F21 + F31 + F41 + FW1 = 1.2;
F12 + F22 + F32 + F42 + FW2 = 5.8;

!Process constraints;
F11 = 0; !Forbidden match between SR1 and SK1;
F12 = 0.8; !Compulsory match between SR1 and SK2;

!Impurity constraints for process sinks;
F11*0 + F21*14 + F31*25 + F41*34 + FW1*0 <= 1.2*0;
F12*0 + F22*14 + F32*25 + F42*34 + FW2*0 <= 5.8*10;

!Flowrate balance for process sources;
F11 + F12 + F1W = 0.8;
F21 + F22 + F2W = 5;
F31 + F32 + F3W = 5.9;
F41 + F42 + F4W = 1.4;

!Total wastewater flowrate (WW);
WW = F1W + F2W + F3W + F4W;

!All variables should take non-negative values;
F11 >= 0; F21 >= 0; F31 >= 0; F41 >= 0;
F12 >= 0; F22 >= 0; F32 >= 0; F42 >= 0;
FW1 >= 0; FW2 >= 0;
F1W >= 0; F2W >= 0; F3W >= 0; F4W >= 0;
END
```

The following is the solution report generated by LINGO:

```
Global optimal solution found.
Objective value:                    2.057143
Total solver iterations:                   3

            Variable        Value
            FW              2.057143
            FW1             1.200000
            FW2             0.8571429
```

		F_{SKj}	1.2	5.8	8.16
		C_{SKj}	0	10	25.4
F_{SRi}	C_{SRi}	SK_j SR_i	SK1	SK2	WW
2.06	0	FW	1.2	0.86	
0.8	0	SR1		0.8	
5	14	SR2		4.14	0.86
5.9	25	SR3			5.9
1.4	34	SR4			1.4

FIGURE 12.5
Network structure for RCN in Example 12.2 (all flowrate terms in kg/s; concentration in ppm).

```
F11                     0.000000
F21                     0.000000
F31                     0.000000
F41                     0.000000
F12                     0.8000000
F22                     4.142857
F32                     0.000000
F42                     0.000000
F1W                     0.000000
F2W                     0.8571429
F3W                     5.900000
F4W                     1.400000
WW                      8.157143
```

It is observed that the minimum freshwater and wastewater flowrates are determined as 2.06 and 8.15 kg/s, respectively, which is identical for the case in Example 12.1. This means that *degenerate* solutions exist for this problem, in which more than one network structure achieves the same minimum water flowrates. The optimized variables are included in the revised matching matrix in Figure 12.5. Similar to Example 12.1, the concentration for wastewater stream (C_{WW}) is determined after the optimization model is solved.

Comparing the network structure in Figures 12.4 and 12.5, we observe that only flowrates that involve matches among freshwater feed (FW), SR1, SK1, and SK2 have been changed, while flowrates for the rest of the matches remain identical.

12.3 Capital and Total Cost Estimations

Another advantage of the superstructural approach is its ability to estimate the capital cost of various RCN elements, e.g., piping connection and storage tank (for batch processes). The estimation of capital cost element then enables us to calculate the total annual cost (*TAC*) of the RCN.

Capital cost (*CC*) correlation often includes variation and fixed items, such as those given in Equation 12.11:

$$CC = aF^b + c \qquad (12.11)$$

where *a* and *b* are parameters for the variation terms and are always expressed as a function of flowrate *F*, while *c* is the parameter for the fixed term.

In order to incorporate capital (e.g., piping) cost correlation, we need to activate the fixed cost component by introducing a binary variable (*BIN*) in the optimization model. Hence, Equation 12.11 is revised as follows:

$$CC = aF^b + c\ BIN \qquad (12.12)$$

For instance, in a direct material reuse/recycle problem where the piping capital cost is to be minimized, binary term $B_{SRi,SKj}$ [0, 1] may be used to define the presence of a material recovery stream between source SR_i and sink SK_j, with flowrate $F_{SRi,SKj}$. The binary term is activated with Equation 12.13:

$$\frac{F_{SRi,SKj}}{M} \le B_{SRi,SKj} \qquad (12.13)$$

where *M* is a parameter with arbitrary large number. Any value that is bigger than the potential flowrate term in Equation 12.13 may be used. When the optimization model determines that the flowrate term takes a value of zero, the fixed cost term in Equation 12.12 is forced to be zero, in order to minimize capital cost. In contrast, when the optimization model determines that the $F_{SRi,SKj}$ term has a nonzero value, the binary term will be forced as unity, thus activating the fixed cost component in Equation 12.12. Apart from piping cost, Equations 12.12 and 12.13 may also be used to determine mass storage cost for batch RCN problems (see Chapter 17).

To determine the *TAC*, the capital cost is to be annualized so that it can be added to the annual operating cost (*AOC*):

$$TAC = AOC + CC\ (AF) \qquad (12.14)$$

where *AF* is the annualizing factor for the capital cost. A commonly used expression for *AF* is given by

$$AF = \frac{IR(1+IR)^{YR}}{(1+IR)^{YR} - 1} \qquad (12.15)$$

where
  IR is the annual fractional interest rate
  YR is the number of years considered for the analysis (Smith, 2005)

The *AOC* in Equation 12.14 is a product of operating cost (*OC*) and annual operating time (*AOT*), i.e.,

$$AOC = OC\left(AOT\right) \tag{12.16}$$

where *OC* is given as the costs of fresh resource and waste discharge. These cost terms are the product of their respective flowrates with their unit costs, which is given as in Equation 12.17:

$$OC = F_R CT_R + F_D CT_D \tag{12.17}$$

where F_R and F_D are the flowrates of fresh resource and waste discharge, respectively, with unit costs of CT_R and CT_D.

Note, however, that introducing the binary term (Equations 12.12 and 12.13) leads the LP model (direct reuse/recycle scheme) into a mixed-integer linear problem (MILP). Most commercial optimization software (e.g., LINGO, GAMS, etc.) may be used to solve the MILP model easily.

Example 12.3 Minimum cost solutions for direct reuse/recycle scheme

Example 12.2 is revisited. Determine the operating, capital, and the minimum total annual costs for a direct water reuse/recycle network. Additional information for relevant cost calculation is given as follows.

1. For operating cost calculation, both freshwater (CT_{FW}) and wastewater (CT_{WW}) are assumed to have a unit cost of $0.001/kg ($1/ton).
2. Capital cost of the RCN is mainly contributed by the piping cost of reuse/recycle streams as well as the connections between freshwater feed and the sinks (wastewater streams are omitted). Capital cost of piping (*PC*, in USD) for a stream of flowrate *F* is given as the revised form of Equation 12.11 (Kim and Smith, 2004):

$$PC = \left(2F + 250\right) DIST \tag{12.18}$$

   where *DIST* is the individual distance between water sinks (SK1, SK2) and water sources (SR1 – SR4), and freshwater feed (FW), respectively, given as in Table 12.2.
3. For calculation of the total annual cost, the annualizing factor (*AF*) in Equation 12.15 is to be used. The piping cost is annualized to a period of 5 years, with an interest rate of 5%. Assume an annual operation of 8000 h.

Solution

To determine the operating cost (*OC*), Equation 12.17 is used:

$$OC(\$/s) = F_{FW} CT_{FW} + F_{WW} CT_{WW} = \left(F_{FW} + F_{WW}\right)\left(\$0.001/kg\right)$$

TABLE 12.2

Distance (m) between
Water Sources/Freshwater
Feed with Water Sinks

	SK1	SK2
FW	20	30
SR1	15	40
SR2	30	50
SR3	40	20
SR4	30	10

Next, the *AOC* is calculated using a revised Equation 12.16 that incorporates the AOT, given as follows:

$$AOC(\$/\text{year}) = (F_{FW} + F_{WW})\ (\$0.001/\text{kg})\ (3600\ \text{s/h})\ (8000\ \text{h})$$

To determine the piping cost, Equation 12.18 is used in conjunction with Equation 12.13, to examine the presence/absence of a stream connection. These are given by the next 10 lines of constraint sets. Piping costs for matches between various sources (SR1-SR4) with sink SK1 are given in lines 1–4, while those with SK2 are given in lines 5–8, respectively. The constraint sets in lines 9 and 10 next determine the piping costs for matches between freshwater feed with both sinks.

$$PC_{SR1,SK1} = (2F_{SR1,SK1} + 250B_{SR1,SK1})15; \qquad F_{SR1,SK1} \leq 1000\,B_{SR1,SK1}$$

$$PC_{SR2,SK1} = (2F_{SR2,SK1} + 250B_{SR2,SK1})30; \qquad F_{SR2,SK1} \leq 1000B_{SR2,SK1}$$

$$PC_{SR3,SK1} = (2F_{SR3,SK1} + 250B_{SR3,SK1})40; \qquad F_{SR3,SK1} \leq 1000B_{SR3,SK1}$$

$$PC_{SR4,SK1} = (2F_{SR4,SK1} + 250B_{SR4,SK1})30; \qquad F_{SR4,SK1} \leq 1000B_{SR4,SK1}$$

$$PC_{SR1,SK2} = (2F_{SR1,SK2} + 250B_{SR1,SK2})40; \qquad F_{SR1,SK2} \leq 1000B_{SR1,SK2}$$

$$PC_{SR2,SK2} = (2F_{SR2,SK2} + 250B_{SR2,SK2})50; \qquad F_{SR2,SK2} \leq 1000B_{SR2,SK2}$$

$$PC_{SR3,SK2} = (2F_{SR3,SK2} + 250B_{SR3,SK2})20; \qquad F_{SR3,SK2} \leq 1000B_{SR3,SK2}$$

$$PC_{SR4,SK2} = (2F_{SR4,SK2} + 250B_{SR4,SK2})10; \qquad F_{SR4,SK2} \leq 1000B_{SR4,SK2}$$

$$PC_{FW,SK1} = (2F_{FW,SK1} + 250B_{FW,SK1})20; \qquad F_{FW,SK1} \leq 1000B_{FW,SK1}$$

$$PC_{FW,SK2} = (2F_{FW,SK2} + 250B_{FW,SK2})30; \qquad F_{FW,SK2} \leq 1000B_{FW,SK2}$$

Note that the arbitrary term *M* in Equation 12.13 takes the value of 1000 in this case. Note also that the binary terms in Equation 12.13 for all reuse/recycle streams as well as those between freshwater feed and the sinks are represented in Figure 12.6.

		F_{SKj}	1.2	5.8	F_{WW}
		C_{SKj}	0	10	
F_{SRi}	C_{SRi}	SK_j / SR_i	SK1	SK2	WW
F_{FW}	0	FW	$B_{FW,SK1}$	$B_{FW,SK2}$	
0.8	0	SR1	$B_{SR1,SK1}$	$B_{SR1,SK2}$	
5	14	SR2	$B_{SR2,SK1}$	$B_{SR2,SK2}$	
5.9	25	SR3	$B_{SR3,SK1}$	$B_{SR3,SK2}$	
1.4	34	SR4	$B_{SR4,SK1}$	$B_{SR4,SK2}$	

FIGURE 12.6
Binary variables in Example 12.3.

The total piping cost (*PC*), and hence the capital cost (*CC*) for the RCN, is contributed by the summation of piping cost for those individual matches, i.e.,

$$CC = PC = PC_{SR1,SK1} + PC_{SR2,SK1} + PC_{SR3,SK1} + PC_{SR4,SK1} + PC_{SR1,SK2}$$
$$+ PC_{SR2,SK2} + PC_{SR3,SK2} + PC_{SR4,SK2} + PC_{FW,SK1} + PC_{FW,SK2}$$

Next, the annualizing factor (*AF*) in Equation 12.15 is calculated as

$$AF = \frac{0.05(1+0.05)^5}{(1+0.05)^5 - 1} = 0.231$$

To determine the *TAC*, Equation 12.14 is used:

$$TAC = AOC + CC(0.231)$$

In this case, it is appropriate to set the optimization objective to minimize the *TAC*, i.e.,

$$\text{Minimize} \, TAC \qquad\qquad (12.19)$$

Since this problem is essentially an extension of that in Example 12.2, all other constraints remain the same as in Example 12.2. LINGO formulation for this MILP problem is given as follows (basic formulation remains the same as in Example 12.2, too)*:

Model:
```
!Optimisation objective: To minimise total annual cost
(TAC);
min = TAC;
```

---

* The command of @ BIN (variable) is used to define binary variables in LINGO.

```
TAC=AOC + PC*0.231;
AOC=OC*3600*8000;
OC=(FW + WW)*0.001; !Unit cost of FW and WW=$1/ton;
PC=PFW1 + PFW2 + P11 + P21 + P31 + P41 + P12
+ P22 + P32 + P42;

!Total freshwater flowrate (FW);
FW=FW1 + FW2;

!Flowrate balance for process sinks;
F11 + F21 + F31 + F41 + FW1=1.2;
F12 + F22 + F32 + F42 + FW2=5.8;

!Process constraints;
F11=0; !Forbidden match between SR1 and SK1;
F12=0.8; !Compulsory match between SR1 and SK2;

!Impurity constraints for process sinks;
F11*0 + F21*14 + F31*25 + F41*34 + FW1*0 <= 1.2*0;
F12*0 + F22*14 + F32*25 + F42*34 + FW2*0 <= 5.8*10;

!Flowrate balance for process sources;
F11 + F12 + F1W=0.8;
F21 + F22 + F2W=5;
F31 + F32 + F3W=5.9;
F41 + F42 + F4W=1.4;

!Total wastewater flowrate (WW);
WW=F1W + F2W + F3W + F4W;

!All variables should take non-negative values;
F11 >= 0; F21 >= 0; F31 >= 0; F41 >= 0;
F12 >= 0; F22 >= 0; F32 >= 0; F42 >= 0;
FW1 >= 0; FW2 >= 0;
F1W >= 0; F2W >= 0; F3W >= 0; F4W >= 0;

!Capital cost estimation;
P11=(2*F11 + 250*B11)*15;
P21=(2*F21 + 250*B21)*30;
P31=(2*F31 + 250*B31)*40;
P41=(2*F41 + 250*B41)*30;
P12=(2*F12 + 250*B12)*40;
P22=(2*F22 + 250*B22)*50;
P32=(2*F32 + 250*B32)*20;
P42=(2*F42 + 250*B42)*10;
PFW1=(2*FW1 + 250*BFW1)*20;
PFW2=(2*FW2 + 250*BFW2)*30;

!Binary variables that represent present/absent of a
streams;
F11 <= M*B11; F21 <= M*B21; F31 <= M*B31; F41 <= M*B41;
F12 <= M*B12; F22 <= M*B22; F32 <= M*B32; F42 <= M*B42;
FW1 <= M*BFW1; FW2 <= M*BFW2;
M=1000; !Arbitrary value;

!Defining binary variables;
@BIN (B11); @BIN (B21); @BIN (B31); @BIN (B41);
@BIN (B12); @BIN (B22); @BIN (B32); @BIN (B42);
@BIN (BFW1); @BIN (BFW2);
```

The following is the solution report generated by LINGO:

```
Global optimal solution found.
Objective value:                        302389.9
Extended solver steps:                         0
Total solver iterations:                       4

              Variable      Value
              TAC           302389.9
              AOC           294171.4
              PC             35577.71
              OC              0.1021429E-01
              FW              2.057143
              WW              8.157143
              PFW1         5048.000
              PFW2         7551.429
              P11             0.000000
              P21             0.000000
              P31             0.000000
              P41             0.000000
              P12         10064.00
              P22         12914.29
              P32             0.000000
              P42             0.000000
              FW1             1.200000
              FW2             0.8571429
              F11             0.000000
              F21             0.000000
              F31             0.000000
              F41             0.000000
              F12             0.8000000
              F22             4.142857
              F32             0.000000
              F42             0.000000
              F1W             0.000000
              F2W             0.8571429
              F3W             5.900000
              F4W             1.400000
              B11             0.000000
              B21             0.000000
              B31             0.000000
              B41             0.000000
              B12             1.000000
              B22             1.000000
              B32             0.000000
              B42             0.000000
              BFW1            1.000000
              BFW2            1.000000
              M            1000.000
```

The various cost elements for the RCN are summarized in Table 12.3. Note that all process variables (i.e., reuse/recycle, freshwater, and waste-water flowrates) remain the same as those in Example 12.2. Hence, the RCN takes the same structure as in Figure 12.5.

TABLE 12.3

Cost Elements of RCN in Example 12.3

Operating cost (OC)	$0.01/s
Annual operating cost (AOC)	$294,171/year
Capital/piping cost (PC)	$35,577
Total annual cost (TAC)	$302,389/year

12.4 Reducing Network Complexity

In some cases, it is desired to reduce network complexity by having less piping connections. Hence, the maximum number of connections may be imposed in the optimization problem. For instance, the maximum number of reuse/recycle connections between source SR_i and sink SK_j may be given by the following constraint:

$$\Sigma_j B_{SRi,SKj} \leq B^{Max} \quad \forall i \tag{12.20}$$

where B^{Max} is the maximum number of connections.

Another realistic consideration in reducing network complexity is to impose minimum flowrate constraint for the piping connections. For instance, the reuse/recycle connections may be forced to take a flowrate that is bigger than the minimum value (F^{Min}), given by Equation 12.21:

$$F^{Min}B_{SRi,SKj} \leq F_{SRi,SKj} \quad \forall i, j \tag{12.21}$$

In Equation 12.21, the binary variable will be forced to take a zero value, if the reuse/recycle flowrate is smaller than the minimum flowrate value.

Example 12.4 Reduced network complexity

Example 12.3 is revisited. For this case, the forbidden match between any process sources with boiler feed water (SK1) is to be maintained, while the compulsory between distillation bottoms (SR1) and scrubber inlet (SK2) is to be relaxed. Determine the various cost elements such as those in Example 12.3, with maximum reuse/recycle connection limited to one; and the minimum flowrates in the reuse/recycle connections are to be bigger than 1 kg/s.

Solution

In order to impose a maximum number of reuse/recycle connections in the RCN, Equation 12.20 takes the following form:

$$B_{SR1,SK1}+B_{SR2,SK1}+B_{SR3,SK1}+B_{SR4,SK1}+B_{SR1,SK2}+B_{SR2,SK2}+B_{SR3,SK2}+B_{SR4,SK2} \leq 1$$

Besides, all reuse/recycle connections in the RCN should have a minimum flowrate that is larger than 1 kg/s, which are given as follows:

$$F^{Min}B_{SR1,SK1} \leq F_{SR1,SK1}; F^{Min}B_{SR2,SK1} \leq F_{SR2,SK1}$$

$$F^{Min}B_{SR3,SK1} \leq F_{SR3,SK1}; F^{Min}B_{SR4,SK1} \leq F_{SR4,SK1}$$

$$F^{Min}B_{SR1,SK2} \leq F_{SR1,SK2}; F^{Min}B_{SR2,SK2} \leq F_{SR2,SK2}$$

$$F^{Min}B_{SR3,SK2} \leq F_{SR3,SK2}; F^{Min}B_{SR4,SK2} \leq F_{SR4,SK2}$$

where F^{Min} takes a value of 1 kg/s.

Note that the rest of the constraints remain identical with those in Example 12.3 (except the compulsory match between SR1 and SK2 is now relaxed). LINGO formulation for this MILP problem is written as follows:

```
Model:
!Optimisation objective: To minimise total annual cost
(TAC);
min=TAC;
TAC=AOC + PC'0.231;
AOC=OC'3600'8000;
OC=(FW + WW)'0.001; !Unit cost of FW and WW=$1/ton;
PC=PFW1 + PFW2 + P11 + P21 + P31 + P41 + P12 + P22 + P32 +
  P42;

!Total freshwater flowrate (FW);
FW=FW1 + FW2;

!Flowrate balance for process sinks;
F11 + F21 + F31 + F41 + FW1=1.2;
F12 + F22 + F32 + F42 + FW2=5.8;

!Process constraints;
F11=0; !Forbidden match between SR1 and SK1;

!Impurity constraints for process sinks;
F11'0 + F21'14 + F31'25 + F41'34 + FW1'0 <= 1.2'0;
F12'0 + F22'14 + F32'25 + F42'34 + FW2'0 <= 5.8'10;

!Flowrate balance for process sources;
F11 + F12 + F1W=0.8;
F21 + F22 + F2W=5;
F31 + F32 + F3W=5.9;
F41 + F42 + F4W=1.4;

!Total wastewater flowrate (WW);
WW=F1W + F2W + F3W + F4W;

!All variables should take non-negative values;
F11 >= 0; F21 >= 0; F31 >= 0; F41 >= 0;
F12 >= 0; F22 >= 0; F32 >= 0; F42 >= 0;
FW1 >= 0; FW2 >= 0;
F1W >= 0; F2W >= 0; F3W >= 0; F4W >= 0;

!Capital cost estimation;
P11=(2'F11 + 250'B11)'15;
P21=(2'F21 + 250'B21)'30;
```

```
P31 = (2*F31 + 250*B31)*40;
P41 = (2*F41 + 250*B41)*30;
P12 = (2*F12 + 250*B12)*40;
P22 = (2*F22 + 250*B22)*50;
P32 = (2*F32 + 250*B32)*20;
P42 = (2*F42 + 250*B42)*10;
PFW1 = (2*FW1 + 250*BFW1)*20;
PFW2 = (2*FW2 + 250*BFW2)*30;

!Binary variables that represent present/absent of a
streams;
F11 <= M*B11; F21 <= M*B21; F31 <= M*B31; F41 <= M*B41;
F12 <= M*B12; F22 <= M*B22; F32 <= M*B32; F42 <= M*B42;
FW1 <= M*BFW1; FW2 <= M*BFW2;
M = 1000; !Arbitrary value;

!Minimum flowrate constraint in piping connections;
FMin*B11 <= F11 ; FMin*B21 <= F21 ; FMin*B31 <= F31 ;
FMin*B41 <= F41 ;
FMin*B12 <= F12 ; FMin*B22 <= F22 ; FMin*B32 <= F32 ;
FMin*B42 <= F42 ;
FMin = 1; !Minimum flowrate constraint;

!Maximum piping connections;
B11 + B21 + B31 + B41 + B12 + B22 + B32 + B42 <= 1;

!Defining binary variables;
@BIN (B11); @BIN (B21); @BIN (B31); @BIN (B41);
@BIN (B12); @BIN (B22); @BIN (B32); @BIN (B42);
@BIN (BFW1); @BIN (BFW2);
END
```

The following is the solution report generated by LINGO:

```
Global optimal solution found.
Objective value:                      346156.2
Extended solver steps:                       0
Total solver iterations:                     4

                Variable        Value
                TAC          346156.2
                AOC          340251.4
                PC           25561.71
                OC               0.1181429E-01
                FW               2.857143
                WW               8.957143
                PFW1          5048.000
                PFW2          7599.429
                P11              0.000000
                P21              0.000000
                P31              0.000000
                P41              0.000000
                P12              0.000000
                P22          12914.29
                P32              0.000000
                P42              0.000000
                B11              0.000000
                B21              0.000000
```

```
B31                    0.000000
B41                    0.000000
B12                    0.000000
B22                    1.000000
B32                    0.000000
B42                    0.000000
FW1                    1.200000
FW2                    1.657143
F11                    0.000000
F21                    0.000000
F31                    0.000000
F41                    0.000000
F12                    0.000000
F22                    4.142857
F32                    0.000000
F42                    0.000000
F1W                    0.8000000
F2W                    0.8571429
F3W                    5.900000
F4W                    1.400000
BFW1                   1.000000
BFW2                   1.000000
M               1000.000
FMIN                   1.000000
```

The various cost elements for the RCN are summarized in Table 12.4. As compared with Example 12.3 (Table 12.3), there is a significant increase in all cost elements for this case. The *AOC* and *TAC* were increased by 15.7% and 14.5%, respectively.

The network structure of the RCN is shown in Figure 12.7. As compared with cases in Examples 12.1 through 12.3, reuse/recycle streams have been reduced from two to one connection. This leads to an increase in both freshwater and wastewater flowrates, i.e., a rise of 38.8% (from 2.06 to 2.86 kg/s) and 9.8% (from 8.16 to 8.96 kg/s), respectively. Note that the reuse/recycle connection has a flowrate that is more than 1 kg/s.

12.5 Superstructural Model for Material Regeneration Network

To synthesize a material regeneration network (with interception unit of single pass type), we may make use of the superstructural representation

TABLE 12.4

Cost Elements of RCN in Example 12.4

Operating cost (OC)	$0.012/s
Annual operating cost (AOC)	$340,251/year
Capital/piping cost (PC)	$25,562
Total annual cost (TAC)	$346,156/year

		F_{SKj}	1.2	5.8	8.96
		C_{SKj}	0	10	5.3
F_{SRi}	C_{SRi}	SK_j SR_i	SK1	SK2	WW
2.86	0	FW	1.2	1.66	
0.8	0	SR1			0.8
5	14	SR2		4.14	0.86
5.9	25	SR3			5.9
1.4	34	SR4			1.4

FIGURE 12.7
Network structure for RCN in Example 12.4 (all flowrate terms in kg/s; concentration in ppm).

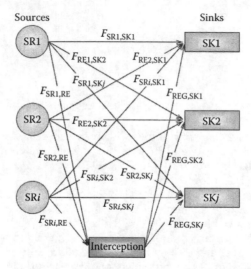

FIGURE 12.8
Superstructural model for material regeneration network.

in Figure 12.8 (reproduced from Figure 1.18) or alternatively its matching matrix form in Figure 12.9. Similar to the case in Figures 12.1 and 12.2, all process sources are matched to the process sinks via connections. Besides, the interception unit acts as both sink and source. When source is sent for regeneration (with flowrate $F_{SRi,RE}$), the inlet of the interception unit acts as a sink. The outlet stream of the unit is the regenerated source of higher quality, which will be sent for reuse/recycle in the process sinks (with flowrate $F_{REG,SKj}$). Note however that no match should be built between the fresh resource and the interception unit as well as between the regenerated source and waste discharge, i.e., $F_{R,RE} = F_{REG,D} = 0$.

The optimization model for a material regeneration network can be extended from that of the reuse/recycle network, which takes the revised form of Equations 12.4 through 12.7. Note that in each constraint set, a new

		F_{SKj}	F_{SK1}	F_{SK2}	F_{SK3}	F_{SKj}	F_{RE}	F_D
		q_{SKj}	q_{SK1}	q_{SK2}	q_{SK3}	q_{SKj}	q_{Rin}	q_D
F_{SRi}	q_{SRi}	SK_j / SR_i	SK1	SK2	SK3	SK_j	RE	WD
F_R	q_R	FR	$F_{R,SK1}$	$F_{R,SK2}$	$F_{R,SK3}$	$F_{R,SKj}$		
F_{Rout}	q_{REG}	REG	$F_{REG,SK1}$	$F_{REG,SK2}$	$F_{REG,SK3}$	$F_{REG,SKj}$		
F_{SR1}	q_{SR1}	SR1	$F_{SR1,SK1}$	$F_{SR1,SK2}$	$F_{SR1,SK3}$	$F_{SR1,SKj}$	$F_{SR1,RE}$	$F_{SR1,D}$
F_{SR2}	q_{SR2}	SR2	$F_{SR2,SK1}$	$F_{SR2,SK2}$	$F_{SR2,SK3}$	$F_{SR2,SKj}$	$F_{SR2\,RE}$	$F_{SR2,D}$
F_{SR3}	q_{SR3}	SR3	$F_{SR3,SK1}$	$F_{SR3,SK2}$	$F_{SR3,SK3}$	$F_{SR3,SKj}$	$F_{SR3,RE}$	$F_{SR3,D}$
F_{SRi}	q_{SRi}	SR_i	$F_{SRi,SK1}$	$F_{SRi,SK2}$	$F_{SRi,SK3}$	$F_{SRi,SKj}$	$F_{SRi\,RE}$	$F_{SRi,D}$

FIGURE 12.9
Superstructural model for material regeneration network in matching matrix form.

flowrate term is added for the interception unit (with outlet quality of q_{ROUT}), to account for the regenerated source that is sent to process sink ($F_{REG,SKj}$) or the source to be sent for regeneration ($F_{SRi,RE}$):

$$\Sigma_i F_{SRi,SKj} + F_{REG,SKj} + F_{R,SKj} = F_{SKj} \quad \forall j \tag{12.22}$$

$$\Sigma_i F_{SRi,SKj}\, q_{SRi} + F_{REG,SKj}\, q_{Rout} + F_{R,SKj}\, q_R \geq F_{SKj}\, q_{SKj} \quad \forall j \tag{12.23}$$

$$\Sigma_j F_{SRi,SKj} + F_{SRi,RE} + F_{SRi,D} = F_{SRi} \quad \forall i \tag{12.24}$$

$$F_{SRi,SKj},\, F_{R,SKj},\, F_{SRi,D},\, F_{REG,SKj},\, F_{SRi,RE} \geq 0 \quad \forall i, j \tag{12.25}$$

Note that a flowrate balance equation is needed for the interception unit (for single pass type), given as follows:

$$\Sigma_i F_{SRi,RE} = \Sigma_j F_{REG,SKj} \tag{12.26}$$

The objective of the optimization problem can be set to minimize the operating costs of the fresh resource and waste discharge (with Equation 12.17) or the TAC of the RCN (with Equation 12.14). Alternatively, we may also minimize the flowrates of both fresh resource (F_R) and regenerated source (F_{REG}). However, since there are two objectives to be minimized, a two-stage optimization approach is needed. In the first stage, the fresh resource flowrate is minimized (with Equation 12.1), subject to the constraints in Equations 12.22 through 12.26. Once the minimum fresh resource flowrate is identified, it is then included as a new constraint in the second stage (Equation 12.27), where the regenerated source flowrate (F_{REG}) is minimized, with the optimization objective in Equation 12.28. Similar to the earlier case, this is an LP model for which any solution found is globally optimal.

$$F_{R,min} = F_R \tag{12.27}$$

$$\text{Minimize } F_{REG} \tag{12.28}$$

Example 12.5 Water regeneration network

The classical water minimization case (Wang and Smith, 1994) in Example 8.2 is revisited here. Use the superstructural approach to synthesize a water regeneration network when an interception unit with outlet concentration (C_{Rout}) of 5 ppm is used to purify water sources for further water recovery. Pure freshwater (0 ppm) is available for use in the RCN. Limiting data are given in Table 12.5 (reproduced from Table 8.1). Solve for the following cases:

1. Determine the minimum operating cost (OC) of the RCN. The unit costs of freshwater, wastewater, and regenerated sources are given as \$1/ton (CT_{FW}), \$1/ton (CT_{WW}), and \$0.8/ton (CT_{RW}), respectively.
2. Minimum flowrates for both freshwater and regenerated source.
3. Redo Case 2, where each process sink can only accept a maximum of two reuse/recycle connections (for the ease of process control).

Solution

A superstructure for the problem is first constructed, as shown in Figure 12.10. All sources including freshwater (FW) regenerated source (REG) are arranged as rows, while sinks including regeneration inlet (RE) and wastewater discharge (WW) are arranged as columns.

The constraints of the problem are given as follows. Equation 12.22 describes the flowrate balances for process sinks, which may be derived by observing flowrate balances in columns 4–7 of Figure 12.10:

$$F_{SR1,SK1} + F_{SR2,SK1} + F_{SR3,SK1} + F_{SR4,SK1} + F_{REG,SK1} + F_{FW,SK1} = 20$$

TABLE 12.5

Limiting Water Data for Example 12.5

Sinks, SK_j	F_{SKj} (t/h)	C_{SKj} (ppm)	Sources, SR_i	F_{SRi} (t/h)	C_{SRi} (ppm)
SK1	20	0	SR1	20	100
SK2	100	50	SR2	100	100
SK3	40	50	SR3	40	800
SK4	10	400	SR4	10	800
$\Sigma_j F_{SKj}$	170		$\Sigma_i F_{SRi}$	170	

F_{SRi}	C_{SRi}	F_{SKj}	20	100	40	10	F_{RE}	F_{WW}
		C_{SKj}	0	50	50	400		
		SK_j / SR_i	SK1	SK2	SK3	SK4	RE	WW
F_{FW}	0	FW	$F_{FW,SK1}$	$F_{FW,SK2}$	$F_{FW,SK3}$	$F_{FW,SK4}$		
F_{REG}	5	REG	$F_{REG,SK1}$	$F_{REG,SK2}$	$F_{REG,SK3}$	$F_{REG,SKj}$		
20	100	SR1	$F_{SR1,SK1}$	$F_{SR1,SK2}$	$F_{SR1,SK3}$	$F_{SR1,SK4}$	$F_{SR1,RE}$	$F_{SR1,WW}$
100	100	SR2	$F_{SR2,SK1}$	$F_{SR2,SK2}$	$F_{SR2,SK3}$	$F_{SR2,SK4}$	$F_{SR2,RE}$	$F_{SR2,WW}$
40	800	SR3	$F_{SR3,SK1}$	$F_{SR3,SK2}$	$F_{SR3,SK3}$	$F_{SR3,SK4}$	$F_{SR3,RE}$	$F_{SR3,WW}$
10	800	SR4	$F_{SR4,SK1}$	$F_{SR4,SK2}$	$F_{SR4,SK3}$	$F_{SR4,SK4}$	$F_{SRi,RE}$	$F_{SR4,WW}$

FIGURE 12.10
Matching matrix for water regeneration network in Example 12.5.

$$F_{SR1,SK2} + F_{SR2,SK2} + F_{SR3,SK2} + F_{SR4,SK2} + F_{REG,SK2} + F_{FW,SK2} = 100$$

$$F_{SR1,SK3} + F_{SR2,SK3} + F_{SR3,SK3} + F_{SR4,SK3} + F_{REG,SK3} + F_{FW,SK3} = 40$$

$$F_{SR1,SK4} + F_{SR2,SK4} + F_{SR3,SK4} + F_{SR4,SK4} + F_{REG,SK4} + F_{FW,SK4} = 10$$

The quality requirement for each sink is described by Equation 12.23, as follows. Note that the freshwater term is omitted as it does not contain any impurity (i.e., $C_{FW} = 0$ ppm). Note also that the inequality sign takes a reverse form as that in Equation 12.23, as the quality index is described by impurity concentration.

$$F_{SR1,SK1}(100) + F_{SR2,SK1}(100) + F_{SR3,SK1}(800) + F_{SR4,SK1}(800)$$
$$+ F_{REG,SK1}(5) \leq 20 \ (0)$$

$$F_{SR1,SK2}(100) + F_{SR2,SK2}(100) + F_{SR3,SK2}(800) + F_{SR4,SK2}(800)$$
$$+ F_{REG,SK2}(5) \leq 100 \ (50)$$

$$F_{SR1,SK3}(100) + F_{SR2,SK3}(100) + F_{SR3,SK3}(800) + F_{SR4,SK3}(800)$$
$$+ F_{REG,SK3}(5) \leq 40 \ (50)$$

$$F_{SR1,SK4}(100) + F_{SR2,SK4}(100) + F_{SR3,SK4}(800) + F_{SR4,SK4}(800)$$
$$+ F_{REG,SK4}(5) \leq 10 \ (400)$$

Equation 12.24 next describes the flowrate balance of the process sources. As shown, each process source may be allocated to any process sinks, while the unutilized source will be sent for wastewater discharge ($F_{SRi,WW}$):

$$F_{SR1,SK1} + F_{SR1,SK2} + F_{SR1,SK3} + F_{SR1,SK4} + F_{SR1,RE} + F_{SR1,WW} = 20$$

$$F_{SR2,SK1} + F_{SR2,SK2} + F_{SR2,SK3} + F_{SR2,SK4} + F_{SR2,RE} + F_{SR2,WW} = 100$$

$$F_{SR3,SK1} + F_{SR3,SK2} + F_{SR3,SK3} + F_{SR3,SK4} + F_{SR3,RE} + F_{SR3,WW} = 40$$

$$F_{SR4,SK1} + F_{SR4,SK2} + F_{SR4,SK3} + F_{SR4,SK4} + F_{SR4,RE} + F_{SR4,WW} = 10$$

A flowrate balance of the regeneration unit is given by Equation 12.26 as follows:

$$F_{SR1,RE} + F_{SR2,RE} + F_{SR3,RE} + F_{SR4,RE} = F_{REG,SK1} + F_{REG,SK2} + F_{REG,SK3} + F_{REG,SK4}$$

Equation 12.25 indicates that all flowrate variables of the optimization model take non-negative values:

$$F_{SR1,SK1}, F_{SR2,SK1}, F_{SR3,SK1}, F_{SR4,SK1} \geq 0$$

$$F_{SR1,SK2}, F_{SR2,SK2}, F_{SR3,SK2}, F_{SR4,SK2} \geq 0$$

$$F_{SR1,SK3}, F_{SR2,SK3}, F_{SR3,SK3}, F_{SR4,SK3} \geq 0$$

$$F_{SR1,SK4}, F_{SR2,SK4}, F_{SR3,SK4}, F_{SR4,SK4} \geq 0$$

$$F_{FW,SK1}, F_{FW,SK2}, F_{FW,SK3}, F_{FW,SK4} \geq 0$$

$$F_{SR1,WW}, F_{SR2,WW}, F_{SR3,WW}, F_{SR4,WW} \geq 0$$

$$F_{SR1,RE} + F_{SR2,RE} + F_{SR3,RE} + F_{SR4,RE} \geq 0$$

$$F_{REG,SK1} + F_{REG,SK2} + F_{REG,SK3} + F_{REG,SK4} \geq 0$$

It is also convenient to relate the total flowrates of the freshwater (F_{FW}), wastewater (F_{WW}), as well as the regenerated sources with their individual flowrate contributions:

$$F_{FW} = F_{FW,SK1} + F_{FW,SK2} + F_{FW,SK3} + F_{FW,SK4}$$

$$F_{WW} = F_{SR1,WW} + F_{SR2,WW} + F_{SR3,WW} + F_{SR4,WW}$$

$$F_{REG} = F_{SR1,RE} + F_{SR2,RE} + F_{SR3,RE} + F_{SR4,RE}$$

1. Case 1—Determine the minimum OC of the RCN.

   For this case, Equation 12.3 takes the revised form as follows:

$$\text{Minimize} OC \ (\$/h)$$

where

$$OC = F_{FW}CT_{FW} + F_{WW}CT_{WW} + F_{REG}CT_{RW}$$

$$= (F_{FW} + F_{WW})(\$1/\text{ton}) + F_{REG}(\$0.8/\text{ton})$$

The LINGO formulation for this LP model can be written as follows:

```
Model:
min = (FW + WW)*1 + REG*0.8;

!Flowrate balance for freshwater;
FW = FW1 + FW2 + FW3 + FW4;

!Flowrate balance for interception units;
```

```
REG=RE1 + RE2 + RE3 + RE4;
REG=REG1 + REG2 + REG3 + REG4;

!Flowrate balance for process sinks;
F11 + F21 + F31 + F41 + REG1 + FW1=20;
F12 + F22 + F32 + F42 + REG2 + FW2=100;
F13 + F23 + F33 + F43 + REG3 + FW3=40;
F14 + F24 + F34 + F44 + REG4 + FW4=10;

!Impurity constraints for process sinks;
F11*100 + F21*100 + F31*800 + F41*800 + REG1*5 <=
20*0;
F12*100 + F22*100 + F32*800 + F42*800 + REG2*5 <=
100*50;
F13*100 + F23*100 + F33*800 + F43*800 + REG3*5 <=
40*50;
F14*100 + F24*100 + F34*800 + F44*800 + REG4*5 <=
10*400;

!Flowrate balance for process sources;
F11 + F12 + F13 + F14 + RE1 + F1W=20;
F21 + F22 + F23 + F24 + RE2 + F2W=100;
F31 + F32 + F33 + F34 + RE3 + F3W=40;
F41 + F42 + F43 + F44 + RE4 + F4W=10;

!Total wastewater flowrate (WW);
WW=F1W + F2W + F3W + F4W;

!All variables should take non-negative values;
F11 >= 0; F21 >= 0; F31 >= 0; F41 >= 0;
F12 >= 0; F22 >= 0; F32 >= 0; F42 >= 0;
F13 >= 0; F23 >= 0; F33 >= 0; F43 >= 0;
F14 >= 0; F24 >= 0; F34 >= 0; F44 >= 0;
FW1 >= 0; FW2 >= 0; FW3 >= 0; FW4 >= 0;
F1W >= 0; F2W >= 0; F3W >= 0; F4W >= 0;
END
```

The following is the solution report generated by LINGO:

```
Global optimal solution found.
Objective value:              98.94737
Total solver iterations:            11


            Variable          Value
            FW             20.00000
            WW             20.00000
            REG            73.68421
            FW1            20.00000
            FW2             0.000000
            FW3             0.000000
            FW4             0.000000
            RE1             0.000000
            RE2            23.68421
            RE3            40.00000
            RE4            10.00000
            REG1            0.000000
            REG2           52.63158
```

```
       REG3                21.05263
       REG4                 0.000000
       F11                  0.000000
       F21                  0.000000
       F31                  0.000000
       F41                  0.000000
       F12                  0.000000
       F22                 47.36842
       F32                  0.000000
       F42                  0.000000
       F13                  0.000000
       F23                 18.94737
       F33                  0.000000
       F43                  0.000000
       F14                 10.00000
       F24                  0.000000
       F34                  0.000000
       F44                  0.000000
       F1W                 10.00000
       F2W                 10.00000
       F3W                  0.000000
       F4W                  0.000000
```

The model determines that the operating cost of this RCN as $98.95/h. The network structure of the RCN takes the form as in Figure 12.11. Note again that the wastewater concentration (C_{WW}) is determined after the model is solved.

2. Case 2—Minimum flowrates for both freshwater and regenerated source

For Case 2 where the flowrates of both freshwater and regenerated source are to be minimized, the two-stage optimization approach is adopted. In Stage 1, the optimization objective is set to minimize the freshwater flowrate (F_{FW}):

		F_{SKj}	20	100	40	10	73.68	20
		C_{SKj}	0	50	50	400	326.73	800
F_{SRi}	C_{SRi}	SK$_j$ SR$_i$	SK1	SK2	SK3	SK4	RE	WW
20	0	FW	20					
73.68	5	REG		52.63	21.05			
20	100	SR1			10	10		
100	100	SR2		47.37	8.95		43.68	
40	800	SR3					30	10
10	800	SR4						10

FIGURE 12.11
Network structure for RCN in Example 12.5 (Cases 1 and 2) (all flowrate terms in t/h; concentration in ppm).

$$\text{Minimize } F_{FW} \qquad\qquad (12.28)$$

subject to all constraints described earlier. The LINGO formulation for this optimization objective is written as follows (similar to Examples 12.1 and 12.2):

```
min = FW;
```

Note that all other constraints are exactly the same as those in Case 1. The following is the solution report generated by LINGO. Note that the results for all optimization variables are omitted, as they are not important for Stage 1.

```
Global optimal solution found.
Objective value:              20.00000
Total solver iterations:            12
```

The result indicates that the minimum freshwater flowrate for this RCN is 20 t/h.

In Stage 2 of the optimization, where regenerated source flowrate is to be minimized, the minimum freshwater flowrate is added as a new constraint to the problem, i.e.,

$$F_{FW} = 20 \qquad\qquad (12.29)$$

The optimization objective is then set to minimize the regenerated source flowrate (F_{REG}):

$$\text{Minimize } F_{REG} \qquad\qquad (12.30)$$

subject to all constraints used in Stage 1 as well as the new constraint in Equation 12.29. In terms of LINGO formulation, Equations 12.29 and 12.30 take the following form:

```
min = REG;
FW = 20;
```

The following is the solution report generated by LINGO:

```
Global optimal solution found.
Objective value:              73.68421
Total solver iterations:            10
```

Note that the results for all optimization variables remain the same as those in Case 1 (i.e. the RCN has the same structure for both minimum cost and minimum flowrate solution). Hence, the network structure of the RCN takes the same form as in Figure 12.11. Note also that for both stages, the inlet concentration of the interception unit and the wastewater concentration are not determined as part of the optimization model. This is to avoid the formation of an NLP model. Once the optimization model is solved, these concentrations are then determined via mass balance and are included in Figure 12.11.

3. Case 3—Redo Case 2 with limited piping connection

Similar to Examples 12.3 and 12.4, the presence of a material recovery stream between sources (including regenerated sources) and sinks is to be activated with the binary term* in Equation 12.13, given as follows:

$$F_{SR1,SK1} \leq 1000\, B_{SR1,SK1}; F_{SR2,SK1} \leq 1000\, B_{SR2,SK1}$$

$$F_{SR3,SK1} \leq 1000\, B_{SR3,SK1}; F_{SR4,SK1} \leq 1000\, B_{SR4,SK1}; F_{REG,SK1} \leq 1000\, B_{REG,SK1}$$

$$F_{SR1,SK2} \leq 1000\, B_{SR1,SK2}; F_{SR2,SK2} \leq 1000\, B_{SR2,SK2}$$

$$F_{SR3,SK2} \leq 1000\, B_{SR3,SK2}; F_{SR4,SK2} \leq 1000\, B_{SR4,SK2}; F_{REG,SK2} \leq 1000\, B_{REG,SK2}$$

$$F_{SR1,SK3} \leq 1000\, B_{SR1,SK3}; F_{SR2,SK3} \leq 1000\, B_{SR2,SK3}$$

$$F_{SR3,SK3} \leq 1000\, B_{SR3,SK3}; F_{SR4,SK3} \leq 1000\, B_{SR4,SK3}; F_{REG,SK3} \leq 1000\, B_{REG,SK3}$$

$$F_{SR1,SK4} \leq 1000\, B_{SR1,SK4}; F_{SR2,SK4} \leq 1000\, B_{SR2,SK4}$$

$$F_{SR3,SK4} \leq 1000 B_{SR3,SK4}; F_{SR4,SK4} \leq 1000\, B_{SR4,SK4}; F_{REG,SK4} \leq 1000\, B_{REG,SK4}$$

Since each process sink can accept only a maximum of two reuse/recycle connections, we make use of Equation 12.20:

$$B_{SR1,SK1} + B_{SR2,SK1} + B_{SR3,SK1} + B_{SR4,SK1} + B_{REG,SK1} \leq 2$$

$$B_{SR1,SK2} + B_{SR2,SK2} + B_{SR3,SK2} + B_{SR4,SK2} + B_{REG,SK2} \leq 2$$

$$B_{SR1,SK3} + B_{SR2,SK3} + B_{SR3,SK3} + B_{SR4,SK3} + B_{REG,SK3} \leq 2$$

$$B_{SR1,SK4} + B_{SR2,SK4} + B_{SR3,SK4} + B_{SR4,SK4} + B_{REG,SK4} \leq 2$$

In terms of LINGO formulation, the aforementioned inequalities of this MILP model take the following form:

```
F11 <= M*B11; F21 <= M*B21; F31 <= M*B31;
F41 <= M*B41; REG1 <= M*BR1;
F12 <= M*B12; F22 <= M*B22; F32 <= M*B32;
F42 <= M*B42; REG2 <= M*BR2;
F13 <= M*B13; F23 <= M*B23; F33 <= M*B33;
F43 <= M*B43; REG3 <= M*BR3;
F14 <= M*B14; F24 <= M*B24; F34 <= M*B34;
F44 <= M*B44; REG4 <= M*BR4;
M=1000;
B11 + B21 + B31 + B41 + BR1 <= 2;
B12 + B22 + B32 + B42 + BR2 <= 2;
B13 + B23 + B33 + B43 + BR3 <= 2;
B14 + B24 + B34 + B44 + BR4 <= 2;
@BIN(B11); @BIN(B21); @BIN(B31); @BIN(B41);
@BIN(BR1);
@BIN(B12); @BIN(B22); @BIN(B32); @BIN(B42);
@BIN(BR2);
```

```
@BIN(B13);  @BIN(B23);  @BIN(B33);  @BIN(B43);
@BIN(BR3);
@BIN(B14);  @BIN(B24);  @BIN(B34);  @BIN(B44);
@BIN(BR4);
```

Note that the rest of the constraints are exactly the same as in Cases 1 and 2. We shall employ the two-stage optimization approach in Case 2 to determine the minimum flowrates of freshwater and regenerated source. The following is the solution report generated by LINGO, while the RCN structure is given in Figure 12.12:

```
Global optimal solution found.
Objective value:                    73.68421
Extended solver steps:                     0
Total solver iterations:                  10

                Variable            Value
                REG              73.68421
                FW               20.00000
                F11              0.000000
                M                1000.000
                B11              0.000000
                F21              0.000000
                B21              0.000000
                F31              0.000000
                B31              0.000000
                F41              0.000000
                B41              0.000000
                REG1             0.000000
                BR1              0.000000
                F12              0.000000
                B12              0.000000
                F22              47.36842
                B22              1.000000
                F32              0.000000
                B32              0.000000
                F42              0.000000
```

F_{SRi}	C_{SRi}	F_{SKj}	20	100	40	10	73.68	20
		C_{SKj}	0	50	50	400	326.73	800
		SK_j SR_i	SK1	SK2	SK3	SK4	RE	WW
20	0	FW	20					
73.68	5	REG		52.63	21.05			
20	100	SR1			18.95			1.05
100	100	SR2		47.37		10	23.68	18.95
40	800	SR3					40	
10	800	SR4					10	

FIGURE 12.12
Network structure for RCN in Example 12.5 (Case 3) (all flowrate terms in t/h; concentration in ppm).

B42	0.000000
REG2	52.63158
BR2	1.000000
F13	18.94737
B13	1.000000
F23	0.000000
B23	0.000000
F33	0.000000
B33	0.000000
F43	0.000000
B43	0.000000
REG3	21.05263
BR3	1.000000
F14	0.000000
B14	0.000000
F24	10.00000
B24	1.000000
F34	0.000000
B34	0.000000
F44	0.000000
B44	0.000000
REG4	0.000000
BR4	0.000000
FW1	20.00000
FW2	0.000000
FW3	0.000000
FW4	0.000000
RE1	0.000000
RE2	23.68421
RE3	40.00000
RE4	10.00000
F1W	1.052632
F2W	18.94737
F3W	0.000000
F4W	0.000000
WW	20.00000

From Figure 12.12, we notice that the 3 connections for SK3 in Figure 12.11 has been reduced to 2. This means that a degenerate solution exists for the problem, since the RCNs have the same flowrates, but with different network structure.

12.6 Superstructural Model for Inter-Plant Resource Conservation Networks

The synthesis of IPRCNs with insight-based techniques has been discussed in Chapter 10. In this section, we shall make use of the superstructural approach in synthesizing the IPRCNs. Figures 12.13 (reproduced from Figure 1.19) and 12.14 show the superstructural representation of an IPRCN with *direct integration scheme* between two individual RCNs. We can easily observe

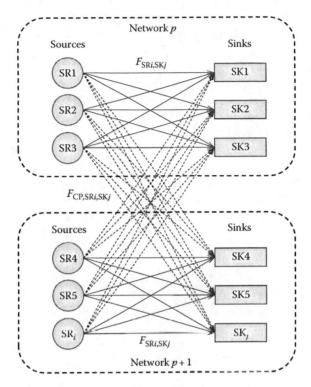

FIGURE 12.13
Superstructural model for IPRCN. (From Chew, I.M.L., Ng, D.K.S., Foo, D.C.Y., Tan, R.R., Majozi, T., and Gouws, J., Synthesis of direct and indirect inter-plant water network, *Ind. Eng. Chem. Res.*, 47, 9485, Copyright 2008 American Chemical Society. With permission.)

		F_{SKj}	F_{SK1}	F_{SK2}	F_{Dp}	F_{SK3}	F_{SKj}	F_{Dp}
		C_{SKj}	q_{SK1}	q_{SK2}	q_D	q_{SK3}	q_{SKj}	q_D
F_{SRi}	C_{SRi}	SK_j	SK1	SK2	WDp	SK3	SK_j	WDp
		SR_i						
F_{Rp}	q_R	FRp	$F_{Rp,SK1}$	$F_{Rp,SK2}$				
F_{SR1}	q_{SR1}	SR1	$F_{SR1,SK1}$	$F_{SR1,SK2}$	$F_{SR1,D}$	$F_{CP,SR1,SK3}$	$F_{CP,SR1,SKj}$	
F_{SR2}	q_{SR2}	SR2	$F_{SR2,SK1}$	$F_{SR2,SK2}$	$F_{SR2,D}$	$F_{CP,SR2,SK3}$	$F_{CP,SR2,SKj}$	
F_{Rp}	q_R	FRp				$F_{Rp,SK3}$	$F_{Rp,SKj}$	
F_{SR3}	q_{SR3}	SR3	$F_{CP,SR3,SK1}$	$F_{CP,SR3,SK2}$		$F_{SR3,SK3}$	$F_{SR3,SKj}$	$F_{SR3,D}$
F_{SRi}	q_{SRi}	SR_i	$F_{CP,SRi,SK1}$	$F_{CP,SRi,SK2}$		$F_{SRi,SK3}$	$F_{SRi,SKj}$	$F_{SRi,D}$

FIGURE 12.14
Superstructural model for IPRCN in matrix form.

that the representation has some features similar to those of direct reuse/recycle scheme for single RCNs (Figure 12.1). Note however that apart from the connections among process sources and sinks for in-plant recovery (i.e. within the same RCNs, with flowrate $F_{SRi,SKj}$), cross-plant stream connections also exist among sources and sinks of different RCNs. The latter are indicated by the gray area in the matching matrix in Figure 12.14 (with flowrate $F_{CP,SRi,SKj}$). Similar to earlier cases, no connection should exist between fresh resource with waste discharge in each RCN.

Similar to the case of material regeneration network, the optimization model for the IPRCN is extended from that of the in-plant reuse/recycle network, with constraints given as in Equations 12.31 through 12.34. Note that in each constraint set, a new flowrate term is added for the cross-plant streams ($F_{CP, SRi, SKj}$).

$$\Sigma_i\, F_{SRi,SKj} + \Sigma_i\, F_{CP,SRi,SKj} + F_{R,SKj} = F_{SKj} \quad \forall j \tag{12.31}$$

$$\Sigma_i\, F_{SRi,SKj}\, q_{SRi} + \Sigma_i\, F_{CP,SRi,SKj}\, q_{SRi} + F_{R,SKj}\, q_R \geq F_{SKj}\, q_{SKj} \quad \forall j \tag{12.32}$$

$$\Sigma_j\, F_{SRi,SKj} + \Sigma_i\, F_{CP,SRi,SKj} + F_{SRiD} = F_{SRi} \quad \forall i \tag{12.33}$$

$$F_{SRi,SKj}, F_{CP,SRi,SKj}, F_{R,SKj}, F_{SRi,D} \geq 0 \quad \forall i, j \tag{12.34}$$

The objective of the optimization model may be set to minimize the various cost elements of the IPRCN, which may take the operating cost (OC) that is contributed by the overall by fresh resources, and waste discharge flowrates (i.e. a revised form of Equation 12.17); or the total annual cost (TAC), which includes the capital costs for in-plant and cross-plant stream connections. Alternatively, we may also set optimization objective to minimize the overall minimum fresh resource (F_R) and the total flowrate of all cross-plant streams (F_{CP}) with the two-stage optimization approach introduced in an earlier section. In this case, the fresh resource flowrate is minimized in the first stage (with Equation 12.35), subject to the constraints in Equations 12.31 through 12.34. Once the overall minimum fresh resource flowrate ($F_{R,min}$) is identified, it is then included as a new constraint in the second stage (Equation 12.36), where the cross-plant stream flowrate is minimized with the optimization objective in Equation 12.37. Similar to earlier cases, this is an LP model for which any solution found is globally optimal:

$$\text{Minimize}\, \Sigma_i\, F_{R,p} \tag{12.35}$$

$$F_{R,min} = \Sigma_i\, F_{R,p} \tag{12.36}$$

$$\text{Minimize}\, F_{CP} \tag{12.37}$$

Example 12.6 Inter-plant water network

The inter-plant water network problem in Example 10.1 is revisited here. Limiting data of two individual RCNs in this IPRCN are given in Table 12.6 (reproduced from Table 10.1). Use the superstructural approach to synthesize an IPRCN that features maximum water recovery, i.e., minimum overall freshwater and wastewater flowrates, assuming that pure freshwater (0 ppm) is available for use. Is there a *degenerate* solution for this case (i.e., different network structures that achieve the same water flowrates)?

Solution

Figure 12.15 shows the superstructure for the IPRCN problem in matrix form.

The constraints of the IPRCN problem are given as follows. Equation 12.31 describes the flowrate balances for all process sinks in Network A, which may be derived by observing flowrate balances in columns 4–7 of Figure 12.15:

$$F_{FWA,SK1} + F_{SR1,SK1} + F_{SR2,SK1} + F_{SR3,SK1} + F_{SR4,SK1} + F_{CP,SR5,SK1}$$

$$+ F_{CP,SR6,SK1} + F_{CP,SR7,SK1} + F_{CP,SR8,SK1} = 20$$

$$F_{FWA,SK2} + F_{SR1,SK2} + F_{SR2,SK2} + F_{SR3,SK2} + F_{SR4,SK2} + F_{CP,SR5,SK2}$$

$$+ F_{CP,SR6,SK2} + F_{CP,SR7,SK2} + F_{CP,SR8,SK2} = 100$$

$$F_{FWA,SK3} + F_{SR1,SK3} + F_{SR2,SK3} + F_{SR3,SK3} + F_{SR4,SK3} + F_{CP,SR5,SK3}$$

$$+ F_{CP,SR6,SK3} + F_{CP,SR7,SK3} + F_{CP,SR8,SK3} = 40$$

TABLE 12.6

Limiting Water Data for Individual RCNs in Example 10.1

Sinks, SK_j	F_{SKj} (t/h)	C_{SKj} (ppm)	Sources, SR_i	F_{SRi} (t/h)	C_{SRi} (ppm)
Network A					
SK1	20	0	SR1	20	100
SK2	100	50	SR2	100	100
SK3	40	50	SR3	40	800
SK4	10	400	SR4	10	800
Network B					
SK5	50	20	SR5	50	50
SK6	100	50	SR6	100	100
SK7	80	100	SR7	70	150
SK8	70	200	SR8	60	250
$\Sigma_j F_{SKj}$	470		$\Sigma_i F_{SRi}$	450	

		F_{SKj}	20	100	40	10	F_{WWA}	50	100	80	70	F_{WWB}
		C_{SKj}	0	50	50	400	WWA	20	50	100	200	WWB
F_{SRi}	C_{SRi}	SKj \ SRi	SK1	SK2	SK3	SK4	WWA	SK5	SK6	SK7	SK8	WWB
F_{FWA}	0	FWA	$F_{FWA,SK1}$	$F_{FWA,SK2}$	$F_{FWA,SK3}$	$F_{FWA,SK4}$						
20	100	SR1	$F_{SR1,SK1}$	$F_{SR1,SK2}$	$F_{SR1,SK3}$	$F_{SR1,SK4}$	$F_{SR1,WWA}$	$F_{CPSR1,SK5}$	$F_{CPSR1,SK6}$	$F_{CPSR1,SK7}$	$F_{CPSR1,SK8}$	
100	100	SR2	$F_{SR2,SK1}$	$F_{SR2,SK2}$	$F_{SR2,SK3}$	$F_{SR2,SK4}$	$F_{SR2,WWA}$	$F_{CPSR2,SK5}$	$F_{CPSR2,SK6}$	$F_{CPSR2,SK7}$	$F_{CPSR2,SK8}$	
40	800	SR3	$F_{SR3,SK1}$	$F_{SR3,SK2}$	$F_{SR3,SK3}$	$F_{SR3,SK4}$	$F_{SR3,WWA}$	$F_{CPSR3,SK5}$	$F_{CPSR3,SK6}$	$F_{CPSR3,SK7}$	$F_{CPSR3,SK8}$	
10	800	SR4	$F_{SR4,SK1}$	$F_{SR4,SK2}$	$F_{SR4,SK3}$	$F_{SR4,SK4}$	$F_{SR4,WWA}$	$F_{CPSR4,SK5}$	$F_{CPSR4,SK6}$	$F_{CPSR4,SK7}$	$F_{CPSR4,SK8}$	
F_{FWB}	0	FWB						$F_{FWB,SK5}$	$F_{FWB,SK6}$	$F_{FWB,SK7}$	$F_{FWB,SK8}$	
50	50	SR5	$F_{CPSR5,SK1}$	$F_{CPSR5,SK2}$	$F_{CPSR5,SK3}$	$F_{CPSR5,SK4}$		$F_{SR5,SK5}$	$F_{SR5,SK6}$	$F_{SR5,SK7}$	$F_{SR5,SK8}$	$F_{SR5,WWB}$
100	100	SR6	$F_{CPSR6,SK1}$	$F_{CPSR6,SK2}$	$F_{CPSR6,SK3}$	$F_{CPSR6,SK4}$		$F_{SR6,SK5}$	$F_{SR6,SK6}$	$F_{SR6,SK7}$	$F_{SR6,SK8}$	$F_{SR6,WWB}$
70	150	SR7	$F_{CPSR7,SK1}$	$F_{CPSR7,SK2}$	$F_{CPSR7,SK3}$	$F_{CPSR7,SK4}$		$F_{SR7,SK5}$	$F_{SR7,SK6}$	$F_{SR7,SK7}$	$F_{SR7,SK8}$	$F_{SR7,WWB}$
60	250	SR8	$F_{CPSR8,SK1}$	$F_{CPSR8,SK2}$	$F_{CPSR8,SK3}$	$F_{CPSR8,SK4}$		$F_{SR8,SK5}$	$F_{SR8,SK6}$	$F_{SR8,SK7}$	$F_{SR8,SK8}$	$F_{SR8,WWB}$

FIGURE 12.15
Matching matrix for IPRCN in Example 12.6.

$$F_{FWA,SK4} + F_{SR1,SK4} + F_{SR2,SK4} + F_{SR3,SK4} + F_{SR4,SK4} + F_{CP,SR5,SK4}$$

$$+ F_{CP,SR6,SK4} + F_{CP,SR7,SK4} + F_{CP,SR8,SK4} = 10$$

Similarly, flowrate balances for all process sinks in Network B may be derived by observing flowrate balances in columns 9–12 of Figure 12.15, given by Equation 12.31:

$$F_{CP,SR1,SK5} + F_{CP,SR2,SK5} + F_{CP,SR3,SK5} + F_{CP,SR4,SK5} + F_{FWB,SK5}$$

$$+ F_{SR5,SK5} + F_{SR6,SK5} + F_{SR7,SK5} + F_{SR8,SK5} = 50$$

$$F_{CP,SR1,SK6} + F_{CP,SR2,SK6} + F_{CP,SR3,SK6} + F_{CP,SR4,SK6} + F_{FWB,SK6}$$

$$+ F_{SR5,SK6} + F_{SR6,SK6} + F_{SR7,SK6} + F_{SR8,SK6} = 100$$

$$F_{CP,SR1,SK7} + F_{CP,SR2,SK7} + F_{CP,SR3,SK7} + F_{CP,SR4,SK7} + F_{FWB,SK7}$$

$$+ F_{SR5,SK7} + F_{SR6,SK7} + F_{SR7,SK7} + F_{SR8,SK7} = 80$$

$$F_{CP,SR1,SK8} + F_{CP,SR2,SK8} + F_{CP,SR3,SK8} + F_{CP,SR4,SK8} + F_{FWB,SK8}$$

$$+ F_{SR5,SK8} + F_{SR6,SK8} + F_{SR7,SK8} + F_{SR8,SK8} = 70$$

The quality requirement for all sinks in Network A is described by Equation 12.32. Note that the freshwater source contains zero impurity (i.e., $C_{FW} = 0\,ppm$):

$$F_{SR1,SK1}(100) + F_{SR2,SK1}(100) + F_{SR3,SK1}(800) + F_{SR4,SK1}(800) + F_{CP,SR5,SK1}(50)$$

$$+ F_{CP,SR6,SK1}(100) + F_{CP,SR7,SK1}(150) + F_{CP,SR8,SK1}(250) \le 20\,(0)$$

$$F_{SR1,SK2}(100) + F_{SR2,SK2}(100) + F_{SR3,SK2}(800) + F_{SR4,SK2}(800) + F_{CP,SR5,SK2}(50)$$

$$+ F_{CP,SR6,SK2}(100) + F_{CP,SR7,SK2}(150) + F_{CP,SR8,SK2}(250) \le 100\,(50)$$

$$F_{SR1,SK3}(100) + F_{SR2,SK3}(100) + F_{SR3,SK3}(800) + F_{SR4,SK3}(800) + F_{CP,SR5,SK3}(50)$$

$$+ F_{CP,SR6,SK3}(100) + F_{CP,SR7,SK3}(150) + F_{CP,SR8,SK3}(250) \le 40\,(50)$$

$$F_{SR1,SK4}(100) + F_{SR2,SK4}(100) + F_{SR3,SK4}(800) + F_{SR4,SK4}(800) + F_{CP,SR5,SK4}(50)$$

$$+ F_{CP,SR6,SK4}(100) + F_{CP,SR7,SK4}(150) + F_{CP,SR8,SK4}(250) \le 10\,(400)$$

Similarly, the quality requirements for all sinks in Network B are described by Equation 12.32, given as follows:

$$F_{CP,SR1,SK5}(100) + F_{CP,SR2,SK5}(100) + F_{CP,SR3,SK5}(800) + F_{CP,SR4,SK5}(800)$$

$$+ F_{SR5,SK5}(50) + F_{SR6,SK5}(100) + F_{SR7,SK5}(150) + F_{SR8,SK5}(250) \le 50\,(20)$$

$$F_{CP,SR1,SK6}(100) + F_{CP,SR2,SK6}(100) + F_{CP,SR3,SK6}(800) + F_{CP,SR4,SK6}(800)$$

$$+ F_{SR5,SK6}(50) + F_{SR6,SK6}(100) + F_{SR7,SK6}(150) + F_{SR8,SK6}(250) \le 100 \ (50)$$

$$F_{CP,SR1,SK7}(100) + F_{CP,SR2,SK7}(100) + F_{CP,SR3,SK7}(800) + F_{CP,SR4,SK7}(800)$$

$$+ F_{SR5,SK7}(50) + F_{SR6,SK7}(100) + F_{SR7,SK7}(150) + F_{SR8,SK7}(250) \le 80 \ (100)$$

$$F_{CP,SR1,SK8}(100) \ + F_{CP,SR2,SK8}(100) + F_{CP,SR3,SK8}(800) + F_{CP,SR4,SK8}(800)$$

$$+ F_{SR5,SK8}(50) + F_{SR6,SK8}(100) + F_{SR7,SK8}(150) + F_{SR8,SK8}(250) \le 70 \ (200)$$

Equation 12.33 next describes the flowrate balance for the process sources. Each of these sources may be allocated to any process sinks via in-plant ($F_{SRi,SKj}$) or inter-plant connections ($F_{CP,SRi,SKj}$), while the unutilized source will be sent for wastewater discharge ($F_{SRi,WWp}$):

$$F_{SR1,SK1} + F_{SR1,SK2} + F_{SR1,SK3} + F_{SR1,SK4} + F_{SR1,WWA} + F_{CP,SR1,SK5}$$

$$+ F_{CP,SR1,SK6} + F_{CP,SR1,SK7} + F_{CP,SR1,SK8} = 20$$

$$F_{SR2,SK1} + F_{SR2,SK2} + F_{SR2,SK3} + F_{SR2,SK4} + F_{SR2,WWA} + F_{CP,SR2,SK5}$$

$$+ F_{CP,SR2,SK6} + F_{CP,SR2,SK7} + F_{CP,SR2,SK8} = 100$$

$$F_{SR3,SK1} + F_{SR3,SK2} + F_{SR3,SK3} + F_{SR3,SK4} + F_{SR3,WWA} + F_{CP,SR3,SK5}$$

$$+ F_{CP,SR3,SK6} + F_{CP,SR3,SK7} + F_{CP,SR3,SK8} = 40$$

$$F_{SR4,SK1} + F_{SR4,SK2} + F_{SR4,SK3} + F_{SR4,SK4} + F_{SR4,WWA} + F_{CP,SR4,SK5}$$

$$+ F_{CP,SR4,SK6} + F_{CP,SR4,SK7} + F_{CP,SR4,SK8} = 10$$

$$F_{CP,SR5,SK1} + F_{CP,SR5,SK2} + F_{CP,SR5,SK3} + F_{CP,SR5,SK4} + F_{SR5,SK5}$$

$$+ F_{SR5,SK6} + F_{SR5,SK7} + F_{SR5,SK8} + F_{SR5,WWB} = 50$$

$$F_{CP,SR6,SK1} + F_{CP,SR6,SK2} + F_{CP,SR6,SK3} + F_{CP,SR6,SK4} + F_{SR6,SK5}$$

$$+ F_{SR6,SK6} + F_{SR6,SK7} + F_{SR6,SK8} + F_{SR6,WWB} = 100$$

$$F_{CP,SR7,SK1} + F_{CP,SR7,SK2} + F_{CP,SR7,SK3} + F_{CP,SR7,SK4} + F_{SR7,SK5}$$

$$+ F_{SR7,SK6} + F_{SR7,SK7} + F_{SR7,SK8} + F_{SR7,WWB} = 70$$

$$F_{CP,SR8,SK1} + F_{CP,SR8,SK2} + F_{CP,SR8,SK3} + F_{CP,SR8,SK4} + F_{SR8,SK5}$$

$$+ F_{SR8,SK6} + F_{SR8,SK7} + F_{SR8,SK8} + F_{SR8,WWB} = 60$$

Equation 12.34 indicates that all flowrate variables of the optimization model take non-negative values:

$$F_{SR1,SK1}, F_{SR2,SK1}, F_{SR3,SK1}, F_{SR4,SK1} \ge 0$$

$$F_{SR1,SK2}, F_{SR2,SK2}, F_{SR3,SK2}, F_{SR4,SK2} \ge 0$$

$$F_{SR1,SK3}, F_{SR2,SK3}, F_{SR3,SK3}, F_{SR4,SK3} \ge 0$$

$$F_{SR1,SK4}, F_{SR2,SK4}, F_{SR3,SK4}, F_{SR4,SK4} \geq 0$$

$$F_{SR5,SK5}, F_{SR5,SK6}, F_{SR5,SK7}, F_{SR5,SK8} \geq 0$$

$$F_{SR6,SK5}, F_{SR6,SK6}, F_{SR6,SK7}, F_{SR6,SK8} \geq 0$$

$$F_{SR7,SK5}, F_{SR7,SK6}, F_{SR7,SK7}, F_{SR7,SK8} \geq 0$$

$$F_{SR8,SK5}, F_{SR8,SK6}, F_{SR8,SK7}, F_{SR8,SK8} \geq 0$$

$$F_{CP,SR1,SK5}, F_{CP,SR1,SK6}, F_{CP,SR1,SK7}, F_{CP,SR1,SK8} \geq 0$$

$$F_{CP,SR2,SK5}, F_{CP,SR2,SK6}, F_{CP,SR2,SK7}, F_{CP,SR2,SK8} \geq 0$$

$$F_{CP,SR3,SK5}, F_{CP,SR3,SK6}, F_{CP,SR3,SK7}, F_{CP,SR3,SK8} \geq 0$$

$$F_{CP,SR4,SK5}, F_{CP,SR4,SK6}, F_{CP,SR4,SK7}, F_{CP,SR4,SK8} \geq 0$$

$$F_{CP,SR5,SK1}, F_{CP,SR5,SK2}, F_{CP,SR5,SK3}, F_{CP,SR5,SK4} \geq 0$$

$$F_{CP,SR6,SK1}, F_{CP,SR6,SK2}, F_{CP,SR6,SK3}, F_{CP,SR6,SK4} \geq 0$$

$$F_{CP,SR7,SK1}, F_{CP,SR7,SK2}, F_{CP,SR7,SK3}, F_{CP,SR7,SK4} \geq 0$$

$$F_{CP,SR8,SK1}, F_{CP,SR8,SK2}, F_{CP,SR8,SK3}, F_{CP,SR8,SK4} \geq 0$$

$$F_{FWA,SK1}, F_{FWA,SK2}, F_{FWA,SK3}, F_{FWA,SK4} \geq 0$$

$$F_{FWB,SK5}, F_{FWB,SK6}, F_{FWB,SK7}, F_{FWB,SK8} \geq 0$$

$$F_{SR1,WWA}, F_{SR2,WWA}, F_{SR3,WWA}, F_{SR4,WWA} \geq 0$$

$$F_{SR5,WWB}, F_{SR6,WWB}, F_{SR7,WWB}, F_{SR8,WWB} \geq 0$$

It is also convenient to relate the total flowrates of freshwater (F_{FWA}, F_{FWB}) and wastewater (F_{WWA}, F_{WWB}) for both Networks A and B:

$$F_{FWA} = F_{FWA,SK1} + F_{FWA,SK2} + F_{FWA,SK3} + F_{FWA,SK4}$$

$$F_{FWB} = F_{FWB,SK5} + F_{FWB,SK6} + F_{FWB,SK7} + F_{FWB,SK8}$$

$$F_{WWA} = F_{SR1,WWA} + F_{SR2,WWA} + F_{SR3,WWA} + F_{SR4,WWA}$$

$$F_{WWB} = F_{SR5,WWB} + F_{SR6,WWB} + F_{SR7,WWB} + F_{SR8,WWB}$$

Since both freshwater and inter-plant flowrates are to be minimized, the two-stage optimization approach is adopted. In Stage 1, the optimization objective is set to minimize the overall freshwater flowrates for both Networks A and B:

$$\text{Minimize } (F_{FWA} + F_{FWB}) \tag{12.38}$$

LINGO formulation for Stage 1 optimization is written as follows:

Model:
```
! Optimisation objective: to minimise overall freshwater
flowrates ;
min = FWA + FWB;
```

```
WW = WWA + WWB;

!Constraints for Network A;

!Flowrate balance for freshwater for Network A;
FWA = FW1 + FW2 + FW3 + FW4;

!Flowrate balance for process sinks in Network A;
FW1 + F11 + F21 + F31 + F41 + CP51 + CP61 + CP71 + CP81 = 20;
FW2 + F12 + F22 + F32 + F42 + CP52 + CP62 + CP72 +
CP82 = 100;
FW3 + F13 + F23 + F33 + F43 + CP53 + CP63 + CP73 + CP83 = 40;
FW4 + F14 + F24 + F34 + F44 + CP54 + CP64 + CP74 + CP84 = 10;

!Impurity constraints for process sinks in Network A;
F11*100 + F21*100 + F31*800 + F41*800 + CP51*50 + CP61*100 +
CP71*150 + CP81*200 <= 20*0;
F12*100 + F22*100 + F32*800 + F42*800 + CP52*50 + CP62*100 +
CP72*150 + CP82*200 <= 100*50;
F13*100 + F23*100 + F33*800 + F43*800 + CP53*50 + CP63*100 +
CP73*150 + CP83*200 <= 40*50;
F14*100 + F24*100 + F34*800 + F44*800 + CP54*50 + CP64*100 +
CP74*150 + CP84*200 <= 10*400;

!Flowrate balance for process sources in Network A;
F11 + F12 + F13 + F14 + F1W + CP15 + CP16 + CP17 + CP18 = 20;
F21 + F22 + F23 + F24 + F2W + CP25 + CP26 + CP27 +
CP28 = 100;
F31 + F32 + F33 + F34 + F3W + CP35 + CP36 + CP37 + CP38 = 40;
F41 + F42 + F43 + F44 + F4W + CP45 + CP46 + CP47 + CP48 = 10;

!Total wastewater flowrate (WW) in Network A;
WWA = F1W + F2W + F3W + F4W;

!Constraints for Network B;

!Flowrate balance for freshwater for Network B;
FWB = FW5 + FW6 + FW7 + FW8;

!Flowrate balance for process sinks in Network B;
CP15 + CP25 + CP35 + CP45 + FW5 + F55 + F65 + F75 + F85 = 50;
CP16 + CP26 + CP36 + CP46 + FW6 + F56 + F66 + F76 +
F86 = 100;
CP17 + CP27 + CP37 + CP47 + FW7 + F57 + F67 + F77 + F87 = 80;
CP18 + CP28 + CP38 + CP48 + FW8 + F58 + F68 + F78 + F88 = 70;

!Impurity constraints for process sinks in Network B;
CP15*100 + CP25*100 + CP35*800 + CP45*800 + F55*50 + F65*100 +
F75*150 + F85*200 <= 50*20;
CP16*100 + CP26*100 + CP36*800 + CP46*800 + F56*50 + F66*100 +
F76*150 + F86*200 <= 100*50;
CP17*100 + CP27*100 + CP37*800 + CP47*800 + F57*50 + F67*100 +
F77*150 + F87*200 <= 80*100;
CP18*100 + CP28*100 + CP38*800 + CP48*800 + F58*50 + F68*100 +
F78*150 + F88*200 <= 70*200;

!Flowrate balance for process sources in Network B;
CP51 + CP52 + CP53 + CP54 + F55 + F56 + F57 + F58 + F5W = 50;
```

```
CP61 + CP62 + CP63 + CP64 + F65 + F66 + F67 + F68 + F6W=100;
CP71 + CP72 + CP73 + CP74 + F75 + F76 + F77 + F78 + F7W= 70;
CP81 + CP82 + CP83 + CP84 + F85 + F86 + F87 + F88 + F8W= 60;

!Total wastewater flowrate (WW) in Network B;
WWB= F5W + F6W + F7W + F8W;

!All variables should take non-negative values;
FW1 >= 0; FW2 >= 0; FW3 >= 0; FW4 >= 0;
FW5 >= 0; FW6 >= 0; FW7 >= 0; FW8 >= 0;
F11 >= 0; F21 >= 0; F31 >= 0; F41 >= 0;
F12 >= 0; F22 >= 0; F32 >= 0; F42 >= 0;
F13 >= 0; F23 >= 0; F33 >= 0; F43 >= 0;
F14 >= 0; F24 >= 0; F34 >= 0; F44 >= 0;
F55 >= 0; F65 >= 0; F75 >= 0; F85 >= 0;
F56 >= 0; F66 >= 0; F76 >= 0; F86 >= 0;
F57 >= 0; F67 >= 0; F77 >= 0; F87 >= 0;
F58 >= 0; F68 >= 0; F78 >= 0; F88 >= 0;
F1W >= 0; F2W >= 0; F3W >= 0; F4W >= 0;
F5W >= 0; F6W >= 0; F7W >= 0; F8W >= 0;
CP15 >= 0; CP25 >= 0; CP35 >= 0; CP45 >= 0;
CP16 >= 0; CP26 >= 0; CP36 >= 0; CP46 >= 0;
CP17 >= 0; CP27 >= 0; CP37 >= 0; CP47 >= 0;
CP18 >= 0; CP28 >= 0; CP38 >= 0; CP48 >= 0;
CP51 >= 0; CP61 >= 0; CP71 >= 0; CP81 >= 0;
CP52 >= 0; CP62 >= 0; CP72 >= 0; CP82 >= 0;
CP53 >= 0; CP63 >= 0; CP73 >= 0; CP83 >= 0;
CP54 >= 0; CP64 >= 0; CP74 >= 0; CP84 >= 0;
END
```

The following is the solution report generated by LINGO. Note that the results for other optimization variables have not been included, as they are not important for Stage 1:

```
Global optimal solution found.
Objective value:                        155.0000
Total solver iterations:                      35
```

The result indicates that the minimum overall freshwater flowrates for this RCN is 155 t/h.

For Stage 2 optimization, where the total inter-plant flowrate is to be minimized, the minimum overall freshwater flowrates obtained in Stage 1 is added as a new constraint to the problem, i.e.,

$$F_{FWA} + F_{FWB} = 155 \qquad (12.39)$$

The optimization objective is then set to minimize the total inter-plant flowrate (F_{REG}):

$$\text{Minimize } F_{CP} \qquad (12.40)$$

where

$$F_{CP} = F_{CP,SR1,SK5} + F_{CP,SR1,SK6} + F_{CP,SR1,SK7} + F_{CP,SR1,SK8}$$
$$+ F_{CP,SR2,SK5} + F_{CP,SR2,SK6} + F_{CP,SR2,SK7} + F_{CP,SR2,SK8}$$
$$+ F_{CP,SR3,SK5} + F_{CP,SR3,SK6} + F_{CP,SR3,SK7} + F_{CP,SR3,SK8}$$
$$+ F_{CP,SR4,SK5} + F_{CP,SR4,SK6} + F_{CP,SR4,SK7} + F_{CP,SR4,SK8}$$
$$+ F_{CP,SR5,SK1} + F_{CP,SR5,SK2} + F_{CP,SR5,SK3} + F_{CP,SR5,SK4}$$
$$+ F_{CP,SR6,SK1} + F_{CP,SR6,SK2} + F_{CP,SR6,SK3} + F_{CP,SR6,SK4}$$
$$+ F_{CP,SR7,SK1} + F_{CP,SR7,SK2} + F_{CP,SR7,SK3} + F_{CP,SR7,SK4}$$
$$+ F_{CP,SR8,SK1} + F_{CP,SR8,SK2} + F_{CP,SR8,SK3} + F_{CP,SR8,SK4}$$

Note that the Stage 2 optimization is subjected to all constraints in Stage 1 as well as the new constraint in Equation 12.39. In terms of LINGO formulation, Equations 12.39 and 12.40 take the following form:

```
min=CP;
FWA + FWB=155;
```

The following is the solution report generated by LINGO for Stage 2 optimization:

```
Global optimal solution found.
Objective value:                    15.00000
Total solver iterations:                  35


            Variable         Value
            CP               15.00000
            CP51              0.000000
            CP61              0.000000
            CP71              0.000000
            CP81              0.000000
            CP52              0.000000
            CP62              0.000000
            CP72              0.000000
            CP82              0.000000
            CP53              0.000000
            CP63              0.000000
            CP73              0.000000
            CP83              0.000000
            CP54              0.000000
            CP64              0.000000
            CP74              0.000000
            CP84              0.000000
            CP15              0.000000
            CP25              0.000000
            CP35              0.000000
            CP45              0.000000
```

CP16	0.000000
CP26	15.00000
CP36	0.000000
CP46	0.000000
CP17	0.000000
CP27	0.000000
CP37	0.000000
CP47	0.000000
CP18	0.000000
CP28	0.000000
CP38	0.000000
CP48	0.000000
FWA	90.00000
FWB	65.00000
WW	135.0000
WWA	75.00000
WWB	60.00000
FW1	20.00000
FW2	50.00000
FW3	20.00000
FW4	0.000000
F11	0.000000
F21	0.000000
F31	0.000000
F41	0.000000
F12	0.000000
F22	50.00000
F32	0.000000
F42	0.000000
F13	0.000000
F23	20.00000
F33	0.000000
F43	0.000000
F14	0.000000
F24	10.00000
F34	0.000000
F44	0.000000
F1W	20.00000
F2W	5.000000
F3W	0.00000
F4W	10.00000
FW5	40.00000
FW6	25.00000
FW7	0.000000
FW8	0.000000
F55	0.000000
F65	10.00000
F75	0.000000
F85	0.000000
F56	50.00000
F66	10.00000
F76	0.000000
F86	0.000000
F57	0.000000
F67	80.00000
F77	0.000000

F87	0.000000
F58	0.000000
F68	0.000000
F78	70.00000
F88	0.000000
F5W	0.000000
F6W	0.000000
F7W	0.000000
F8W	60.00000

Hence, results of the superstructural model indicate that a minimum cross-plant stream flowrate of 15 t/h is needed to achieve the overall freshwater flowrate of 155 t/h. The structure of the IPRCN is given in Figure 12.16. As shown, the cross-plant stream is indicated by the sink–source match in the gray-highlighted area, connecting between source SR2 of Network A with sink SK6 of Network B. Note that similar to earlier cases, the concentration for wastewater discharge is determined after the ATM is solved.

We next examine if there exist degenerate solutions for this IPRCN. Since the network structure in Figure 12.16 shows that an inter-plant connection exists between SR2 and SK6, we can add a constraint to forbid this connection and then examine if there exists another network structure that features the same water flowrates. The LINGO formulation for this constraint is given as follows:

```
F26 = 0;
```

Solving the superstructural model again with the newly added constraint generates an alternative IPRCN that achieves the same flowrates, i.e., a degenerate solution does exist for this case. The structure for the alternative IPRCN is shown in Figure 12.17. We then observed that the in-plant connections of Network A are essentially the same as those in Figure 12.16. However, two cross-plant streams exist in this case connecting between source SR2 in Network A with sinks SK5 and SK7 in Network B. Note that Network B has one in-plant connection less than that in Figure 12.16.

Finally, note that the model may also be extended to determine for cases with minimum TAC, reduced piping connection or network complexity, as has been discussed in earlier sections. However, note that the LP model will then become an MILP model with the use of binary terms.

12.7 Additional Readings

The superstructural approach discussed in this chapter is the basic model for RCN synthesis. Many works have shown that the basic model can be

	F_{SRi}	C_{SRi}	SK_j / SR_i	SK1	SK2	SK3	SK4	WWA	SK5	SK6	SK7	SK8	WWB
			F_{SKj}	20	100	40	10	$F_{WWA}=75$	50	100	80	70	$F_{WWB}=60$
			C_{SKj}	0	50	50	400	567	20	50	100	200	250
				SK1	SK2	SK3	SK4	WWA	SK5	SK6	SK7	SK8	WWB
$F_{FWA}=90$	20	0	FWA	20	50	20							
	100	100	SR1					20					
	100	100	SR2		50	20	10	5		15			
	40	800	SR3					40					
	10	800	SR4					10					
$F_{FWB}=65$	50	0	FWB						40	25			
	50	50	SR5							50			
	100	100	SR6						10	10	80		
	70	150	SR7									70	
	60	250	SR8										60

FIGURE 12.16
Network structure for IPRCN in Example 12.6 (all flowrate terms in t/h; concentration in ppm).

		F_{SKj}	20	100	40	10	$F_{WWA}=75$	50	100	80	70	$F_{WWB}=60$
		C_{SKj}	0	50	50	400	567	50	50	100	200	250
F_{SRi}	C_{SRi}	SK$_j$ / SR$_i$	SK1	SK2	SK3	SK4	WWA	SK5	SK6	SK7	SK8	WWB
$F_{FWA}=90$	0	FWA	20	50	20							
20	100	SR1					20					
100	100	SR2		50		10	5			5		
40	800	SR3					40					
10	800	SR4					10					
$F_{FWB}=65$	0	FWB						40	25			
50	50	SR5							50			
100	100	SR6							25	75		
70	150	SR7									70	
60	250	SR8										60

FIGURE 12.17
Alternative network structure for IPRCN in Example 12.6, with forbidden match between SR2 and SK6 (all flowrate terms in t/h; concentration in ppm).

extended to many other RCN cases, e.g., material regeneration networks with single pass interception unit of removal ratio type (Gabriel and El-Halwagi, 2005); or with partitioning interception units (Tan et al., 2009); total water network (Karuppiah and Grossmann, 2006); as well as batch material network (Kim and Smith, 2004; Majozi, 2005; Ng et al., 2008; Shoaib et al., 2008; Majozi, and Gouws, 2009). Readers may refer to the respective references for the extended models.

For IPRCN synthesis, we may also consider *indirect integration* scheme, where different RCNs are integrated via *centralized utility facility*, as well as the use of interception units for source regeneration and treatment prior to environmental waste discharge (Chew et al., 2008). Besides, the incorporation of property integration for IPRCN synthesis has also been reported (Lovelady et al., 2009; Rubio-Castro et al., 2010, 2011).

Problems

Water Minimization Problems

12.1 For the classical water minimization case in Example 12.5 (Wang and Smith, 1994), synthesize a direct reuse/recycle network (i.e. no interception unit) with minimum freshwater flowrate. Compare the network structure with that obtained using the nearest neighbor algorithm (note to readers: solve Problem 7.16 to compare results).

12.2 Synthesize an RCN for direct water reuse/recycle scheme for Example 7.1 (Polley and Polley, 2000) for the following cases:

   a. RCN with minimum freshwater flowrate, with zero impurity freshwater feed. Compare the resulting network structure with that obtained using the nearest neighbor algorithm (Example 7.1).

   b. Redo case (a) when freshwater feed has an impurity of 10 ppm. Compare the network structure with that obtained using the nearest neighbor algorithm (note to readers: solve Problem 7.1 to compare results).

   c. Redo case (a) when the number of reuse/recycle streams is limited to three connections.

   d. Determine the piping and water costs for the RCN in cases (a) and (c). Capital cost of piping is given by Equation 12.18, while the distances between the water sources and sinks are given in Table 12.7. Unit costs for freshwater and wastewater are given as $1/ton, each.

   e. RCN with minimum TAC. Make use of the cost elements in case (d). The piping cost is annualized to a period of 5 years, with an interest rate of 5%. Assume an annual operation of 8000 h.

TABLE 12.7

Distance (m) between Water
Sources with Sinks in Problem 12.2

		Sinks		
Sources	SK1	SK2	SK3	SK4
SR1	0	30	150	65
SR2	30	0	45	120
SR3	150	45	0	80
SR4	65	120	80	0

12.3 Synthesize a minimum freshwater network for the textile plant in Problem 2.4, assuming pure freshwater is used. Is the freshwater and wastewater flowrates identical with that obtained by targeting technique? (Note to readers: solve problem 3.2 to compare result).

12.4 Synthesize a water reuse/recycle network for the bulk chemical production case in Problem 3.5, assuming pure freshwater is used. Use the limiting data in Table 3.11. The RCN should feature the following characteristics:

a. Minimum freshwater flowrate. Compare the network structure with that obtained using the nearest neighbor algorithm (note to readers: solve Problem 7.6 to compare results).

b. Minimum total annual cost. Unit cost for freshwater and wastewater are given as $1/ton each. Capital cost of piping is given by Equation 12.18, while the distances between the water sources and sinks are given in Table 12.8. The piping cost is annualized to a period of 5 years, with an interest rate of 5%. Assume an annual operation of 8000 h.

12.5 Synthesize a water reuse/recycle network for the specialty chemical production process in Problem 4.6, with minimum freshwater flowrate. Assume that pure freshwater is used. Compare the network structure with that obtained using the nearest neighbor algorithm (note to readers: solve Problem 7.7 to compare results).

TABLE 12.8

Distance (m) between Water Sources with Sinks in Problem 12.4

		Sinks		
Sources	Reactor	Boiler Makeup	Cooling Tower Makeup	Cleaning
Decanter	100	100	100	40
Cooling tower blowdown	100	30	12	50
Boiler blowdown	110	0	40	75

12.6 Synthesize a water reuse/recycle network for the palm oil mill in Problem 4.8, with minimum freshwater flowrate. Assume pure freshwater is used. Compare the network structure with that obtained using the nearest neighbor algorithm (note to readers: solve Problem 7.4 to compare results).

12.7 For the water minimization case in Problem 6.4 (Sorin and Bédard, 1999), synthesize the water networks for the following cases (assume that pure freshwater is used):

   a. Direct reuse/recycle network with minimum freshwater (0 ppm) flowrate. Compare the solution with that obtained via nearest neighbor algorithm (Note to reader: solve Problem 7.12a to c compare results).

   b. Water regeneration network with minimum operating cost. Unit costs for freshwater (CT_{FW}), wastewater (CT_{WW}), and regenerated water (CT_{RW}) are given as \$1/ton, \$0.5/ton, and \$0.5/ton, respectively. The interception unit has a fixed outlet concentration of 10 ppm (C_{Rout}).

12.8 The paper milling process in Problem 4.5 is revisited here. Synthesize a water reuse/recycle network for this case, assuming that pure freshwater is used. The limiting data are given in Table 4.23, while the distances between the water sources and sinks are given in Table 12.9. Unit costs for freshwater and wastewater are given as \$1/ton each. Capital cost of piping is given by Equation 12.18. The piping cost is annualized to a period of 5 years, with an interest rate of 5%. Assume an annual operation of 8000 h. Determine the minimum *TAC* for the RCN.

Utility Gas Recovery Problems

12.9 Synthesize an RCN with minimum fresh oxygen supply for oxygen recovery case in Problem 3.9. The fresh oxygen supply has an impurity of 5% (note to readers: solve Problem 7.19 to compare results).

12.10 Synthesize a reuse/recycle network for the refinery hydrogen network in Example 2.4 with minimum fresh hydrogen supply. The fresh

TABLE 12.9

Distance (m) between Water Sources with Sinks in Problem 12.8

Distance	Storage Tank	Pressing Section	Forming Section	De-Inking Pulper (DIP) and Others	DIP
Pressing section	82		5	72	103
Forming section	77	5		67	98
DIP and others	20	67	72		40
DIP	68	98	103	40	
Chemical preparation (CP)	60	53	58	40	85
Approach flow	75	8	8	85	98

hydrogen feed has 1% impurity. Compare the network structure with that obtained via nearest neighbor algorithm (note to readers: solve Problem 7.20 to compare results).

12.11 Synthesize a reuse/recycle network for the refinery hydrogen network in Problems 2.6 and 4.15 (see limiting data in Table 4.30). The fresh hydrogen feed has an impurity content of 5%. Compare the network structure with that obtained using nearest neighbor algorithm (note to readers: solve Problem 7.21 to compare results).

Property Integration Problems

12.12 Synthesize an RCN for fiber recovery in the paper making process in Example 2.6 in order to minimize fresh fiber use. Compare the RCN with that synthesized using the nearest neighbor algorithm (note to readers: solve Problem 7.24 to compare results).

12.13 Synthesize an RCN for the microelectronics manufacturing facility in Problem 3.11, to minimize the use of the ultrapure water. The fresh ultrapure water has a resistivity value of $18,000 \, k\Omega/cm$. Compare the solution with the RCN synthesized using the nearest neighbor algorithm (note to readers: solve Problem 7.25 to compare results).

IPRCN Problems

12.14 Revisit the inter-plant water network problem in Example 10.3. Is that possible to generate several other IPRCN alternatives with the same flowrate targets?

12.15 Figure 12.18 shows the process flow diagram of an integrated pulp mill and a bleached paper plant that are geographically located next to each other (Lovelady et al., 2007). Table 12.10 summarizes the limiting data

TABLE 12.10

Limiting Water Data for Individual RCNs in Problem 12.15

Sinks, SKj	F_{SKj} (t/h)	C_{SKj} (ppm)			Sources, SRi	F_{SRi} (t/h)	C_{SRi} (ppm)		
		Cl^-	K^+	Na^+			Cl^-	K^+	Na^+
Pulp mill									
Washer (SK1)	13,995	34.4	12.7	89.4	Stripper 1 (SR1)	8,901	0	0	0
Screening (SK2)	1,450	241.0	2.4	241.0	Screening (SR2)	1,450	308.6	115.5	840.3
Washer/ filter (SK3)	5,762	0	0	0	Stripper 2 (SR3)	1,024	0	0	0
Bleached paper plant									
Bleaching (SK4)	30,990	3.7	1.1	3.6	Bleaching (SR4)	30,990	500.0	5.0	500.0

Synthesis of Resource Conservation Networks: A Superstructural Approach **353**

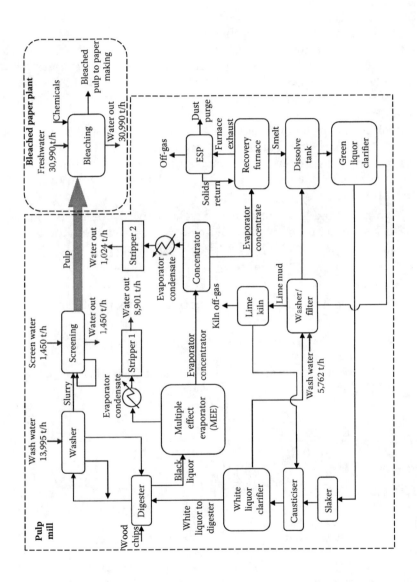

FIGURE 12.18
Process flow diagram of an integrated pulp mill and a bleached paper plant. (From Chew, I.M.L., Ng, D.K.S., Foo, D.C.Y., Tan, R.R., Majozi, T., and Gouws, J., Synthesis of direct and indirect inter-plant water network, *Ind. Eng. Chem. Res.*, 47, 9485, Copyright 2008 American Chemical Society. With permission.)

for both plants (Chew et al., 2008). Note that the recovery of water is dictated by the concentration of three ions, i.e., chlorine (Cl⁻), potassium (K⁺), and sodium (Na⁺). The concentrations of these ions in the freshwater source are given as 3.7, 1.1, and 3.6 ppm, respectively. Synthesize an IPRCN for this case with minimum freshwater and cross-plant flowrates (hint: since this is a multiple impurities problem, Equation 12.5 is written for each impurity term).

12.16 Assume that the two refinary hydrogen networks in Problems 4.15 and 4.16 form an IPRCN with direct integration scheme. Synthesize the IPRCN for this case. Fresh hydrogen feed are assumed the same as in the original problems. Limiting data is summarized in Tables 4.30 and 4.31.

12.17 Revisit the inter-plant water recovery problem between the two wafer fabrication plants in Problem 10.4a. Synthesize an IPRCN for this case with minimum ultrapure water and cross-plant flowrates.

12.18 Revisit the inter-plant water network problem in Example 10.2. Solve the following tasks:

    a. Synthesize an in-plant reuse/recycle network for all individual RCNs with minimum overall freshwater flowrates without any cross-plant stream (i.e., no IPRCN is implemented).

    b. Synthesize an IPRCN with minimum freshwater and cross-plant flowrates.

    c. Determine the minimum numbers of cross-plant connections needed for the IPRCN in (b).

    d. Determine the freshwater and cross-plant flowrates when only two cross-plant connections are allowed for the IPRCN.
       Hint: binary variables are needed for tasks (c) and (d), which leads to MILP models.

References

Chew, I. M. L., Ng, D. K. S., Foo, D. C. Y., Tan, R. R., Majozi, T., and Gouws, J. 2008. Synthesis of direct and indirect inter-plant water network. *Industrial and Engineering Chemistry Research*, 47, 9485–9496.

El-Halwagi, M. M. 1997. *Pollution Prevention through Process Integration: Systematic Design Tools*. San Diego, CA: Academic Press.

Gabriel, F. B. and El-Halwagi, M. M. 2005. Simultaneous synthesis of waste interception and material reuse networks: Problem reformulation for global optimization. *Environmental Progress*, 24(2), 171–180.

Karuppiah, R. and Grossmann, I. E. 2006. Global optimization for the synthesis of integrated water systems in chemical processes. *Computers and Chemical Engineering*, 30(4), 650–673.

Kim, J.-K. and Smith, R. 2004. Automated design of discontinuous water systems. *Process Safety and Environmental Protection*, 82, 238–248.

Lovelady, E. M., El-Halwagi M. M., Chew, I, M. L., Ng, D. K. S., Foo, D. C. Y., and Tan, R. R. 2009. A property-integration approach to the design and integration of eco-industrial parks. *Seventh International Conference on the Foundations of Computer-Aided Process Design (FOCAPD 2009)*, Breckenridge, CO, June 7–12, 559–567.

Lovelady, E. M., El-Halwagi, M. M., and Krishnagopalan, G. A. 2007. An integrated approach to the optimisation of water usage and discharge in pulp and paper plants. *International Journal of Environment and Pollution*, 29(1/2/3), 274–307.

Majozi, T. 2005. Wastewater minimization using central reusable storage in batch plants. *Computers and Chemical Engineering*, 29, 1631–1646.

Majozi, T. and Gouws, J. 2009. A mathematical optimization approach for wastewater minimization in multiple contaminant batch plants. *Computers and Chemical Engineering*, 33, 1826–1840.

Ng, D. K. S., Foo, D. C. Y., Rabie, A., and El-Halwagi, M. M. 2008. Simultaneous synthesis of property-based water reuse/recycle and interception networks for batch processes. *AIChE Journal*, 54(10), 2624–2632.

Polley, G. T. and Polley, H. L. 2000. Design better water networks. *Chemical Engineering Progress*, 96(2), 47–52.

Rubio-Castro, E., Ponce-Ortega, J. M., Nápoles-Rivera, F., El-Halwagi, M. M., Serna-González, M., and Jiménez-Gutiérrez, A. 2010. Water integration of eco-industrial parks using a global optimization approach. *Industrial and Engineering Chemistry Research*, 49, 9945–9960.

Rubio-Castro, E., Ponce-Ortega, J. M., Serna-González, M., Jiménez-Gutiérrez, A., and El-Halwagi, M. M. 2011. A global optimal formulation for the water integration in eco-industrial parks considering multiple pollutants. *Computers and Chemical Engineering*, 35, 1558–1574.

Shoaib, A. M., Aly, S. M., Awad, M. E., Foo, D. C. Y., and El-Halwagi, M. M. 2008. A hierarchical approach for the synthesis of batch water network. *Computers and Chemical Engineering*, 32(3), 530–539.

Smith, R. 2005. *Chemical Process Design and Integration*. New York: John Wiley & Sons.

Sorin, M. and Bédard, S. 1999. The global pinch point in water reuse networks. *Process Safety and Environmental Protection*, 77, 305–308.

Tan, R. R., Ng, D. K. S., Foo, D. C. Y., and Aviso, K. B. 2009. A superstructure model for the synthesis of single-contaminant water networks with partitioning regenerators. *Process Safety and Environmental Protection*, 87, 197–205.

Wang, Y. P. and Smith, R. 1994. Wastewater minimisation. *Chemical Engineering Science*, 49, 981–1006.

13

Automated Targeting Model for Direct Reuse/Recycle Networks

In this chapter, the basic framework for *automated targeting model* (ATM) is introduced. The basic ATM framework is based on the *material cascade analysis* (MCA) technique described in Chapter 4. In this chapter, the ATM is demonstrated for basic resource conservation problems, i.e., direct reuse/recycle scheme. A new variant of direct reuse/recycle scheme, i.e., *bilateral problem*, is also introduced in this chapter. Both concentration- and property-based resource conservation networks (RCNs) are considered.

13.1 Basic Framework and Mathematical Formulation of ATM

The basic framework of the ATM takes the form of a cascade diagram, as shown in Figure 13.1. Steps to formulate the ATM are given as follows (Ng et al., 2009a,b,c):

1. Construct the cascade diagram, with all quality levels (q_k) of process sinks and sources arranged in descending order. When the quality levels for fresh resources do not coincide with any of the process sinks and sources that are present, additional levels are added. An arbitrary value is also added at the final quality level q_n (lowest among all quality levels) to allow the calculation of residual impurity/property load.

2. Flowrates of all process sinks (F_{SKj}) and sources (F_{SRi}) are located at their respective quality levels in the *material cascade*. When there are more than one sink/source at the same quality levels, their flowrates are added. For RCNs with multiple fresh resources, the fresh resources may also be located at the quality levels like other process sources.

3. The *impurity/property load cascade* is constructed next to the material cascade.

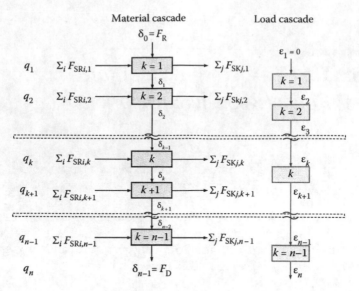

FIGURE 13.1
Basic framework of the ATM. (From Ng, D.K.S. et al., *Ind. Eng. Chem. Res.*, 48(16), 7646, 2009a. With permission.)

4. Formulate the optimization model. The optimization objective can be set to determine the minimum fresh resource flowrate (F_R) or the minimum cost of the RCN, given as in Equations 13.1 and 13.2, respectively (reproduced from Equations 12.1 and 12.3):

$$\text{Minimize } F_R \tag{13.1}$$

$$\text{Minimize cost} \tag{13.2}$$

subject to

$$\delta_k = \delta_{k-1} + (\Sigma_i F_{SRi} - \Sigma_j F_{SKj})_k \quad k = 1, 2, \ldots, n-1 \tag{13.3}$$

$$\varepsilon_k = \begin{cases} 0 & k = 1 \\ \varepsilon_{k-1} + \delta_{k-1}(q_k - q_{k-1}) & k = 2, 3, \ldots, n \end{cases} \tag{13.4}$$

$$\delta_{k-1} \geq 0 \quad k = 1, n \tag{13.5}$$

$$\varepsilon_k \geq 0 \quad k = 2, 3, \ldots, n. \tag{13.6}$$

Similar to the superstructural model in Chapter 12, the RCN cost in Equation 13.2 may take the form of minimum operating cost (MOC) or annual operating cost (AOC). Constraints in Equations 13.3 and 13.4 describe the material and impurity/property load cascades across all quality levels, respectively. As shown in Equation 13.3, the *net material flowrate* of level k (δ_k) is determined from the summation of the net material flowrate cascaded from an earlier quality level (δ_{k-1}) with the flowrate balance at operator level k, i.e., $(\Sigma_i F_{SRi} - \Sigma_j F_{SKj})_k$. Equation 13.4 indicates that the *residual impurity/property load* (ε_k) at the first quality level has zero value, i.e., $\varepsilon_1 = 0$, while residual loads in other quality levels are contributed by the residual load cascaded from previous level (ε_{k-1}), as well as that contributed by the net material flowrate within each quality interval. The latter is given by the product of the net material flowrate from previous level k (δ_{k-1}) and the difference between two adjacent quality levels ($q_k - q_{k-1}$), represented by the last term in Equation 13.4. Equation 13.5 next indicates that the net material flowrate that enters first (δ_0) and last levels (δ_{n-1}) should take non-negative values. Apart from the residual load at level 1 (which is set to zero), all residual loads must also take non-negative values, as indicated by Equation 13.6. Both Equations 13.5 and 13.6 are set to fulfill the feasibility condition of an RCN (Ng et al., 2009a,b,c).

In Equations 13.1 through 13.6, the flowrates of fresh resource (F_R) as well as the net material (δ_k) and residual load at each level k (ε_k) are process variables, which will be determined when the optimization model is solved. On the other hand, the quality level (q_k) and sink (F_{SRi}) and source (F_{SKj}) flowrates are parameters that make Equations 13.1 through 13.6 in linear form. In other words, the ATM is a linear program (LP) that can be solved to obtain a global optimum solution.

Note that the ATM has similar characteristics as the MCA technique in Chapter 4. The material cascade represents the cumulative flowrate of the MCA (column 5 of Table 4.1), while the load cascade is identical to the cumulative load of the MCA (column 7 of Table 4.1). Hence, upon solving the optimization model, the first and last entries of the material cascade represent the minimum flowrates of fresh resource ($\delta_0 = F_R$; for single fresh resource problem) and waste discharge ($\delta_{n-1} = F_D$) of the RCN, respectively. On the other hand, the residual load that vanishes ($\varepsilon_k = 0$) indicates the pinch point of the RCN.

 Example 13.1 Water minimization for acrylonitrile (AN) production

Water minimization case study of the AN process (El-Halwegi, 1997) in Examples 2.1, 3.1, and 4.1 is revisited here. The limiting water data for the problem are given in Table 13.1 (reproduced from Table 4.2). Assuming that pure freshwater is used, determine the following RCN targets using the ATM:

1. Minimum freshwater and wastewater flowrates
2. The pinch concentration and the pinch-causing source
3. The flowrate allocation targets of the pinch-causing source to the higher (HQR) and lower quality (LQR) regions

TABLE 13.1

Limiting Water Data for AN Production

Water Sinks, SK_j		Flowrate F_{SKj} (kg/s)	NH_3 Concentration C_{SKj} (ppm)
j	**Stream**		
1	BFW	1.2	0
2	Scrubber inlet	5.8	10

Water Sources, SR_i		Flowrate F_{SRi} (kg/s)	NH_3 Concentration C_{SRi} (ppm)
i	**Stream**		
1	Distillation bottoms	0.8	0
2	Off-gas condensate	5	14
3	Aqueous layer	5.9	25
4	Ejector condensate	1.4	34

Solution

The four-step procedure is followed to formulate the ATM for the AN case study:

1. *Construct the cascade diagram, with all quality levels (q_k) of process sinks and sources arranged in descending order. When the quality levels for fresh resources do not coincide with any of the process sinks and sources that are present, additional levels are added. An arbitrary value is also added at the final level q_n (lowest among all quality levels) to allow the calculation of the residual impurity/property load.*
2. *Flowrates of the process sinks and sources are located at their respective quality levels in the material cascade.*
3. *The impurity/property load cascade is constructed next to the material cascade.*

   Following Steps 1–3, the cascade diagram for Example 13.1 is shown in Figure 13.2. Note that the highest possible concentration (i.e., 1,000,000 ppm) is added as the final level.
4. *Formulate the optimization model.*

   To determine the minimum freshwater flowrate (FFW) of the water network, we shall make use of a revised form of Equation 13.1:

$$\text{Minimize } F_{FW} \tag{13.7}$$

The constraints of the optimization problem are given as follows. Based on Equation 13.3, the material (water) cascades across all quality levels are given as follows:

$$\delta_1 = \delta_0 + (0.8 - 1.2)$$
$$\delta_2 = \delta_1 + (0 - 5.8)$$
$$\delta_3 = \delta_2 + (5.0 - 0)$$

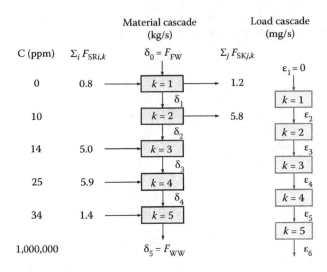

FIGURE 13.2
Cascade diagram for Example 13.1.

$$\delta_4 = \delta_3 + (5.9 - 0)$$
$$\delta_5 = \delta_4 + (1.4 - 0)$$

Since the net material flowrates that enter the first and last concentration levels correspond to the minimum freshwater (F_{FW}) and wastewater (F_{WW}) flowrates, respectively, the following equations may be added:

$$\delta_0 = F_{FW}$$
$$\delta_5 = F_{WW}$$

Based on Equation 13.4, the impurity load cascades across all quality levels are given as follows:

$$\varepsilon_1 = 0$$
$$\varepsilon_2 = \varepsilon_1 + \delta_1(10 - 0)$$
$$\varepsilon_3 = \varepsilon_2 + \delta_2(14 - 10)$$
$$\varepsilon_4 = \varepsilon_3 + \delta_3(25 - 14)$$
$$\varepsilon_5 = \varepsilon_4 + \delta_4(34 - 25)$$
$$\varepsilon_6 = \varepsilon_5 + \delta_5(1000,000 - 34)$$

Equation 13.5 indicates that net material flowrate that enters first and last levels should take non-negative values:

$$\delta_0 \geq 0$$
$$\delta_5 \geq 0$$

Based on Equation 13.6, all residual loads (except for that at level 1) should take non-negative values:

$$\varepsilon_2 \geq 0$$
$$\varepsilon_3 \geq 0$$
$$\varepsilon_4 \geq 0$$
$$\varepsilon_5 \geq 0$$
$$\varepsilon_6 \geq 0$$

In terms of LINGO formulation, the previous model can be written as follows*:

Model:

```
! Optimisation objective: to minimise freshwater (FW)
flowrate ;
min = FW;

! Water cascade ;
D1 = D0 - 0.4 ;
D2 = D1 - 5.8 ;
D3 = D2 + 5.0 ;
D4 = D3 + 5.9 ;
D5 = D4 + 1.4 ;

! The first and last net material flowrates correspond to
minimum freshwater (FW) and wastewater (WW) flowrates ;
D0 = FW ;
D5 = WW ;

! Impurity load cascade ;
E1 = 0 ;
E2 = E1 + 10*D1 ;
E3 = E2 + 4*D2 ;
E4 = E3 + 11*D3 ;
E5 = E4 + 9*D4 ;
E6 = E5 + 999966*D5 ;

! D0 and D5 should take non-negative values ;
D0 >= 0 ;
D5 >= 0 ;

! E2-E6 should take non-negative values ;
E2 >= 0 ;
E3 >= 0 ;
E4 >= 0 ;
E5 >= 0 ;
E6 >= 0 ;

! D1-D4 may take negative values ;
@free (D1) ;
@free (D2) ;
@free (D3) ;
@free (D4) ;
END
```

---

* The LINGO command of @ free (variable) indicates the variables may take negative value.

Note that the last set of constraints (net material flowrate of all levels may take negative flowrate; see discussion in Chapter 4) are only needed when the model is formulated in LINGO, as LINGO assumes that all variables have non-negative values by default.

The following is the solution report generated by LINGO:

```
Global optimal solution found.
Objective value:                          2.057143
Total solver iterations:                          0


            Variable            Value
            FW               2.057143
            D1               1.657143
            D0               2.057143
            D2              -4.142857
            D3               0.8571429
            D4               6.757143
            D5               8.157143
            WW               8.157143
            E1               0.000000
            E2              16.57143
            E3               0.000000
            E4               9.428571
            E5              70.24286
            E6           8156936.
```

The solution report indicates that a global optimum solution is found for the given ATM, with the optimization objective achieved at 2.06 kg/s (freshwater flowrate). The optimized variables may also be incorporated into the cascade diagram for ease of interpretation, as shown in Figure 13.3.

Since the ATM has characteristics similar to those of the MCA technique, insights for MCA may be extended here for the ATM. From Figure 13.3, the minimum wastewater flowrate for the water network is determined as 8.16 kg/s. The pinch corresponds to the concentration where the residual load vanishes ($\varepsilon_3 = 0$), i.e., 14 ppm. This indicates that SR2 (5 kg/s; see Table 13.1 for its limiting concentration) is the pinch-causing source of the problem. Its allocation flowrates to the higher and lower quality regions are identified from the net material flowrates that are just lower and just higher than the pinch, i.e., δ_2 (4.14 kg/s) and δ_3 (0.86 kg/s), respectively (see Sections 3.2 and 4.1 for more details on *flowrate allocation targets*).

Finally, it should be noted that the results of ATM in Figure 13.3 are indeed a numerical representation of the cascade table in the MCA technique, given in Table 4.5. One easily notes that the material cascade is found in both Table 4.5 and Figure 13.3; while the residual impurity/property loads in Figure 13.3 correspond to the cumulative load (Cum. Δm_k) column in Table 4.5. The water network that achieves the established RCN targets was shown in Figure 4.1.

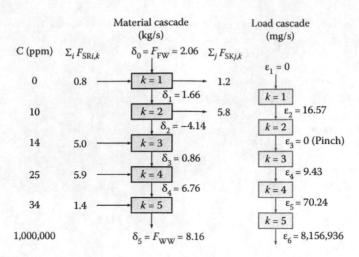

FIGURE 13.3
Results of ATM for Example 13.1.

13.2 Incorporation of Process Constraints into ATM

Since the ATM is constructed on an optimization framework, additional process constraints (e.g., flowrate requirement of certain process sinks, etc.) may be included in the model. However, care should be taken as adding certain constraints may entail a nonlinear program (NLP) for the ATM. However, most commercial optimization software (e.g., LINGO, GAMS) have the ability to solve an NLP problem with a global solution.

> **Example 13.2 Flowrate constraints for water minimization problem in AN production**
>
> Water minimization of the AN process in Example 13.1 is revisited. Instead of using the limiting water data in Table 13.1, one may also make use of the original process constraints given as in Example 2.2. One of the relevant constraints indicates the flowrate requirement of process sink SK2, i.e., scrubber, as follows:
>
> $$5.8 \leq \text{flowrate of wash feed (kg/s)} \leq 6.2 \qquad (13.8)$$
>
> Use the ATM to determine the optimum flowrate for the scrubber wash water (SK2) as well as the freshwater and wastewater flowrate targets for the case study.
>
> **Solution**
>
> Since only an additional flowrate constraint is added to the problem, the same ATM model in Example 13.1 may be extended for use here.

The only change on the ATM is found in Equation 13.3 (material cascade), where the flowrate of SK2 (F_{SK2}) is added as a variable instead of its actual flowrate value:

$$\delta_2 = \delta_1 + (0 - F_{SK2})$$

where F_{SK2} is bound with the given operating range in Equation 13.8. The revised model is still an LP problem, which may be solved to obtain global optimum solution.

The complete LINGO formulation for the model is written as follows:

```
Model:
! Optimisation objective: to minimise freshwater (FW)
flowrate ;
min = FW;

! Water cascade (note: flowrate of sink SK2 is given as an
operating range) ;
D1 = D0 - 0.4 ;
D2 = D1 - SK2 ;
D3 = D2 + 5.0 ;
D4 = D3 + 5.9 ;
D5 - D4 + 1.4 ;
SK2 >= 5.8 ;
SK2 <= 6.2 ;

! The first and last net material flowrates correspond to
minimum freshwater (FW) and wastewater (WW) flowrates ;
D0 = FW ;
D5 = WW ;

! Impurity load cascade ;
E1 = 0 ;
E2 = E1 + 10*D1 ;
E3 = E2 + 4*D2 ;
E4 = E3 + 11*D3 ;
E5 = E4 + 9*D4 ;
E6 = E5 + 999966*D5 ;

! D0 and D5 should take non-negative values ;
D0 >= 0 ;
D5 >= 0 ;

! E2-E6 should take non-negative values ;
E2 >= 0 ;
E3 >= 0 ;
E4 >= 0 ;
E5 >= 0 ;
E6 >= 0 ;

! D1-D4 may take negative values ;
@free (D1) ;
@free (D2) ;
@free (D3) ;
@free (D4) ;
END
```

The following is the solution report generated by LINGO:

```
Global optimal solution found.
Objective value:                        2.057143
Total solver iterations:                       2


            Variable          Value
            FW            2.057143
            D1            1.657143
            D0            2.057143
            D2           -4.142857
            SK2           5.800000
            D3            0.8571429
            D4            6.757143
            D5            8.157143
            WW            8.157143
            E1            0.000000
            E2           16.57143
            E3            0.000000
            E4            9.428571
            E5           70.24286
            E6       8156936.
```

The solution report indicates that the optimization objective (i.e., freshwater flowrate for the RCN) remains the same as in Example 13.1. The model also determines that SK2 should be operated at the optimum flowrate of 5.8 kg/s. This justifies the use of Heuristic 2 in Chapter 2 in identifying limiting data for an RCN (see detailed discussion in Section 2.2). The cascade diagram for this case takes the same form as in Figure 13.3.

13.3 ATM for Property Integration with Inferior Operator Level

Since the ATM is based on the MCA technique, modification is hence needed when it is used for cases when fresh resource has an inferior a property operator value among all material sources (see detailed discussion in Section 2.5). Step 1 to formulate the ATM is modified as follows:

1. Construct the cascade diagram, with all quality levels (q_k) of process sinks and sources arranged in *ascending* order. When the quality levels for fresh resources do not coincide with any of the process sinks and sources that are present, additional levels are added. An arbitrary value is also added at the final level q_n (*highest* among all quality levels) to allow the calculation of the residual property load.

TABLE 13.2

Limiting Data for Fiber Recovery Case Study

Sink, SK_j	F_{SKj} (t/h)	ψ_{SKj} (Dimensionless)	Source, SR_i	F_{SRi} (t/h)	ψ_{SRi} (Dimensionless)
Machine I (SK1)	100	0.382	Bleached fiber (SR1)	90	0.469
Machine II (SK2)	40	0.536	Broke (SR2)	60	0.182
			Fresh fiber (FF)	To be determined	0.738

Besides, Equation 13.4 of the ATM is also modified as follows:

$$\varepsilon_k = \begin{cases} 0 & k = 1 \\ \varepsilon_{k-1} + \delta_{k-1}\left(q_{k-1} - q_k\right) & k = 2, 3, \ldots, n \end{cases} \quad (13.9)$$

Example 13.3 Targeting for property integration—fresh resource with inferior operator value

Table 13.2 shows the limiting data for the fiber recovery case study in Examples 2.6, 3.8, and 4.5. Determine the minimum fresh and waste fiber targets using the ATM.

Solution

The four-step procedure is followed to formulate the ATM for the AN case study, with revised step 1 and Equation 13.9:

1. *Construct the cascade diagram, with all quality levels (q_k) of process sinks and sources arranged in ascending order. When the quality levels for fresh resources do not coincide with any of the process sinks and sources that are present, additional levels are added. An arbitrary value is also added at the final level q_n (highest among all quality levels) to allow the calculation of the residual property load.*
2. *Flowrates of the process sinks and sources are located at their respective quality levels in the material cascade.*
3. *The property load cascade is constructed next to the material cascade.*

   Since the property operator of the fresh resource is found in the inferior level, the operator levels are arranged in the descending order (ascending order of quality), as shown in the cascade diagram in Figure 13.4. Note that the lowest possible operator value (zero) is also added at the final level.
4. *Formulate the optimization model.*

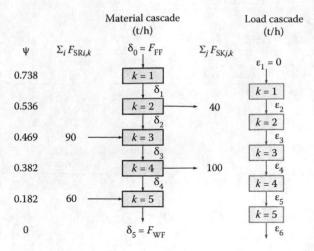

FIGURE 13.4
Cascade diagram for Example 13.3.

To determine the minimum fresh fiber (F_{FF}) of the RCN, a revised form of optimization objective in Equation 13.1 is used:

$$\text{Minimize } F_{FF} \tag{13.10}$$

The constraints of the optimization problem are given as follows. Equation 13.3 describes the fiber cascades across all quality levels:

$$\delta_1 = \delta_0$$
$$\delta_2 = \delta_1 + (0 - 40)$$
$$\delta_3 = \delta_2 + (90 - 0)$$
$$\delta_4 = \delta_3 + (0 - 100)$$
$$\delta_5 = \delta_4 + (60 - 0)$$

The following equations are added to relate the net material flowrates that enter the first and last operator levels to the minimum fresh (F_{FF}) and waste fiber (F_{WF}) flowrates, respectively:

$$\delta_0 = F_{FF}$$
$$\delta_5 = F_{WF}$$

Equation 13.9 next gives the property load cascades across all operator levels:

$$\varepsilon_1 = 0$$
$$\varepsilon_2 = \varepsilon_1 + \delta_1(0.738 - 0.536)$$

$$\varepsilon_3 = \varepsilon_2 + \delta_2(0.536 - 0.469)$$

$$\varepsilon_4 = \varepsilon_3 + \delta_3(0.469 - 0.382)$$

$$\varepsilon_5 = \varepsilon_4 + \delta_4(0.382 - 0.182)$$

$$\varepsilon_6 = \varepsilon_5 + \delta_5(0.182 - 0)$$

Equation 13.5 indicates that net material flowrate that enters first and last levels should take non-negative values:

$$\delta_0 \geq 0$$

$$\delta_5 \geq 0$$

Equation 13.6 indicates that all residual loads (except that at level 1) should take non-negative values:

$$\varepsilon_2 \geq 0$$

$$\varepsilon_3 \geq 0$$

$$\varepsilon_4 \geq 0$$

$$\varepsilon_5 \geq 0$$

$$\varepsilon_6 \geq 0$$

The LINGO formulation for this LP model can be written as follows:

Model:

```
! Optimisation objective: to minimise fresh fibre (FF)
flowrate ;
min = FF;

! Fibre cascade ;
D1 = D0 ;
D2 = D1 - 40 ;
D3 = D2 + 90 ;
D4 = D3 - 100 ;
D5 = D4 + 60;

! The first and last net material flowrates correspond to
minimum fresh fibre (FF) and waste fibre (WF) flowrates ;
D0 = FF ;
D5 = WF ;

! Property load cascade ;
E1 = 0 ;
E2 = E1 + D1*(0.738 - 0.536) ;
E3 = E2 + D2*(0.536 - 0.469) ;
E4 = E3 + D3*(0.469 - 0.382) ;
E5 = E4 + D4*(0.382 - 0.182) ;
E6 = E5 + D5*(0.182 - 0) ;
```

```
! D0 and D5 should take non-negative values ;
D0 >= 0 ;
D5 >= 0 ;

! E2-E6 should take non-negative values ;
E2 >= 0 ;
E3 >= 0 ;
E4 >= 0 ;
E5 >= 0 ;
E6 >= 0 ;

! D1-D4 may take negative values ;
@free (D1) ;
@free (D2) ;
@free (D3) ;
@free (D4) ;
END
```

The following is the solution report generated by LINGO:

```
Global optimal solution found.
Objective value:                      14.98201
Total solver iterations:                     0


             Variable           Value
             FF              14.98201
             D1              14.98201
             D0              14.98201
             D2             -25.01799
             D3              64.98201
             D4             -35.01799
             D5              24.98201
             WF              24.98201
             E1               0.000000
             E2               3.026367
             E3               1.350162
             E4               7.003597
             E5               0.000000
             E6               4.546727
```

The solution report indicates that a global optimum solution is found, with the objective value of 14.98 t/h (i.e., minimum fresh fiber flowrate). The revised cascade diagram with optimized variables takes the form as in Figure 13.5. From Figure 13.5, it is observed that the minimum waste fiber (F_{WF}) for the RCN is determined as 24.98 t/h. The pinch operator is identified at the operator of 0.182 (dimensionless), corresponding to the residual load that vanishes ($\varepsilon_5 = 0$). The pinch-causing source is hence identified for the source at the operator level of 0.182. The allocated flowrates to the LQR and HQR are identified from the net material flowrates at δ_4 (35.02 t/h) and δ_5 (24.98 t/h), respectively. Note that Figure 13.5 corresponds to the cascade table of the MCA technique in Table 4.19.

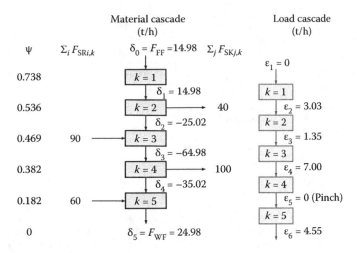

FIGURE 13.5
Results of ATM for Example 13.3.

13.4 ATM for Bilateral Problems

Another special case in property integration is *bilateral problem*, where multiple fresh resources are used, and the operator values of these fresh resources are at the superior and inferior levels among all process sources, respectively. Hence, we will have to formulate the ATM to handle both superior and inferior cases simultaneously. This is shown with the following example.

Example 13.4 Material reuse/recycle in palm oil milling process

Figure 13.6 shows the clay bath operation of a palm oil milling process (Ng et al., 2009c), which is used to separate the palm kernel from its shell and the uncracked nuts (from the cracked mixture stream) based on flotation principle. Due to density difference, the clay bath suspension is operated with a density of $1120\,kg/m^3$. This allows the kernel to emit as the overflow stream, while shell fractions and the uncracked nuts with higher density are withdrawn as bottom stream. Freshwater and clay (fresh resources) are fed to the premixing tank to achieve the desired density of the clay suspension, before it is fed to the clay bath. Sieving is used in both top and bottom streams to separate the solids (kernel, shell, and uncracked nuts) from the liquid portion.

To reduce fresh resource usage, the liquid portion from the overflow stream (SR1) is considered for recovery to the premixing tank (SK1). However, the recovery of wastewater from the bottom flow stream is not considered due to its high impurity content. Freshwater and clay may be used to supplement the recovery of overflow stream in order to meet its flowrate and density requirement of SK1. The limiting data of the sinks

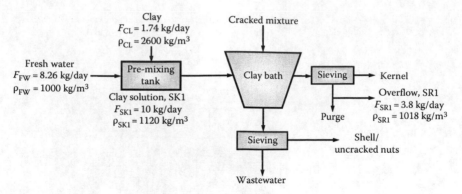

FIGURE 13.6

Clay bath operation in palm oil milling process. (From *Chem. Eng. J.*, 149, Ng, D.K.S., Foo, D.C.Y., Tan, R.R., Pau, C.H., and Tan, Y.L., Automated targeting for conventional and bilateral property-based resource conservation network, 87–101, Copyright 2009c, with permission from Elsevier.)

and sources (including fresh resources) are given in Table 13.3. Note that the density values and the corresponding property operators of the sinks and sources are given in columns 3 and 4, respectively.

Complete the following tasks:

1. Determine the minimum MOC of the RCN, assuming that freshwater (CT_{FW}) and clay (CT_{CL}) have unit costs of \$0.001/kg and \$0.25/kg, respectively. Assume an annual operating time (AOT) of 330 days.
2. Develop the material recovery scheme for this RCN based on targeting insights.

Solution

From Table 13.3, it is observed that the fresh resources available for this case have operator values at the superior (clay) and inferior (freshwater) levels, respectively. In other words, one of the two fresh resources will be placed at the lowest level of the material cascade, for either case where the superior or inferior cascade is formulated with the ATM. Hence, in

TABLE 13.3

Limiting Data for Palm Oil Milling Case Study

Sink, SK_j	F_{SKj} (kg/day)	Density, ρ_{SKj} (kg/m³)	$\psi_{SKj} \times 10^{-4}$ (m³/kg)
Clay solution (SK1)	10.0	1120	8.93
Source, SR_i	**F_{SRi} (kg/day)**	**Density, ρ_{SRi} (kg/m³)**	**$\psi_{SRi} \times 10^{-4}$ (m³/kg)**
Overflow (SR1)	3.8	1018	9.82
Freshwater (FW)	To be determined	1000	10
Clay (CL)	To be determined	2600	3.85

order to optimize the use of both resources simultaneously, both superior and inferior cascades are to be formulated within the same ATM.

The four-step procedure may be followed to formulate the superior or inferior cascade for the ATM.

1. *Construct the cascade diagram, with all quality levels (q_k) of process sinks and sources arranged in descending (superior cascade)/ascending order (inferior cascade). When the quality levels for fresh resources do not coincide with any of the process sinks and sources that are present, additional levels are added. An arbitrary value is also added at the final level q_n (lowest/highest among all quality levels) to allow the calculation of the residual property load.*
2. *Flowrates of the process sinks and sources are located at their respective quality levels in the material cascade.*
3. *The property load cascade is constructed next to the material cascade.*

   The operator levels for the superior cascade are arranged in ascending order (Figure 13.7a), while the reverse is applied for operator levels of the inferior cascade (Figure 13.7b). An arbitrary large value of 100 is added at the lowest operator level in the superior cascade, while zero property operator is added at the final level of the inferior cascade. Note that for both cascades, one of the fresh resources is placed at the highest level, while the other fresh resource is treated as a process source at its corresponding operator level (the same strategy applies to all cases with multiple fresh resources). Note also that to avoid duplication of variables, all variables for the superior cascade are represented with subscript A, while those of the inferior cascade with subscript B. Note however that the flowrate variables for both resources, i.e., freshwater (F_{FW}) and clay (F_{CL}) remain identical in both cascades.

4. *Formulate the optimization model.*

   The minimum AOC of the RCN is given as the product of AOT with the cost of fresh resources, and the latter is determined from the product of the flowrates of freshwater (FFW) and clay (FCL) with their individual unit cost, respectively. Hence, the optimization objective in Equation 13.2 takes the form as in Equation 13.11:

$$\text{Minimize AOC} \tag{13.11a}$$

where

$$AOC = (F_{CL}CT_{CL} + F_{FW}CT_{FW})AOT \tag{13.11b}$$

For this example, the variables of CT_{FW} and CT_{CL} correspond to \$0.001/kg and \$0.25/kg, respectively, while AOT takes the value of 330 days. Since both superior and inferior cascades are needed for this problem, the

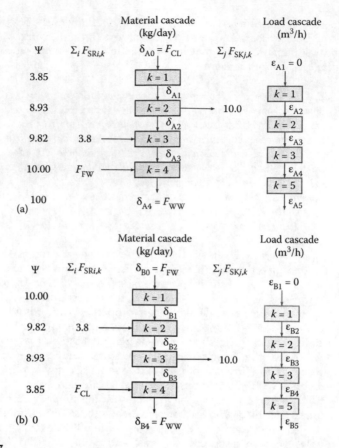

FIGURE 13.7
Cascade diagram for Example 13.4: (a) superior cascade and (b) inferior cascade. (From *Chem. Eng. J.*, 149, Ng, D.K.S., Foo, D.C.Y., Tan, R.R., Pau, C.H., and Tan, Y.L., Automated targeting for conventional and bilateral property-based resource conservation network, 87–101, Copyright 2009c, with permission from Elsevier.)

model is constructed for both cascades. In the following section, constraints on the left are those associated with the superior cascade, while those on the right are for the inferior cascade.

First, Equation 13.3 describes the material cascades across all quality levels:

$$\delta_{A1} = \delta_{A0} \qquad\qquad \delta_{B1} = \delta_{B0}$$

$$\delta_{A2} = \delta_{A1} + (0 - 10.0) \quad \delta_{B2} = \delta_{B1} + (3.8 - 0)$$

$$\delta_{A3} = \delta_{A2} + (3.8 - 0) \quad \delta_{B3} = \delta_{B2} + (0 - 10)$$

$$\delta_{A4} = \delta_{A3} + (F_{FW} - 0) \quad \delta_{B4} = \delta_{B3} + (F_{CL} - 0)$$

The following equations are added to relate the net material flowrates that enter the first and the last operator levels to the minimum fresh resource and wastewater (FWW) flowrates, respectively. Note that the fresh resource at the first operator level is different for both cascades, i.e., clay (with flowrate FCL), for superior cascade, while freshwater (with flowrate FFW) for inferior cascade. Note however that both cascades share the same flowrate for wastewater (FWW) at the final operator level:

$$\delta_{A0} = F_{CL} \qquad \delta_{B0} = F_{FW}$$
$$\delta_{A4} = F_{WW} \qquad \delta_{B4} = F_{WW}$$

Equations 13.4 and 13.9 next give the property load cascades across all operator levels in the superior (left) and inferior (right) cascades, respectively:

$$\varepsilon_{A1} = 0 \qquad\qquad\qquad\quad \varepsilon_{B1} = 0$$
$$\varepsilon_{A2} = \varepsilon_{A1} + \delta_{A1}(8.93 - 3.85) \quad \varepsilon_{B2} = \varepsilon_{B1} + \delta_{B1}(10 - 9.82)$$
$$\varepsilon_{A3} = \varepsilon_{A2} + \delta_{A2}(9.82 - 8.93) \quad \varepsilon_{B3} = \varepsilon_{B2} + \delta_{B2}(9.82 - 8.93)$$
$$\varepsilon_{A4} = \varepsilon_{A3} + \delta_{A3}(10 - 9.82) \quad \varepsilon_{B4} = \varepsilon_{B3} + \delta_{B3}(8.93 - 3.85)$$
$$\varepsilon_{A5} = \varepsilon_{A4} + \delta_{A4}(100 - 10) \quad \varepsilon_{B5} = \varepsilon_{B4} + \delta_{B4}(3.85 - 0)$$

Equation 13.5 indicates that net material flowrate that enters the first and last levels should take non-negative values:

$$\delta_{A0} \geq 0 \quad \delta_{B0} \geq 0$$
$$\delta_{A4} \geq 0 \quad \delta_{B4} \geq 0$$

Equation 13.6 indicates that all residual loads (except that at level 1) should take non-negative values:

$$\varepsilon_{A2} \geq 0 \quad \varepsilon_{B2} \geq 0$$
$$\varepsilon_{A3} \geq 0 \quad \varepsilon_{B3} \geq 0$$
$$\varepsilon_{A4} \geq 0 \quad \varepsilon_{B4} \geq 0$$
$$\varepsilon_{A5} \geq 0 \quad \varepsilon_{B5} \geq 0$$

The LINGO formulation for the previous model can be written as follows:

Model:

```
! Optimisation objective: to minimise annual operating cost
(AOC);
min = AOC;
AOC = (FW*0.001 + CL*0.25)*330;

! Material and load cascade for SUPERIOR cascade!
! Material cascade ;
DA1 = DA0 ;
DA2 = DA1 - 10 ;
DA3 = DA2 + 3.8 ;
DA4 = DA3 + FW ;
```

```
! The first and last net material flowrates correspond
to minimum fresh clay (CL) and wastewater (WW) flowrates ;
DA0 = CL ;
DA4 = WW ;

! Property load cascade ;
EA1 = 0 ;
EA2 = EA1 + DA1*(8.93 - 3.85) ;
EA3 = EA2 + DA2*(9.82 - 8.93) ;
EA4 = EA3 + DA3*(10 - 9.82) ;
EA5 = EA4 + DA4*(100 - 10) ;

! DA0 and DA4 should take non-negative values ;
DA0 >= 0 ;
DA4 >= 0 ;

! EA2-EA5 should take non-negative values ;
EA2 >= 0 ;
EA3 >= 0 ;
EA4 >= 0 ;
EA5 >= 0 ;

! DA1-DA3 may take negative values ;
@free (DA1) ;
@free (DA2) ;
@free (DA3) ;

! Material and load cascade for INFERIOR cascade!
! Material cascade ;
DB1 = DB0 ;
DB2 = DB1 + 3.8 ;
DB3 = DB2 - 10;
DB4 = DB3 + CL ;

! The first and last net material flowrates correspond
to minimum freshwater (FW) and wastewater (WW) flowrates ;
DB0 = FW ;
DB4 = WW ;

! Property load cascade ;
EB1 = 0 ;
EB2 = EB1 + DB1*(10 - 9.82) ;
EB3 = EB2 + DB2*(9.82 - 8.93) ;
EB4 = EB3 + DB3*(8.93 - 3.85) ;
EB5 = EB4 + DB4*(3.85 - 0) ;

! DA0 and DA4 should take non-negative values ;
DB0 >= 0 ;
DB4 >= 0 ;

! EB2-EB5 should take non-negative values ;
EB2 >= 0 ;
EB3 >= 0 ;
EB4 >= 0 ;
EB5 >= 0 ;

! DB1-DB3 may take negative values ;
@free (DB1) ;
@free (DB2) ;
@free (DB3) ;
END
```

The following is the solution report generated by LINGO:

```
Global optimal solution found.
Objective value:                      135.8695
Total solver iterations:                     0

            Variable          Value
            AOC             135.8695
            FW              4.571382
            CL              1.628618
            DA1             1.628618
            DA0             1.628618
            DA2            -8.371382
            DA3            -4.571382
            DA4             0.000000
            WW              0.000000
            EA1             0.000000
            EA2             8.273379
            EA3             0.8228488
            EA4             0.000000
            EA5             0.000000
            DB1             4.571382
            DB0             4.571382
            DB2             8.371382
            DB3            -1.628618
            DB4             0.000000
            EB1             0.000000
            EB2             0.8228488
            EB3             8.273379
            EB4             0.000000
            EB5             0.000000
```

The solution report of LINGO indicates that the AOC of the RCN is determined as $135.87/year. Next, the revised cascade diagram takes the form as in Figure 13.8. As shown, the clay and freshwater flowrates are determined as 1.63 and 4.57 kg/day, respectively. Note that each cascade has its respective pinch operator and also the pinch-causing source.

Since the ATM is based on the MCA technique, insights to develop the material recovery scheme in the MCA can also be used for the ATM. For this case, the material recovery scheme for this RCN may be derived from either of the cascade diagrams in Figure 13.8. As shown in Figure 13.8a, 1.63 kg/day of clay is cascaded from the first level to the sink (SK1, 10 kg/day) at level 2. Since the flowrate of clay is insufficient to fulfill that of the sink, sources of lower levels are used. Hence, the source at level 3 (SR1, 3.8 kg/day) and freshwater at level 4 (4.57 kg/day) are sent to fulfill the flowrate requirement of the sink. The allocation of sources from a lower quality level to sink at a higher quality level is indicated by the negative net material flowrate between levels 2 and 4 (see Example 4.1 for discussion on other similar problems).*

The RCN that achieves the established flowrate targets is shown in Figure 13.9.

---

* Note that derivation of RCN structure from targeting insights only applies to cases where single process sink (with non-zero quality level) that exists in the HQR.

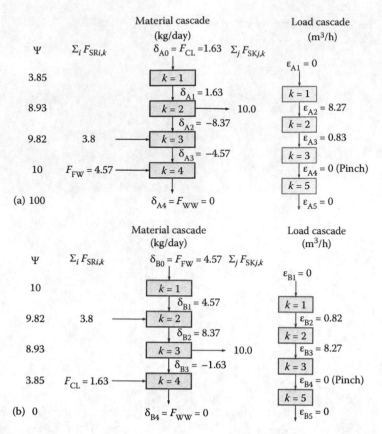

FIGURE 13.8
Results of ATM for Example 13.4: (a) superior cascade and (b) inferior cascade. (From *Chem. Eng. J.*, 149, Ng, D.K.S., Foo, D.C.Y., Tan, R.R., Pau, C.H., and Tan, Y.L., Automated targeting for conventional and bilateral property-based resource conservation network, 87–101, Copyright 2009c, with permission from Elsevier.)

Problems

Water Minimization Problems

13.1 Determine the minimum flowrates of freshwater and wastewater for the tire-to-fuel process in Problem 2.1. Note that the operating flow-rates of the water sinks and sources may be added as process constraints in the ATM. Note also that the source flowrate of the seal pot has to be equal to its sink flowrate.

13.2 Revisit Example 3.6 with ATM. Determine the minimum freshwater and wastewater flowrate targets for this RCN, assuming that pure

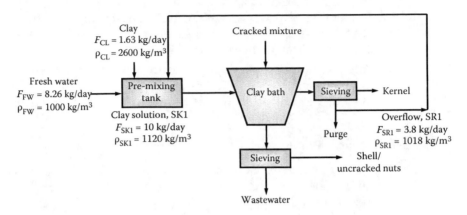

FIGURE 13.9
RCN for clay bath operation. (From *Chem. Eng. J.*, 149, Ng, D.K.S., Foo, D.C.Y., Tan, R.R., Pau, C.H., and Tan, Y.L., Automated targeting for conventional and bilateral property-based resource conservation network, 87–101, Copyright 2009c, with permission from Elsevier.)

freshwater is used. Limiting data for the case are given in Table 3.5 (note: this is a threshold problem; see answer in Figure 3.17).

13.3 Revisit the water minimization example in Problem 6.4 (Sorin and Bédard, 1999). Determine the minimum flowrate targets for its freshwater and wastewater requirement, assuming that pure freshwater is used. Limiting data for the case are given in Table 6.28.

13.4 For the bulk chemical production case in Problem 3.5, solve the following tasks:

  a. Determine the minimum freshwater and wastewater flowrates for the process, assuming that pure freshwater is used.

  b. Synthesize an RCN based on the insights obtained from the ATM (refer to Section 3.2). Compare the network structure from that obtained using the nearest neighbor algorithm and superstructural approach (note to readers: solve Problems 7.6 and 12.4 to compare results).

13.5 Revisit the AN case in Example 13.1. Determine the minimum water cost ($/s) for the RCN, when a secondary impure freshwater source with 5 ppm ammonia content is available for use apart from the primary pure freshwater source. Unit costs for the pure (CT_{FW1}) and impure freshwater sources (CT_{FW2}) are given as \$0.001/kg and \$0.0005/kg, respectively (see answer in Example 3.4).

13.6 Revisit Problem 3.4 (Polley and Polley, 2000) with ATM.

13.7 Revisit the classical water minimization problem in Problem 3.3. Determine the following targets for a direct water reuse/recycle scheme:

  a. Minimum freshwater and wastewater flowrates, when a single pure freshwater feed is available (note: see answers in Example 8.1).

b. Total water cost ($/h), when two freshwater sources of 0 and 10 ppm are used. These freshwater sources have unit costs of $1.0/ton (0 ppm) and $0.8/ton (10 ppm), respectively.

c. Perform a sensitivity analysis to analyze the effect of unit cost of impure freshwater source (range between $0.8/ton and $1.0/ton) on the total water cost and the flowrates of individual freshwater sources.

Utility Gas Recovery Problems

13.8 For the refinery hydrogen network in Examples 2.4 and 3.3, identify the minimum flowrates of the fresh hydrogen feed and purge stream for the RCN with ATM. The limiting data are found in Table 3.2.

13.9 For the refinery hydrogen network in Problems 2.6 and 4.15, identify the minimum flowrates of the fresh hydrogen feed and purge stream for the RCN. The limiting data are found in Table 4.30.

Property Integration Problems

13.10 For the metal degreasing case in Examples 2.5 and 5.2, solve the following cases:

a. Determine the minimum fresh solvent and solvent waste flowrates for the RCN, with limiting data in Table 2.4.

b. Determine the optimum operating temperature of the thermal processing unit that allows the maximum recovery of the process solvent sources. Other information on the problem is given in Example 5.2 (note: the ATM becomes an NLP model).

13.11 Determine the minimum flowrates of fresh resources (ultrapure water and municipal freshwater) and wastewater for both scenarios of the wafer fabrication process in Problem 4.19. Use the limiting data in Table 4.32.

References

El-Halwagi, M. M. 1997. *Pollution Prevention through Process Integration: Systematic Design Tools*. San Diego, CA: Academic Press.

Ng, D. K. S., Foo, D. C. Y., and Tan, R. R. 2009a. Automated targeting technique for single-component resource conservation networks—Part 1: Direct reuse/recycle. *Industrial and Engineering Chemistry Research*, 48(16), 7637–7646.

Ng, D. K. S., Foo, D. C. Y., and Tan, R. R. 2009b. Automated targeting technique for single-component resource conservation networks—Part 2: Single pass and partitioning waste interception systems. *Industrial and Engineering Chemistry Research*, 48(16), 7647–7661.

Ng, D. K. S., Foo, D. C. Y., Tan, R. R., Pau, C. H., and Tan, Y. L. 2009c. Automated targeting for conventional and bilateral property-based resource conservation network. *Chemical Engineering Journal*, 149, 87–101.

Polley, G. T. and Polley, H. L. 2000. Design better water networks. *Chemical Engineering Progress*, 96(2), 47–52.

Sorin, M. and Bédard, S. 1999. The global pinch point in water reuse networks. *Process Safety and Environmental Protection*, 77, 305–308.

14

Automated Targeting Model for Material Regeneration and Pretreatment Networks

As discussed in Chapters 6 and 9, interception units are broadly categorized into *single pass* and *partitioning* types (Ng et al., 2009a; Tan et al., 2009). However, the algebraic targeting procedure given in Chapter 6 is applicable only for a material regeneration network that utilizes interception units of the single pass type; while the graphical approach given in Chapter 9 is restricted to the pretreatment network with partitioning units. In this chapter, the automated targeting model (ATM) outlined in Chapter 13 is extended for various resource conservation networks (RCNs) with both types of interception units. This includes the material regeneration network (where process source is purified for recovery); as well as the pretreatment network (where fresh resource is purified).

14.1 Types of Interception Units and Their Characteristics

Figure 14.1 illustrates the two general types of interception units: single pass and partitioning units. In the former type (Figure 14.1a), it may be assumed that the feed and purified streams have the same flowrates, i.e., $F_{Rin} = F_{Rout}$. In other words, flowrate loss is negligible. The purified stream is of better quality compared to the feed stream, i.e., $q_{Rout} > q_{Rin}$.

In the partitioning type interception units, however, flowrate loss is significant. As shown in Figure 14.1b, the interception unit accepts a feed stream with quality q_{Rin} and produces two outlet streams: *purified* stream of higher quality q_{RP} (with flowrates F_{RP}) and *reject* stream of lower quality q_{RJ} (with flowrate F_{RJ}). Note that the quality of the streams are denoted as $q_{RP} > q_{Rin} > q_{RJ}$.

Both types of interception units may be classified as *fixed outlet quality* or *removal ratio*. As discussed in Chapter 6, interception units of the former type purify feed streams (with quality q_{Rin}) to a constant outlet quality, i.e., q_{Rout} or q_{RP} (e.g., 20 ppm, 5 wt%, etc.). On the other hand, interception units of the removal ratio type purify a given feed stream by removing a portion of its impurity/property load.

In the following sections, the ATM for an RCN with different types of interception units is outlined.

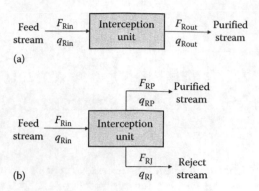

FIGURE 14.1
Types of interception units: (a) single pass and (b) partitioning units. (From Ng, D.K.S., Foo, D.C.Y., and Tan, R.R., Automated targeting technique for single-component resource conservation networks—Part 2: Single pass and partitioning waste interception systems, *Ind. Eng. Chem. Res.*, 48(16), 7647. Copyright 2009a American Chemical Society. With permission.)

14.2 ATM for RCN with Single Pass Interception Unit of Fixed Outlet Quality Type

The basic ATM framework for this RCN takes the revised form as that for the reuse/recycle network given in Chapter 13, as shown in Figure 14.2. The same procedure may be followed to formulate the ATM for the RCN (see detailed steps in Chapter 13). However, two minor modifications are needed:

1. The outlet quality of the interception unit (q_{Rout}) is added in the cascade diagram, if it is not present among the quality levels of the process sinks and sources.

2. For all quality levels that are lower than q_{Rout}, wherever a process source is found, a new sink is added to represent the potential flow-rate to be sent for regeneration ($F_{RE,r}$, $r = 1, 2, ..., RG$). Also, a new source is added at q_{Rout} level which represents the regenerated source (F_{REG}).

The formulation of the ATM also differs slightly from that of the direct reuse/recycle network. The following constraints describe the material and load cascades in the ATM framework:

$$\delta_k = \delta_{k-1} + (\Sigma_i F_{SRi} - \Sigma_j F_{SKj})_k + F_{REG} - F_{RE,r} \quad k = 1, 2, ..., n-1 \quad (14.1)$$

$$F_{REG} = \Sigma_r F_{RE,r} \quad (14.2)$$

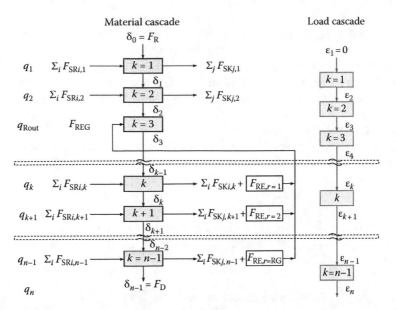

FIGURE 14.2
ATM framework for RCN with single pass interception unit(s) of fixed outlet quality. (From Ng, D.K.S., Foo, D.C.Y., and Tan, R.R., Automated targeting technique for single-component resource conservation networks—Part 2: Single pass and partitioning waste interception systems, *Ind. Eng. Chem. Res.*, 48(16), 7647. Copyright 2009a American Chemical Society. With permission.)

$$F_{RE,r} \leq \Sigma_i F_{SRi,k} \quad q_k < q_{Rout}; \quad \forall r \tag{14.3}$$

$$\varepsilon_k = \begin{cases} 0 & k = 1 \\ \varepsilon_{k-1} + \delta_{k-1}(q_k - q_{k-1}) & k = 2, 3, \ldots, n \end{cases} \tag{14.4}$$

$$\delta_{k-1} \geq 0 \quad k = 1, n \tag{14.5}$$

$$\varepsilon_k \geq 0 \quad k = 2, 3, \ldots, n \tag{14.6}$$

Equation 14.1 is an extended form of Equation 13.3 (reuse/recycle network), which describes the *net material flowrate* of level k (δ_k) in the material cascade. Apart from the net material flowrate cascaded from an earlier quality level (δ_{k-1}), the equation also takes into account the flowrate difference of process sources and sinks at level k ($\Sigma_i F_{SRi,k} - \Sigma_j F_{SKj,k}$), as well as the regeneration flowrate (F_{REG}) and individual flowrate to be sent for regeneration ($F_{RE,r}$). Note that Equation 14.1 is a generic equation that is applicable for all quality levels. For quality levels other than q_{Rout}, the F_{REG} term is set to zero. Similarly, the potential regeneration flowrate terms ($F_{RE,r}$) are set to zero for quality levels where no sources exist, as well as for levels of higher quality than the regeneration outlet quality, i.e., $q_k > q_{Rout}$. Equation 14.2 next describes the flowrate balance for the interception unit(s). Since the regeneration flowrate originates from the process source, the maximum flowrate to be regenerated is bound by the maximum available flowrate, as described by Equation 14.3. Equations 14.4 through 14.6 are essentially the same as those used to describe the ATM for direct reuse/recycle network in Chapter 13.

Since there are two variables to be optimized in this case, i.e., fresh resource and regeneration flowrates, we may set the optimization objective to minimize for RCN cost (e.g. operating or total annual cost (TAC)). Alternatively, we can also minimize the flowrates of both fresh resource and regeneration source with the two-stage optimization approach introduced in Chapter 12 (see Section 12.5 for details).

Since the model is an extended form of that given in Chapter 13, it is a linear program (LP) model that can be solved to obtain a global optimum solution.

 Example 14.1 Water regeneration network

The classical water minimization case study in Example 8.2 (Wang and Smith, 1994) is revisited here. The limiting water data are given in Table 14.1 (reproduced from Table 8.1). A single pass interception unit with outlet concentration (C_{Rout}) of 5 ppm is used to regenerate water for further water recovery. Determine the minimum flowrates of freshwater (0 ppm) and regenerated sources for this *regeneration network*. Also identify the structure of the RCN using the insights of the ATM.

Solution

The cascade diagram for this example is shown in Figure 14.3. Note that a new concentration level of 5 ppm is needed in the cascade diagram, which corresponds to the outlet concentration of the interception unit. Also, a new source is added here for the regenerated water and two new sinks are added at 100 and 800 ppm, to allow the water sources to be sent for regeneration from these levels.

TABLE 14.1

Limiting Water Data for Example 14.1

Sinks, SK$_j$	F_{SKj} (t/h)	C_{SKj} (ppm)	Sources, SR$_i$	F_{SRi} (t/h)	C_{SRi} (ppm)
SK1	20	0	SR1	20	100
SK2	100	50	SR2	100	100
SK3	40	50	SR3	40	800
SK4	10	400	SR4	10	800
$\Sigma_j F_{SKj}$	170		$\Sigma_i F_{SRi}$	170	

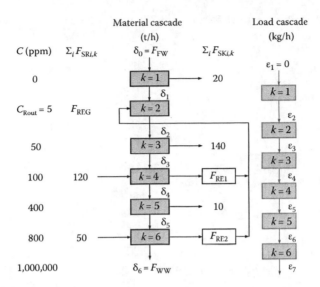

FIGURE 14.3
Cascade diagram for Example 14.1.

Since both freshwater and regenerated flowrates are to be minimized, the two-stage optimization approach is used. In this case, we will minimize the freshwater flowrate in the first stage, while the regenerated flowrate is minimized at the second stage. To determine the minimum freshwater flowrate (F_{FW}), the optimization objective in Equation 14.7 is used:

$$\text{Minimize } F_{FW} \tag{14.7}$$

subject to the following constraints:
Water cascade across all quality levels is given by Equation 14.1:

$$\delta_1 = \delta_0 + (0 - 20)$$
$$\delta_2 = \delta_1 + (F_{REG} - 0)$$
$$\delta_3 = \delta_2 + (0 - 140)$$
$$\delta_2 = \delta_3 + (120 - F_{RE1})$$
$$\delta_2 = \delta_4 + (0 - 10)$$
$$\delta_2 = \delta_5 + (50 - F_{RE2})$$

Since the net material flowrates that enter the first and the last concentration levels correspond to the minimum freshwater (F_{FW}) and wastewater (F_{WW}) flowrates, respectively, the following equations may be added:

$$\delta_0 = F_{FW}$$
$$\delta_6 = F_{WW}$$

Equation 14.2 describes the flowrate balance of the interception unit:

$$F_{REG} = F_{RE1} + F_{RE2}$$

Next, the source flowrate to be regenerated should not be more than what is available, as in Equation 14.3:

$$F_{RE1} \leq 120$$
$$F_{RE2} \leq 50$$

Based on Equation 14.4, the impurity load cascades (in kg/h) across all quality levels are given as follows:

$$\varepsilon_1 = 0$$
$$\varepsilon_2 = \varepsilon_1 + \delta_1(5 - 0)/1,000$$
$$\varepsilon_3 = \varepsilon_2 + \delta_2(50 - 5)/1,000$$
$$\varepsilon_4 = \varepsilon_3 + \delta_3(100 - 50)/1,000$$
$$\varepsilon_5 = \varepsilon_4 + \delta_4(400 - 100)/1,000$$
$$\varepsilon_6 = \varepsilon_5 + \delta_5(800 - 400)/1,000$$
$$\varepsilon_7 = \varepsilon_6 + \delta_6(1,000,000 - 800)/1,000$$

Equation 14.5 indicates that the net material flowrates that enter the first and last levels should assume non-negative values:

$$\delta_0 \geq 0$$
$$\delta_6 \geq 0$$

Based on Equation 14.6, all residual loads (except for level 1) should take non-negative values:

$$\varepsilon_2 \geq 0; \varepsilon_3 \geq 0$$
$$\varepsilon_4 \geq 0; \varepsilon_5 \geq 0$$
$$\varepsilon_6 \geq 0; \varepsilon_7 \geq 0$$

In terms of LINGO formulation, the previous model can be written as follows:

Model:
```
! Optimisation objective: to minimise freshwater (FW)
flowrate ;
min = FW ;

! Water cascade ;
D1 = D0 - 20 ;
D2 = D1 + REG ;
```

```
D3 = D2 - 140 ;
D4 = D3 + ( 120 - RE1 ) ;
D5 = D4 - 10 ;
D6 = D5 + ( 50 - RE2 ) ;

! The first and last net material flowrates correspond to
minimum freshwater (FW) and wastewater (WW) flowrates ;
D0 = FW ;
D6 = WW ;

! Flowrate balance for interception unit(s) ;
REG = RE1 + RE2 ;

! Flowrate constraints of regenerated water ;
RE1 <= 120 ;
RE2 <= 50 ;

! Impurity load cascade (in kg/s) ;
E1 = 0 ;
E2 = E1 + D1'(5 - 0)/1000 ;
E3 = E2 + D2'(50 - 5)/1000 ;
E4 = E3 + D3'(100 - 50)/1000 ;
E5 = E4 + D4'(400 - 100)/1000 ;
E6 = E5 + D5'(800 - 400)/1000 ;
E7 = E6 + D6'(1000000 - 800)/1000 ;

! D0 and D6 should take non-negative values ;
D0 >= 0 ;
D6 >= 0 ;

! E2-E7 should take non-negative values ;
E2 >= 0 ; E3 >= 0 ; E4 >= 0 ;
E5 >= 0 ; E6 >= 0 ; E7 >= 0 ;

! D1-D5 may take negative values ;
@free (D1) ; @free (D2) ; @free (D3) ;
@free (D4) ; @free (D5) ;
END
```

As mentioned in the previous chapter, the last set of constraints is only necessary for LINGO, as it assumes all variables to have non-negative values by default. For the same reason, the constraints for non-negativity (given by Equations 14.5 and 14.6) need not be coded in LINGO. The following is the solution report generated by LINGO for Stage 1 optimization (the results for optimization variables are omitted, as they are not necessary right now):

```
Global optimal solution found.
Objective value:                      20.00000
Total solver iterations:                     4
```

Note that the optimization result indicates that the minimum freshwater flowrate for the water network is determined to be 20 t/h. Hence, the minimum flowrate is included as a new constraint in Stage 2 optimization:

$$F_{FW} = 20 \qquad (14.8)$$

while the optimization objective is revised to minimize the regenerated flowrate:

$$\text{Minimize } F_{REG} \qquad (14.9)$$

The LINGO formulation for the second optimization stage remains identical to that in the first stage, except that the new optimization objective (given by Equation 14.9) and the new constraint for freshwater flowrate (given by Equation 14.8) are added, as follows:

Model:
```
! Optimisation objective: to minimise regenerated water
(REG) flowrate ;
min = REG ;
FW = 20 ;

! Water cascade ;
D1 = D0 - 20 ;
D2 = D1 + REG ;
D3 = D2 - 140 ;
D4 = D3 + ( 120 - RE1 ) ;
D5 = D4 - 10 ;
D6 = D5 + ( 50 - RE2 ) ;

! The first and last net material flowrates corresponds to
minimum freshwater (FW) and wastewater (WW) flowrates ;
D0 = FW ;
D6 = WW ;

! Flowrate balance for interception unit(s) ;
REG = RE1 + RE2 ;

! Flowrate constraints of regenerated water ;
RE1 <= 120 ;
RE2 <= 50 ;

! Impurity load cascade (in kg/s) ;
E1 = 0 ;
E2 = E1 + D1*(5 - 0)/1000 ;
E3 = E2 + D2*(50 - 5)/1000 ;
E4 = E3 + D3*(100 - 50)/1000 ;
E5 = E4 + D4*(400 - 100)/1000 ;
E6 = E5 + D5*(800 - 400)/1000 ;
E7 = E6 + D6*(1000000 - 800)/1000 ;

! D0 and D6 should take non-negative values ;
D0 >= 0 ;
D6 >= 0 ;

! E2-E7 should take non-negative values ;
E2 >= 0 ; E3 >= 0 ; E4 >= 0 ;
E5 >= 0 ; E6 >= 0 ; E7 >= 0 ;

! D1-D5 may take negative values ;
@free (D1) ; @free (D2) ; @free (D3) ;
@free (D4) ; @free (D5) ;
END
```

The following is the solution report generated by LINGO for Stage 2 optimization:

```
Global optimal solution found.
Objective value:                     73.68421
Total solver iterations:                    3

                Variable        Value
                REG          73.68421
                FW           20.00000
                D1            0.000000
                D0           20.00000
                D2           73.68421
                D3          -66.31579
                D4            5.714286
                RE1          47.96992
                D5           -4.285714
                D6           20.00000
                RE2          25.71429
                WW           20.00000
                E1            0.000000
                E2            0.000000
                E3            3.315789
                E4            0.000000
                E5            1.714286
                E6            0.000000
                E7        19984.00
```

As shown, the optimization result indicates that the minimum regenerated water flowrate is determined as 73.68 t/h. The revised cascade diagram that incorporates all optimized variables is shown in Figure 14.4. Note that the results are comparable to those obtained using the algebraic targeting technique (see Table 8.5 in Example 8.2). Also note that even though the flowrates for fresh and regenerated water remain the same, the individual sources that are sent for regeneration are slightly different for both approaches, which indicates that a degenerate solution exists for this case.

For a simple case such as this problem, the structure for the RCN can be determined from the result of the ATM in Figure 14.4.* At 0 ppm level, there exists a sink (SK1) that requires 20 t/h of freshwater feed. Hence, the entire freshwater feed is consumed by this sink, which leaves zero net water flowrate from the first concentration level, i.e., $\delta_1 = 0$.

We then move to the next concentration level of 5 ppm, where 73.68 t/h of regenerated water source is available. Since no water sink is found at this level, the regenerated source is cascaded to the following level

---

* There should only be sink that exists in each pinch region, with non-zero quality level, in order to apply this technique.

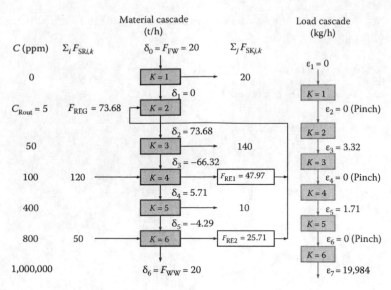

FIGURE 14.4
Result of ATM for Example 14.1.

(50 ppm), i.e., $\delta2 = 73.68$ t/h. Sinks SK2 and SK3 are at 50 ppm with a flow-rate requirement of 140 t/h, for which it is insufficient to use the regenerated source alone. In order to fulfill the flowrate requirement, a flowrate of 66.32 t/h is drawn from the lower concentration level of 100 ppm (indicated by negative net water flowrate, i.e., $\delta3 = -66.32$ t/h). The latter may be contributed by either SR1 or SR2, since both sources are found at 100 ppm. To minimize piping connections, we shall utilize SR2 which has sufficient flowrates for both sinks. The allocation of regenerated water and SR2 to both SK2 and SK3 are then determined based on the flowrate ratio of the sinks.

We then move to next concentration level, i.e., 100 ppm where sources SR1 and SR2 are found. Apart from supplying water to the sinks at 50 ppm, a flowrate of 47.97 t/h (F_{RE1}) is also sent for regeneration from these sources. We also observe that 5.71 t/h of these sources are cascaded down to the next concentration level at 400 ppm, from either of these sources ($\delta_4 = 5.71$ t/h). At 400 ppm level, the cascaded flowrate from 100 ppm ($\delta_4 = 5.71$ t/h) is mixed with 4.29 t/h water from the lower concentration level (800 ppm, indicated by negative δ_5 value), to fulfill the 10 t/h flowrate requirement of sink SK4.

In the concentration level of 800 ppm, we observe the presence of the sources SR3 and SR4. Apart from sending water to the sink at 400 ppm, 25.71 t/h (F_{RE2}) of these sources was also drawn for water regeneration. The leftover flowrate from these sources, i.e., 20 t/h will then leave as wastewater ($\delta_6 = F_{WW}$) from the RCN. A complete structure of the water regeneration network is shown in Figure 14.5. Note that the network

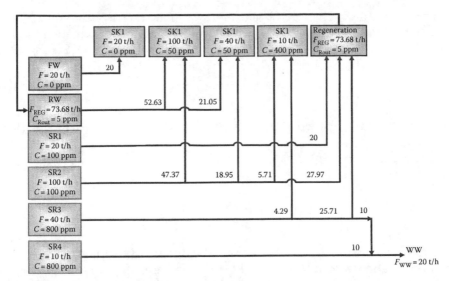

FIGURE 14.5
RCN for Example 14.1 (all flowrate terms in t/h).

structure shows us that the targeting insights may be different from those obtained from the nearest neighbor algorithm,[*] indicating that a degenerate solution exists for the problem.

14.3 ATM for RCN with Single Pass Interception Unit of Removal Ratio Type

Apart from the interception units that have a fixed outlet quality, another type of interception unit are those that are rated in terms of its *removal ratio index* (*RR*), which describes the fractional removal of impurity/property load from the feed stream of the interception unit. This parameter has been discussed previously for partitioning interception units in a pretreatment network (see Section 9.1). In this section, the equation for this parameter takes a revised form, given as follows:

$$RR = \frac{F_{Rin}\left(q^{max} - q_{Rin}\right) - F_{Rout}\left(q^{max} - q_{Rout}\right)}{F_{Rin}\left(q^{max} - q_{Rin}\right)} \tag{14.10}$$

---

[*] Solve Problem 7.16b to compare results.

where q^{max} is the highest possible value for a given type of purity index, e.g., fraction of 1, 100%, etc. A more commonly used version of Equation 14.10 is given for the quality index of concentration, as follows (Wang and Smith, 1994):

$$RR = \frac{F_{Rin}C_{Rin} - F_{Rout}C_{Rout}}{F_{Rin}C_{Rin}} \qquad (14.11)$$

where C_{Rin} and C_{Rout} are the impurity concentrations of the feed and purified streams.

For a single pass interception unit without flowrate loss, i.e., $F_{Rin} = F_{Rout}$, Equation 14.11 may be simplified as follows:

$$RR = \frac{C_{Rin} - C_{Rout}}{C_{Rin}} \qquad (14.12)$$

From Equation 14.12, we observe that the quality of the purified stream is dependent on the feed stream quality. For instance, for an interception unit of $RR = 0.9$, feed streams of 100 ppm will be purified to 10 ppm, while that of 200 ppm will end up at 20 ppm. In other words, new quality levels are to be included in the cascade diagram of the ATM for every source that is present, as all sources may be sent for regeneration (see Figure 14.6). Identifying the quality levels for the interception units helps to maintain the ATM as LP model that ensure global optimum solution.

The formulation for the ATM of this case is similar to that of the interception unit of fixed outlet quality, except that the constraints which involve flowrate terms of the interception units (i.e., Equations 14.1 and 14.2) are revised slightly as follows:

$$\delta_k = \delta_{k-1} + (\Sigma_i F_{SRi} - \Sigma_j F_{SKj})_k + F_{REG,r} - F_{RE,r} \quad k = 1, 2, \ldots, n-1 \qquad (14.13)$$

$$F_{RE,r} = F_{REG,r} \quad r = 1, 2, \ldots, RG \qquad (14.14)$$

where $F_{RE,r}$ and $F_{REG,r}$ are the inlet and outlet rates of the interception unit r.

Example 14.2 Water regeneration network

Example 14.1 is revisited here. A single pass interception unit with $RR = 0.9$ is used to purify water source for further recovery. Determine the minimum operating cost (MOC) for the water regeneration network, assuming that the unit costs for fresh and regenerated water are given as \$1/ton (CT_{FW}) and \$0.8/ton (CT_{RW}), respectively. Wastewater treatment cost is assumed to be insignificant here.

Solution

From Table 14.1, it is observed that the water sources that may be sent for regeneration originate from 100 and 800 ppm. Equation 14.12 hence

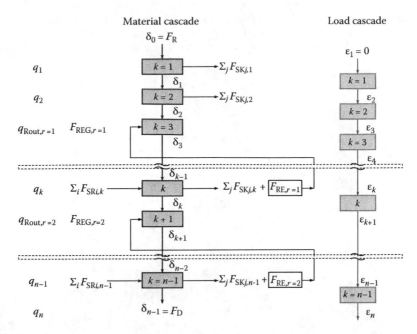

FIGURE 14.6
ATM framework for RCN with single pass interception unit(s) of removal ratio type. (From Ng, D.K.S., Foo, D.C.Y., and Tan, R.R., Automated targeting technique for single-component resource conservation networks—Part 2: Single pass and partitioning waste interception systems, *Ind. Eng. Chem. Res.*, 48(16), 7647. Copyright 2009a American Chemical Society. With permission.)

determines that the concentrations of the regenerated sources correspond to 10 and 80 ppm, respectively. Hence, two new concentration levels are added to the cascade diagram for the regenerated sources. The cascade diagram for this example is shown in Figure 14.7.

The optimization objective is set to determine the MOC for the water regeneration network:

$$\text{Minimize } F_{FW}\, CT_{FW} + (F_{RE1} + F_{RE2})\, CT_{RW} \tag{14.15}$$

where $F_{RE1} + F_{RE2}$ are the two possible water sources that may be sent for regeneration. The optimization problem is subject to the following constraints.

The water cascades across all quality levels are described by Equation 14.13 as follows:

$$\delta_1 = \delta_0 + (0 - 20)$$
$$\delta_2 = \delta_1 + (F_{REG1} - 0)$$
$$\delta_3 = \delta_2 + (0 - 140)$$
$$\delta_4 = \delta_3 + (F_{REG2} - 0)$$

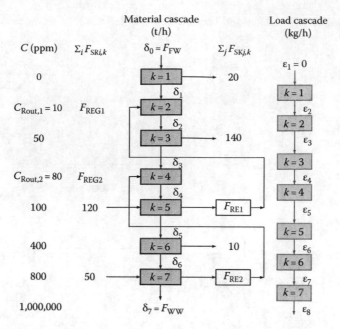

FIGURE 14.7
Cascade diagram for Example 14.2.

$$\delta_5 = \delta_4 + (120 - F_{RE1})$$
$$\delta_6 = \delta_5 + (0 - 10)$$
$$\delta_7 = \delta_6 + (50 - F_{RE2})$$

Since the net material flowrates that enter the first and the last concentration levels correspond to the minimum freshwater (F_{FW}) and wastewater (F_{WW}) flowrates, respectively, the following equations may be added:

$$\delta_0 = F_{FW}$$
$$\delta_7 = F_{WW}$$

Equation 14.14 describes the flowrate balance of the interception unit:

$$F_{RE1} = F_{REG1}$$
$$F_{RE2} = F_{REG2}$$

Next, the source flowrate to be regenerated should not be more than what is available, according to Equation 14.3:

$$F_{RE1} \leq 120$$
$$F_{RE2} \leq 50$$

Based on Equation 14.4, the impurity load cascades (in kg/s) across all quality levels are given as follows:

$$\varepsilon_1 = 0$$
$$\varepsilon_2 = \varepsilon_1 + \delta_1(10 - 0)/1,000$$
$$\varepsilon_3 = \varepsilon_2 + \delta_2(50 - 10)/1,000$$
$$\varepsilon_4 = \varepsilon_3 + \delta_3(80 - 50)/1,000$$
$$\varepsilon_5 = \varepsilon_4 + \delta_4(100 - 80)/1,000$$
$$\varepsilon_6 = \varepsilon_5 + \delta_5(400 - 100)/1,000$$
$$\varepsilon_7 = \varepsilon_6 + \delta_6(800 - 400)/1,000$$
$$\varepsilon_8 = \varepsilon_7 + \delta_7(1,000,000 - 800)/1,000$$

Equation 14.5 indicates that the net material flowrates that enter the first and last levels should take non-negative values:

$$\delta_0 \geq 0$$
$$\delta_7 \geq 0$$

Based on Equation 14.6, all residual loads (except at level 1) should take non-negative values:

$$\varepsilon_2 \geq 0 \quad \varepsilon_3 \geq 0$$
$$\varepsilon_4 \geq 0 \quad \varepsilon_5 \geq 0$$
$$\varepsilon_6 \geq 0 \quad \varepsilon_7 \geq 0$$
$$\varepsilon_8 \geq 0$$

The LINGO formulation for the previous model is written as follows:

```
Model:
! Optimisation objective: to determine minimum operating
cost (MOC) ;
min = FW*1 + (RE1 + RE2)*0.8 ;
! Water cascade ;
D1 = D0 - 20 ;
D2 = D1 + REG1 ;
D3 = D2 - 140 ;
D4 = D3 + REG2 ;
D5 = D4 + ( 120 - RE1 ) ;
D6 = D5 - 10 ;
D7 = D6 + ( 50 - RE2 ) ;

! The first and last net material flowrates correspond to
minimum freshwater (FW) and wastewater (WW) flowrates
;
D0 = FW ;
D7 = WW ;
```

```
! Flowrate balance for interception unit(s) ;
RE1 = REG1 ;
RE2 = REG2 ;

! Flowrate constraints of regenerated water ;
RE1 <= 120 ;
RE2 <= 50 ;

! Impurity load cascade ;
E1 = 0 ;
E2 = E1 + D1*(10 - 0)/1000 ;
E3 = E2 + D2*(50 - 10)/1000 ;
E4 = E3 + D3*(80 - 50)/1000 ;
E5 = E4 + D4*(100 - 80)/1000 ;
E6 = E5 + D5*(400 - 100)/1000 ;
E7 = E6 + D6*(800 - 400)/1000 ;
E8 = E7 + D7*(1000000 - 800)/1000 ;

! D0 and D7 should take non-negative values ;
D0 >= 0 ;
D7 >= 0 ;

! E2-E8 should take non-negative values ;
E2 >= 0 ; E3 >= 0 ; E4 >= 0 ; E5 >= 0 ;
E6 >= 0 ; E7 >= 0 ; E8 >= 0 ;

! D1-D6 may take negative values ;
@free (D1) ; @free (D2) ; @free (D3) ;
@free (D4) ; @free (D5) ; @free (D6) ;
END
```

Note that the non-negativity constraints are not included, as LINGO assumes all variables take non-negative values by default (see discussion in Example 14.1). The following is the solution report generated by LINGO:

```
Global optimal solution found.
Objective value:                    85.07937
Total solver iterations:                   4

                Variable        Value
                FW            45.71429
                RE1           49.20635
                RE2            0.000000
                D1            25.71429
                D0            45.71429
                D2            74.92063
                REG1          49.20635
                D3           -65.07937
                D4           -65.07937
                REG2           0.000000
                D5            5.714286
                D6           -4.285714
                D7            45.71429
                WW            45.71429
```

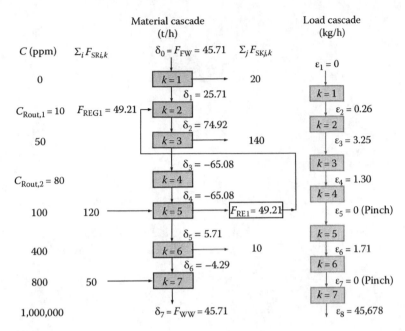

FIGURE 14.8
Result of ATM for Example 14.2.

E1	0.000000
E2	0.2571429
E3	3.253968
E4	1.301587
E5	0.000000
E6	1.714286
E7	0.000000
E8	45677.71

The optimization result indicates that the MOC is determined as \$85/h. The result also indicates that only 49.21 t/h of water source at 100 ppm is to be regenerated to 10 ppm, in order to reduce the freshwater and wastewater flowrates to 45.71 t/h. The revised cascade diagram is shown in Figure 14.8. Note that the result is different from that in Example 14.1 (see Figure 14.4), as different optimization objectives are used (minimum flowrates for Example 14.1; and minimum cost for this example).

14.4 Modeling for Partitioning Interception Unit(s) of Fixed Outlet Quality Type

Basic modeling for a partitioning interception unit has been previously described in Chapter 9 for a pretreatment network. In this section, a more

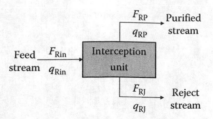

FIGURE 14.9
A partitioning interception unit.

detailed model is described, which takes into account the impurity/property load of the RCN. Figure 14.9 shows a partitioning type interception unit reproduced from Figures 11.1 and 14.1b.

The flowrate and material balances of the interception unit are given in Equations 14.16 and 14.17 (reproduced from Equations 9.1 and 9.2):

$$F_{Rin} = F_{RP} + F_{RJ}$$ (14.16)

$$F_{Rin}\, q_{Rin} = F_{RP}\, q_{RP} + F_{RJ}\, q_{RJ}$$ (14.17)

where
F_{RP} and q_{RP} are the flowrate and quality index of the higher quality *purified* stream, respectively
F_{RJ} and q_{RJ} are flowrate and quality index of the lower quality *reject* streams respectively

Since the interception unit possesses two outlet streams, an overall *product recovery factor (RC)* is used to define the recovery of the desired product from the inlet stream, as given in Equation 14.18. Note that the latter is an extended form of Equation 9.3 (pretreatment network), where stream quality is considered:

$$RC = \frac{F_{RP}\, q_{RP}}{F_{Rin}\, q_{Rin}}$$ (14.18)

Combining Equations 14.16 through 14.18 leads to Equation 14.19:

$$q_{RJ} = \frac{q_{RP}(RC-1)}{RC-QI}$$ (14.19)

where *QI* is the *quality improvement* index for a given interception unit, given as follows:

$$QI = \frac{q_{RP}}{q_{Rin}}$$ (14.20)

Similar expressions may also be developed from the perspective of impurity concentration. In this case, Equation 14.17 takes the following form:

$$F_{Rin}\, C_{Rin} = F_{RP}\, C_{RP} + F_{RJ}\, C_{RJ} \tag{14.21}$$

where C_{RP} and C_{RJ} are the impurity concentrations of the purified and reject streams. Similarly, the product recovery factor in Equation 14.18 takes the revised form as follows:

$$RC = \frac{F_{RP}\left(C^{max} - C_{RP}\right)}{F_{Rin}\left(C^{max} - C_{Rin}\right)} \tag{14.22}$$

where C^{max} is the highest possible value for a given concentration index, e.g., 1,000,000 ppm, 100 wt%, etc. Combining Equations 14.16, 14.21, and 14.22 leads to Equation 14.23:

$$C_{RJ} = \frac{C_{RP}RC - C_{Rin}QI}{RC - QI} \tag{14.23}$$

Note that the quality improvement (QI) index in Equation 14.23 has been defined in Equation 14.20 from quality perspective. A revised form of this equation in impurity concentration is given as follows:

$$QI = \frac{\left(C^{max} - C_{RP}\right)}{\left(C^{max} - C_{Rin}\right)} \tag{14.24}$$

From Equations 14.19 and 14.23, it may be observed that the reject stream quality of a partitioning interception unit is dependent on the quality/concentration of its inlet and purified streams, as well as the product recovery factor (RC), but not on the stream flowrates. Hence, for a given interception unit with purified stream quality and product recovery factor, the quality of the reject stream can be determined, provided that inlet stream quality is also known.

14.5 Modeling for Partitioning Interception Unit(s) of Removal Ratio Type

Apart from the product recovery factor (RC), a partitioning interception unit may also be rated by removal ratio index (RR), as given by Equations 14.10

and 14.11. It will be useful to relate the purified stream quality (q_{RP}) to both RC and RR values for a given unit. Hence, combining Equations 14.10 and 14.18 leads to the following correlation:

$$q_{RP} = \frac{q_{Rin} q^{max} RC}{q^{max}(1-RR) + q_{Rin}(RR+RC-1)} \qquad (14.25)$$

Similarly, combining Equations 14.11 and 14.22 leads to the revised version of Equation 14.26 as follows:

$$C_{RP} = \frac{C_{Rin} C^{max}(1-RR)}{(C^{max} - C_{Rin})RC + C_{Rin}(1-RR)} \qquad (14.26)$$

From Equations 14.25 and 14.26, we observe that the quality/concentration of the purified stream is dependent on the inlet stream quality, as well as the values of RR and RC of the interception unit. Once the purified stream quality is known, we can use Equation 14.19 or 14.23 to determine the reject stream quality/concentration.

Example 14.3 Mass balance calculation for a partitioning interception unit

Figure 14.10 shows a partitioning interception unit that regenerate an inlet stream of 100 kg/s (F_{Rin}) and 20% impurity concentration (C_{Rin}), to a purified stream with flowrate F_{RP} and impurity concentration C_{RP}, and a reject stream of flowrate F_{RJ} and impurity concentration C_{RJ}. Determine the missing flowrate and/or concentration parameters of the purified and stream streams for the following cases:

1. The interception unit is of the fixed outlet quality type, which produces 80 kg/s (F_{RP}) purified stream of 10% impurity (C_{RP}).
2. The interception unit has a removal ratio (RR) of 0.9, and product recovery factor (RC) of 0.8.

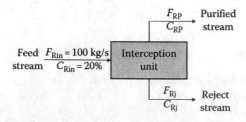

FIGURE 14.10
A partitioning interception unit for Example 14.3.

Solution

Case 1—$F_{RP}=80\,kg/s$; $C_{RP}=10\%$

For the fixed outlet quality type interception unit, the product recovery factor and quality improvement index are determined using Equations 14.22 and 14.24:

$$RC = \frac{80(100-10)}{100(100-20)} = 0.9$$

$$QI = \frac{(100-10)}{(100-20)} = 1.125$$

In other words, the interception unit recovers 90% of the desired product from the inlet stream and improves the stream quality by 12.5%.

To determine the reject stream concentration (C_{RJ}), we can perform a mass balance calculation using Equation 14.21, since the other parameters are known values (F_{RJ} can be easily determined as 20 kg/s using Equation 14.16). Alternatively, we can also use Equation 14.23 to determine the value of C_{RJ}, without having to perform mass balance calculation of the streams:

$$C_{RJ} - \frac{10(0.9)-20(1.125)}{0.9-1.125} = 60\%$$

Case 2—$RR=0.9$, $RC=0.8$.

For this case, Equation 14.26 is used to determine that the purified stream concentration (C_{RP}) of the interception unit:

$$C_{RP} = \frac{20(100)(1-0.9)}{(100-20)0.8 + 20(1-0.9)} = 3.03\%$$

Next, instead of using the mass balance equation, we can make use of Equations 14.23 and 14.24 to determine the reject stream concentration (C_{RJ}):

$$QI = \frac{(100-3.03)}{(100-20)} = 1.212$$

$$C_{RJ} = \frac{3.03(0.8)-20(1.212)}{0.8-1.212} = 52.94\%$$

Then, we can determine the flowrates of the purified (F_{RP}) and reject (F_{RJ}) streams by solving simultaneous equations with Equations 14.16 and 14.17:

$$100\ kg/s = F_{RP} + F_{RJ}$$

$$100\ kg/s(20\%) = F_{RP}(3.03\%) + F_{RJ}(52.94\%)$$

which leads to $F_{RP}=66\,kg/s$ and $F_{RJ}=34\,kg/s$.

14.6 ATM for RCN with Partitioning Interception Unit(s)

To incorporate interception units into the ATM, the quality levels of the purified and reject streams are first determined using the model described in the previous sections. Similar to the earlier cases with a single pass interception unit, the quality levels of the purified and reject streams of the interception unit are added to the ATM framework, if they are not readily present among the quality levels of the process sinks and sources. An ATM framework for the incorporation of a partitioning interception unit is shown in Figure 14.11. Note that the ATM frameworks for RCNs with both types of partitioning interception units are quite similar. However, the quality level of the purified streams of all interception units are identical for the case of fixed outlet quality type units (see Example 14.3 for illustration), while the RCN with interception units of removal ratio type will have a unique quality level for each of their purified and reject streams (depending on its inlet stream quality).

Most constraints in the ATM formulation remain similar to the earlier cases, except that the constraints that involve purified and reject flowrates of

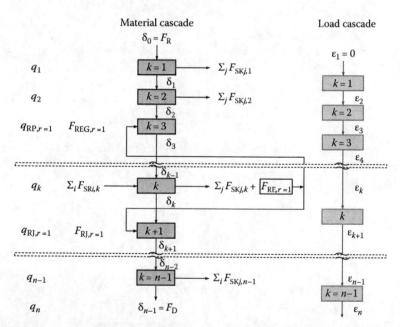

FIGURE 14.11
ATM framework for RCN with partitioning interception unit(s). (From Ng, D.K.S., Foo, D.C.Y., and Tan, R.R., Automated targeting technique for single-component resource conservation networks—Part 2: Single pass and partitioning waste interception systems, *Ind. Eng. Chem. Res.*, 48(16), 7647. Copyright 2009a American Chemical Society. With permission.)

the interception units (i.e., Equations 14.1 and 14.2) take the revised version as follows:

$$\delta_k = \delta_{k-1} + (\Sigma_i F_{SRi} - \Sigma_j F_{SKj})_k + F_{REG,r} + F_{RJ,r} - F_{RE,r} \quad k = 1, 2, \ldots, n-1 \quad (14.27)$$

$$F_{RE,r} = F_{REG,r} + F_{RJ,r} \quad r = 1, 2, \ldots, RG \quad (14.28)$$

where $F_{RJ,r}$ is the reject stream flowrate of interception unit r.

Also, the product recovery factor in Equations 14.18 and 14.22 should be rearranged so as to avoid nonlinearity (division of variables $F_{REG,r}$ and $F_{RE,r}$) in the ATM. The revised forms of the equations are given as follows, in quality and concentration indices, respectively:

$$F_{RE,r} \, q_{RE,r} \, RC = F_{REG,r} \, q_{REG,r} \quad r = 1, 2, \ldots, RG \quad (14.29)$$

$$F_{RE,r} (C^{max} - C_{RE,r}) \, RC = F_{REG,r} (C^{max} - C_{REG,r}) \quad r = 1, 2, \ldots, RG \quad (14.30)$$

where

$q_{RE,r}$ and $C_{RE,r}$ are the quality and concentration of the inlet stream of interception unit r

$q_{REG,r}$ and $C_{REG,r}$ are the quality and concentration of its purified stream

Example 14.4 Hydrogen network with partitioning interception unit of fixed outlet quality type

The hydrogen recovery problem in Examples 2.4 and 3.3 is revisited here, with the limiting data shown in Table 14.2 (reproduced from Table 2.2). The fresh hydrogen feed has an impurity concentration of 1%. A pressure swing adsorption (PSA) unit is used to purify the process sources for further recovery. The PSA unit can purify a hydrogen stream to 5% impurity (C_{RP}) and has a hydrogen recovery factor (RC) of 0.9 (Agrawal and Shenoy, 2006). Determine the MOC of the RCN given that the unit costs for fresh hydrogen (CT_{FH}) and the regenerated hydrogen sources (CT_{RH}) are $2000/MMscf and $1000/MMscf, respectively (cost of purged stream is ignored).

TABLE 14.2

Limiting Data for Example 14.4

Sinks, SK_j	F_{SKj} (MMscfd)	C_{SKj} (%)	Sources, SR_i	F_{SRi} (MMscfd)	C_{SRi} (%)
SK1	400	7.20	SR1	350	9.00
SK2	600	12.43	SR2	500	15.00

Solution

Since two process sources exist in the problem, both sources may be sent for purification in the PSA units, for further recovery to the sinks. Hence, the concentrations of the purified (5%) and reject streams of the PSA units are to be added as new quality levels in the ATM. The quality improvement index and reject stream concentrations for both sources are calculated using Equations 14.23 and 14.24 as follows:

SR1:

$$QI = \frac{(100-5)}{(100-9)} = 1.044$$

$$C_{RJ} = \frac{5(0.9)-9(1.044)}{0.9-1.044} = 34\%$$

SR2:

$$QI = \frac{(100-5)}{(100-15)} = 1.118$$

$$C_{RJ} = \frac{5(0.9)-15(1.118)}{0.9-1.118} = 56.28\%$$

Hence, two new concentration levels (34% and 56.28%) are added to the cascade diagram, along with the concentration of the purified stream (5%), as shown in Figure 14.12.

The optimization objective is set to determine the MOC for the hydrogen network:

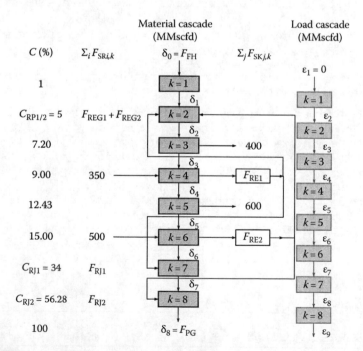

FIGURE 14.12
Cascade diagram for Example 14.3.

$$\text{Minimize } F_{FH}\, CT_{FH} + (F_{RE1} + F_{RE2})\, CT_{RH} \qquad (14.31)$$

where F_{FH} is the fresh hydrogen feed of the RCN. The optimization problem is subjected to the following constraints:

Hydrogen cascade across all quality levels, as given by Equation 14.27:

$$\delta_1 = \delta_0$$
$$\delta_2 = \delta_1 + (F_{REG1} + F_{REG2})$$
$$\delta_3 = \delta_2 + (0 - 400)$$
$$\delta_4 = \delta_3 + (350 - F_{RE1})$$
$$\delta_5 = \delta_4 + (0 - 600)$$
$$\delta_6 = \delta_5 + (500 - F_{RE2})$$
$$\delta_7 = \delta_6 + F_{RJ1}$$
$$\delta_8 = \delta_7 + F_{RJ2}$$

The following equations are added to relate the net material flowrates that enter the first and the last concentration levels to the minimum fresh hydrogen (F_{FH}) and purge (F_{PG}) flowrates:

$$\delta_0 = F_{FH}$$
$$\delta_8 = F_{PG}$$

Equation 14.28 describes the flowrate balance of the PSA units:

$$F_{RE1} = F_{REG1} + F_{RJ1}$$
$$F_{RE2} = F_{REG2} + F_{RJ2}$$

The recovery of hydrogen ($RC = 0.9$) to the purified stream is described by Equation 14.30:

$$F_{RE1}(0.9)(100 - 9) = F_{REG1}(100 - 5)$$
$$F_{RE2}(0.9)(100 - 15) = F_{REG2}(100 - 5)$$

Source flowrate to be regenerated should not be more than what is available according to Equation 14.3:

$$F_{RE1} \leq 350$$
$$F_{RE2} \leq 500$$

Impurity load cascades across all quality levels are given by Equation 14.4:

$$\varepsilon_1 = 0$$
$$\varepsilon_2 = \varepsilon_1 + \delta_1 (5 - 1) / 100$$

$$\varepsilon_3 = \varepsilon_2 + \delta_2(7.2 - 5)/100$$

$$\varepsilon_4 = \varepsilon_3 + \delta_3(9 - 7.2)/100$$

$$\varepsilon_5 = \varepsilon_4 + \delta_4(12.43 - 9)/100$$

$$\varepsilon_6 = \varepsilon_5 + \delta_5(15 - 12.43)/100$$

$$\varepsilon_7 = \varepsilon_6 + \delta_6(34 - 15)/100$$

$$\varepsilon_8 = \varepsilon_7 + \delta_7(56.28 - 34)/100$$

$$\varepsilon_9 = \varepsilon_8 + \delta_8(100 - 56.28)/100$$

Net material flowrates that enter the first and last levels should take non-negative values (Equation 14.5):

$$\delta_0 \geq 0$$

$$\delta_8 \geq 0$$

Finally, all residual loads (except at level 1) should take non-negative values (Equation 14.6):

$$\varepsilon_2 \geq 0 \quad \varepsilon_3 \geq 0$$

$$\varepsilon_4 \geq 0 \quad \varepsilon_5 \geq 0$$

$$\varepsilon_6 \geq \quad \varepsilon_7 \geq 0$$

$$\varepsilon_8 \geq 0 \quad \varepsilon_9 \geq 0$$

The LINGO formulation for this LP model is written as follows:

```
Model:
! Optimisation objective: to determine the minimum operating
cost (MOC) ;
min = FH*2000 + (RE1 + RE2)*1000 ;

! Hydrogen cascade ;
D1 = D0 ;
D2 = D1 + REG1 + REG2 ;
D3 = D2 - 400 ;
D4 = D3 + ( 350 - RE1 ) ;
D5 = D4 - 600 ;
D6 = D5 + ( 500 - RE2 ) ;
D7 = D6 + RJ1 ;
D8 = D7 + RJ2 ;

! The first and last net material flowrates corresponds to
minimum fresh hydrogen (FH) and purge (PG) flowrates ;
D0 = FH ;
D8 = PG ;

! Flowrate balance for interception unit(s) ;
RE1 = REG1 + RJ1 ;
RE2 = REG2 + RJ2 ;
```

```
RE1*0.9*(100 - 9) = REG1*(100 - 5) ;
RE2*0.9*(100 - 15) = REG2*(100 - 5) ;

! Flowrate constraints of regenerated hydrogen ;
RE1 <= 350 ;
RE2 <= 500 ;

! Impurity load cascade ;
E1 = 0 ;
E2 = E1 + D1*(5 - 1)/100 ;
E3 = E2 + D2*(7.2 - 5)/100 ;
E4 = E3 + D3*(9 - 7.2)/100 ;
E5 = E4 + D4*(12.43 - 9)/100 ;
E6 = E5 + D5*(15 - 12.43)/100 ;
E7 = E6 + D6*(34 - 15)/100 ;
E8 = E7 + D7*(56.28 - 34)/100 ;
E9 = E8 + D8*(100 - 56.28)/100 ;

! D1-D7 may take negative values ;
@free (D1) ; @free (D2) ; @free (D3) ;
@free (D4) ; @free (D5) ; @free (D6) ; @free (D7) ;
END
```

Similar to Example 14.2, the non-negativity constraints are excluded here, as LINGO assumes all variables take non-negative values by default (see discussion in Example 14.1). The following is the solution report generated by LINGO:

```
Global optimal solution found.
Objective value:                        359554.7
Total solver iterations:                       2


                Variable        Value
                FH              158.3467
                RE1               0.000000
                RE2              42.86133
                D1              158.3467
                D0              158.3467
                D2              192.8613
                REG1              0.000000
                REG2             34.51465
                D3             -207.1387
                D4              142.8613
                D5             -457.1387
                D6                0.000000
                D7                0.000000
                RJ1               0.000000
                D8                8.346680
                RJ2               8.346680
                PG                8.346680
                E1                0.000000
                E2                6.333867
                E3               10.57682
                E4                6.848320
```

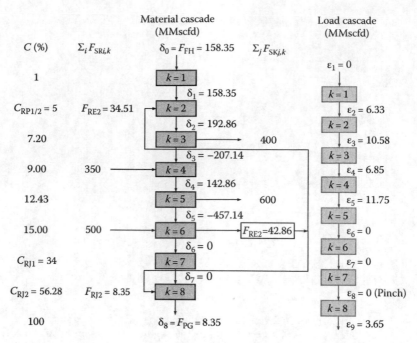

FIGURE 14.13
Result of ATM for Example 14.3.*

E5	11.74846
E6	0.000000
E7	0.000000
E8	0.000000
E9	3.649168

The optimization result indicates that the MOC is determined as $359,5551day. The revised cascade diagram is shown in Figure 14.13. The results indicate that only the hydrogen source at 15% level will be sent for purification in the PSA unit ($F_{RE}=42.86$ MMscfd); the source at 9% level will not be purified. Also, the minimum fresh hydrogen (F_{FH}) and purge flowrate (F_{PG}) corresponding to this scenario are identified as 158.35 and 8.35 MMscfd, respectively.

14.7 ATM for Pretreatment Networks

In Chapter 9, flowrate targeting for a pretreatment network has been carried out using insight-based technique. In this section, the targeting task

---

* Note that the targeting insights should not be used to drive the RCN structure here, as there are more then 1 sink in the pinch region (of different quality index).

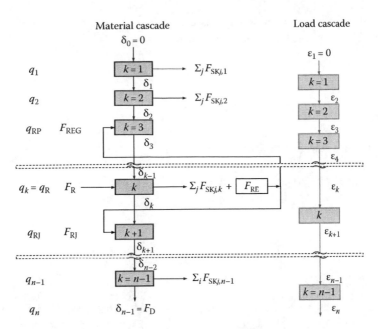

FIGURE 14.14
ATM framework for pretreatment network.

is carried out with the ATM. Since a partitioning interception unit is normally used in a pretreatment network for water purification, the ATM for a material regeneration network with partitioning unit (Section 14.6) may be extended for use here, with the revised framework given in Figure 14.14. It is observable that the latter resembles the framework in Figure 14.11. The main difference between the two frameworks is that the ATM framework for the pretreatment network in Figure 14.14 has only one interception unit, with its inlet located at the level where the fresh resource (with flowrate F_R) is found. Hence, a new sink is added to represent the potential flowrate to be sent for purification (F_{RE}) at this level. Also, the flowrate terms of the purified (F_{REG}) and reject (F_{RJ}) streams of the interception unit are added as sources in the ATM framework. Note also that the flowrate term of fresh resource is added at a lower level as it possesses a lower quality than some of the process sinks.

Most constraints in the ATM formulation remain similar to the case in Section 14.6. In the absence of process sources, and with a single interception unit, the net material flowrates as determined by Equation 14.27 take revised forms as follows:

$$\delta_k = \delta_{k-1} - \Sigma_j F_{SKj,k} + F_{REG} + F_{RJ} - F_{RE} \quad k = 1, 2, \ldots, n-1 \quad (14.32)$$

Since fresh resources do not exist at the highest quality level, no net material flowrate should enter the first quality level. On the other hand, the net material flowrate that enters the last level should take non-negative value. Hence, Equation 14.5 is revised as follows:

$$\delta_{k-1} = \begin{cases} 0 & k=1 \\ \geq 0 & k=n \end{cases} \qquad (14.33)$$

Also, Equations 14.28 and 14.29 that determine the flowrate balance and allocation of the interception unit are revised as Equations 14.34 and 14.35, respectively, as only one interception unit is present in the RCN. Note that Equation 14.35 is a simplified form of Equation 14.18, which is applicable for a dilute system such as in a pretreatment network (see discussion in Section 9.1):

$$F_{RE} = F_{REG} + F_{RJ} \qquad (14.34)$$

$$F_{RE}RC = F_{REG} \qquad (14.35)$$

Equation 14.3 is also revised to impose a constraint where the flowrate for purification should not be greater than the fresh resource flowrate:

$$F_{RE} \leq F_R \qquad (14.36)$$

Apart from the earlier-revised constraints, other constraints given by Equations 14.4 and 14.6 remain identical as in the earlier cases.

Example 14.5 Water pretreatment network

The water pretreatment network problem in Example 9.1 is revisited. A freshwater feed with an impurity concentration of 25 ppm (C_R) is to be supplied to four water sinks, with their limiting data given in Table 14.3 (reproduced from Table 9.1). A partitioning interception unit is used in the pretreatment network to purify the freshwater feed in order to fulfill sink requirements. The unit purifies the inlet stream to an impurity concentration of 5 ppm (C_{RP}) and has a recovery factor (RC) of 0.75. Determine the minimum freshwater flowrate needed for this pretreatment network.

TABLE 14.3

Limiting Data for Example 14.3

Sinks, SK$_j$	F_{SKj} (t/h)	C_{SKj} (ppm)
SK1	4	10
SR2	1	20
SR3	2	30
SR4	1	40

Solution

Example 9.1 demonstrated that when a 25 ppm freshwater feed is purified with an interception unit with purified stream concentration (C_{RP}) of 5 ppm and $RC = 0.75$, the reject stream of the unit has a concentration of 85 ppm (C_{RJ}). We can now construct the cascade diagram for the problem as given in Figure 14.15.

To determine the minimum freshwater flowrate (F_{FW}), the optimization objective in Equation 14.7 is used:

$$\text{Minimize } F_{FW} \tag{14.7}$$

subject to the constraints that follow:

Water cascade across all quality levels is given by Equation 14.32:

$$\delta_1 = \delta_0 + F_{REG}$$
$$\delta_2 = \delta_1 - 4$$
$$\delta_3 = \delta_2 - 1$$
$$\delta_4 = \delta_3 + (F_{FW} - F_{RE})$$
$$\delta_5 = \delta_4 - 2$$
$$\delta_6 = \delta_5 - 1$$
$$\delta_7 = \delta_6 + F_{RJ}$$

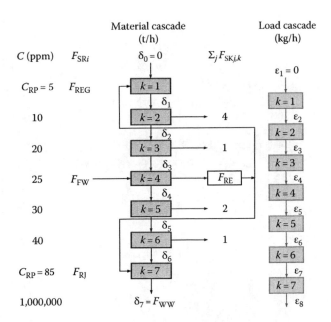

FIGURE 14.15
Cascade diagram for Example 14.5.

Equation 14.33 indicates that the net material flowrates that enter the first concentration level are set to zero, while those in the last level should have a value greater than zero. Also, it is useful to note that the latter corresponds to the minimum wastewater flowrate (F_{WW}) of the pretreatment network:

$$\delta_0 = 0$$
$$\delta_7 \geq 0$$
$$\delta_7 = F_{WW}$$

Equations 14.34 and 14.35 indicate the flowrate balance and allocation of the interception unit as follows (with $RC = 0.75$):

$$F_{RE} = F_{REG} + F_{RJ}$$
$$F_{RE}(0.75) = F_{REG}$$

The source flowrate to be purified should not be more than the freshwater flowrate (F_{FW}) available for the pretreatment network, according to Equation 14.35:

$$F_{RE} \leq F_{FW}$$

Impurity load cascades (in kg/h) across all quality levels are given by Equation 14.4 as follows:

$$\varepsilon_1 = 0$$
$$\varepsilon_2 = \varepsilon_1 + \delta_1(10 - 5) / 1,000$$
$$\varepsilon_3 = \varepsilon_2 + \delta_2(20 - 10) / 1,000$$
$$\varepsilon_4 = \varepsilon_3 + \delta_3(25 - 20) / 1,000$$
$$\varepsilon_5 = \varepsilon_4 + \delta_4(30 - 25) / 1,000$$
$$\varepsilon_6 = \varepsilon_5 + \delta_5(40 - 30) / 1,000$$
$$\varepsilon_7 = \varepsilon_6 + \delta_6(85 - 40) / 1,000$$
$$\varepsilon_8 = \varepsilon_7 + \delta_7(1,000,000 - 85) / 1,000$$

Finally, all residual loads (except at level 1) should take non-negative values (Equation 14.6):

$$\varepsilon_2 \geq 0 \quad \varepsilon_3 \geq 0$$
$$\varepsilon_4 \geq 0 \quad \varepsilon_5 \geq 0$$
$$\varepsilon_6 \geq 0 \quad \varepsilon_7 \geq 0$$
$$\varepsilon_8 \geq 0$$

The LINGO formulation for this LP model is written as follows:

Model:

```
! Optimisation objective: to minimise freshwater flowrate
(FW) ;
min = FW ;

! Water cascade ;
D1 = D0 + REG ;
D2 = D1 - 4 ;
D3 = D2 - 1 ;
D4 = D3 + ( FW - RE ) ;
D5 = D4 - 2 ;
D6 = D5 - 1;
D7 = D6 + RJ ;

! No net flowrate enters the first level, net flowrate that
enters the last level should have non-negative value;
D0 = 0 ;
D7 >= 0;

! The last net material flowrates corresponds to the minimum
wastewater (WW) flowrate ;
D7 = WW ;

! Flowrate balance and allocation for interception unit ;
RE = REG + RJ ;
RE*0.75 = REG ;
! Flowrate constraints of regenerated source ;
RE <= FW ;

! Impurity load cascade ;
E1 = 0 ;
E2 = E1 + D1*(10 - 5)/1000 ;
E3 = E2 + D2*(20 - 10)/1000 ;
E4 = E3 + D3*(25 - 20)/1000 ;
E5 = E4 + D4*(30 - 25)/1000 ;
E6 = E5 + D5*(40 - 30)/1000 ;
E7 = E6 + D6*(85 - 40)/1000 ;
E8 = E7 + D7*(1000000 - 85)/1000 ;

! E2-E8 should take non-negative values ;
E2 >= 0 ; E3 >= 0 ; E4 >= 0 ; E5 >= 0 ;
E6 >= 0 ; E7 >= 0 ; E8 >= 0 ;
! D1-D6 may take negative values ;
@free (D1) ; @free (D2) ; @free (D3) ;
@free (D4) ; @free (D5) ; @free (D6) ;
END
```

Similar to earlier cases, the non-negativity constraints are excluded here, as LINGO assumes all variables take non-negative values by default. The following is the solution report generated by LINGO:

```
Global optimal solution found.
Objective value:                      8.666667
Total solver iterations:                     0
            Variable          Value
            FW                8.666667
            D1                3.250000
            D0                0.000000
```

REG	3.250000
D2	-0.7500000
D3	-1.750000
D4	2.583333
RE	4.333333
D5	0.5833333
D6	-0.4166667
D7	0.6666667
RJ	1.083333
WW	0.6666667
E1	0.000000
E2	0.1625000E-01
E3	0.8750000E-02
E4	0.000000
E5	0.1291667E-01
E6	0.1875000E-01
E7	0.000000
E8	666.6100

The optimization result indicates that the minimum freshwater flow-rate needed for the pretreatment network is 8.67 t/h. The revised cascade diagram is shown in Figure 14.16. The results show that 4.33 t/h of the freshwater feed is sent for purification in the partitioning unit (F_{RE}). In other words, 4.34 t/h (= 8.67 − 4.33 t/h) of freshwater will bypass the interception unit. With the RC value of 0.75, the interception unit produces a purified stream with a flowrate of 3.25 t/h (F_{REG}) at 5% level, while a flowrate of 1.08 t/h is found in the reject stream (F_{RJ}). Also, the

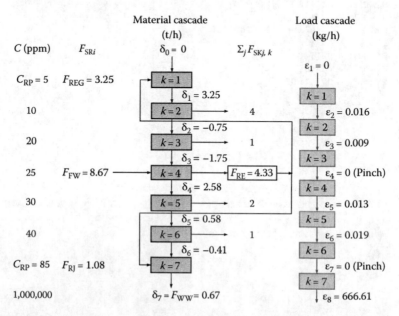

FIGURE 14.16
Result of ATM for Example 14.5.

wastewater flowrate (F_{WW}) is determined as 0.67 t/h. Note that all flow-rate targets are identical to those determined using the insight-based technique in Chapter 9 (see Example 9.1—Case 1 for comparison).

14.8 Additional Readings

For material regeneration networks that employ removal ratio type interception units (single pass or partitioning interception units), there may be a possibility that more than one interception unit is to be used, as a result of the ATM. For cases where only one interception unit is desired (e.g., due to space consideration, additional constraints may be added to the ATM. In most cases, the ATM will become a nonlinear program (NLP). Readers can refer to the work of Ng et al. (2009a) for an extended model of this case.

Problems

Water Minimization Problems

14.1 Rework Example 14.2, assuming that the unit cost for freshwater is $1/ton ($CT_{FW}$) and that for the regenerated water (CT_{RW}) are given in terms of impurity load removal, i.e., $0.8/kg impurity load removed. Impurity load removal for the interception unit $r(\Delta m_r)$ is given as in Equation 14.37 (revised form of Equation 6.4):

$$\Delta m_r = F_{RE,r}(C_{Rin,r} - C_{Rout,r}) \qquad (14.37)$$

Use the insights from the ATM to identify the structure of the RCN.

14.2 A water minimization case study (Gabriel and El-Halwagi, 2005) is analyzed here, with the limiting water data given in Table 14.4. To enhance water recovery, stream stripping is used to regenerate water sources for further recovery. The unit cost for water regeneration (CT_{RW}) is given as a function of impurity load removal and varies for

TABLE 14.4

Limiting Water Data for Problem 14.2

Sinks, SK$_j$	F_{SKj} (t/h)	C_{SKj} (ppm)	Sources, SR$_i$	F_{SRi} (t/h)	C_{SRi} (ppm)
SK1	200	20	SR1	150	10
SK2	80	75	SR2	60	50
			SR3	100	85

TABLE 14.5

Cost Data for Interception Unit in Problem 14.2

Sources, SR_i	Removal Ratio (RR)	CT_{RW} ($\$$/kg Impurity Removed)
SR1	0.1	0.68
	0.5	1.46
	0.9	2.96
SR2	0.1	0.54
	0.5	1.16
	0.9	2.36
SR3	0.1	0.45
	0.5	0.97
	0.9	1.97

different removal ratios for each water source, as shown in Table 14.5. The unit costs for freshwater (CT_{FW}) and wastewater (CT_{WW}) are taken as $\$0.13$/ton and $\$0.22$/ton, respectively. Determine the MOC for this RCN. Use the superstructural approach in Chapter 13 to synthesize the RCN that achieves the network targets.

14.3 The water minimization case in Problems 6.4 and 7.12 is revisited. The limiting data for this case are given in Table 6.25. To further recover water sources to the sinks, single pass interception units are used for source regeneration. The unit costs for freshwater (CT_{FW}), wastewater (CT_{WW}), and regenerated water (CT_{RW}) are given as $\$1$/ton, $\$0.5$/ton, and $\$0.5$/ton, respectively. Determine the MOC for the following cases:

   a. The interception unit has a fixed outlet concentration (C_{Rout}) of 10 ppm. Compare the result with that obtained using the superstructural approach (note to readers: solve Problem 12.7b to compare results).

   b. The interception units have a removal ratio of 0.9.

14.4 The paper milling process in Problem 4.5 is revisited here. A dissolved air flotation unit (DAF) is employed to remove the total suspended solid (TSS) content in the water sources, which is the most significant quality factor for water recovery in this process. The DAF unit takes the form of a partitioning interception unit, with a water recovery factor (RC) of 0.98 and TTS removal ratio (RR) of 0.9 (Ng et al., 2009a). The unit costs for fresh and regenerated water are taken as $\$1$/ton and $\$0.75$/ton, respectively. Determine the MOC for this RCN, which is mainly contributed by fresh and regenerated water. Use the limiting water data in Table 4.24.

14.5 Rework the water minimization problem of the acrylonitrile (AN) problem in Example 6.3. For this case, determine the MOC of the RCN when resin adsorption is used as an interception unit to purify water sources for further recovery. The equilibrium relation for adsorbing the ammonia content from the water sources with this mass exchange unit is given in Equation 14.38 (see Section 6.3 for the use of mass exchange operation as interception unit). Unit costs for freshwater and waste-water are given as $0.001/kg each, while that for the resin adsorption unit is given in terms of its impurity removal, i.e., 91×10^{-6}/mg NH_3 removed (El-Halwagi, 1997).*

$$C_{SRi} = 0.01 \, C_{MSA} \tag{14.38}$$

14.6 Rework the water minimization problem of the Kraft pulping process in Problem 4.1. Air stripping is used as an interception unit to regenerate water sources for further recovery, by removing its methanol content. The equilibrium relation for stripping of methanol using this mass exchange unit is given as follows:

$$C_{SRi} = 0.38 \, C_{MSA} \tag{14.39}$$

where C_{MSA} is the impurity (methanol) concentration in the MSA (stripping air). A minimum allowable concentration difference of 4.275 ppm should be incorporated to operate the mass exchange unit (El-Halwagi, 1997). It may be assumed that the ambient air is used as the mass separating agent in this case, and hence has an inlet concentration of 0 ppm (refer to Section 6.3 for the use of mass exchange unit as interception unit).

The operating (OC_{AS}) and the capital costs (CC_{AS}) of the stripping unit are given as follows (El-Halwagi, 1997; Ng et al., 2010):

$$OC_{AS}(\$/h) = 1.5 \, F_{AS} \, RR_{AS} \tag{14.40}$$

$$CC_{AS}(\$) = 1317(F_{AS})^{0.65} \tag{14.41}$$

Note that the operating cost of the stripping unit is a function of its regeneration flowrate (F_{AS}), as well as the removal ratio of the methanol load from the water sources (RR_{AS}). Note further that only one stripping unit is to be used; hence, the removal ratio of the stripping unit is determined by the following equation:

$$RR_{AS} = \frac{\sum_r F_{RE,r} C_{RE,r} - F_{REG} C_{REG}}{\sum_r F_{RE,r} C_{RE,r}} \tag{14.42}$$

Note that the unknown terms in Equations 14.40 and 14.42 lead to an NLP of the model.

---

* The outlet concentration of the regenerated source may also be set as a variable, which leads to an NLP model.

Determine the total annualized cost (TAC) of the RCN, which consists of the freshwater and wastewater costs, as well as the operating and capital costs of the interception unit. The unit costs of freshwater and wastewater are both taken as $1/ton. Piping costs may be ignored in this case. The annual operating time (AOT) for the RCN is assumed as 8000h. To annualize the capital cost, the annualizing factor in Equation 12.15 can be used, assuming an annual fractional interest rate of 10%, for a total period of 5 years (note: refer to Example 12.3 for a sample of TAC calculation).

Utility Gas Recovery Problems

14.7 For the refinery hydrogen network in Problems 4.15 and 13.9 (Alves and Towler, 2002), a gas membrane unit with a hydrogen recovery factor (RC) of 0.95 is used for source rereneration. The membrane unit produces a purified stream of 2 mol % impurity (C_{RP}). The unit costs for fresh and regenerated hydrogen are given as 1.6×10^{-3}/mol and 1.0×10^{-3}/mol, respectively. The limiting data are given in Table 4.30. Determine the MOC for the RCN.

14.8 Consider the magnetic tape process in Problem 3.8. Three MSAs are available to purify nitrogen sources for further recovery. Data for the MSAs are given in Table 14.6 (El-Halwagi, 1997). The unit cost for fresh nitrogen gas (CT_{N2}) is given as $0.20/kg. Determine the MOC for the RCN, which is mainly contributed by the fresh and regenerated nitrogen sources (note: see Section 6.3 for the use of mass exchange units as interception units).

Property Integration Problems

14.9 Revisit the wafer fabrication process in Problems 4.19 and 13.11. In order to enhance water recovery, a membrane treatment unit is installed. The membrane unit purifies water sources to a permeate quality (purified stream) of 15 MΩ m (hint: convert this to operator using Equation 2.64). The membrane unit is treated as a partitioning unit, with a water recovery factor of 0.95.

    a. When water minimization is only considered for the wafer fabrication section (FAB), determine the minimum UPW and regeneration flowrates. Use the limiting data in Table 4.32a.

TABLE 14.6

Data for MSAs in Problem 14.8

MSAl	Inlet Concentration, $C_{MSA,in}$ (wt%)	Outlet Concentration, $C_{MSA,out}$ (wt%)	α_l	Minimum Driving Force, ξ_l (wt%)	Unit Cost, CT_{MSAl} ($/kg MSA)
1	0.014	0.040	0.4	0.001	0.002
2	0.020	0.080	1.5	0.001	0.001
3	0.001	0.010	0.1	0.001	0.002

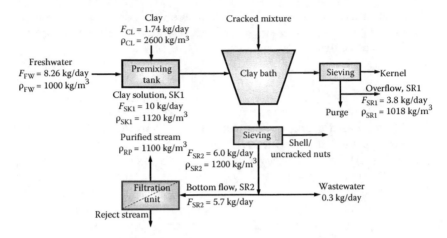

FIGURE 14.17
RCN for clay bath operation in palm oil milling process. (From Ng, D.K.S. et al., *Chem. Eng. J.*, 149, 87, 2009c.)

  b. To consider water recovery in the entire plant, a MOC solution is to be identified. The unit costs for ultrapure (UPW), fresh, wastewater, and regenerated water are taken as $2/t, $1/t, $0.5/t, and $1.5/t, respectively (Ng et al., 2010). Use the limiting data in Table 4.32b.

14.10 The RCN problem of palm oil milling process in Example 13.4 is revisited. A filtration unit is installed to purify the bottom flow of the clay bath that is currently sent to wastewater treatment. The filtration unit is modeled as a partitioning interception unit of the fixed outlet quality type. It purifies the bottom flow to 1100 kg/m³, and produces a reject stream of 1600 kg/m³ (Ng et al., 2009b). However, to avoid accumulation of impurity load, 5% of wastewater flowrate is purged from the bottom stream, as shown in Figure 14.17. Besides, the reject stream of the filtration unit will not be recycled (i.e. sent as wastewater discharge). The revised limiting data are shown in Table 14.7 (reproduced from Table 13.3). Determine the MOC of the RCN, given that

TABLE 14.7

Limiting Data for Problem 14.10

Sink, SK_j	F_{SK_j} (kg/day)	Density, ρ (kg/m³)	$\psi_{SR_i} \times 10^{-4}$ (m³/kg)
Clay solution (SK1)	10.0	1120	8.93
Source, SR_i	F_{SR_i} (kg/day)	Density, ρ (kg/m³)	$\psi_{SK_j} \times 10^{-4}$ (m³/kg)
Overflow (SR1)	3.8	1018	9.82
Bottom flow (SR2)	5.7	1200	8.33
Freshwater (FW)	To be determined	1000	10
Clay (CL)	To be determined	2600	3.85

the unit costs for clay, freshwater, wastewater, and regenerated water are $0.25/kg, $0.001/kg, $0.0008/kg, and $0.0008/kg, respectively (Ng et al., 2010).

Pretreatment Network Problems

14.11 Rework the water pretreatment network problem in Example 14.5, when a partitioning interception unit with an *RR* of 0.9 is used. The interception unit produces a reject stream of 100 ppm. The purified stream quality is unknown; however, it is expected to be of a higher quality than all the sinks (note: this leads to NLP model). Determine the flowrate targets for freshwater, inlet, purified and reject streams of the interception unit, as well as the water recovery factor (*RC*) of the latter (hint: since the purified stream concentration is an unknown, it is best to include Equation 14.21 as a constraint in the ATM).

14.12 Rework Problem 9.2 with an ATM. Compare the flowrate targets with those obtained by the insight-based technique (Chapter 9).

References

Agrawal, V. and Shenoy U. V. 2006. Unified conceptual approach to targeting and design of water and hydrogen networks. *AIChE Journal*, 52(3), 1071–1081.

Alves, J. J. and Towler, G. P. 2002. Analysis of refinery hydrogen distribution systems. *Industrial and Engineering Chemistry Research*, 41, 5759–5769.

El-Halwagi, M. M. 1997. *Pollution Prevention through Process Integration: Systematic Design Tools*. San Diego, CA: Academic Press.

Gabriel, F. B. and El-Halwagi, M. M. 2005. Simultaneous synthesis of waste interception and material reuse networks: Problem reformulation for global optimization. *Environmental Progress*, 24(2), 171–180.

Ng, D. K. S., Foo, D. C. Y., and Tan, R. R. 2009a. Automated targeting technique for single-component resource conservation networks—Part 2: Single pass and partitioning waste interception systems. *Industrial and Engineering Chemistry Research*, 48(16), 7647–7661.

Ng, D. K. S., Foo, D. C. Y., Tan, R. R., and El-Halwagi, M. M. 2010. Automated targeting technique for concentration- and property-based total resource conservation network. *Computers and Chemical Engineering*, 34(5), 825–845.

Ng, D. K. S., Foo, D. C. Y., Tan, R. R., Pau, C. H., and Tan, Y. L. 2009b. Automated targeting for conventional and bilateral property-based resource conservation network. *Chemical Engineering Journal*, 149, 87–101.

Tan, R. R., Ng, D. K. S., Foo, D. C. Y., and Aviso, K. B. 2009. A superstructure model for the synthesis of single-contaminant water networks with partitioning regenerators. *Process Safety and Environmental Protection*, 87, 197–205.

Wang, Y. P. and Smith, R. 1994. Wastewater minimisation. *Chemical Engineering Science*, 49, 981–1006.

15

Automated Targeting Model for Waste Treatment and Total Material Networks

As discussed in Chapter 8, waste treatment is part of an overall framework called the *total material network* (TMN). Two other important elements that form the TMN are direct reuse/recycle and material regeneration. The automated targeting model (ATM) outlined for the material regeneration network in Chapter 14 is further extended in this chapter for a waste treatment network (WTN) as well as for a TMN.

15.1 ATM for Waste Treatment Network

Similar to interception units (see discussion in Chapters 6, 9, and 14), waste treatment units may also be classified as *fixed outlet quality* or *fixed removal ratio types*, based on their performance. Hence, the basic ATM framework for a WTN takes the same form as for the material regeneration network outlined in Chapter 14, for both fixed outlet quality (Figure 15.1) and fixed removal ratio (Figure 15.2) waste treatment units. However, some minor modifications are required:

1. All quality levels (q_k) of waste streams, discharge quality of the terminal waste (q_D), as well as the outlet quality of the treatment unit(s) (q_T) are arranged in descending order in the cascade diagram. An arbitrary value is also added at the final level (lowest among all quality levels) for the calculation of the residual impurity/property load.

2. Flowrates of individual waste streams (F_{Wy}) are located at their respective quality levels, on the left side of the *waste cascade*. When there are more than one waste streams at the same quality level, their flowrates are added. The *terminal waste discharge flowrate* (F_D) is added at the quality level of the terminal waste discharge (q_D), on the right side of the waste cascade. Conceptually, the individual waste streams and the terminal waste discharge act as sources and sink in the waste cascade, respectively (corresponding to the process sources and sinks in the material cascade in Chapters 13 and 14).

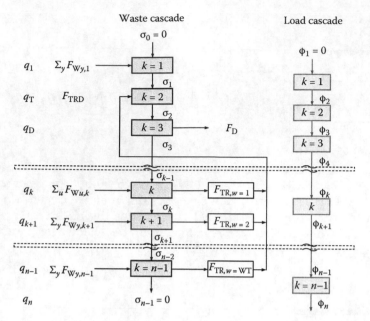

FIGURE 15.1
ATM framework for WTN with waste treatment unit of fixed outlet quality. (From Ng, D.K.S. et al., *Comput. Chem. Eng.*, 34(5), 825, 2010.)

3. A new sink is added at the same quality level in the ATM frame-work wherever an individual waste stream (source) exists. This new sink represents the potential waste to be sent for treatment (with flowrate $F_{TR,w}$). Besides, a new source is added at outlet quality level of the treatment unit (q_T), representing the treated waste source.

For a WTN with a fixed outlet quality type treatment unit (ATM framework in Figure 15.1), the ATM formulation is described by the following constraints:

$$\sigma_k = \sigma_{k-1} + \Sigma_y F_{Wy,k} - F_D + F_{TRD} - F_{TR,w} \quad k = 1, 2, \ldots, n-1 \tag{15.1}$$

$$\sigma_{k-1} = 0 \quad k = 1, n \tag{15.2}$$

$$F_{TRD} = \Sigma_w F_{TR,w} \tag{15.3}$$

$$F_{TR,w} \leq F_{Wy} \quad w = 1, 2, \ldots, WT \tag{15.4}$$

$$\varphi_k = \begin{cases} 0 & k = 1 \\ \varphi_{k-1} + \sigma_{k-1}\left(q_k - q_{k-1}\right) & k = 2, 3, \ldots, n \end{cases} \tag{15.5}$$

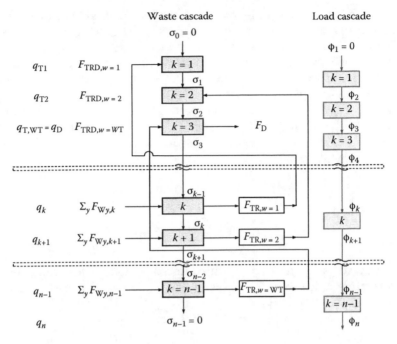

FIGURE 15.2
ATM framework for WTN with waste treatment unit of removal ratio type. (From Ng, D.K.S. et al., *Comput. Chem. Eng.*, 34(5), 825, 2010.)

$$\varphi_k \geq 0 \quad k = 2, 3, \ldots, n \tag{15.6}$$

Equation 15.1 is a revised form of Equation 14.1 (material regeneration network), which describes the *net waste flowrate* of level k (σ_k) in the waste cascade. The net waste flowrate takes into account the net waste flowrate cascaded from an earlier quality level (σ_{k-1}), the individual waste flowrate ($\Sigma_y F_{Wy,k}$), the terminal waste flowrate (F_D), as well as the treated waste flowrate (F_{TRD}) and the individual treatment flowrate w ($F_{TR,w}$) that present at each level k (their values are set to zero if absent). Equation 15.2 describes that no net waste flowrate enters the first (σ_0) and last levels (σ_{n-1}) of the waste cascade. The flowrate balance for the treatment unit is described by Equation 15.3. Since the treatment flowrate originates from the individual waste stream (source), the maximum flowrate to be treated is bound with the maximum available flowrate, as described by Equation 15.4. Equation 15.5 indicates that the *residual waste load* (φ_k) entering the first quality level has zero value, i.e., $\varphi_1 = 0$, while residual loads in other quality levels are calculated based on the residual load cascaded from the previous levels (φ_{k-1}) as well as the net waste flowrate within each quality interval. The latter is given by the product of the net waste flowrate from previous level (σ_{k-1}) and the difference between two adjacent quality levels $(q_k - q_{k-1})$, represented by the last term in Equation 15.5. Also, all other residual loads (except from σ_1 that enters to level 1, which is set to zero in Equation 15.5) must take non-negative values, as indicated by Equation 15.6. Note that the constraints in both Equations 15.5 and 15.6 are similar to the load constraints for direct reuse/recycle (Chapter 13) and material regeneration networks (Chapter 14).

For a WTN with a removal ratio type treatment unit (ATM framework in Figure 15.2), the constraints in Equations 15.1 and 15.3 are revised to cater to the individual treated $(F_{TRD,w})$ and treatment flowrate w $(F_{TR,w})$. The revised equations are given as follows:

$$\sigma_k = \sigma_{k-1} + \Sigma_y F_{Wy,k} - F_D + F_{TRD,w} - F_{TR,w} \quad k = 1, 2, \dots, n-1 \qquad (15.7)$$

$$F_{TRD,w} = F_{TR,w} \quad w = 1, 2, \dots, WT \qquad (15.8)$$

For both cases, we can set the optimization objective to determine the minimum treatment flowrate.

$$\text{Minimize } \Sigma_w F_{TR,w} \qquad (15.9)$$

Similar to the ATM in Chapter 14, the extended ATM for WTN is also a linear program (LP) that can be solved for global optimum solution, if it exists.

Example 15.1 Targeting minimum wastewater treatment

Example 8.3 is revisited here. Data for the three wastewater streams to be treated before environmental discharge are given in Table 15.1 (reproduced from Table 8.7). Determine the minimum treatment flowrate needed for these wastewater streams for the following cases. Derive the WTNs for both cases based on the insights from the ATM.

1. When a wastewater treatment unit with a fixed outlet concentration (C_T) of 10 ppm is used. The environmental discharge limit (C_D) is given as 20 ppm.
2. When a wastewater treatment unit with a fixed removal ratio (RR) of 0.9 is used, for an environmental discharge limit (C_D) of 100 ppm.

TABLE 15.1

Data for Wastewater Streams in Example 15.1

Wastewater, WWy	F_{WW_y} (t/h)	C_{WW_y} (ppm)
WW1	3.69	100
WW2	6.31	800
WW3	10.00	800

Solution

1. Case 1—$C_T = 10\,$ppm; $C_D = 20\,$ppm.

Following the three-step procedure, the cascade diagram for this case is shown in Figure 15.3. The final concentration level of 1,000,000 ppm is added in the cascade diagram, according to Step 1 in the procedure. The treated wastewater (with flowrate F_{TRD}) is added as a new source at $C_T = 10\,$ppm. Also, two new sinks are added at 100 and 800 ppm (with flowrates F_{TR1} and F_{TR2} respectively), to allow the wastewater to be sent for treatment from these levels.

To determine the minimum wastewater treatment flowrate, the optimization objective in Equation 15.9 is used. The optimization problem is subjected to the constraints in Equations 15.1 through 15.6. Equation 15.1 describes the waste cascade across all quality levels:

$$\sigma_1 = \sigma_0$$
$$\sigma_2 = \sigma_1 + F_{TRD}$$
$$\sigma_3 = \sigma_2 - F_{WW}$$
$$\sigma_4 = \sigma_3 + (3.69 - F_{TR1})$$
$$\sigma_5 = \sigma_4 + (16.31 - F_{TR2})$$

According to Equation 15.2, no net wastewater flowrate enters the first and last levels of the waste cascade:

$$\sigma_0 = 0$$
$$\sigma_5 = 0$$

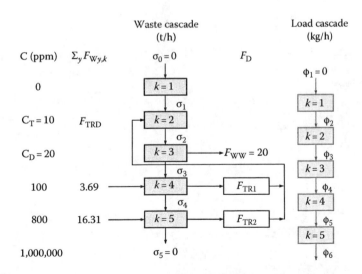

FIGURE 15.3
Cascade diagram for Example 15.1 (wastewater treatment unit with fixed outlet concentration of 10 ppm).

Equation 15.3 describes the flowrate balance of the wastewater treatment unit:

$$F_{TRD} = F_{TR1} + F_{TR1}$$

Next, the value for the wastewater source to be treated should not be more than what is available, according to Equation 15.4:

$$F_{TR1} \leq 3.69$$

$$F_{TR2} \leq 16.31$$

Equation 15.5 describes the waste load cascades (in kg/s) across all concentration levels:

$$\varphi_1 = 0$$

$$\varphi_2 = \varphi_1 + \delta_1(10 - 0)/1,000$$

$$\varphi_3 = \varphi_2 + \delta_2(20 - 10)/1,000$$

$$\varphi_4 = \varphi_3 + \delta_3(100 - 20)/1,000$$

$$\varphi_5 = \varphi_4 + \delta_4(800 - 100)/1,000$$

$$\varphi_6 = \varphi_5 + \delta_5(1,000,000 - 800)/1,000$$

Based on Equation 15.6, all residual loads (except that at level 1) should take non-negative values:

$$\varphi_2 \geq 0; \quad \varphi_3 \geq 0$$

$$\varphi_4 \geq 0; \quad \varphi_5 \geq 0$$

$$\varphi_6 \geq 0$$

In terms of LINGO formulation, this LP model can be written as follows:

Model:

```
! Optimisation objective: to minimise wastewater treatment
(TRD) flowrate ;
min = TR1 + TR2 ;

! Waste cascade ;
S1 = S0 ;
S2 = S1 + TRD ;
S3 = S2 - 20 ;
S4 = S3 + ( 3.69 - TR1 ) ;
S5 = S4 + ( 16.31 - TR2 ) ;

! No net waste flowrate enters the first and the last
levels;
S0 = 0 ;
S5 = 0 ;

! Flowrate balance(s) for treatment unit(s) ;
TRD = TR1 + TR2 ;
```

```
! Flowrate constraints of treated wastewater source ;
TR1 <= 3.69 ;
TR2 <= 16.31 ;

! Waste load cascade (in kg/s) ;
J1 = 0 ;
J2 = J1 + S1'(10 - 0)/1000 ;
J3 = J2 + S2'(20 - 10)/1000 ;
J4 = J3 + S3'(100 - 20)/1000 ;
J5 = J4 + S4'(800 - 100)/1000 ;
J6 = J5 + S5'(1000000 - 800)/1000 ;

! J2-J6 should take non-negative values ;
J2 >= 0 ; J3 >= 0 ; J4 >= 0 ;
J5 >= 0 ; J6 >= 0 ;

! S1-S4 may take negative values ;
@free (S1) ; @free (S2) ;
@free (S3) ; @free (S4) ;
END
```

As in the previous chapters, the last set of constraints is only necessary when LINGO is used to solve the optimization model, as this software assumes all variables to have non-negative values by default. For the same reason, the constraints for non-negativity (given by Equation 15.6) need not be coded in LINGO.

The following is the solution report generated by LINGO:

```
Global optimal solution found.
Objective value:                      17.77778
Total solver iterations:                     0


              Variable          Value
              TR1               1.467778
              TR2              16.31000
              G1                0.000000
              G0                0.000000
              G2               17.77778
              TRD              17.77778
              G3               -2.222222
              G4                0.000000
              G5                0.000000
              J1                0.000000
              J2                0.000000
              J3                0.1777778
              J4                0.000000
              J5                0.000000
              J6                0.000000
```

A revised cascade diagram is shown in Figure 15.4, with the process variables obtained from the optimization model. As shown, a total flowrate of 17.78 t/h of wastewater is sent for wastewater treatment (F_{TRD}). A large portion of this treated wastewater originates from the 800 ppm sources (WW2 and WW3 with $F_{TQ2}=16.31$ t/h), with the remaining flowrate ($F_{TR1}=1.47$ t/h) being drawn from the 100 ppm wastewater source

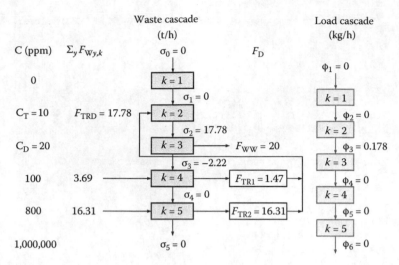

FIGURE 15.4
Result of ATM for Example 15.1.

(WW1). In other words, wastewater sources WW2 and WW3 are fully treated, while a large portion of 100 ppm source bypasses (with flowrate $F_{BP} = 2.22\,t/h$) the wastewater treatment unit.

A WTN can then be derived based on the insights from the cascade diagram, as shown in Figure 15.5. Note that the network structure is exactly the same as the one derived based on the insights from the waste treatment pinch diagram (Figure 8.12). Also note that small flowrate discrepancies between the two networks are acceptable.

2. Case 2—$RR = 0.9$; $C_D = 100\,ppm$

The cascade diagram for this case is shown in Figure 15.6. The main difference of this cascade diagram as compared with the one in Figure 15.3 is that, each wastewater source is treated to a different concentration. Also, it is worth noting that the discharge concentration (C_D) coincides with the concentration for one of the wastewater sources, i.e., 100 ppm.

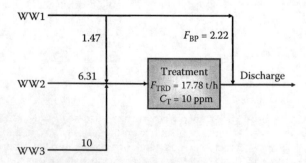

FIGURE 15.5
Waste treatment network for Example 15.1 (Case 1 flowrate terms in t/h).

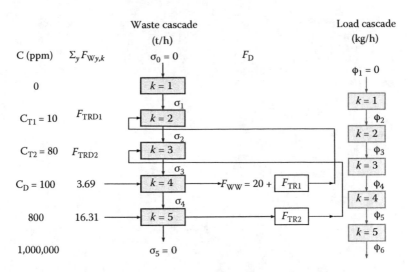

FIGURE 15.6
Cascade diagram for Example 15.1 (wastewater treatment unit with RR of 0.9).

Similar to the previous case, the optimization objective is set to determine the minimum wastewater treatment flowrate using Equation 15.9 and is subject to the constraints in Equations 15.2 and 15.4 through 15.8.

Note that since most constraints are similar to those in Example 15.1, only Equations 15.7 and 15.8 are shown here for the sake of simplicity. Equation 15.7 describes the waste cascade across all quality levels for the problem:

$$\sigma_1 = \sigma_0$$
$$\sigma_2 = \sigma_1 + \Gamma_{TRD1}$$
$$\sigma_3 = \sigma_2 + F_{TRD2}$$
$$\sigma_4 = \sigma_3 - F_{WW} + (3.69 - F_{TR1})$$
$$\sigma_5 = \sigma_4 + (16.31 - F_{TR2})$$

Equation 15.8 describes the flowrate balance of the wastewater treatment unit:

$$F_{TRD1} = F_{TR1}$$
$$F_{TRD2} = F_{TR2}$$

The LINGO formulation for this LP model can be written as follows:

Model:
```
! Optimisation objective: to minimise total wastewater
treatment flowrate ;
min = TR1 + TR2 ;
```

```
! Waste cascade ;
S1 = S0 ;
S2 = S1 + TRD1 ;
S3 = S2 + TRD2 ;
S4 = S3 - 20 + ( 3.69 - TR1 ) ;
S5 = S4 + ( 16.31 - TR2 ) ;

! No net waste flowrate enters the first and the last
levels;
S0 = 0 ;
S5 = 0 ;

! Flowrate balance(s) for treatment unit(s) ;
TRD1 = TR1 ;
TRD2 = TR2 ;

! Flowrate constraints of treated wastewater source ;
TR1 <= 3.69 ;
TR2 <= 16.31 ;

! Waste load cascade (in kg/s) ;
J1 = 0 ;
J2 = J1 + S1*(10 - 0)/1000 ;
J3 = J2 + S2*(80 - 10)/1000 ;
J4 = J3 + S3*(100 - 80)/1000 ;
J5 = J4 + S4*(800 - 100)/1000 ;
J6 = J5 + S5*(1000000 - 800)/1000 ;

! J2-J6 should take non-negative values ;
J2 >= 0 ; J3 >= 0 ; J4 >= 0 ;
J5 >= 0 ; J6 >= 0 ;

! S1-S4 may take negative values ;
@free (S1) ; @free (S2) ;
@free (S3) ; @free (S4) ;
END
```

The following is the solution report generated by LINGO:

```
Global optimal solution found.
Objective value:                    15.85694
Total solver iterations:                   0

                 Variable          Value
                 TR1           0.000000
                 TR2          15.85694
                 S1            0.000000
                 S0            0.000000
                 S2            0.000000
                 TRD1          0.000000
                 S3           15.85694
                 TRD2         15.85694
                 S4           -0.4530556
                 S5            0.000000
                 J1            0.000000
```

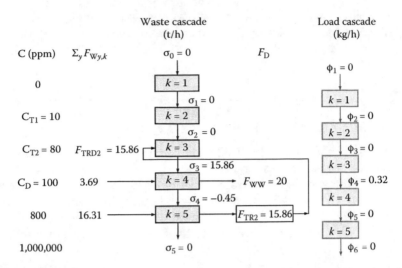

FIGURE 15.7
Result of ATM for Example 15.1 (Case 2).

J2	0.000000
J3	0.000000
J4	0.3171389
J5	0.000000
J6	0.000000

A revised cascade diagram is shown in Figure 15.7. It is observed that only 15.86 t/h from the 800 ppm wastewater source will be sent for wastewater treatment; the remaining flowrate (0.45 t/h) will be discharged directly (i.e., without treatment) together with the 100 ppm wastewater source. In other words, the total bypass flowrate (F_{BP}) is determined as 4.14 t/h (=0.45 + 3.69 t/h).

A WTN is then derived from these insights and represented in Figure 15.8. Note that the network structure is exactly the same as the

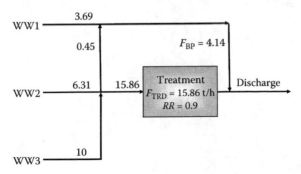

FIGURE 15.8
Waste treatment network for Example 15.1 (Case 2 flowrate terms in t/h).

one derived based on the insights of the waste treatment pinch diagram (Figure 8.14). Also note that small flowrate discrepancies between the two networks are negligible.

15.2 ATM for TMN without Waste Recycling

As discussed in Chapter 8, a TMN consists of the individual elements of direct reuse/recycle, material regeneration and waste treatment. In this section, we will extend the ATM framework for material regeneration network in Chapter 14 for TMNs. In principle, the extended ATM enables the identification of individual waste streams that emit at their respective quality levels. These identified waste streams can then be handled by a separate ATM framework for WTN, as described in the earlier sections. Since there exist several variants of material regeneration network (due to different types and characteristics of the interception units; see Chapter 14 for details), the extended ATM for the TMNs hence takes several forms. Each of these will be described in the following sections.

15.2.1 TMN with Single Pass Interception Unit of Fixed Outlet Quality Type

When a TMN has a single pass interception unit of the fixed outlet quality type, its ATM framework takes the revised form of that used for a material regeneration network (Figure 14.2), as given in Figure 15.9. As shown, the discharge of waste from the final quality level (such as that in Figure 14.2) is forbidden (i.e., $\delta_{n-1} = 0$). Instead, the individual waste sources are added as new sinks at each quality level wherever a source exists. This enables the tracking of individual waste streams that may emit from the quality level where the process sources exit.

The optimization constraints for the ATM framework in Figure 15.9 are given as follows:

$$\delta_k = \delta_{k-1} + (\Sigma_i F_{\text{SR}i} - \Sigma_j F_{\text{SK}j})_k + F_{\text{REG}} - F_{\text{RE},r} - F_{\text{W}y} \quad k = 1, 2, \ldots, n-1 \quad (15.10)$$

$$F_{\text{REG}} = \Sigma_r F_{\text{RE},r} \quad (15.11)$$

$$F_{\text{RE},r} + F_{\text{W}y} \le F_{\text{SR}i} \quad r = 1, 2, \ldots, RG \quad (15.12)$$

$$\varepsilon_k = \begin{cases} 0 & k = 1 \\ \varepsilon_{k-1} + \delta_{k-1}(q_k - q_{k-1}) & k = 2, 3, \ldots, n \end{cases} \quad (15.13)$$

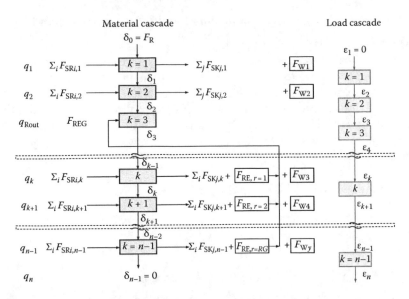

FIGURE 15.9
ATM framework for TMN with fixed outlet quality type single pass interception unit (cascade for WTN is not shown). (From Ng, D.K.S. et al., *Comput. Chem. Eng.*, 34(5), 825, 2010.)

$$\delta_{k-1} = \begin{cases} \geq 0 & k = 1 \\ 0 & k = n \end{cases} \tag{15.14}$$

$$\varepsilon_k \geq 0 \quad k = 2, 3, \ldots, n \tag{15.15}$$

Note that the constraints in Equations 15.10 through 15.15 resemble those of the material regeneration network described in Chapter 14 (i.e., Equations 14.1 through 14.6), with the addition of the waste flowrate terms (F_{WY}). The main difference between the two different sets of constraints is found in Equations 15.10, 15.12, and 15.14. In Equation 15.10 (extended form of Equation 14.1), a waste flowrate is added to the *net material flowrate* at level k (δ_k) in the material cascade. In Equation 15.12, the regeneration and waste flowrates are bound with the maximum flowrate of the process source, where they originate from. Equation 15.14 describes that the net material flowrates that enter the first quality level (δ_0) should take non-negative values; while those entering the last levels (δ_{n-1}) are set to zero. The latter will forbid the emission of total waste discharge from the final quality level. Doing this enables the tracking of the individual waste streams in the RCN, as discussed earlier. The remaining constraints are essentially the same as those for the material regeneration network described in Chapter 14. Hence, the linearity of the ATM is preserved.

15.2.2 TMN with Removal Ratio Type Single Pass Interception Unit

Apart from a fixed outlet quality type interception unit, one may also utilize a removal ratio type interception unit for the TMN. Similar to the earlier case, the ATM framework takes the revised form as that for a material regeneration network (Figure 14.6), as given in Figure 15.10.

Most optimization constraints for this ATM framework remain identical to those in Figure 15.9. However, Equations 15.10 and 15.11 are revised to cater to the flowrate terms of the interception unit of the removal ratio type, given as follows:

$$\delta_k = \delta_{k-1} + (\Sigma_i F_{SRi} - \Sigma_j F_{SKj})_k + F_{REG,r} - F_{RE,r} - F_{Wy} \quad k = 1, 2, \ldots, n-1 \quad (15.16)$$

$$F_{RE,r} = F_{REG,r} \quad r = 1, 2, \ldots, RG \quad (15.17)$$

15.2.3 TMN with Partitioning Interception Unit

Apart from single pass unit, we can also utilize a partitioning interception unit in the TMN. The ATM framework for this case may take two different forms. In Figure 15.11 (revised form of Figure 14.11), both the purified ($F_{REG,r}$)

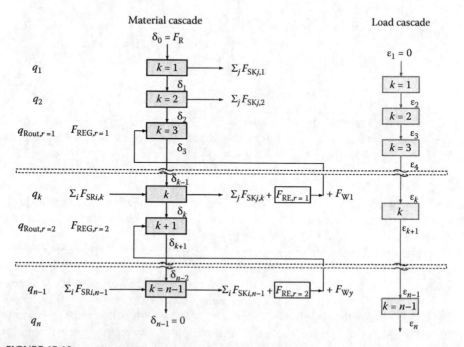

FIGURE 15.10
ATM framework for TMN with removal ratio type single pass interception unit (cascade for WTN is not shown).

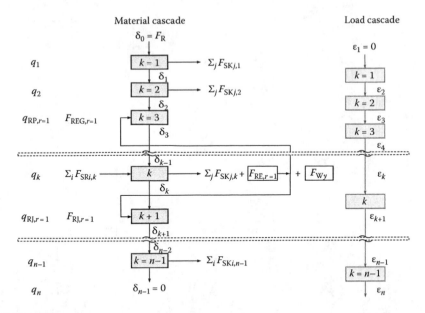

FIGURE 15.11
ATM framework for TMN with reject steam of partitioning interception unit being recovered within the RCN (cascade for WTN is not shown).

and reject ($F_{RJ,r}$) flowrate terms are included in the material cascade of the ATM framework. This means that both the purified and reject streams may be reused/recycled to the sink after regeneration.

Similar to the case in Figure 15.10, most optimization constraints for this ATM framework remain identical, with Equations 15.16 and 15.17 being replaced by Equations 15.18 and 15.19 so as to cater to the flowrate terms of the partitioning interception unit. In addition, an equation that defines product recovery factor (RC) is also needed for the partitioning interception unit (see Section 14.4 for more details), as given by Equation 15.20:

$$\delta_k = \delta_{k-1} + (\Sigma_i F_{SRi} - \Sigma_j F_{SKj})_k + \Sigma_r F_{REG,r} + F_{RJ,r} - F_{RE,r} - F_{Wy} \quad k = 1, 2, \ldots, n-1 \quad (15.18)$$

$$F_{RE,r} = F_{REG,r} + F_{RJ,r} \quad r = 1, 2, \ldots, RG \quad (15.19)$$

$$F_{RE,r} \, q_{RE,r} \, RC = F_{REG,r} \, q_{REG,r} \quad r = 1, 2, \ldots, RG \quad (15.20)$$

It is also worth noting that there are cases where the recovery of the reject stream of the partitioning interception unit is not desired. This may be due to the fact that there are certain impurities/properties that are not permitted by the process. In these cases, the reject stream will be sent to the WTN along with other individual waste streams (i.e., $F_{RJ,r} = F_{Wy}$). The ATM framework for this case is shown in Figure 15.12.

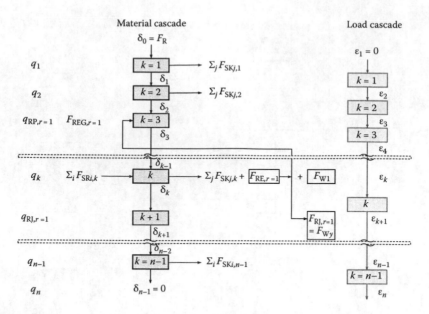

FIGURE 15.12
ATM framework for TMN with reject steam of partitioning interception unit being sent for WTN (cascade for WTN is not shown).

For the case in Figure 15.12, Equation 15.18 is revised to determine the various net material flowrate terms within the material cascade:

$$\delta_k = \delta_{k-1} + (\Sigma_i F_{SRi} - \Sigma_j F_{SKj})_k + \Sigma_r F_{REG,r} - F_{RE,r} - F_{Wy} \quad k = 1, 2, \ldots, n-1 \quad (15.21)$$

For both cases with or without the recovery of reject stream, the ATM remains in its linear form.

Finally, note that the cascade diagrams for the WTN have not been included during the discussion of Figures 15.9 through 15.12. In principle, all identified waste streams from the RCN will be sent to the WTN. As discussed in the previous section, different sets of constraints are to be included when different types of waste treatment units are used in the WTN. For instance, the constraints in Equations 15.1 through 15.6 are to be included when the WTN employs a fixed outlet quality type treatment unit (the ATM framework is given in Figure 15.1). Alternatively, Equations 15.2 and 15.4 through 15.8 are needed when a removal ratio type treatment unit is used (for the ATM framework in Figure 15.2). Also note that when the individual waste streams are handled by a separate ATM framework, the treated waste streams will be discharged to the environment and not recycled to the RCN.

Example 15.2 Total water network

Examples 8.4 and 14.1 is revisited here, with the limiting water data given in Table 15.2 (reproduced from Table 14.1). A single pass type interception unit with outlet concentration (C_{Rout}) of 5 ppm is used for water regeneration (see Example 14.1). Wastewater streams emitted from the water regeneration network are to be sent for wastewater treatment before final discharge. The waste treatment unit has a fixed outlet concentration (C_T) of 10 ppm, while the environmental discharge limit (C_D) is given as 20 ppm. Determine the minimum flowrates for freshwater as well as the associated regeneration and waste treatment flowrates needed for the water network. Note that the treated wastewater will not be recycled in the process.

Solution

The ATM framework in Figure 15.9 takes the revised form as in Figure 15.13a. Note that the ATM framework in Figure 15.13a resembles that described in Figure 14.3, with two potential wastewater streams added at 100 and 800 ppm (with flowrates of F_{WW1} and F_{WW2} respectively). In other words, apart from being sent for water regeneration, the two water sources at 100 and 800 ppm may also emerge as individual wastewater streams. These wastewater streams are sent for treatment prior to environmental discharge. Since the treated wastewater will not be recycled into the RCN, they are handled by a separate WTN. The cascade diagram for the latter is given in Figure 15.13b and is similar to that of Figure 15.3 (Example 15.1, case 1), except that the flowrates of the wastewater sources are now given as variables.

The optimization objective is to determine the minimum freshwater flowrate (F_{FW}) of the RCN, i.e.,

$$\text{Minimize } F_{FW} \tag{15.22}$$

subject to the constraints in Equations 15.10 through 15.15 for material recovery as well as Equations 15.1 through 15.6 for WTN. Note that the constraints in Equations 15.10 through 15.15 are essentially the revised form of those in Example 14.1.

TABLE 15.2

Limiting Water Data for Example 15.2

Sinks, SK_j	F_{SKj} (t/h)	C_{SKj} (ppm)	Sources, SR_i	F_{SRi} (t/h)	C_{SRi} (ppm)
SK1	20	0	SR1	20	100
SK2	100	50	SR2	100	100
SK3	40	50	SR3	40	800
SK4	10	400	SR4	10	800
$\Sigma_j F_{SKj}$	170		$\Sigma_i F_{SRi}$	170	

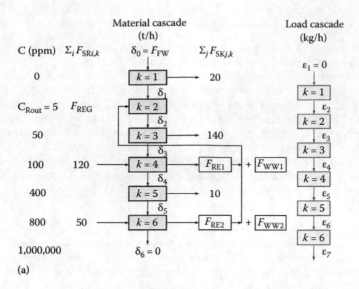

(a)

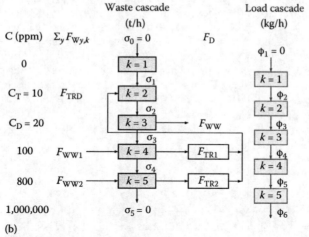

(b)

FIGURE 15.13
Cascade diagram for total water network in Example 15.2: (a) cascade diagram for water recovery and (b) cascade diagram for wastewater treatment.

Constraints for material recovery cascade are first discussed. Water cascade across all quality levels is given by Equation 15.10:

$$\delta_1 = \delta_0 + (0 - 20)$$
$$\delta_2 = \delta_1 + (F_{REG} - 0)$$
$$\delta_3 = \delta_2 + (0 - 140)$$
$$\delta_4 = \delta_3 + (120 - F_{RE1} - F_{WW1})$$

$$\delta_5 = \delta_4 + (0 - 10)$$
$$\delta_6 = \delta_5 + (50 - F_{RE2} - F_{WW2})$$

The net material flowrate entering the first concentration level corresponds to the minimum freshwater (F_{FW}) flowrate of the RCN, i.e.,

$$\delta_0 = F_{FW}$$

Equation 15.11 describes the flowrate balance of the interception unit:

$$F_{REG} = F_{RE1} + F_{RE2}$$

Next, the source flowrate to be sent for regeneration and wastewater treatment should not be more than what is available, according to Equation 15.12:

$$F_{RE1} + F_{WW1} \leq 120$$
$$F_{RE2} + F_{WW2} \leq 50$$

Based on Equation 15.13, the impurity load cascades (in kg/s) across all quality levels are given as follows:

$$\varepsilon_1 = 0$$
$$\varepsilon_2 = \varepsilon_1 + \delta_1(5 - 0)/1,000$$
$$\varepsilon_3 = \varepsilon_2 + \delta_2(50 - 5)/1,000$$
$$\varepsilon_4 = \varepsilon_3 + \delta_3(100 - 50)/1,000$$
$$\varepsilon_5 = \varepsilon_4 + \delta_4(400 - 100)/1,000$$
$$\varepsilon_6 = \varepsilon_5 + \delta_5(800 - 400)/1,000$$
$$\varepsilon_7 = \varepsilon_6 + \delta_6(1,000,000 - 800)/1,000$$

Equation 15.14 indicates that the net material flowrate that enters the first concentration level should take non-negative values, while no flowrate should enter the last level:

$$\delta_0 \geq 0$$
$$\delta_6 = 0$$

Based on Equation 15.15, all residual loads (except at level 1) should take non-negative values:

$$\varepsilon_2 \geq 0; \quad \varepsilon_3 \geq 0$$
$$\varepsilon_4 \geq 0; \quad \varepsilon_5 \geq 0$$
$$\varepsilon_6 \geq 0; \quad \varepsilon_7 \geq 0$$

Next, we discuss the constraints for the WTN, which resemble the constraints in Example 15.1 (Case 1). Equation 15.1 describes the waste cascade across all quality levels:

$$\sigma_1 = \sigma_0$$
$$\sigma_2 = \sigma_1 + F_{TRD}$$
$$\sigma_3 = \sigma_2 - F_{WW}$$
$$\sigma_4 = \sigma_3 + (F_{WW1} - F_{TR1})$$
$$\sigma_5 = \sigma_4 + (F_{WW2} - F_{TR2})$$

According to Equation 15.2, no net wastewater flowrate enters the first and last levels of the waste cascade:

$$\sigma_0 = 0$$
$$\sigma_5 = 0$$

Equation 15.3 describes the flowrate balance of the wastewater treatment unit:

$$F_{TRD} = F_{TR1} + F_{TR1}$$

Next, the wastewater source to be treated should not be more than what is available, according to Equation 15.4:

$$F_{TR1} \leq F_{WW1}$$
$$F_{TR2} \leq F_{WW2}$$

Equation 15.5 describes the waste load cascades (in kg/s) across all concentration levels:

$$\varphi_1 = 0$$
$$\varphi_2 = \varphi_1 + \delta_1(10 - 0)/1{,}000$$
$$\varphi_3 = \varphi_2 + \delta_2(20 - 10)/1{,}000$$
$$\varphi_4 = \varphi_3 + \delta_3(100 - 20)/1{,}000$$
$$\varphi_5 = \varphi_4 + \delta_4(800 - 100)/1{,}000$$
$$\varphi_6 = \varphi_5 + \delta_5(1{,}000{,}000 - 800)/1{,}000$$

Based on Equation 15.6, all residual loads (except at level 1) should take non-negative values:

$$\varphi_2 \geq 0; \quad \varphi_3 \geq 0$$
$$\varphi_4 \geq 0; \quad \varphi_5 \geq 0$$
$$\varphi_6 \geq 0$$

In terms of LINGO formulation, the LP model can be written as follows.

```
Model:
! Optimisation objective 1: to minimise freshwater (FW)
flowrate ;
min = FW ;

! Constraints for water recovery ;
! Water cascade ;
D1 = D0 - 20 ;
D2 = D1 + REG ;
D3 = D2 - 140 ;
D4 = D3 + ( 120 - RE1 - WW1 ) ;
D5 = D4 - 10 ;
D6 = D5 + ( 50 - RE2 - WW2 ) ;

! The first net material flowrate corresponds to minimum
freshwater (FW) flowrate ;
D0 = FW ;

! Flowrate balance for interception unit(s) ;
REG = RE1 + RE2 ;

! Flowrate constraints of regenerated and treated water ;
RE1 + WW1 <= 120 ;
RE2 + WW2 <= 50 ;

! Impurity load cascade (in kg/s) ;
E1 = 0 ;
E2 = E1 + D1*(5 - 0)/1000 ;
E3 = E2 + D2*(50 - 5)/1000 ;
E4 = E3 + D3*(100 - 50)/1000 ;
E5 = E4 + D4*(400 - 100)/1000 ;
E6 = E5 + D5*(800 - 400)/1000 ;
E7 = E6 + D6*(1000000 - 800)/1000 ;

! D0 should take non-negative values, and D6 is set to zero;
D0 >= 0 ;
D6 = 0 ;

! E2-E7 should take non-negative values ;
E2 >= 0 ; E3 >= 0 ; E4 >= 0 ;
E5 >= 0 ; E6 >= 0 ; E7 >= 0 ;

! D1-D5 may take negative values ;
@free (D1) ; @free (D2) ; @free (D3) ;
@free (D4) ; @free (D5) ;

! Constraints for WTN ;
! Waste cascade ;
S1 = S0 ;
S2 = S1 + TRD ;
S3 = S2 - 20 ;
S4 = S3 + ( WW1 - TR1 ) ;
S5 = S4 + ( WW2 - TR2 ) ;

! No net waste flowrate enters the first and the last
levels;
S0 = 0 ;
S5 = 0 ;
```

```
! Flowrate balance(s) for treatment unit(s) ;
TRD = TR1 + TR2 ;

! Flowrate constraints of treated wastewater source ;
TR1 <= WW1 ;
TR2 <= WW2 ;

! Waste load cascade (in kg/s) ;
J1 = 0 ;
J2 = J1 + S1*(10 - 0)/1000 ;
J3 = J2 + S2*(20 - 10)/1000 ;
J4 = J3 + S3*(100 - 20)/1000 ;
J5 = J4 + S4*(800 - 100)/1000 ;
J6 = J5 + S5*(1000000 - 800)/1000 ;

! J2-J6 should take non-negative values ;
J2 >= 0 ; J3 >= 0 ; J4 >= 0 ;
J5 >= 0 ; J6 >= 0 ;

! S1-S4 may take negative values ;
@free (S1) ; @free (S2) ;
@free (S3) ; @free (S4) ;
END
```

The following is the solution report generated by LINGO; the revised cascade diagrams for the total water network are shown in Figure 15.14.

```
Global optimal solution found.
Objective value:                        20.00000
Total solver iterations:                       5

          Variable            Value
          FW               20.00000
          D1               0.000000
          D0               20.00000
          D2               73.68421
          REG              73.68421
          D3              -66.31579
          D4               5.714286
          RE1              27.96992
          WW1              20.00000
          D5              -4.285714
          D6               0.000000
          RE2              45.71429
          WW2              0.000000
          E1               0.000000
          E2               0.000000
          E3               3.315789
          E4               0.000000
          E5               1.714286
          E6               0.000000
          E7               0.000000
          S1               0.000000
          S0               0.000000
          S2               17.77778
          TRD              17.77778
```

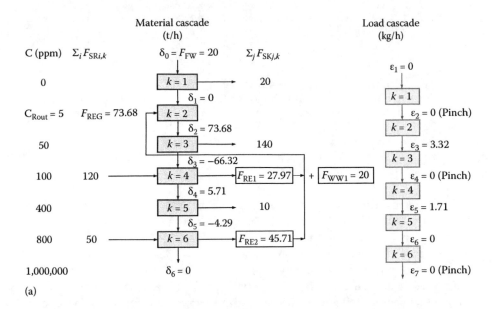

(a)

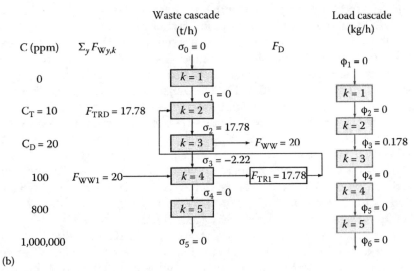

(b)

FIGURE 15.14
Result of ATM for Example 15.2: (a) cascade diagram for material recovery and (b) cascade diagram for WTN.

S3	−2.222222
S4	0.000000
TR1	17.77778
S5	0.000000
TR2	0.000000
J1	0.000000
J2	0.000000

J3	0.1777778
J4	0.000000
J5	0.000000
J6	0.000000

The optimization result shows that 20 t/h of freshwater flowrate (F_{FW}) is needed for the RCN. To enhance water recovery, both water sources are sent for regeneration, with their respective flowrates calculated as 27.97 (F_{RE1}) and 45.71 t/h (F_{RE2}). Besides, the model also determines that 20 t/h of wastewater WW1 (F_{WW1}) is emitted from the 100 ppm water source. Figure 15.14b indicates that 17.78 t/h of wastewater source WW1 is treated by the wastewater treatment unit.

Similar to the earlier cases, the structure for the RCN can be determined from the result of the ATM in Figure 15.14.* Note that the cascade diagram for water recovery in Figure 15.14a resembles that of Example 14.1 (see Figure 14.4). However, there are differences between these two cascade diagrams. First, it is observed that wastewater flowrate is not emitted from the last concentration interval in Figure 15.14a, i.e., $\delta_6 = 0$. Instead, the individual wastewater is observed to be emitted at the levels where the water source(s) exists. In this case, the latter corresponds to SR1 at 100 ppm, as it has an exact flowrate of 20 t/h. Second, even though the total regeneration flowrate for both examples remain identical, the individual water sources that are sent for water regeneration are of different flowrates. In Example 14.1 (Figure 14.4), 47.97 t/h of 100 ppm and 25.71 t/h of 800 ppm sources are sent for regeneration. In contrast, a bigger flowrate (45.71 t/h) is drawn from the 800 ppm source in Figure 15.14a, with the smaller flowrate (27.97 t/h) contributed by the 100 ppm source. Hence, the network structure for water recovery between water sources and sinks takes a form similar to that in Example 14.1 (Figure 14.5), with differences in the water source flowrates that are sent for regeneration and wastewater treatment.

Next, a wastewater treatment scheme can be developed from the insights provided by Figure 15.14b. We observe that a treatment flowrate of 17.78 t/h is needed for the 20 t/h wastewater stream (100 ppm). In other words, 2.22 t/h of this 100 ppm wastewater source will bypass the treatment unit.

A complete total water network is found in Figure 15.15. Note that this network is an extended form of that in Figure 14.5. However, it has a different structure as compared to that in designed using the insight based technique in Chapter 8 (see Figure 8.13), which means that this case is a degenerate problem, where more than one solution exists (for the same flowrate targets).

Example 15.3 Total hydrogen network

The hydrogen recovery problem in Example 14.4 is revisited, with the limiting data shown in Table 15.3 (reproduced from Table 14.2). The fresh

---

* The method is applicable for cases where only one sink (with non zero quality level) exists in each pinch region.

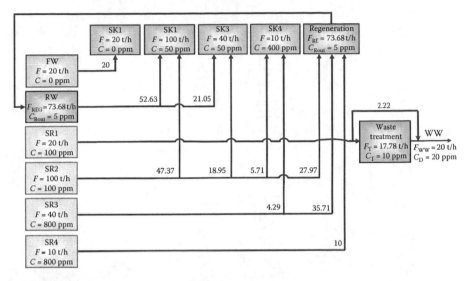

FIGURE 15.15
Total water network for Example 15.2 (flowrate terms in t/h). (From Ng, D.K.S. et al., *Comput. Chem. Eng.*, 34(5), 825, 2010.)

TABLE 15.3

Limiting Data for Example 15.3

Sinks, SK$_j$	F_{SKj} (MMscfd)	C_{SKj} (%)	Sources, SR$_i$	F_{SRi} (MMscfd)	C_{SRi} (%)
SK1	400	7.20	SR1	350	9.00
SK2	600	12.43	SR2	500	15.00

hydrogen feed has an impurity concentration of 1% (C_R). A pressure swing adsorption (PSA) unit with hydrogen recovery factor (RC) of 0.9 and purified stream concentration of 5% impurity (C_{RP}) is used to purify the process sources for further recovery. The reject stream from the PSA unit will not be recovered and hence be sent to the purge gas system (i.e. the WTN), along with other purge source(s). The purge gas system consists of a steam boiler and a flare. The steam boiler can utilize the purge gas as fuel, which will generate some savings for the RCN. However, it has a limited capacity of taking 5 MMscfd of gas stream, which should not exceed 50% impurity. On the other hand, no saving is possible for the flare. Determine the minimum operating cost (MOC) of the RCN. The unit costs for the fresh hydrogen (CT_{FH}) and regenerated hydrogen sources (CT_{RH}) are given as $2000/MMscf and $1000/MMscf, respectively; while the fuel savings can be determined from the value of $850/MMscf hydrogen gas.

Solution

The ATM framework for this RCN is given in Figure 15.16. Note that Figure 15.16a is a revised form of Figure 14.12, with the reject streams of

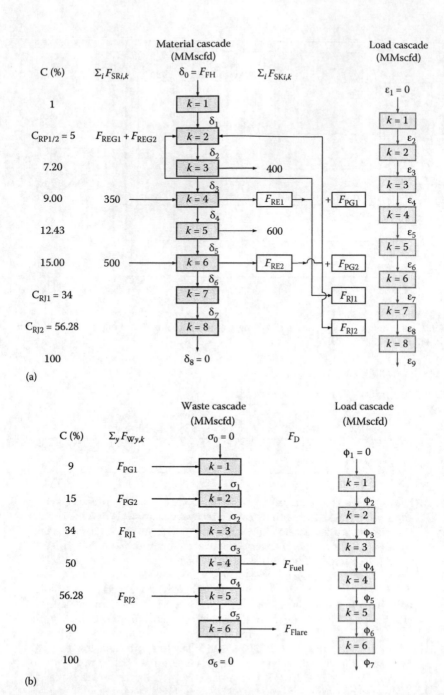

FIGURE 15.16
Cascade diagram for total hydrogen network in Example 15.3: (a) cascade diagram for hydrogen recovery and (b) cascade diagram for purge gas system.

the PSA units (RJ1 and RJ2) and individual purge streams (PG1 and PG2) being sent to the purge gas system. The cascade diagram for the latter is shown in Figure 15.16b.

The optimization objective in Equation 14.31 in Example 14.4 is revised to include the fuel savings term in order to minimize the MOC for the hydrogen network:

$$\text{Minimize } F_{FH} \, CT_{FH} + (F_{RE1} + F_{RE2}) \, CT_{RH} - \text{SAVING} \qquad (15.23)$$

where F_{FH} is the fresh hydrogen feed of the RCN. The optimization problem is subjected to the constraints in Equations 15.12 through 15.15 and 15.19 through 15.21 for hydrogen recovery, as well as Equations 15.1 through 15.6 for the purge gas system (WTN).

First, the constraints for hydrogen recovery are described. Hydrogen cascade across all quality levels are given by Equation 15.21:

$$\delta_1 = \delta_0$$
$$\delta_2 = \delta_1 + (F_{REG1} + F_{REG2})$$
$$\delta_3 = \delta_2 + (0 - 400)$$
$$\delta_4 = \delta_3 + (350 - F_{RE1} - F_{PG1})$$
$$\delta_5 = \delta_4 + (0 - 600)$$
$$\delta_6 = \delta_5 + (500 - F_{RE2} - F_{PG2})$$
$$\delta_7 = \delta_6$$
$$\delta_8 = \delta_7$$

Note that the individual purge flowrates for both sources, i.e., PG1 (F_{PG1}) and PG2 (F_{PG2}), are now included in the hydrogen cascade. Note, however, that the flowrates of reject stream RJ1 and RJ2 (with flowrates F_{RJ1} and F_{RJ2} respectively) are excluded; the latter will be included in the waste cascade (to be discussed later).

The net material flowrate that enters the first concentration level corresponds to the minimum fresh hydrogen (F_{FH}) for the RCN:

$$\delta_0 = F_{FH}$$

Equation 15.19 describes the flowrate balance of the PSA units:

$$F_{RE1} = F_{REG1} + F_{RJ1}$$
$$F_{RE2} = F_{REG2} + F_{RJ2}$$

The recovery of hydrogen ($RC = 0.9$) to the purified stream is described by Equation 15.20 (note: quality index takes the form of concentration here):

$$F_{RE1}(0.9)(100 - 9) = F_{REG1}(100 - 5)$$
$$F_{RE2}(0.9)(100 - 15) = F_{REG2}(100 - 5)$$

The source flowrate to be sent for regeneration and purge should not be more than what is available, according to Equation 15.12:

$$F_{RE1} + F_{PG1} \leq 350$$
$$F_{RE2} + F_{PG2} \leq 500$$

Impurity load cascades across all quality levels are given by Equation 15.13 as follows:

$$\varepsilon_1 = 0$$
$$\varepsilon_2 = \varepsilon_1 + \delta_1(5-1)/100$$
$$\varepsilon_3 = \varepsilon_2 + \delta_2(7.2-5)/100$$
$$\varepsilon_4 = \varepsilon_3 + \delta_3(9-7.2)/100$$
$$\varepsilon_5 = \varepsilon_4 + \delta_4(12.43-9)/100$$
$$\varepsilon_6 = \varepsilon_5 + \delta_5(15-12.43)/100$$
$$\varepsilon_7 = \varepsilon_6 + \delta_6(34-15)/100$$
$$\varepsilon_8 = \varepsilon_7 + \delta_7(56.28-34)/100$$
$$\varepsilon_9 = \varepsilon_8 + \delta_8(100-56.28)/100$$

The net material flowrates that enter the first level should take non-negative values, while that to the last level is set to zero. (Equation 15.14):

$$\delta_0 \geq 0$$
$$\delta_8 = 0$$

Finally, all residual loads (except at level 1) should take non-negative values (Equation 15.15):

$$\varepsilon_2 \geq 0; \quad \varepsilon_3 \geq 0$$
$$\varepsilon_4 \geq 0; \quad \varepsilon_5 \geq 0$$
$$\varepsilon_6 \geq 0; \quad \varepsilon_7 \geq 0$$
$$\varepsilon_8 \geq 0; \quad \varepsilon_9 \geq 0$$

We next discuss the constraints for the purge gas system, which acts as the WTN. Equation 15.1 describes the waste cascade across all quality levels:

$$\sigma_1 = \sigma_0 + F_{PG1}$$
$$\sigma_2 = \sigma_1 + F_{PG2}$$
$$\sigma_3 = \sigma_2 + F_{RJ1}$$
$$\sigma_4 = \sigma_3 - F_{Fuel}$$
$$\sigma_5 = \sigma_4 + F_{RJ2}$$
$$\sigma_6 = \sigma_5 + F_{Flare}$$

According to Equation 15.2, no net wastewater flowrate enters the first and the last levels of the waste cascade:

$$\sigma_0 = 0$$
$$\sigma_6 = 0$$

Equation 15.3 describes the flowrate balance of the purge gas system:

$$F_{PG1} + F_{PG2} + F_{RJ1} + F_{RJ2} = F_{Fuel} + F_{Flare}$$

In addition, the boiler has a flowrate constraint of 5 MMscfd of feed:

$$F_{Fuel} \leq 5$$

Note that the source availability described by Equation 15.4 is not applicable for this case. Equation 15.5 describes the waste load cascades (in kg/s) across all concentration levels:

$$\varphi_1 = 0$$
$$\varphi_2 = \varphi_1 + \delta_1(15 - 9)/100$$
$$\varphi_3 = \varphi_2 + \delta_2(34 - 15)/100$$
$$\varphi_4 = \varphi_3 + \delta_3(50 - 34)/100$$
$$\varphi_5 = \varphi_4 + \delta_4(56.28 - 50)/100$$
$$\varphi_6 = \varphi_5 + \delta_5(90 - 56.28)/100$$
$$\varphi_7 = \varphi_6 + \delta_6(100 - 90)/100$$

Based on Equation 15.6, all residual loads (except at level 1) should take non-negative values:

$$\varphi_2 \geq 0; \quad \varphi_3 \geq 0$$
$$\varphi_4 \geq 0; \quad \varphi_5 \geq 0$$
$$\varphi_6 \geq 0; \quad \varphi_7 \geq 0$$

To determine the fuel savings from the boiler, we shall make use of the cascade diagram in Figure 15.16b. In principle, the boiler can take a mix of purge gases that has a hydrogen content of at least 50%. This means that both purge streams PG1 and PG2, as well as the reject stream RJ1 with lower impurity concentrations can be used in the boiler. The optimization model will determine if the reject stream RJ2 is suitable for use. Since the fuel savings term is given as $850/MMscf hydrogen gas, we can calculate the potential savings from the hydrogen content of the purge and reject streams, as given by following equation:

$$\text{SAVING} = 850 \sum_y F_{Wy} \left(\frac{100 - C_{Wy}}{100} \right)$$

The LINGO formulation for this LP model is written as follows:

Model:

```
! Optimisation objective: to minimise operating cost (MOC) ;
min = FH*2000 + (RE1 + RE2)*800 - SAVING ;
SAVING = 850*(PG1*(100 - 5)/100 + PG2*(100 - 15)/100 +
RJ1*(100 - 34)/100 - S4*(100 - 56.28)/100) ;

! Constraints for hydrogen recovery ;

! Hydrogen cascade ;
D1 = D0 ;
D2 = D1 + REG1 + REG2 ;
D3 = D2 - 400 ;
D4 = D3 + ( 350 - RE1 - PG1 ) ;
D5 = D4 - 600 ;
D6 = D5 + ( 500 - RE2 - PG2 ) ;
D7 = D6 ;
D8 = D7 ;

! The first net material flowrates corresponds to minimum
fresh hydrogen (FH) flowrate ;
D0 = FH ;

! Flowrate balance for interception unit(s) ;
RE1 = REG1 + RJ1 ;
RE2 = REG2 + RJ2 ;
RE1*0.9*(100 - 9) = REG1*(100 - 5) ;
RE2*0.9*(100 - 15) = REG2*(100 - 5) ;

! Flowrate constraints of regenerated source and purge
streams ;
RE1 + PG1 <= 350 ;
RE2 + PG2 <= 500 ;

! Impurity load cascade ;
E1 = 0 ;
E2 = E1 + D1*(5 - 1)/100 ;
E3 = E2 + D2*(7.2 - 5)/100 ;
E4 = E3 + D3*(9 - 7.2)/100 ;
E5 = E4 + D4*(12.43 - 9)/100 ;
E6 = E5 + D5*(15 - 12.43)/100 ;
E7 = E6 + D6*(34 - 15)/100 ;
E8 = E7 + D7*(56.28 - 34)/100 ;
E9 = E8 + D8*(100 - 56.28)/100 ;

! D0 should take non-negative values, and D8 is set to zero;
D0 >= 0 ;
D8 = 0 ;

! E2-E9 should take non-negative values ;
E2 >= 0 ; E3 >= 0 ; E4 >= 0 ; E5 >= 0 ;
```

```
E6 >= 0 ; E7 >= 0 ; E8 >= 0 ; E9 >= 0 ;

! D1-D7 may take negative values ;
@free (D1) ; @free (D2) ; @free (D3) ; @free (D4) ;
@free (D5) ; @free (D6) ; @free (D7) ;
```

! Constraints for WTN ;

```
! Waste cascade ;
S1 = S0 + PG1 ;
S2 = S1 + PG2 ;
S3 = S2 + RJ1 ;
S4 = S3 - Fuel;
S5 = S4 + RJ2 ;
S6 = S5 - Flare ;

! No net waste flowrate enters the first and the last
levels;
S0 = 0 ;
S6 = 0 ;

! Flowrate balance(s) for treatment unit(s) ;
PG1 + PG2 + RJ1 + RJ2 = Fuel + Flare ;

! Flowrate constraints of boiler ;
Fuel <= 5 ;

! Waste load cascade (in kg/s) ;
J1 = 0 ;
J2 = J1 + S1*(15 - 9)/100 ;
J3 = J2 + S2*(34 - 15)/100 ;
J4 = J3 + S3*(50 - 34)/100 ;
J5 = J4 + S4*(56.28 - 50)/100 ;
J6 = J5 + S5*(90 - 56.28 )/100 ;
J7 = J6 + S6*(100- 90)/100 ;

! J2-J7 should take non-negative values ;
J2 >= 0 ; J3 >= 0 ; J4 >= 0 ;
J5 >= 0 ; J6 >= 0 ; J7 >= 0 ;

! S1-S5 may take negative values ;
@free (S1) ; @free (S2) ; @free (S3) ;
@free (S4) ; @free (S5) ;
END
```

The following is the solution report generated by LINGO:

```
Global optimal solution found.
Objective value:                        349203.6
Total solver iterations:                      10


               Variable          Value
               FH             158.9149
               RE1            0.000000
               RE2            41.87336
               SAVING         2125.000
               PG1            0.000000
```

PG2	0.7606589
RJ1	0.000000
S4	-4.239341
D1	158.9149
D0	158.9149
D2	192.6340
REG1	0.000000
REG2	33.71908
D3	-207.3660
D4	142.6340
D5	-457.3660
D6	0.000000
D7	0.000000
D8	0.000000
RJ2	8.154286
E1	0.000000
E2	6.356598
E3	10.59455
E4	6.861959
E5	11.75431
E6	0.000000
E7	0.000000
E8	0.000000
E9	0.000000
S1	0.000000
S0	0.000000
S2	0.7606589
S3	0.7606589
FUEL	5.000000
S5	3.914945
S6	0.000000
FLARE	3.914945
J1	0.000000
J2	0.000000
J3	0.1445252
J4	0.2662306
J5	0.000000
J6	1.320120
J7	1.320120

The ATM indicates that the MOC for the RCN is \$349,000/day. The revised cascade diagrams for the total hydrogen network are shown in Figure 15.17. As compared to Example 14.4, the fresh hydrogen flowrate increases slightly to 158.91 MMscfd (from 158.35 MMscfd; see Figure 14.13). It is also observed that a small flowrate is purged from source SR2, which does not happen in Example 14.4.

On the other hand, for the purge gas system, Figure 15.17b indicates that the entire purge stream PG2 (15%) will be sent to the boiler, along with 4.24 MMscfd of 56.28% reject stream RJ2 (indicated by the absolute value of σ_4 in the cascade diagram). A simple mass balance calculation indicates that the boiler feed has an impurity content of 5%. The remaining flowrate of RJ2 (3.91 MMscfd) is then sent to the flare.

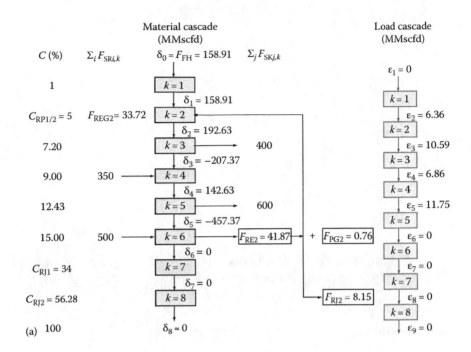

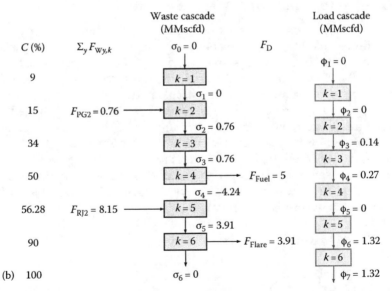

FIGURE 15.17
Result of ATM for Example 15.3: (a) cascade diagram for material recovery and (b) cascade diagram for purge system (WTN).

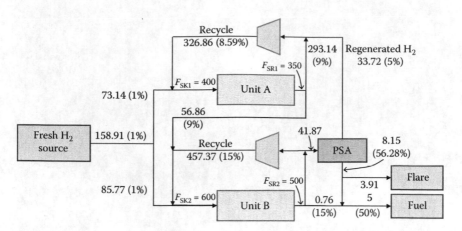

FIGURE 15.18
Total hydrogen network for Example 15.3 (flowrate in MMscfd).

A total hydrogen network that achieves the targets is given in Figure 15.18, designed using *nearest neighbor algorithm* introduceed in Chapter 7.*

15.3 ATM for TMN with Waste Recycling

In the previous section, the waste treatment was handled by a separate ATM framework. In such an arrangement, the treated waste streams are discharged to the environment directly, i.e., not recycled to the process sink in the RCN. However, there are cases where the recycling of treated waste streams may be permitted. In such cases, the treatment unit performs as an interception unit in regenerating the process sources. The purified sources may be discharged to the environment and/or be recycled within the RCN. The ATM framework for this case is shown in Figure 15.19, when both interception and waste treatment units are of the fixed outlet quality type.

The optimization constraints for the ATM framework in Figure 15.16 are given as the revised version of those in Figure 15.19. Note that Equations 15.10 and 15.12 are revised as follows:

$$\delta_k = \delta_{k-1} + (\Sigma_i F_{SRi} - \Sigma_j F_{SKj})_k + F_{REG} - F_{RE,r} + F_{TRD} - F_{TR,w} - F_D \quad k = 1, 2, \ldots, n-1$$

(15.24)

$$F_{RE,r} + F_{TR,w} \le F_{SRi} \quad r = 1, 2, \ldots, RG$$

(15.25)

---

* Readers are adviced to revisit Chapter 7 and try to design the RCN in Figure 15.18 as an additional exercise (hint: flowrate targets should be incorporated in design).

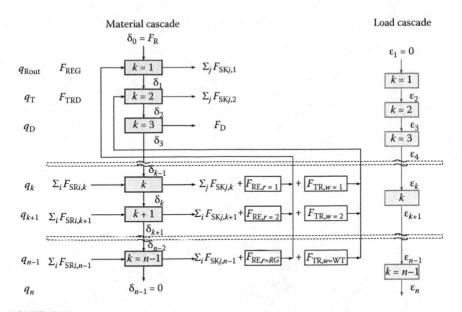

FIGURE 15.19
ATM framework for TMN with waste recycling. (From Ng, D.K.S. et al., *Comput. Chem. Eng.*, 34(5), 825, 2010.)

Also, the constraints described by Equations 15.13 through 15.15 are also needed for the ATM framework here. Flowrate balances for the waste treatment and source interception units are defined by Equations 15.3 and 15.11, respectively. Similar to earlier case, the ATM remains as an LP model.

Example 15.4 Total water network with waste recycling

Example 15.2 is revisited. The earlier restriction on wastewater recycling is relaxed here. In other words, both regenerated source(s) from the interception unit ($C_{Rout} = 5\,\text{ppm}$) and the treated wastewater stream(s) from wastewater treatment unit ($C_T = 10\,\text{ppm}$) may be recovered to the RCN. Synthesize a total water network for the following cases:

1. Minimum freshwater flowrate.
2. MOC solution. For this case, the unit cost for freshwater (CT_{FW}) is given as \$1/ton. Also, the unit costs for water regeneration (CT_{REG}) and wastewater treatment (CT_{TRD}) are given as a function of impurity load removal, i.e., 0.75 and 0.5 \$/kg impurity removed.

Solution

The ATM framework in Figure 15.19 takes the revised form as in Figure 15.20. In this case, both water sources at 100 and 800 ppm may be

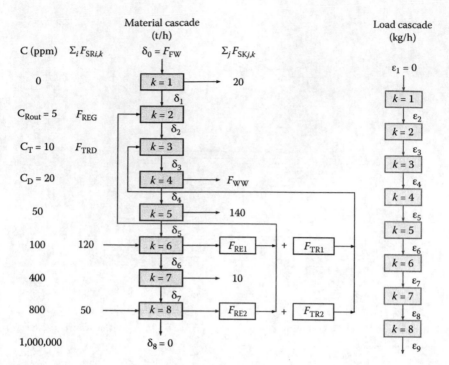

FIGURE 15.20
Cascade diagram for total water network in Example 15.4. (From Ng, D.K.S. et al., *Comput. Chem. Eng.*, 34(5), 825, 2010.)

sent for water regeneration as well as wastewater treatment. Similar to the regenerated source(s), the treated wastewater source(s) may be recovered to the RCN.

1. Case 1—Minimum freshwater flowrate

   To determine the minimum freshwater flowrate, the optimization objective in Equation 15.22 is used:

$$\text{Minimize } F_{FW} \qquad\qquad (15.22)$$

subject to the constraints in Equations 15.3, 15.11 through 15.15, 15.24, and 15.25. These are described as follows.

Water cascade across all quality levels is given by Equation 15.24:

$$\delta_1 = \delta_0 + (0 - 20)$$
$$\delta_2 = \delta_1 + F_{REG}$$
$$\delta_3 = \delta_2 + F_{TRD}$$
$$\delta_4 = \delta_3 + (0 - F_{WW})$$

$$\delta_5 = \delta_4 + (0 - 140)$$
$$\delta_6 = \delta_5 + (120 - F_{RE1} - F_{TR1})$$
$$\delta_7 = \delta_6 + (0 - 10)$$
$$\delta_8 = \delta_7 + (50 - F_{RE2} - F_{TR2})$$

The net material flowrate that enters the first concentration level corresponds to the minimum freshwater (F_{FW}) flowrate of the RCN:

$$\delta_0 = F_{FW}$$

Equations 15.3 and 15.11 describe the flowrate balance of the water treatment and water regeneration units, respectively:

$$F_{TRD} = F_{TR1} + F_{TR2}$$
$$F_{REG} = F_{RE1} + F_{RE2}$$

Next, the source flowrate to be sent for regeneration and wastewater treatment should not be more than what is available, according to Equation 15.25:

$$F_{RE1} + F_{TR1} \le 120$$
$$F_{RE2} + F_{TR2} \le 50$$

Based on Equation 15.13, the impurity load cascades (in kg/s) across all quality levels are given as follows:

$$\varepsilon_1 = 0$$
$$\varepsilon_2 = \varepsilon_1 + \delta_1(5-0)/1,000$$
$$\varepsilon_3 = \varepsilon_2 + \delta_2(10-5)/1,000$$
$$\varepsilon_4 = \varepsilon_3 + \delta_3(20-10)/1,000$$
$$\varepsilon_5 = \varepsilon_4 + \delta_4(50-20)/1,000$$
$$\varepsilon_6 = \varepsilon_5 + \delta_5(100-50)/1,000$$
$$\varepsilon_7 = \varepsilon_6 + \delta_6(400-100)/1,000$$
$$\varepsilon_8 = \varepsilon_7 + \delta_7(800-400)/1,000$$
$$\varepsilon_9 = \varepsilon_8 + \delta_8(1,000,000-800)/1,000$$

Equation 15.14 indicates that the net material flowrates that enter the first concentration level should take non-negative values, while no flowrate should enter the last level:

$$\delta_0 \ge 0$$
$$\delta_8 = 0$$

Based on Equation 15.15, all residual loads (except at level 1) should take non-negative values:

$$\varepsilon_2 \geq 0; \quad \varepsilon_3 \geq 0$$
$$\varepsilon_4 \geq 0; \quad \varepsilon_5 \geq 0$$
$$\varepsilon_6 \geq 0; \quad \varepsilon_7 \geq 0$$
$$\varepsilon_8 \geq 0; \quad \varepsilon_9 \geq 0$$

In terms of LINGO formulation, the LP model can be written as follows.

Model:

```
! Case 1: to minimise freshwater (FW) flowrate ;
min = FW ;

! Water cascade ;
D1 = D0 - 20 ;
D2 = D1 + REG ;
D3 = D2 + TRD ;
D4 = D3 - WW ;
D5 = D4 - 140 ;
D6 = D5 + ( 120 - RE1 - TR1 ) ;
D7 = D6 - 10 ;
D8 = D7 + ( 50 - RE2 - TR2 ) ;

! The first net material flowrates corresponds to minimum
freshwater (FW) flowrate ;
D0 = FW ;

! Flowrate balance(s) for interception unit(s) ;
REG = RE1 + RE2 ;

! Flowrate balance(s) for treatment unit(s) ;
TRD = TR1 + TR2 ;

! Flowrate constraints of regenerated and treated water ;
RE1 + TR1 <= 120 ;
RE2 + TR2 <= 50 ;

! Impurity load cascade (in kg/s) ;
E1 = 0 ;
E2 = E1 + D1*(5 - 0)/1000 ;
E3 = E2 + D2*(10 - 5)/1000 ;
E4 = E3 + D3*(20 - 10)/1000 ;
E5 = E4 + D4*(50 - 20)/1000 ;
E6 = E5 + D5*(100 - 50)/1000 ;
E7 = E6 + D6*(400 - 100)/1000 ;
E8 = E7 + D7*(800 - 400)/1000 ;
E9 = E8 + D8*(1000000 - 800)/1000 ;

! D0 should take non-negative values, and D8 is set to zero;
D0 >= 0 ;
D8 = 0 ;

! E2-E7 should take non-negative values ;
E2 >= 0 ; E3 >= 0 ; E4 >= 0 ; E5 >= 0 ;
E6 >= 0 ; E7 >= 0 ; E8 >= 0 ; E9 >= 0 ;
```

```
! D1-D5 may take negative values ;
@free (D1) ; @free (D2) ; @free (D3) ; @free (D4) ;
@free (D5) ; @free (D6) ; @free (D7) ;
END
```

The following is the solution report generated by LINGO:

```
Global optimal solution found.
Objective value:                          20.00000
Total solver iterations:                          3
```

Variable	Value
FW	20.00000
D1	0.000000
D0	20.00000
D2	90.52632
REG	90.52632
D3	90.52632
TRD	0.000000
D4	70.52632
WW	20.00000
D5	-69.47368
D6	5.714286
RE1	44.81203
TR1	0.000000
D7	-4.285714
D8	0.000000
RE2	45.71429
TR2	0.000000
E1	0.000000
E2	0.000000
E3	0.4526316
E4	1.357895
E5	3.473684
E6	0.000000
E7	1.714286
E8	0.000000
E9	0.000000

The ATM determines that 20 t/h of freshwater is needed for the total water network. The revised cascade diagram for this case is shown in Figure 15.21. It is observed that both water sources are sent for regeneration; and no water source is sent to the wastewater treatment unit. At 20 ppm level (C_D), it is observed that 20 t/h wastewater is discharged from the RCN (F_{WW}). This means that the interception unit has purified the water sources that meet the discharge quality (20 ppm for this case). In other words, the interception unit acts as a wastewater treatment unit in this case.

2. Case 2—MOC solution

The optimization objective is to determine the MOC of the RCN, which is given as follows:

$$MOC = F_{FW}CT_{FW} + CT_{RW}\Sigma_r F_{REG,r}(C_{SRi} - C_{Rout}) + CT_{TRD}\Sigma_w F_{TR,w} (C_{SRi} - C_T)$$

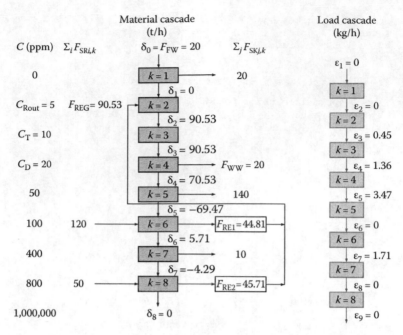

FIGURE 15.21
Result of ATM for Example 15.4 (Case 1—minimum freshwater flowrate).

Note that all optimization constraints remain identical as in Case 1. In terms of LINGO formulation, the optimization objective can be written as follows:

```
! Case 2: to determine the MOC;
min = FW*1 + ( RE1*(100 - 5)/1000 + RE2*(800 - 5)/1000 )*0.75
+ ( TR1*(100 - 10)/1000 + TR2*(800 - 10)/1000 )*0.5;
```

The following is the solution report generated by LINGO:

```
Global optimal solution found.
Objective value:                    60.30000
Total solver iterations:                   3
          Variable          Value
          FW             20.00000
          WW             20.00000
          RE1             0.000000
          RE2             0.000000
          TR1            49.84127
          TR2            45.71429
          D1              0.000000
          D0             20.00000
          D2              0.000000
          REG             0.000000
          D3             95.55556
```

TRD	95.55556
D4	75.55556
D5	-64.44444
D6	5.714286
D7	-4.285714
D8	0.000000
E1	0.000000
E2	0.000000
E3	0.000000
E4	0.9555556
E5	3.222222
E6	0.000000
E7	1.714286
E8	0.000000
E9	0.000000

As shown, the MOC for this RCN is determined as \$60/h. The revised cascade diagram for this case is shown in Figure 15.22. In this case, instead of being sent to the interception unit as in Case 1, both water sources are sent for waste treatment (with flowrates of 49.85 and 45.71 t/h for sources at 100 and 800 ppm respectively). The treated wastewater is being recycled in the water network, while a portion of that leaves as wastewater ($F_{WW} = 20\,t/h$). Note that the flowrate for waste treatment is lower than that for interception, as in Case 1 (due to lower treatment outlet concentration). Note however that the freshwater (F_{FW}) and wastewater (F_{WW}) flowrates remain identical to those in Case 1.

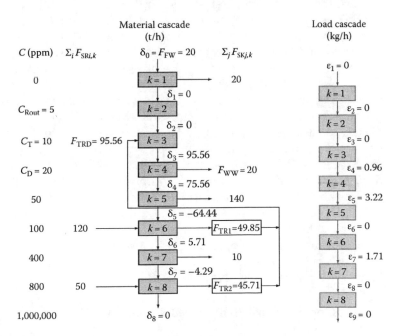

FIGURE 15.22
Result of ATM for Example 15.4 (Case 2—MOC solution).

Finally, note that from cases 1 and 2, we can conclude that both interception and waste treatment units act as purifying unit in upgrading water quality for both reuse/recycle as well as for environmental discharge.

15.4 Additional Readings

Similar to the case in material regeneration network (Chapter 14), where the TMNs employ a waste treatment unit of the removal ratio type, there may be a possibility that more than one treatment unit is needed, as determined by the ATM. Additional constraints can then be added to restrict the number of treatment units, in which the ATM will become a nonlinear program (NLP). Readers may refer to the work of Ng et al. (2010) for an extended model of this case.

Problems

Water Minimization Problems

15.1 Rework the water minimization case in Problems 6.4 and 14.3 (Sorin and Bédard, 1999). The limiting data for this case are given in Table 6.25. In this case, single pass interception units with a removal ratio of 0.9 are used for both water regeneration and wastewater treatment. In other words, apart from being discharged to the environment, the effluent is also being considered for recycling within the water network. Environment discharge limit is set at 50 ppm. The unit costs for pure freshwater (CT_{FW}) and regenerated water (CT_{RW}) are given as $1/ton and $0.5/ton, respectively. Determine the MOC for this case.

15.2 Rework the water minimization problem of the Kraft pulping process in Problem 14.6. In this case, a biotreatment facility is employed for wastewater treatment in order to comply with environmental discharge limit. It is also determined that the maximum inlet and average outlet concentrations of methanol for the biotreatment facility are given as 1000 and 15 ppm, respectively (El-Halwagi, 1997). Note that upon treatment, the effluent from the biotreatment may be recycled to the process. The cost elements of the biotreatment facility are given as follows (El-Halwagi, 1997; Ng et al., 2010):

$$\text{Operating cost } (\$/\text{h}) = 0.00011 \, \Delta m + 1.3 \, \Sigma_w F_{\text{TRD},w} \qquad (15.26)$$

$$\text{Captial cost } (\$) = 190 \, \Sigma_w F_{\text{TRD},w} \qquad (15.27)$$

where Δm is the methanol load removal. The unit cost of freshwater is taken as $1/ton (ignore the wastewater unit cost in Problem 14.6). To determine the annualized capital cost, the annualizing factor in Equation 12.15 is used, assuming an annual fractional interest rate of 10%, for a total period of 5 years. Determine the total annualized cost (TAC) for the RCN, assuming an annual operating time (AOT) of 8000 h.

Utility Gas Recovery Problems

15.3 Rework the refinery hydrogen network problem in Problem 14.7 (Alves and Towler, 2002). The reject stream from the PSA unit will not be recovered to the RCN and is sent to the purge gas system along with other purge sources. The purge gas system consists of a steam boiler with a limited capacity of 80 mol/s and a flare. The steam boiler can utilize the purge gas as fuel. However, its feed stream should not contain more than 40% impurity. Determine the MOC of the RCN. The unit costs for the fresh hydrogen (CT_{FH}) and regenerated hydrogen sources (CT_{RH}) are given as $1.58/kmol and $0.79/kmol, respectively; the fuel savings can be estimated from the value of $0.67/kmol of hydrogen gas.

Property Integration Problems

15.4 Water recovery for the wafer fabrication process (Ng et al., 2010) in Problem 14.9b is revisited. To enhance water recovery, a membrane unit (water recovery factor = 0.95; permeate quality = 15 MΩ m) is used for water regeneration. On the other hand, the environmental discharge limit (C_D) for these wastewater streams is set as 2 ppm for their heavy metal content. Hence, a wastewater treatment facility that is able to purify wastewater to an outlet concentration (C_T) of 0.5 ppm heavy metal content is employed for use. Note that the treated wastewater will not be recycled to the RCN. The unit costs for the ultrapure water (UPW), freshwater, wastewater, and regenerated water are taken as $2/ton, $1/ton, $0.5/ton, and $1.5/ton, respectively. Use the limiting data of the water sinks and sources, as well as the heavy metal content of the latter as shown in Table 15.4. The resistivity values of the municipal freshwater and UPW are given as 0.02 and 18 MΩ m, respectively. Determine the MOC for this RCN.

15.5 Revisit the RCN problem of palm oil milling process (Ng et al., 2010) in Problem 14.10. Apart from the filtration unit that is used to purify the bottom flow of clay bath for recovery, a waste treatment unit is added to purify wastewater before it is discharged to the environment. In this case, the total suspended solid (TSS) content in the wastewater stream is the main concern for final discharge; and its limit is given as 250 ppm. The wastewater treatment unit has an

TABLE 15.4

Limiting Data for Problem 15.4

j	Sinks, SK$_j$	F_{SKj} (t/h)	ψ_{SKj} ([MΩ m]$^{-1}$)	i	Sources, SR$_i$	F_{SRi} (t/h)	ψ_{SRi} ([MΩ m]$^{-1}$)	C_{WW} (ppm)
1	Wet	500	0.143	1	Wet I	250	1	5.00
2	Lithography	450	0.125	2	Wet II	200	0.5	4.50
3	CMP	700	0.100	3	Lithography	350	0.333	5.00
4	Etc.	350	0.200	4	CMP I	300	10	10.00
5	Cleaning	200	125.000	5	CMP II	200	0.5	4.50
6	Cooling tower makeup	450	50.000	6	Etc	280	2	5.00
7	Scrubber	300	100.000	7	Cleaning	180	500	15.00
				8	Scrubber	300	200	10.00
				9	UF reject	30% of UF inlet	100	1.50
				10	RO reject	30% of RO inlet	200	10.00

outlet concentration of 20 ppm (C_T). Besides, the treated effluent from WTN cannot be recycled to the RCN as it does not improve the density of the effluent.

The schematic for the RCN, as well as the TSS content of the various wastewater streams, is given in Figure 15.23. Determine the MOC for the RCN. The unit costs for all water sources are the same as those in Problem 14.10.

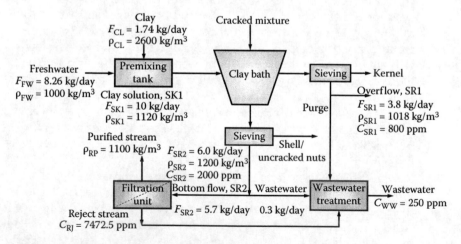

FIGURE 15.23
RCN for palm oil milling process in Problem 15.5.

References

Alves, J. J. and Towler, G. P. 2002. Analysis of refinery hydrogen distribution systems. *Industrial and Engineering Chemistry Research*, 41, 5759–5769.

El-Halwagi, M. M. 1997. *Pollution Prevention through Process Integration: Systematic Design Tools*. San Diego, CA: Academic Press.

Ng, D. K. S., Foo, D. C. Y., Tan, R. R., and El-Halwagi, M. M. 2010. Automated targeting technique for concentration- and property-based total resource conservation network. *Computers and Chemical Engineering*, 34(5), 825–845.

Sorin, M. and Bédard, S. 1999. The global pinch point in water reuse networks. *Process Safety and Environmental Protection*, 77, 305–308.

16

Automated Targeting Model for Inter-Plant Resource Conservation Networks

As discussed in Chapter 10, apart from in-plant recovery within a single resource conservation network (RCN), material recovery can also be carried out across different plants/processes/sections of a plant through *inter-plant resource conservation networks* (IPRCNs). In addition to recovering material from process sinks and sources of different RCNs through *direct integration* scheme, different RCNs may also be integrated *indirectly* through a *centralized utility facility* (CUF). Automated targeting model (ATM) developed for single RCN in the previous chapters is extended for use in both direct and indirect integration schemes of IPRCNs in this chapter.

16.1 ATM for Direct Integration Scheme—Direct Material Reuse/Recycle

The basic ATM framework for direct integration scheme (with direct reuse/recycle) for IPRCN is shown in Figure 16.1. It is easily observable that the framework is similar to the direct reuse/recycle scheme for a single RCN (Figure 13.1), except that each variable term now has an additional subscript p which denotes the set of individual RCNs that are involved in the IPRCN scheme. Also, note that there are two flowrate terms newly added to the framework: *import* and *export flowrates*. These enable the integration between sinks and sources of different RCNs to take place. On the left side of the material cascade, along with the total process source flowrate of network p, the total import flowrate from other network(s) p' to network $p(\Sigma_{p'}F_{CP,p',p})$ is added as a new source. On the other hand, the total export flowrate from network p to other network(s) $p'(\Sigma_{p'}F_{CP,p,p'})$ is added as a new sink, along with the total process sink flowrate of plant p, on the right side of the material cascade. In order to carry out the targeting task for an IPRCN, ATM frameworks are to be constructed for all networks that are involved in the IPRCN schemes.

For the ATM framework in Figure 16.1, the following constraints are needed for each network in the IPRCN:

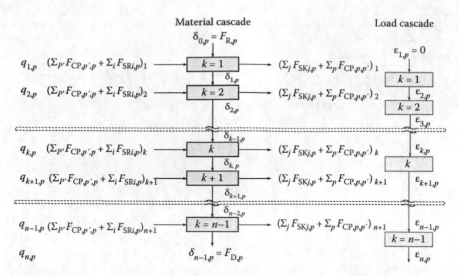

FIGURE 16.1
ATM framework for direct reuse/recycle in direct integration scheme. (From *Chem. Eng. J.*, 153(1–3), Chew, I.M.L. and Foo, D.C.Y., Automated targeting for inter-plant water integration, 23–36, Copyright 2009, with permission from Elsevier.)

$$\delta_{k,p} = \delta_{k-1,p} + (\Sigma_i F_{SRi} - \Sigma_j F_{SKj})_{k,p} + (\Sigma_{p'} F_{CP,p',p} - \Sigma_{p'} F_{CP,p,p'})_k$$

$$k = 1, 2, \ldots, n-1; \quad \forall p; \, p \neq p' \tag{16.1}$$

$$\varepsilon_{k,p} = \begin{cases} 0 & k = 1; \quad \forall p \\ \varepsilon_{k-1,p} + \delta_{k-1,p}(q_{k,p} - q_{k-1,p}) & k = 2,3,\ldots,n; \quad \forall p \end{cases} \tag{16.2}$$

$$\delta_{k-1,p} \geq 0 \quad k = 1, n; \quad \forall p \tag{16.3}$$

$$\varepsilon_{k,p} \geq 0 \quad k = 2,3,\ldots, n; \quad \forall p \tag{16.4}$$

Equation 16.1 describes the material cascades across all quality levels for each network p. The *net material flowrate* of level k in network p ($\delta_{k,p}$) is a result of summation of the net material flowrate cascaded from an earlier quality level ($\delta_{k-1,p}$) with the flowrate balance at quality level k ($\Sigma_i F_{SRi} - \Sigma_j F_{SKj})_{k,p}$, and the net flowrate of the total import and export streams at each level ($\Sigma_{p'} F_{CP,p',p} - \Sigma_{p'} F_{CP,p,p'}$). Equation 16.2 indicates that the *residual impurity/ property load* ($\varepsilon_{k,p}$) at the first quality level of each plant p has zero value, i.e., $\varepsilon_{1,p} = 0$, while those in other quality levels are contributed by the residual load cascaded from previous level ($\varepsilon_{k-1,p}$) and the net material flowrate within each quality interval. The latter is given by the product of the net material flowrate from previous level ($\delta_{k-1,p}$) and the difference between two adjacent quality levels ($q_{k,p} - q_{k-1,p}$). The net material flowrates entering the first ($\delta_{0,p}$) and last levels ($\delta_{n-1,p}$) of each network p should take non-negative values (Equation 16.3). These variables correspond to the fresh resource ($F_{R,p}$) and waste discharge flowrates ($F_{D,p}$) of the individual network p. In Equation 16.4, all residual loads are set to take non-negative values, except at level 1 (which is set to zero according to Equation 16.2). Note that the constraints in Equations 16.2 through 16.4 resemble those of the reuse/recycle scheme for single network in Chapter 13 (i.e., Equations 13.4 through 13.6).

The two-stage optimization approach introduced in Chapters 12 and 14 may be used to determine the minimum total fresh resource and cross-plant flowrates across all RCNs. In this case, the overall fresh resource across all networks ($\Sigma_p F_{R,p}$) is first minimized with the optimization objective in Equation 16.5, subject to the constraints in Equations 16.1 through 16.4. Then, the minimum total cross-plant flowrates are determined using the optimization objective in Equation 16.6, where the total export flowrates are minimized, with the minimum fresh resource determined in Stage 1 added as a new constraint (Equation 16.7), along with other constraints in Stage 1 (Equations 16.1 through 16.4). Note that one may also set Equation 16.6 to minimize for total import flowrates to achieve the same objective value:

$$\text{Minimize } \Sigma_p F_{R,p} \tag{16.5}$$

$$\text{Minimize } \Sigma_k \Sigma_p F_{CP,p,p',k} \tag{16.6}$$

$$F_{R,min} = \Sigma_p F_{R,p} \tag{16.7}$$

Alternatively, the optimization objective may also be set to determine the minimum cost of the IPRCN, which takes the form of the minimum operating cost (MOC), or minimum total annualized cost (TAC), as given in Equation 16.8:

$$\text{Minimize cost} \tag{16.8}$$

Similar to ATM in previous chapters, the ATM for IPRCNs is also a linear program (LP) that can be solved to achieve global optimum solution.

 Example 16.1 Inter-plant water network with direct reuse/recycle

Example 10.1 considering inter-plant water network is revisited here. The limiting water data for both Networks A and B are found in Table 16.1 (reproduced from Table 10.1). The freshwater feed is available at 0 ppm. Determine the minimum overall freshwater and total cross-plant flowrates across all RCNs using the two-stage optimization approach.

TABLE 16.1

Limiting Water Data for Example 16.1

Sinks, SK$_j$	F_{SKj} (t/h)	C_{SKj} (ppm)	Sources, SR$_i$	F_{SRi} (t/h)	C_{SRi} (ppm)
Network A					
SK1	20	0	SR1	20	100
SK2	100	50	SR2	100	100
SK3	40	50	SR3	40	800
SK4	10	400	SR4	10	800
Network B					
SK5	50	20	SR5	50	50
SK6	100	50	SR6	100	100
SK7	80	100	SR7	70	150
SK8	70	200	SR8	60	250
$\Sigma_i F_{SKj}$	470		$\Sigma_i F_{SRi}$	450	

Solution

Based on the ATM framework in Figure 16.1, the cascade diagrams for both networks are shown in Figure 16.2. Note that all the import and export flowrate terms are added at each quality level in the cascade diagrams. The export flowrates from Network A become the import flowrates for Network B. Similarly, the export flowrates from Network B then become the import flowrates for Network A.

Next, we use the two-stage optimization approach to determine the minimum overall freshwater and total cross-plant flowrates. In Stage 1, the optimization objective is set to minimize the overall freshwater flowrates according to Equation 16.5:

$$\text{Minimize } F_{FW,A} + F_{FW,B} \tag{16.9}$$

subject to the constraints given by Equations 16.1 through 16.4 for both Networks A and B. We first describe the constraints for Network A.

Water cascade across all quality levels is given by Equation 16.1:

$$\delta_{1,A} = \delta_{0,A} - 20 + (F_{CP,B,A,1} - F_{CP,A,B,1})$$

$$\delta_{2,A} = \delta_{1,A} + (F_{CP,B,A,2} - F_{CP,A,B,2})$$

$$\delta_{3,A} = \delta_{2,A} - 140 + (F_{CP,B,A,3} - F_{CP,A,B,3})$$

$$\delta_{4,A} = \delta_{3,A} + 120 + (F_{CP,B,A,4} - F_{CP,A,B,4})$$

$$\delta_{5,A} = \delta_{4,A} + (F_{CP,B,A,5} - F_{CP,A,B,5})$$

$$\delta_{6,A} = \delta_{5,A} + (F_{CP,B,A,6} - F_{CP,A,B,6})$$

$$\delta_{7,A} = \delta_{6,A} + (F_{CP,B,A,7} - F_{CP,A,B,7})$$

$$\delta_{8,A} = \delta_{7,A} - 10 + (F_{CP,B,A,8} - F_{CP,A,B,8})$$

$$\delta_{9,A} = \delta_{8,A} + 50 + (F_{CP,B,A,9} - F_{CP,A,B,9})$$

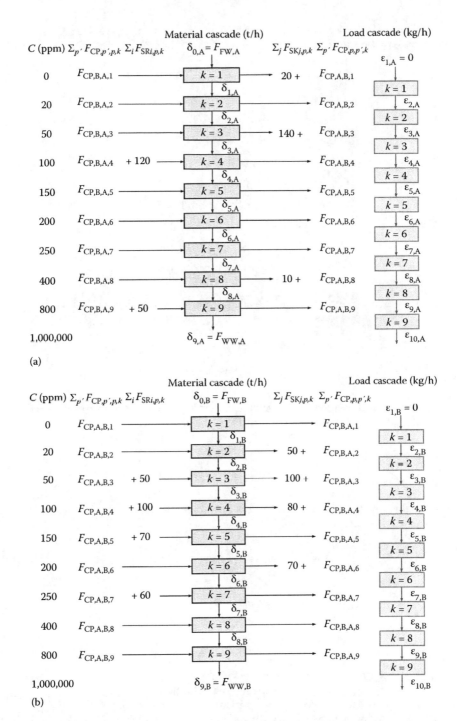

FIGURE 16.2
Cascade diagram for Example 16.1: (a) Network A and (b) Network B.

The impurity load cascades (in kg/s) across all quality levels are given by Equation 16.2:

$$\varepsilon_{1,A} = 0$$

$$\varepsilon_{2,A} = \varepsilon_{1,A} + \delta_{1,A}(20 - 0)/1,000$$

$$\varepsilon_{3,A} = \varepsilon_{2,A} + \delta_{2,A}(50 - 20)/1,000$$

$$\varepsilon_{4,A} = \varepsilon_{3,A} + \delta_{3,A}(100 - 50)/1,000$$

$$\varepsilon_{5,A} = \varepsilon_{4,A} + \delta_{4,A}(150 - 100)/1,000$$

$$\varepsilon_{6,A} = \varepsilon_{5,A} + \delta_{5,A}(200 - 150)/1,000$$

$$\varepsilon_{7,A} = \varepsilon_{6,A} + \delta_{6,A}(250 - 200)/1,000$$

$$\varepsilon_{8,A} = \varepsilon_{7,A} + \delta_{7,A}(400 - 250)/1,000$$

$$\varepsilon_{9,A} = \varepsilon_{8,A} + \delta_{8,A}(800 - 400)/1,000$$

$$\varepsilon_{10,A} = \varepsilon_{9,A} + \delta_{9,A}(1,000,000 - 800)/1,000$$

Equation 16.3 indicates that the net material flowrates of the first and last concentration levels should take non-negative values:

$$\delta_{0,A} \geq 0$$

$$\delta_{9,A} \geq 0$$

Note that these flowrate terms correspond to the minimum freshwater ($F_{FW,A}$) and wastewater flowrates ($F_{WW,A}$) of Network A:

$$\delta_{0,A} = F_{FW,A}$$

$$\delta_{9,A} = F_{WW,A}$$

Equation 16.4 indicates that all residual loads (except at level 1) should take non-negative values:

$$\varepsilon_{2,A} \geq 0; \quad \varepsilon_{3,A} \geq 0$$

$$\varepsilon_{4,A} \geq 0; \quad \varepsilon_{5,A} \geq 0$$

$$\varepsilon_{6,A} \geq 0; \quad \varepsilon_{7,A} \geq 0$$

$$\varepsilon_{8,A} \geq 0; \quad \varepsilon_{9,A} \geq 0$$

$$\varepsilon_{10,A} \geq 0;$$

The constraints in Equations 16.1 through 16.4 are also written for Network B. For brevity, only the net material flowrate in Equation 16.1 is discussed here; the rest of the constraints of Network A are exactly the same as that in Network A (with subscript A changed to B). According to Equation 16.1, the water cascade for Network B is given as follows:

$$\delta_{1,B} = \delta_{0,B} + (F_{CP,A,B,1} - F_{CP,B,A,1})$$

$$\delta_{2,B} = \delta_{1,B} - 50 + (F_{CP,A,B,2} - F_{CP,B,A,2})$$

$$\delta_{3,B} = \delta_{2,B} + 50 - 100 + (F_{CP,A,B,3} - F_{CP,B,A,3})$$

$$\delta_{4,B} = \delta_{3,B} + 100 - 80 + (F_{CP,A,B,4} - F_{CP,B,A,4})$$

$$\delta_{5,B} = \delta_{4,B} + 70 + (F_{CP,A,B,5} - F_{CP,B,A,5})$$

$$\delta_{6,B} = \delta_{5,B} - 70 + (F_{CP,A,B,6} - F_{CP,B,A,6})$$

$$\delta_{7,B} = \delta_{6,B} + 60 + (F_{CP,A,B,7} - F_{CP,B,A,7})$$

$$\delta_{8,B} = \delta_{7,B} + (F_{CP,A,B,8} - F_{CP,B,A,8})$$

$$\delta_{9,B} = \delta_{8,B} + (F_{CP,A,B,9} - F_{CP,B,A,9})$$

The LINGO formulation for the previous LP model can be written as follows.

```
Model:
! Optimisation objective 1: to minimise total freshwater
flowrates in Networks A (FWA) and B (FWB);
min = FWA x FWB;

! Constraints for Network A;
! Water cascade;
D1A = D0A - 20 + (CPBA1 - CPAB1);
D2A = D1A + (CPBA2 - CPAB2);
D3A = D2A - 140 + (CPBA3 - CPAB3);
D4A = D3A + 120 + (CPBA4 - CPAB4);
D5A = D4A + (CPBA5 - CPAB5);
D6A = D5A + (CPBA6 - CPAB6);
D7A = D6A + (CPBA7 - CPAB7);
D8A = D7A - 10 + (CPBA8 - CPAB8);
D9A = D8A + 50 + (CPBA9 - CPAB9);

! Impurity load cascade (in kg/s) ;
E1A = 0 ;
E2A = E1A + D1A*(20 - 0)/1000;
E3A = E2A + D2A*(50 - 20)/1000;
E4A = E3A + D3A*(100 - 50)/1000;
E5A = E4A + D4A*(150 - 100)/1000;
E6A = E5A + D5A*(200 - 150)/1000;
E7A = E6A + D6A*(250 - 200)/1000;
E8A = E7A + D7A*(400 - 250)/1000;
E9A = E8A + D8A*(800 - 400)/1000;
E10A = E9A + D9A*(1000000 - 800)/1000;

! The first and last net material flowrates should take
non-negative values ;
D0A >= 0;
D9A >= 0;

! These net material flowrates correspond to minimum
freshwater (FW) and wastewater (WW) flowrates for each
network;
```

```
D0A = FWA;
D9A = WWA;

! E2A-E10A should take non-negative values ;
E2A >= 0;  E3A >= 0;  E4A >= 0;
E5A >= 0;  E6A >= 0;  E7A >= 0;
E8A >= 0;  E9A >= 0;  E10A >= 0;

! D1A-D8A may take negative values ;
@free (D1A);  @free (D2A);  @free (D3A);
@free (D4A);  @free (D5A);  @free (D6A); @free (D7A);
@free (D8A);

! Constraints for Network B;
! Water cascade;
D1B = D0B + (CPAB1 - CPBA1);
D2B = D1B - 50 + (CPAB2 - CPBA2);
D3B = D2B + (50 - 100) + (CPAB3 - CPBA3);
D4B = D3B + (100 - 80) + (CPAB4 - CPBA4);
D5B = D4B + 70 + (CPAB5 - CPBA5);
D6B = D5B - 70 + (CPAB6 - CPBA6);
D7B = D6B + 60 + (CPAB7 - CPBA7);
D8B = D7B + (CPAB8 - CPBA8);
D9B = D8B + (CPAB9 - CPBA9);

! Impurity load cascade (in kg/s) ;
E1B = 0 ;
E2B = E1B + D1B*(20 - 0)/1000;
E3B = E2B + D2B*(50 - 20)/1000;
E4B = E3B + D3B*(100 - 50)/1000;
E5B = E4B + D4B*(150 - 100)/1000;
E6B = E5B + D5B*(200 - 150)/1000;
E7B = E6B + D6B*(250 - 200)/1000;
E8B = E7B + D7B*(400 - 250)/1000;
E9B = E8B + D8B*(800 - 400)/1000;
E10B = E9B + D9B*(1000000 - 800)/1000;

! The first and last net material flowrates should take
non-negative values ;
D0B >= 0;
D9B >= 0;

! These net material flowrates correspond to minimum
freshwater (FW) and wastewater (WW) flowrates for each
network;
D0B = FWB;
D9B = WWB;

! E2B-E10B should take non-negative values ;
E2B >= 0;  E3B >= 0;  E4B >= 0;
E5B >= 0;  E6B >= 0;  E7B >= 0;
E8B >= 0;  E9B >= 0;  E10B >= 0;

! D1B-D8B may take negative values ;
@free (D1B);  @free (D2B);  @free (D3B);  @free (D4B);
@free (D5B);  @free (D6B);  @free (D7B);  @free (D8B);
END
```

As mentioned in the previous chapters, the last set of constraints are only necessary when LINGO is used to solve the optimization model, as the software assumes all variables to have non-negative values by default. For the same reason, the constraints for non-negativity (given by Equations 16.4) need not be coded in LINGO.

The following is the solution report generated by LINGO for Stage 1 optimization (the results for optimization variables are omitted, as they are not necessary for now):

```
Global optimal solution found.
Objective value:                      155.0000
Total solver iterations:                     0
```

The optimization result indicates that the overall minimum freshwater flowrates for both water networks are determined as 155 t/h. Hence, the minimum flowrate is included as a new constraint in Stage 2 optimization, following Equation 16.6:

$$F_{FW,A} + F_{FW,B} = 155 \tag{16.10}$$

while the optimization objective is revised to minimize the total export flowrates, given by Equation 16.6:

$$\text{Minimize } \Sigma_k F_{CP,B,A,k} + \Sigma_k F_{CP,A,B,k} \tag{16.11}$$

LINGO formulation for the second optimization stage remains similar to that in the first stage, except for the newly added optimization objective (given by Equation 16.6) and freshwater flowrate constraint (Equation 16.10). In this case, the LINGO formulation for the Stage 2 optimization objective and the new constraints are given as follows:

```
! Optimisation objective 2: to minimise cross-plant
flowrates;
min = CPAB1 + CPAB2 + CPAB3 + CPAB4 + CPAB5 + CPAB6 + CPAB7 + CPAB8 +
CPAB9 + CPBA1 + CPBA2 + CPBA3 + CPBA4 + CPBA5 + CPBA6 + CPBA7 + CPBA8 +
CPBA9;
FWA + FWB = 155;
```

The following is the solution report generated by LINGO for Stage 2 optimization:

```
Global optimal solution found.
Objective value:                      15.00000
Total solver iterations:                    17

              Variable       Value
              CPAB1          0.000000
              CPAB2          0.000000
              CPAB3          0.000000
              CPAB4         15.00000
              CPAB5          0.000000
              CPAB6          0.000000
```

CPAB7	0.000000
CPAB8	0.000000
CPAB9	0.000000
CPBA1	0.000000
CPBA2	0.000000
CPBA3	0.000000
CPBA4	0.000000
CPBA5	0.000000
CPBA6	0.000000
CPBA7	0.000000
CPBA8	0.000000
CPBA9	0.000000
FWA	90.00000
FWB	65.00000
D1A	70.00000
D0A	90.00000
D2A	70.00000
D3A	−70.00000
D4A	35.00000
D5A	35.00000
D6A	35.00000
D7A	35.00000
D8A	25.00000
D9A	75.00000
E1A	0.000000
E2A	1.400000
E3A	3.500000
E4A	0.000000
E5A	1.750000
E6A	3.500000
E7A	5.250000
E8A	10.50000
E9A	20.50000
E10A	74960.50
WWA	75.00000
D1B	65.00000
D0B	65.00000
D2B	15.00000
D3B	−35.00000
D4B	0.000000
D5B	70.00000
D6B	0.000000
D7B	60.00000
D8B	60.00000
D9B	60.00000
E1B	0.000000
E2B	1.300000
E3B	1.750000
E4B	0.000000
E5B	0.000000
E6B	3.500000
E7B	3.500000
E8B	12.50000
E9B	36.50000
E10B	59988.50
WWB	60.00000

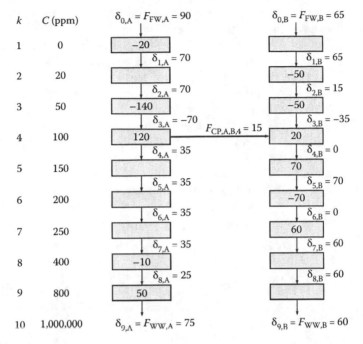

FIGURE 16.3
Result of ATM for Example 16.1 (flowrate in t/h; load cascades are not displayed).

The optimization result indicates that the minimum total cross-plant flowrate is determined as 15 t/h. A simplified cascade diagram that consists of the material cascades for both networks is shown in Figure 16.3 (load cascades are not displayed). We observe that Networks A and B require freshwater flowrates of 90 and 65 t/h, respectively, i.e., a total flowrate of 155 t/h as determined in Stage 1. Only one cross-plant flowrate is needed to connect the two water networks at 100 ppm level ($F_{CP,A,B,4}$). Note that the total flowrates for fresh- and wastewater as well as the cross-plant stream are identical to those obtained using the insight-based targeting technique (see Example 10.1). Hence, the IPRCN for this example has the same structure as that in Figures 10.13 and 10.14.

16.2 ATM for Direct Integration Scheme: RCNs with Individual Interception Unit

As discussed in Chapters 6 and 14, to further explore material recovery potential, interception units may be used to regenerate process sources for further recovery in the process sinks. In the context of IPRCN, the interception unit may be used for regeneration in the individual RCN or for the entire IPRCN. The former is now discussed.

In principle, the ATM framework for this case takes the hybrid form of that in Figure 16.1 and those variants in Chapter 14, e.g., single pass or partitioning interception units of fixed outlet quality or removal ratio types. For instance, for the case where a single pass interception unit of fixed outlet quality type is used for the individual network p, the ATM framework takes the form as outlined in Figure 16.4, which is a hybrid of those in Figures 16.1 and 14.1 (single pass interception unit of fixed outlet quality type for a single network).

For the ATM framework in Figure 16.4, the constraints in Equations 16.2 through 16.4 are to be used, along with the following constraints for each network that consists of an individual interception unit:

$$\delta_{k,p} = \delta_{k-1,p} + (\Sigma_i F_{SRi} - \Sigma_j F_{SKj})_{k,p} + (\Sigma_{p'} F_{CP,p',p} - \Sigma_{p'} F_{CP,p,p'})_k$$
$$+ F_{REG,p,r} - \Sigma_r F_{RE,p,r} \quad k = 1, \ldots, n-1; \quad \forall p; p \neq p' \tag{16.12}$$

$$F_{REG,p} = \Sigma_r F_{RE,r,p} \quad \forall p \tag{16.13}$$

$$F_{RE,r,p} \leq \Sigma_i F_{SRi,k,p} \quad \forall r; q_k < q_{Rout} \quad \forall p \tag{16.14}$$

Equation 16.12 is the revised form of Equation 16.1, where the regeneration flowrate terms are added to the net material flowrate calculation.

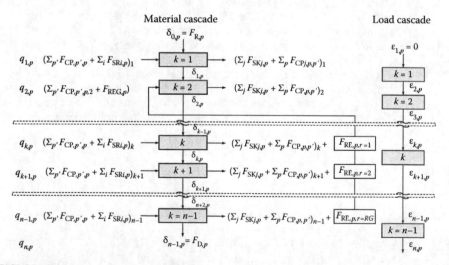

FIGURE 16.4
ATM framework for direct integration scheme—RCNs with individual interception unit. (From *Chem. Eng. J.*, 153(1–3), Chew, I.M.L. and Foo, D.C.Y., Automated targeting for inter-plant water integration, 23–36, Copyright 2009, with permission from Elsevier.)

The flowrate balance of the interception unit is given by Equation 16.13; the flowrate availability for regeneration is given by Equation 16.14. Note that for network(s) in the IPRCN scheme that does not have an interception unit, the ATM framework in Figure 16.1 and the associated constraints in Equations 16.1 through 16.4 are to be used.

Example 16.2 Inter-plant water network with an individual interception unit

Example 16.1 is revisited here. In this case, an interception unit with an outlet concentration of 20 ppm is installed in Network A for regeneration of water sources; Network B does not have an interception unit. Determine the minimum TAC target for the IPRCN. Additional information for cost calculations is given as follows:

1. Unit costs for freshwater (CT_{FW}), regenerated water (CT_{RW}), and wastewater (CT_{WW}) are given as \$1/ton, \$0.8/ton, and \$1/ton, respectively. Note that the unit cost for regenerated water includes capital and operating costs.
2. Capital cost for the IPRCN is mainly contributed by the piping cost of the cross-plant streams $(PC_{CP},$ in USD), which is given by the revised form of Equation 12.18 (cost for stream connections within individual networks may be omitted):

$$PC_{CP} = (2F_{CP,p,p'} + 250)\, D_{p,p'} \qquad (16.15)$$

where

$F_{CP,p,p'}$ is the cross-plant flowrate between network p and p' (import or export stream; in t/h)

$D_{p,p'}$ is the distance between network p and p'

For this example, the distance between Networks A and B is given as 100 m.

3. For calculation of the TAC, the annualizing factor (AF) in Equation 12.15 may be used. The piping cost is annualized to a period of 5 years, with an interest rate of 5%. Assume an annual operation of 8000 h.

Solution

Since only Network A is installed with an interception unit, its cascade diagram takes a revised form as in Figure 16.5; the cascade diagram of Network B remains the same as in Figure 16.2b.

For the cascade diagram in Figure 16.5, the water cascade across all quality levels in Network A is given by Equation 16.12 as follows:

$$\delta_{1,A} = \delta_{0,A} - 20 + (F_{CP,B,A,1} - F_{CP,A,B,1})$$

$$\delta_{2,A} = \delta_{1,A} + (F_{CP,B,A,2} - F_{CP,A,B,2}) + (F_{REG,A})$$

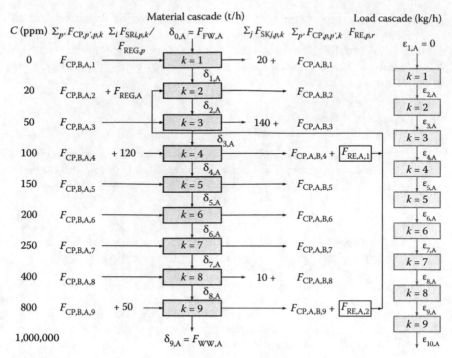

FIGURE 16.5
Cascade diagram for Network A in Example 16.2.

$$\delta_{3,A} = \delta_{2,A} - 140 + (F_{CP,B,A,3} - F_{CP,A,B,3})$$

$$\delta_{4,A} = \delta_{3,A} + 120 + (F_{CP,B,A,4} - F_{CP,A,B,4}) + F_{RE,A,1}$$

$$\delta_{5,A} = \delta_{4,A} + (F_{CP,B,A,5} - F_{CP,A,B,5})$$

$$\delta_{6,A} = \delta_{5,A} + (F_{CP,B,A,6} - F_{CP,A,B,6})$$

$$\delta_{7,A} = \delta_{6,A} + (F_{CP,B,A,7} - F_{CP,A,B,7})$$

$$\delta_{8,A} = \delta_{7,A} - 10 + (F_{CP,B,A,8} - F_{CP,A,B,8})$$

$$\delta_{9,A} = \delta_{8,A} + 50 + (F_{CP,B,A,9} - F_{CP,A,B,9}) + F_{RE,A,2}$$

Next, the flowrate balance of the interception unit is given by Equation 16.13:

$$F_{REG,A} = F_{RE,A,1} + F_{RE,A,2}$$

Equation 16.14 indicates that the flowrate sent for regeneration is bound with its maximum availability:

$$F_{RE,A,1} \leq 120$$

$$F_{RE,A,2} \leq 50$$

The rest of the constraints for Network A (given by Equations 16.2 through 16.4) as well as those for Network B (Equations 16.1 through 16.4) are identical to those in Example 16.1 and hence are omitted here for brevity.

We next discuss the derivation of the optimization objective for the problem, which resembles that in Example 12.3 (see Section 12.3 for more detailed description). In this case, it is desired to determine the TAC target of the IPRCN. Hence, the optimization objective is given as described in Equation 16.16:

$$\text{Minimize TAC} \tag{16.16}$$

We first determine the operating cost (OC) of the IPRCN, which is contributed by the various water costs:

$$OC(\$/h) = \Sigma_p F_{FW,p} CT_{FW} + \Sigma_r F_{RE,A,r} CT_{RW} + \Sigma_p F_{WW,p} CT_{WW}$$

$$= (F_{FW,A} + F_{FW,B} + F_{WW,A} + F_{WW,B})(\$1/\text{ton})$$

$$+ (F_{RE,A,1} + F_{RE,A,2})(\$0.8/\text{ton})$$

Next, the AOC is determined from the product of OC and the annual operating time (AOT), given as follows:

$$AOC(\$/\text{year}) = OC(8000\,\text{h})$$

The total piping cost (PC) and, hence, the capital cost (CC) for the IPRCN is contributed by the cross-plant piping cost at each concentration level:

$$CC = PC = PC_{CP,A,B,1} + PC_{CP,A,B,2} + PC_{CP,A,B,3} + PC_{CP,A,B,4}$$

$$+ PC_{CP,A,B,5} + PC_{CP,A,B,6} + PC_{CP,A,B,7} + PC_{CP,A,B,8} + PC_{CP,A,B,9}$$

$$+ PC_{CP,B,A,1} + PC_{CP,B,A,2} + PC_{CP,B,A,3} + PC_{CP,B,A,4} + PC_{CP,B,A,5}$$

$$+ PC_{CP,B,A,6} + PC_{CP,B,A,7} + PC_{CP,B,A,8} + PC_{CP,B,A,9}$$

To determine the piping cost, Equation 16.15 is revised as Equation 16.17, where a binary term $(B_{CP,p,p',k})$ is added to indicate the presence of cross-plant (export) streams at level k:

$$PC_{CP,p,p',k} = (2F_{CP,p,p',k} + 250\,B_{CP,p,p',k})\,D_{p,p'}; \quad p \neq p' \tag{16.17}$$

Equation 16.17 is used in conjunction with Equation 16.18 to examine the presence/absence of a cross-plant connection (see Section 12.3 and Example 12.3 for the use of these equations in single RCN):

$$\frac{F_{CP,p,p',k}}{M} \leq B_{CP,p,p',k} \quad k = 1,\ldots,n-1; \quad \forall p; p \neq p' \tag{16.18}$$

where M is an arbitrary large number. In this case, the value of M is taken as 1000. Equations 16.17 and 16.18 are applied for all export streams for both Networks A (lines 1–9) and B, respectively (lines 10–18), given as follows:

$$PC_{CP,A,B,1} = (2\,F_{CP,A,B,1} + 250\,B_{CP,A,B,1})\,D_{A,B}; \qquad F_{CP,A,B,1} \le 1000\,B_{CP,A,B,1}$$
$$PC_{CP,A,B,2} = (2\,F_{CP,A,B,2} + 250\,B_{CP,A,B,2})\,D_{A,B}; \qquad F_{CP,A,B,2} \le 1000\,B_{CP,A,B,2}$$
$$PC_{CP,A,B,3} = (2\,F_{CP,A,B,3} + 250\,B_{CP,A,B,3})\,D_{A,B}; \qquad F_{CP,A,B,3} \le 1000\,B_{CP,A,B,3}$$
$$PC_{CP,A,B,4} = (2\,F_{CP,A,B,4} + 250\,B_{CP,A,B,4})\,D_{A,B}; \qquad F_{CP,A,B,4} \le 1000\,B_{CP,A,B,4}$$
$$PC_{CP,A,B,5} = (2\,F_{CP,A,B,5} + 250\,B_{CP,A,B,5})\,D_{A,B}; \qquad F_{CP,A,B,5} \le 1000\,B_{CP,A,B,5}$$
$$PC_{CP,A,B,6} = (2\,F_{CP,A,B,6} + 250\,B_{CP,A,B,6})\,D_{A,B}; \qquad F_{CP,A,B,6} \le 1000\,B_{CP,A,B,6}$$
$$PC_{CP,A,B,7} = (2\,F_{CP,A,B,7} + 250\,B_{CP,A,B,7})\,D_{A,B}; \qquad F_{CP,A,B,7} \le 1000\,B_{CP,A,B,7}$$
$$PC_{CP,A,B,8} = (2\,F_{CP,A,B,8} + 250\,B_{CP,A,B,8})\,D_{A,B}; \qquad F_{CP,A,B,8} \le 1000\,B_{CP,A,B,8}$$
$$PC_{CP,A,B,9} = (2\,F_{CP,A,B,9} + 250\,B_{CP,A,B,9})\,D_{A,B}; \qquad F_{CP,A,B,9} \le 1000\,B_{CP,A,B,9}$$
$$PC_{CP,B,A,1} = (2\,F_{CP,B,A,1} + 250\,B_{CP,B,A,1})\,D_{A,B}; \qquad F_{CP,B,A,1} \le 1000\,B_{CP,B,A,1}$$
$$PC_{CP,B,A,2} = (2\,F_{CP,B,A,2} + 250\,B_{CP,B,A,2})\,D_{A,B}; \qquad F_{CP,B,A,2} \le 1000\,B_{CP,B,A,2}$$
$$PC_{CP,B,A,3} = (2\,F_{CP,B,A,3} + 250\,B_{CP,B,A,3})\,D_{A,B}; \qquad F_{CP,B,A,3} \le 1000\,B_{CP,B,A,3}$$
$$PC_{CP,B,A,4} = (2\,F_{CP,B,A,4} + 250\,B_{CP,B,A,4})\,D_{A,B}; \qquad F_{CP,B,A,4} \le 1000\,B_{CP,B,A,4}$$
$$PC_{CP,B,A,5} = (2\,F_{CP,B,A,5} + 250\,B_{CP,B,A,5})\,D_{A,B}; \qquad F_{CP,B,A,5} \le 1000\,B_{CP,B,A,5}$$
$$PC_{CP,B,A,6} = (2\,F_{CP,B,A,6} + 250\,B_{CP,B,A,6})\,D_{A,B}; \qquad F_{CP,B,A,6} \le 1000\,B_{CP,B,A,6}$$
$$PC_{CP,B,A,7} = (2\,F_{CP,B,A,7} + 250\,B_{CP,B,A,7})\,D_{A,B}; \qquad F_{CP,B,A,7} \le 1000\,B_{CP,B,A,7}$$
$$PC_{CP,B,A,8} = (2\,F_{CP,B,A,8} + 250\,B_{CP,B,A,8})\,D_{A,B}; \qquad F_{CP,B,A,8} \le 1000\,B_{CP,B,A,8}$$
$$PC_{CP,B,A,9} = (2\,F_{CP,B,A,9} + 250\,B_{CP,B,A,9})\,D_{A,B}; \qquad F_{CP,B,A,9} \le 1000\,B_{CP,B,A,9}$$

The annualizing factor in Equation 12.15 is calculated as follows:

$$AF = \frac{0.05(1+0.05)^5}{(1+0.05)^5 - 1} = 0.231$$

Hence, the total annual cost (TAC) in Equation 16.16 is given as follows:

$$TAC = AOC + CC(0.231)$$

Due to the presence of binary variables, the previous problem is a mixed integer linear program (MILP). The LINGO formulation for the previous problem may be written as follows.

```
Model:
! Optimisation objective: to determine minimum total
annualised cost (TAC);
min = TAC;
TAC = AOC + PC*0.231;
AOC = ( (FWA + FWB + WWA + WWB) *1 + (REA1 + REA2) *0.8 ) *8000;
PC = (PAB1 + PAB2 + PAB3 + PAB4 + PAB5 + PAB6 + PAB7 + PAB8 + PAB9 + PBA1 +
PBA2 + PBA3 + PBA4 + PBA5 + PBA6 + PBA7 + PBA8 + PBA9) ;

! Constraints for Network A;
! Water cascade;
D1A = D0A - 20 + (CPBA1 - CPAB1) ;
D2A = D1A + REGA + (CPBA2 - CPAB2) ;
D3A = D2A - 140 + (CPBA3 - CPAB3) ;
D4A = D3A + 120 + (CPBA4 - CPAB4) - REA1;
```

```
D5A = D4A + (CPBA5 - CPAB5);
D6A = D5A + (CPBA6 - CPAB6);
D7A = D6A + (CPBA7 - CPAB7);
D8A = D7A - 10 + (CPBA8 - CPAB8);
D9A = D8A + 50 + (CPBA9 - CPAB9) - REA2;

! Flowrate balance for interception unit;
REGA = REA1 + REA2;

! Source availability for regeneration;
REA1 <= 120;
REA2 <= 50;

! Impurity load cascade (in kg/s) ;
E1A = 0 ;
E2A = E1A + D1A*(20 - 0)/1000;
E3A = E2A + D2A*(50 - 20)/1000;
E4A = E3A + D3A*(100 - 50)/1000;
E5A = E4A + D4A*(150 - 100)/1000;
E6A = E5A + D5A*(200 - 150)/1000;
E7A = E6A + D6A*(250 - 200)/1000;
E8A = E7A + D7A*(400 - 250)/1000;
E9A = E8A + D8A*(800 - 400)/1000;
E10A = E9A + D9A*(1000000 - 800)/1000;

! The first and last net material flowrates should take
non-negative values ;
D0A >= 0;
D9A >= 0;

! These net material flowrates correspond to minimum
freshwater (FW) and
wastewater (WW) flowrates for each network;
D0A = FWA;
D9A = WWA;

! E2A-E10A should take non-negative values ;
E2A >= 0; E3A >= 0; E4A >= 0;
E5A >= 0; E6A >= 0; E7A >= 0;
E8A >= 0; E9A >= 0; E10A >= 0;

! D1A-D8A may take negative values ;
@free (D1A); @free (D2A); @free (D3A); @free (D4A);
@free (D5A); @free (D6A); @free (D7A); @free (D8A);

! Constraints for Network B;
! Water cascade;
D1B = D0B + (CPAB1 - CPBA1);
D2B = D1B - 50 + (CPAB2 - CPBA2);
D3B = D2B + (50 - 100) + (CPAB3 - CPBA3);
D4B = D3B + (100 - 80) + (CPAB4 - CPBA4);
D5B = D4B + 70 + (CPAB5 - CPBA5);
D6B = D5B - 70 + (CPAB6 - CPBA6);
D7B = D6B + 60 + (CPAB7 - CPBA7);
D8B = D7B + (CPAB8 - CPBA8);
D9B = D8B + (CPAB9 - CPBA9);

! Impurity load cascade (in kg/s) ;
E1B = 0 ;
E2B = E1B + D1B*(20 - 0)/1000;
```

```
E3B=E2B+D2B*(50-20)/1000;
E4B=E3B+D3B*(100-50)/1000;
E5B=E4B+D4B*(150-100)/1000;
E6B=E5B+D5B*(200-150)/1000;
E7B=E6B+D6B*(250-200)/1000;
E8B=E7B+D7B*(400-250)/1000;
E9B=E8B+D8B*(800-400)/1000;
E10B=E9B+D9B*(1000000-800)/1000;

! The first and last net material flowrates should take
non-negative values ;
D0B>=0;
D9B>=0;

! These net material flowrates correspond to minimum
freshwater (FW) and
wastewater (WW) flowrates for each network;
D0B=FWB;
D9B=WWB;

! E2B-E10B should take non-negative values ;
E2B>=0;  E3B>=0;  E4B>=0;
E5B>=0;  E6B>=0;  E7B>=0;
E8B>=0;  E9B>=0;  E10B>=0;

! D1B-D8B may take negative values ;
@free (D1B); @free (D2B); @free (D3B); @free (D4B);
@free (D5B); @free (D6B); @free (D7B); @free (D8B);

! Cost estimation for cross-plant piping;
!Capital cost estimation;
!Export streams of Network A;
PAB1=(2*CPAB1+250*BAB1)*D;
PAB2=(2*CPAB2+250*BAB2)*D;
PAB3=(2*CPAB3+250*BAB3)*D;
PAB4=(2*CPAB4+250*BAB4)*D;
PAB5=(2*CPAB5+250*BAB5)*D;
PAB6=(2*CPAB6+250*BAB6)*D;
PAB7=(2*CPAB7+250*BAB7)*D;
PAB8=(2*CPAB8+250*BAB8)*D;
PAB9=(2*CPAB9+250*BAB9)*D;

!Export streams of Network B;
PBA1=(2*CPBA1+250*BBA1)*D;
PBA2=(2*CPBA2+250*BBA2)*D;
PBA3=(2*CPBA3+250*BBA3)*D;
PBA4=(2*CPBA4+250*BBA4)*D;
PBA5=(2*CPBA5+250*BBA5)*D;
PBA6=(2*CPBA6+250*BBA6)*D;
PBA7=(2*CPBA7+250*BBA7)*D;
PBA8=(2*CPBA8+250*BBA8)*D;
PBA9=(2*CPBA9+250*BBA9)*D;
D=100; !Distance=100m;

!Binary variables that represent present/absent of a
streams;
!Network A;
CPAB1<=M*BAB1; CPAB2<=M*BAB2; CPAB3<=M*BAB3;
CPAB4<=M*BAB4; CPAB5<=M*BAB5;
```

```
CPAB6<=M*BAB6; CPAB7<=M*BAB7; CPAB8<=M*BAB8; CPAB9<=M*BAB9;

!Network B;
CPBA1<=M*BBA1; CPBA2<=M*BBA2; CPBA3<=M*BBA3; CPBA4<=M*BBA4;
CPBA5<=M*BBA5;
CPBA6<=M*BBA6; CPBA7<=M*BBA7; CPBA8<=M*BBA8; CPBA9<=M*BBA9;
M=1000;

!Defining binary variables;
!Network A;
@BIN (BAB1); @BIN (BAB2); @BIN (BAB3); @BIN (BAB4); @BIN
(BAB5);
@BIN (BAB6); @BIN (BAB7); @BIN (BAB8); @BIN (BAB9);
!Network B;
@BIN (BBA1); @BIN (BBA2); @BIN (BBA3); @BIN (BBA4); @BIN
(BBA5);
@BIN (BBA6); @BIN (BBA7); @BIN (BBA8); @BIN (BBA9);
END
```

The following is the solution report generated by LINGO:

```
Global optimal solution found.
Objective value:                      1526158.
Extended solver steps:                      3
Total solver iterations:                   278


            Variable           Value
            TAC          1526158.
            AOC          1504000.
            PC             95923.08
            FWA              20.00000
            FWB              16.00000
            WWA               0.000000
            WWB              16.00000
            REA1            120.0000
            REA2             50.00000
            PAB1              0.000000
            PAB2           37461.54
            PAB3              0.000000
            PAB4              0.000000
            PAB5              0.000000
            PAB6              0.000000
            PAB7              0.000000
            PAB8              0.000000
            PAB9              0.000000
            PBA1              0.000000
            PBA2              0.000000
            PBA3              0.000000
            PBA4              0.000000
            PBA5           31461.54
            PBA6              0.000000
            PBA7           27000.00
            PBA8              0.000000
            PBA9              0.000000
            D1A               0.000000
            D0A              20.00000
            CPBA1             0.000000
```

CPAB1	0.000000
D2A	107.6923
REGA	170.0000
CPBA2	0.000000
CPAB2	62.30769
D3A	−32.30769
CPBA3	0.000000
CPAB3	0.000000
D4A	−32.30769
CPBA4	0.000000
CPAB4	0.000000
D5A	0.000000
CPBA5	32.30769
CPAB5	0.000000
D6A	0.000000
CPBA6	0.000000
CPAB6	0.000000
D7A	10.00000
CPBA7	10.00000
CPAB7	0.000000
D8A	0.000000
CPBA8	0.000000
CPAB8	0.000000
D9A	0.000000
CPBA9	0.000000
CPAB9	0.000000
E1A	0.000000
E2A	0.000000
E3A	3.230769
E4A	1.615385
E5A	0.000000
E6A	0.000000
E7A	0.000000
E8A	1.500000
E9A	1.500000
E10A	1.500000
D1B	16.00000
D0B	16.00000
D2B	28.30769
D3B	−21.69231
D4B	−1.692308
D5B	36.00000
D6B	−34.00000
D7B	16.00000
D8B	16.00000
D9B	16.00000
E1B	0.000000
E2B	0.3200000
E3B	1.169231
E4B	0.8461538E−01
E5B	0.000000
E6B	1.800000
E7B	0.1000000
E8B	2.500000
E9B	8.900000
E10B	15996.10

```
BAB1              0.000000
D               100.0000
BAB2              1.000000
BAB3              0.000000
BAB4              0.000000
BAB5              0.000000
BAB6              0.000000
BAB7              0.000000
BAB8              0.000000
BAB9              0.000000
BBA1              0.000000
BBA2              0.000000
BBA3              0.000000
BBA4              0.000000
BBA5              1.000000
BBA6              0.000000
BBA7              1.000000
BBA8              0.000000
BBA9              0.000000
M              1000.000
```

The optimization result indicates that the TAC for the IPRCN in this case is $1.53 million. A simplified cascade diagram that consists of the material cascades of the individual networks is shown in Figure 16.6 (load cascades are omitted). As shown, Networks A and B consume freshwater flowrates of 20 and 16 t/h, respectively. The model also determines

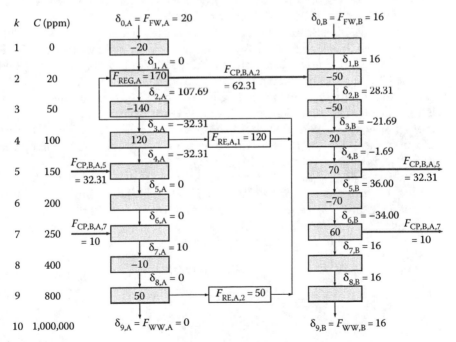

FIGURE 16.6
Result of ATM for Example 16.2 (flowrate in t/h; load cascades are not displayed).

that both water sources at 100 and 800 ppm of Network A are sent for regeneration for in-plant recovery. This leads to the complete removal of wastewater from Network A.

It is also interesting to note that three cross-plant streams are observed between both networks. At 20 ppm level, 62.31 t/h of water is exported from Network A to Network B. On the other hand, Network B also sends two export streams to Network A, at the levels of 150 and 250 ppm, respectively. These cross-plant streams contribute to the piping cost (PC) of $0.96 million (see the solution report of LINGO).

16.3 ATM for IPRCNs with Centralized Utility Facility

As discussed earlier, a CUF enables material recovery to be carried out between different RCNs indirectly. In most cases, the CUF supplies fresh resources to the RCNs, as well as purifies sources from the latter with centralized interception or waste treatment unit(s). The purified sources can be further recovered to the RCNs and/or emit as final waste discharge. Hence, separate ATM frameworks are needed for flowrate targeting in the CUF. Also, a revised ATM framework is needed for the individual RCNs, in order to track the individual sources to be sent for regeneration and/or waste treatment in the CUF. These ATM frameworks are shown in Figure 16.7. We notice that the framework in Figure 16.7a resembles that in Figure 16.1, except that the individual sources are now emitted as waste streams from the RCNs (with flowrate $F_{W,p,y}$), to be sent for regeneration or treatment in the CUF. Also, fresh resource and treated source (with flowrate $F_{TRD,p}$) are supplied from the CUF for each RCN. On the other hand, the ATM framework for the interception/treatment unit in the CUF in Figure 16.7b resembles that for *waste treatment network* in Figure 15.1, with the interception unit of the fixed outlet quality type. Note that the ATM framework for the CUF in Figure 16.7b may also take the form as that in Figure 15.2, where an interception unit of the fixed removal ratio type is used.

We may also view the ATM frameworks in Figure 16.7 as the extended versions of those used for *total material network* for a single RCN (see Sections 15.2 and 15.3). Hence, the various constraints used for the latter may be extended for use here. For the ATM framework in Figure 16.7a, the following constraints are imposed for each network p:

$$\delta_{k,p} = \delta_{k-1,p} + (\Sigma_i F_{SRi} - \Sigma_j F_{SKj})_{k,p} + (\Sigma_{p'} F_{CP,p',p} - \Sigma_{p'} F_{CP,p,p'})_k + F_{TRD,p} - F_{W,p,y}$$

$$k = 1, \ldots, n-1; \quad \forall p; p \neq p' \tag{16.19}$$

$$\delta_{k-1,p} = \begin{cases} \geq 0 & k = 1; \forall p \\ 0 & k = n; \forall p \end{cases} \tag{16.20}$$

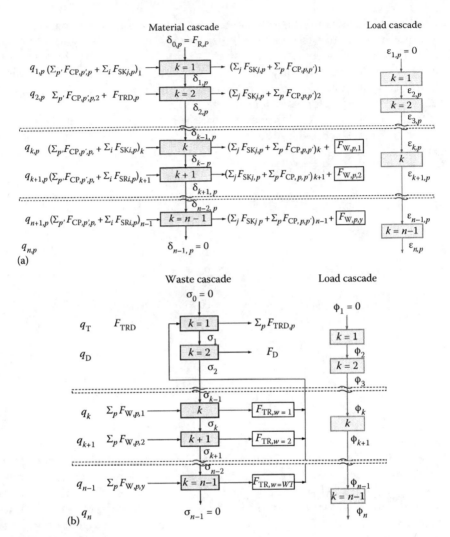

FIGURE 16.7
ATM frameworks for (a) indirect integration scheme and (b) CUF with fixed outlet quality type interception unit. (From *Chem. Eng. J.*, 153(1–3), Chew, I.M.L. and Foo, D.C.Y., Automated targeting for inter-plant water integration, 23–36, Copyright 2009, with permission from Elsevier.)

$$F_{W,p,y} \leq \Sigma_i F_{SRi,k,p} \quad \forall r; \forall p; \forall y \qquad (16.21)$$

Equation 16.19 is the revised form of Equation 16.12, where flowrates of the individual waste streams ($F_{W,p,y}$, added as sinks), as well as that of the purified flowrate from the CUF ($F_{TRD,p}$, added as source), are included for net material flowrate calculation. Next, Equation 16.20 is the extended form of

Equation 15.14, which indicates that the net material flowrates that enter the first quality level of each network p ($\delta_{0,p}$) should take non-negative values, while that to the last levels ($\delta_{n-1,p}$) is set to zero. The latter forbids waste discharge from the final quality level. The flowrate availability of each waste stream y is given by Equation 16.21. The constraints in Equations 16.2 and 16.4 are also needed for each network p.

On the other hand, ATM framework for the interception unit in the CUF in Figure 16.7b is described by the following constraints, which are essentially extended from those of ATM framework for waste treatment network (see Section 15.1, Equations 15.1 through 15.6):

$$\sigma_k = \sigma_{k-1} + F_{\mathrm{TRD}} - F_D - \Sigma_p F_{\mathrm{TRD},p} + \Sigma_p F_{\mathrm{W},p,y} - F_{\mathrm{TR},w} \quad k = 1, 2, \ldots, n-1 \quad (16.22)$$

$$\sigma_{k-1} = 0 \quad k = 1, n \quad (16.23)$$

$$F_{\mathrm{TRD}} = \Sigma_w F_{\mathrm{TR},w} \quad (16.24)$$

$$F_{\mathrm{TR},w} \leq \Sigma_p F_{\mathrm{W},p,y} \quad w = 1, 2, \ldots, WT \quad (16.25)$$

$$\varphi_k = \begin{cases} 0 & k = 1 \\ \varphi_{k-1} + \sigma_{k-1}\left(q_k - q_{k-1}\right) & k = 2, 3, \ldots, n \end{cases} \quad (16.26)$$

$$\varphi_k \geq 0 \quad k = 2, 3, \ldots, n \quad (16.27)$$

Equation 16.22 is a revised form of Equation 15.1, which describes the *net waste flowrate* of level k (σ_k) in the waste cascade (Figure 16.7b). The net waste flowrate takes into account the net waste flowrate cascaded from an earlier quality level (σ_{k-1}), the treated waste flowrate (F_{TRD}), terminal waste flowrate (F_D, if it is present), the treated source to be sent to network p ($F_{\mathrm{TRD},p}$), the total waste flowrates from network p ($\Sigma_p F_{\mathrm{W},p,y}$), as well as the individual treatment flowrate w ($F_{\mathrm{TR},w}$). Note that the treated source to be recycled to network p ($F_{\mathrm{TRD},p}$) is a new flowrate term added in Equation 16.22 (as compared to Equation 15.1). Next, Equation 16.23 states that no net waste flowrate enters the first (σ_0) and the last levels (σ_{n-1}) of the waste cascade. The flowrate balance for the interception/treatment unit in the CUF is described by Equation 16.24. Since the treatment flowrate originates from the individual waste stream (from network p), the maximum flowrate to be treated is bound by the maximum available flowrate, as described by Equation 16.25. Equation 16.26 indicates that the *residual waste load* (φ_k) at the first quality level has zero value, i.e., $\varphi_1 = 0$, while those in other quality levels are contributed by the residual load cascaded from previous level (φ_{k-1}), as well as by the net waste flowrate within each quality interval (given by the product of σ_{k-1} and the difference between two adjacent quality levels, i.e., $q_k - q_{k-1}$), represented by the last term in Equation 16.26. Equation

16.27 indicates that all the residual loads except those from level 1 (which is set to zero following Equation 16.26) must take non-negative values.

Note that for cases where more than one interception/treatment unit is present in the CUF (quite a common situation), the previous model will be written for each interception unit. In other words, all the variables in the ATM model will be tracked for each interception/treatment unit that is present. This is illustrated in the following example.

Example 16.3 Inter-plant water network for wafer fabrication plants

The water minimization problem for two wafer fabrication plants (Networks A and B) in Problem 10.4 is revisited here. In this case, a CUF is available to supply ultrapure water (UPW) for both the plants. The CUF consists of two interception units that can be used to purify process water sources. The intercepted sources can either be recovered to the wafer fabrication plants (regeneration) or be discharged as wastewater streams into the environment. The quality in concern for water recovery is resistivity (R), which is an index that reflects the total ionic content in aqueous streams (see Equation 10.1 for its property mixing rule). On the other hand, the environmental limit for heavy metal content in terminal wastewater discharge (C_{WW}) has been set to 2 ppm. The limiting data for all process sinks and sources of both networks, as well as the heavy metal content of all sources, are summarized in Tables 16.2 and 16.3 (reproduced from Tables 10.31 and 10.32); the performance and unit costs of the interception unit are given in Table 16.4. The UPW supply has a unit cost of \$2/ton (CT_{UPW}), with a resistivity value of 18 MΩ m (i.e., an operator value of 0.0556 (MΩ m)$^{-1}$), and does not contain any heavy metal. Other parameters for capital and total cost calculations (i.e., piping cost, distance, and parameters for annualizing factor) are assumed to be identical to those in Example 16.2. Determine the minimum TAC target for the IPRCN.

Solution

The cascade diagrams for both networks are shown in Figure 16.8. For the cascade diagram for Network A in Figure 16.8a, the water cascade across all quality levels is described by Equation 16.19:

TABLE 16.2

Limiting Data for Example 16.3: Network A

J	Sinks, SK_j	F_{SKj} (t/h)	ψ_{SKj} [MΩ m]$^{-1}$	i	Sources, SR_i	F_{SRi} (t/h)	ψ_{SRi} ([MΩ m]$^{-1}$)	C_{WW} (ppm)
1	Wet	500	0.143	1	Wet I	250	1.0000	5.0
2	Lithography	450	0.125	2	Wet II	200	0.5000	4.5
3	CMP	700	0.100	3	Lithography	350	0.3333	5.0
4	Etc	350	0.200	4	CMP I	300	10.000	10.0
				5	CMP II	200	0.5000	4.5
				6	Etc	280	2.0000	5.0

TABLE 16.3

Limiting Data for Example 16.3: Network B

J	Sinks, SK_j	F_{SKj} (t/h)	ψ_{SKj} ([MΩ m]$^{-1}$)	i	Sources, SR_i	F_{SRi} (t/h)	ψ_{SRi} ([MΩ m]$^{-1}$)	C_{WW} (ppm)
5	Wafer Fab	182	0.0625	7	50% spent rinse	227.12	0.1250	5.0
6	CMP	159	0.1	8	100% spent rinse	227.12	0.5000	11.0

TABLE 16.4

Outlet Stream Quality and Unit Cost for Interception Unit in Example 16.3

Interception Unit	Heavy Metal Content, C_T (ppm)	Resistivity, R_T (MΩ m)	Operator, ψ_T ([MΩ m]$^{-1}$)	Unit Cost, CT_{TRD} ($/ton)
I	2	5	0.2000	0.9
II	2	8	0.1250	1.5

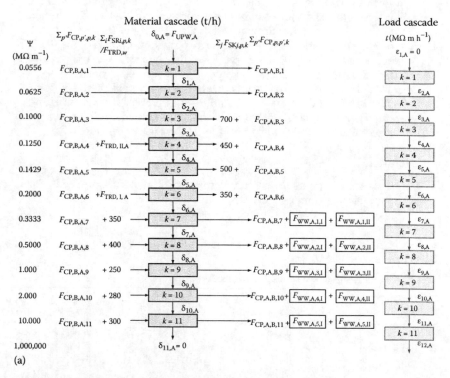

FIGURE 16.8

Cascade diagrams for Example 16.3: (a) Network A and (b) Network B.

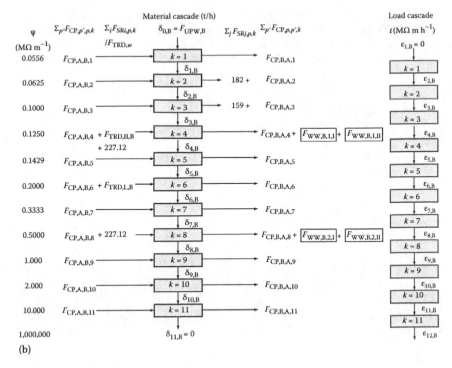

(b)

FIGURE 16.8 (continued)

$$\delta_{1,A} = \delta_{0,A} + (F_{CP,B,A,1} - F_{CP,A,B,1})$$

$$\delta_{2,A} = \delta_{1,A} + (F_{CP,B,A,2} - F_{CP,A,B,2})$$

$$\delta_{3,A} = \delta_{2,A} - 700 + (F_{CP,B,A,3} - F_{CP,A,B,3})$$

$$\delta_{4,A} = \delta_{3,A} - 450 + (F_{CP,B,A,4} - F_{CP,A,B,4}) + F_{TRD,II,A}$$

$$\delta_{5,A} = \delta_{4,A} - 500 + (F_{CP,B,A,5} - F_{CP,A,B,5})$$

$$\delta_{6,A} = \delta_{5,A} - 350 + (F_{CP,B,A,6} - F_{CP,A,B,6}) + F_{TRD,I,A}$$

$$\delta_{7,A} = \delta_{6,A} + 350 + (F_{CP,B,A,7} - F_{CP,A,B,7}) + F_{WW,A,1,I} + F_{WW,A,1,II}$$

$$\delta_{8,A} = \delta_{7,A} + 400 + (F_{CP,B,A,8} - F_{CP,A,B,8}) + F_{WW,A,2,I} + F_{WW,A,2,II}$$

$$\delta_{9,A} = \delta_{8,A} + 250 + (F_{CP,B,A,9} - F_{CP,A,B,9}) + F_{WW,A,3,I} + F_{WW,A,3,II}$$

$$\delta_{10,A} = \delta_{9,A} + 280 + (F_{CP,B,A,10} - F_{CP,A,B,10}) + F_{WW,A,4,I} + F_{WW,A,4,II}$$

$$\delta_{11,A} = \delta_{10,A} + 300 + (F_{CP,B,A,11} - F_{CP,A,B,11}) + F_{WW,A,5,I} + F_{WW,A,5,II}$$

Note that since two interception units (I and II) are available for use in the CUF, different flowrate terms are used to track the wastewater streams that are sent to these units, given by the last two terms in the last five sets of constraints.

Equation 16.20 states that the net material flowrate that enters the first quality level of each network p ($\delta_{0,p}$) should take non-negative values, while that to the last levels ($\delta_{n-1,p}$) is set to zero. Note that the former corresponds to the minimum UPW ($F_{UPW,A}$) needed by Network A:

$$\delta_{0,A} \geq 0$$

$$\delta_{11,A} = 0$$

$$\delta_{0,A} = F_{UPW,A}$$

The flowrate availability of each waste stream y is described by Equation 16.21:

$$F_{WW,A,1,I} + F_{WW,A,1,II} \leq 350$$

$$F_{WW,A,2,I} + F_{WW,A,2,II} \leq 400$$

$$F_{WW,A,3,I} + F_{WW,A,3,II} \leq 250$$

$$F_{WW,A,4,I} + F_{WW,A,4,II} \leq 280$$

$$F_{WW,A,5,I} + F_{WW,A,5,II} \leq 300$$

Likewise, the previous sets of constraints may be written for Network B, given as follows:

$$\delta_{1,B} = \delta_{0,B} + (F_{CP,A,B,1} - F_{CP,B,A,1})$$

$$\delta_{2,B} = \delta_{1,B} - 182 + (F_{CP,A,B,2} - F_{CP,B,A,2})$$

$$\delta_{3,B} = \delta_{2,B} - 159 + (F_{CP,A,B,3} - F_{CP,B,A,3})$$

$$\delta_{4,B} = \delta_{3,B} + 227.12 + (F_{CP,A,B,4} - F_{CP,B,A,4}) + F_{TRD,II,B} + F_{WW,B,1,I} + F_{WW,B,1,II}$$

$$\delta_{5,B} = \delta_{4,B} + (F_{CP,A,B,5} - F_{CP,B,A,5})$$

$$\delta_{6,B} = \delta_{5,B} + (F_{CP,A,B,6} - F_{CP,B,A,6}) + F_{TRD,I,B}$$

$$\delta_{7,B} = \delta_{6,B} + (F_{CP,A,B,7} - F_{CP,B,A,7})$$

$$\delta_{8,B} = \delta_{7,B} + 227.12 + (F_{CP,A,B,8} - F_{CP,B,A,8}) + F_{WW,B,2,I} + F_{WW,B,2,II}$$

$$\delta_{9,B} = \delta_{8,B} + (F_{CP,A,B,9} - F_{CP,B,A,9})$$

$$\delta_{10,B} = \delta_{9,B} + (F_{CP,A,B,10} - F_{CP,B,A,10})$$

$$\delta_{11,B} = \delta_{10,B} + (F_{CP,A,B,11} - F_{CP,B,A,11})$$

$$\delta_{0,B} \geq 0$$

$$\delta_{11,B} = 0$$

$$\delta_{0,B} = F_{UPW,B}$$

$$F_{WW,B,1,I} + F_{WW,B,1,II} \leq 227.12$$

$$F_{WW,B,2,I} + F_{WW,B,2,II} \leq 227.12$$

Apart from the aforementioned constraints, two other sets of constraints given by Equations 16.2 and 16.4 are also needed for each network p.

The constraints for inter-plant piping connections as well as those for cost calculation remain identical to those in Example 16.2. Both of these are omitted here for brevity.

We next describe the constraints needed for the interception units. As described earlier, separate cascade diagrams are needed for different interception units available for service in the CUF. Hence, two cascade diagrams are needed for interception units I and II, and are shown in Figure 16.9. Note that different indices have been added to indicate the variable sets for interception units I and II, respectively. Also, note that heavy metal content is used as the quality index for the waste cascades

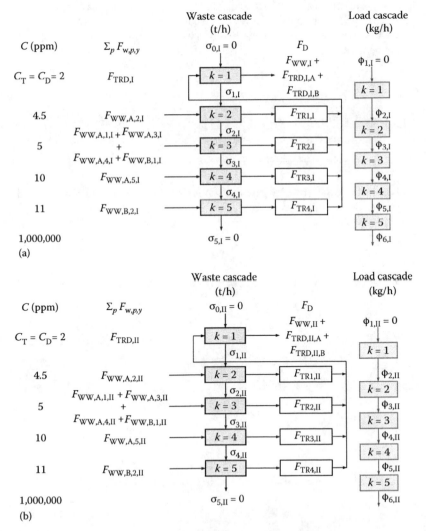

FIGURE 16.9
Cascade diagrams for CUF in Example 16.3 for (a) interception unit I and (b) interception unit II.

in Figure 16.9. In other words, the individual wastewater streams that emit from the material cascades in Figure 16.8 will now be tracked based on their heavy metal content, rather than their resistivity operator values. For instance, wastewater stream with flowrate $F_{WW,A,1,I}$ is observed to emerge from the operator level of 0.3333 (MΩ m)$^{-1}$ in Figure 16.8. From Table 16.2, it is found that this wastewater stream possesses a heavy metal content of 5 ppm. Hence, it is located at its corresponding concentration level in Figure 16.9 (along with other three wastewater streams with the same heavy metal content). Constraints for interception unit I (given by Equations 16.22 through 16.27) are next described.

Equation 16.22 describes the waste cascade across all quality levels. Note that in the first level, apart from the terminal wastewater stream ($F_{WW,I}$), two other flowrate terms are also added to enable the treated sources to be recycled to Networks A ($F_{TRD,I,A}$) and B ($F_{TRD,I,B}$), respectively:

$$\sigma_{1,I} = \sigma_{0,I} + F_{TRD,I} - F_{WW,I} - F_{TRD,I,A} - F_{TRD,I,B}$$

$$\sigma_{2,I} = \sigma_{1,I} + (F_{WW,A,2,I} - F_{TR1,I})$$

$$\sigma_{3,I} = \sigma_{2,I} + (F_{WW,A,1,I} + F_{WW,A,3,I} + F_{WW,A,4,I} + F_{WW,B,1,I} - F_{TR2,I})$$

$$\sigma_{4,I} = \sigma_{3,I} + (F_{WW,A,5,I} - F_{TR3,I})$$

$$\sigma_{5,I} = \sigma_{4,I} + (F_{WW,B,2,I} - F_{TR4,I})$$

Based on Equation 16.23, no net wastewater flowrate enters the first and the last levels of the waste cascade:

$$\sigma_{0,I} = 0$$

$$\sigma_{5,I} = 0$$

Equation 16.24 describes the flowrate balance of the interception unit:

$$F_{TRD,I} = F_{TR1,I} + F_{TR2,I} + F_{TR3,I} + F_{TR4,I}$$

Equation 16.25 indicates that the wastewater sources to be treated should not be more than what is available:

$$F_{TR1,I} \leq F_{WW,A,2,I}$$

$$F_{TR2,I} \leq F_{WW,A,1,I} + F_{WW,A,3,I} + F_{WW,A,4,I} + F_{WW,B,1,I}$$

$$F_{TR3,I} \leq F_{WW,A,5,I}$$

$$F_{TR4,I} \leq F_{WW,B,2,I}$$

Equation 16.26 next describes the waste load cascades (in kg/h) across all concentration levels:

$$\varphi_{1,I} = 0$$

$$\varphi_{2,I} = \varphi_{1,I} + \delta_{1,I}(4.5 - 2)/1,000$$

$$\varphi_{3,I} = \varphi_{2,I} + \delta_{2,I}(5 - 4.5)/1,000$$

$$\varphi_{4,I} = \varphi_{3,I} + \delta_{3,I}(10 - 5)/1,000$$

$$\varphi_{5,I} = \varphi_{4,I} + \delta_{4,I}(11 - 10)/1,000$$

$$\varphi_{6,I} = \varphi_{5,I} + \delta_{5,I}(1,000,000 - 11)/1,000$$

Based on Equation 16.27, all residual loads (except at level 1) should take non-negative values:

$$\varphi_{2,I} \geq 0; \quad \varphi_{3,I} \geq 0$$

$$\varphi_{4,I} \geq 0; \quad \varphi_{5,I} \geq 0$$

$$\varphi_{6,I} \geq 0$$

Constraints for interception unit II may be derived similar to those for unit I and, hence, are omitted for brevity. Note that the optimization objective for this example resembles that for Example 16.2, except that the operating cost (OC) term takes a slightly different form, i.e.,

$$OC(\$/h) = \Sigma_p F_{\text{UPW},p} CT_{\text{UPW}} + \Sigma_w F_{\text{TRD},w} CT_{\text{TRD}}$$

$$= (F_{\text{UPW,A}} + F_{\text{UPW,B}})(\$2/\text{ton}) + F_{\text{TRD,I}}(\$0.9/\text{ton}) + F_{\text{TRD,II}}(\$1.5/\text{ton})$$

In terms of LINGO formulation,* the previous MILP model can be written as follows:

```
Model:
! Optimisation objective: to determine minimum total
annualised cost (TAC);
min=TAC;
TAC=AOC+PC*0.231;
AOC=((UPWA+UPWB)*2+TRDI*0.9+TRDII*1.5)*8000;
PC=(PAB1+PAB2+PAB3+PAB4+PAB5+PAB6+PAB7+PAB8+ PAB9+PAB10
+PAB11+PBA1+PBA2+PBA3+PBA4+PBA5+PBA6+PBA7+PBA8+ PBA9+
PBA10+ PBA11);

! Constraints for Network A;
! Water cascade;
D1A=D0A+(CPBA1-CPAB1);
D2A=D1A+(CPBA2-CPAB2);
D3A=D2A-700+(CPBA3-CPAB3);
D4A=D3A-450+(CPBA4-CPAB4)+TRDIIA;
D5A=D4A-500+(CPBA5-CPAB5);
D6A=D5A-350+(CPBA6-CPAB6)+TRDIA;
D7A=D6A+350+(CPBA7-CPAB7)-WWA1I-WWA1II;
D8A=D7A+400+(CPBA8-CPAB8)-WWA2I-WWA2II;
```

---

* A full LINGO software is needed to solve this model, as the number of constraints has exceeded limit of the demo version.

```
D9A = D8A + 250 + (CPBA9 - CPAB9) - WWA3I - WWA3II;
D10A = D9A + 280 + (CPBA10 - CPAB10) - WWA4I - WWA4II;
D11A = D10A + 300 + (CPBA11 - CPAB11) - WWA5I - WWA5II;

! Net material flowrate that enters the first concentration
level should take
non-negative values, while that to the last level should
take zero value;
D0A >= 0;
D11A = 0;

! The net material flowrate entering the first level
corresponds to minimum ultrapure water (UPW) for the
network;
D0A = UPWA;

!Flowrate availability of wastewater;
WWA1I + WWA1II <= 350;
WWA2I + WWA2II <= 400;
WWA3I + WWA3II <= 250;
WWA4I + WWA4II <= 280;
WWA5I + WWA5II <= 300;

! Property load cascade ;
E1A = 0 ;
E2A = E1A + D1A*(0.0625 - 0.0556);
E3A = E2A + D2A*(0.1 - 0.0625);
E4A = E3A + D3A*(0.125 - 0.1);
E5A = E4A + D4A*(0.1429 - 0.125);
E6A = E5A + D5A*(0.2 - 0.1429);
E7A = E6A + D6A*(0.3333 - 0.2);
E8A = E7A + D7A*(0.5 - 0.3333);
E9A = E8A + D8A*(1 - 0.5);
E10A = E9A + D9A*(2 - 1);
E11A = E10A + D10A*(10 - 2);
E12A = E11A + D11A*(100 - 10);

! E2A-E12A should take non-negative values ;
E2A >= 0; E3A >= 0; E4A >= 0; E5A >= 0;
E6A >= 0; E7A >= 0; E8A >= 0; E9A >= 0;
E10A >= 0; E11A >= 0; E12A >= 0;

! D1A-D10A may take negative values ;
@free (D1A); @free (D2A); @free (D3A); @free (D4A);
@free (D5A);
@free (D6A); @free (D7A); @free (D8A); @free (D9A);
@free (D10A);

! Constraints for Network B;
! Water cascade;
D1B = D0B + (CPAB1 - CPBA1);
D2B = D1B - 182 + (CPAB2 - CPBA2);
D3B = D2B - 159 + (CPAB3 - CPBA3);
D4B = D3B + 227.12 + (CPAB4 - CPBA4) - WWB1I - WWB1II + TRDIIB;
D5B = D4B + (CPAB5 - CPBA5);
D6B = D5B + (CPAB6 - CPBA6) + TRDIB;
D7B = D6B + (CPAB7 - CPBA7);
```

```
D8B = D7B + 227.12 + (CPAB8 - CPBA8) - WWB2I - WWB2II;
D9B = D8B + (CPAB9 - CPBA9);
D10B = D9B + (CPAB10 - CPBA10);
D11B = D10B + (CPAB11 - CPBA11);
```

! Net material flowrate that enters the first concentration level should
take non-negative values, while that to the last level should take zero value;
```
D0B >= 0;
D11B = 0;
```

! The net material flowrate entering the first level corresponds to minimum ultrapure water (UPW) for the network;
```
D0B = UPWB;
```

!Flowrate availability of wastewater;
```
WWB1I + WWB1II <= 227.12;
WWB2I + WWB2II <= 227.12;
```

! Property load cascade ;
```
E1B = 0 ;
E2B = E1B + D1B*(0.0625 - 0.0556);
E3B = E2B + D2B*(0.1 - 0.0625);
E4B = E3B + D3B*(0.125 - 0.1);
E5B = E4B + D4B*(0.1429 - 0.125);
E6B = E5B + D5B*(0.2 - 0.1429);
E7B = E6B + D6B*(0.3333 - 0.2);
E8B = E7B + D7B*(0.5 - 0.3333);
E9B = E8B + D8B*(1 - 0.5);
E10B = E9B + D9B*(2 - 1);
E11B = E10B + D10B*(10 - 2);
E12B = E11B + D11B*(100 - 10);
```

! E2B-E12B should take non-negative values ;
```
E2B >= 0; E3B >= 0; E4B >= 0; E5B >= 0;
E6B >= 0; E7B >= 0; E8B >= 0; E9B >= 0;
E10B >= 0; E11B >= 0; E12B >= 0;
```

! D1B-D10B may take negative values ;
```
@free (D1B); @free (D2B); @free (D3B); @free (D4B);
@free (D5B);
@free (D6B); @free (D7B); @free (D8B); @free (D9B);
@free (D10B);
```

! Cost estimation for cross-plant piping;
!Capital cost estimation;
!Export streams of Network A;
```
PAB1 = (2*CPAB1 + 250*BAB1)*D;
PAB2 = (2*CPAB2 + 250*BAB2)*D;
PAB3 = (2*CPAB3 + 250*BAB3)*D;
PAB4 = (2*CPAB4 + 250*BAB4)*D;
PAB5 = (2*CPAB5 + 250*BAB5)*D;
PAB6 = (2*CPAB6 + 250*BAB6)*D;
PAB7 = (2*CPAB7 + 250*BAB7)*D;
PAB8 = (2*CPAB8 + 250*BAB8)*D;
PAB9 = (2*CPAB9 + 250*BAB9)*D;
```

```
PAB10 = (2*CPAB10 + 250*BAB10)*D;
PAB11 = (2*CPAB11 + 250*BAB11)*D;

!Export streams of Network B;
PBA1 = (2*CPBA1 + 250*BBA1)*D;
PBA2 = (2*CPBA2 + 250*BBA2)*D;
PBA3 = (2*CPBA3 + 250*BBA3)*D;
PBA4 = (2*CPBA4 + 250*BBA4)*D;
PBA5 = (2*CPBA5 + 250*BBA5)*D;
PBA6 = (2*CPBA6 + 250*BBA6)*D;
PBA7 = (2*CPBA7 + 250*BBA7)*D;
PBA8 = (2*CPBA8 + 250*BBA8)*D;
PBA9 = (2*CPBA9 + 250*BBA9)*D;
PBA10 = (2*CPBA10 + 250*BBA10)*D;
PBA11 = (2*CPBA11 + 250*BBA11)*D;
D = 100;  !Distance = 100m;

!Binary variables that represent present/absent of a
streams;
!Network A;
CPAB1 <= M*BAB1; CPAB2 <= M*BAB2; CPAB3 <= M*BAB3; CPAB4 <= M*BAB4;
CPAB5 <= M*BAB5; CPAB6 <= M*BAB6; CPAB7 <= M*BAB7; CPAB8 <= M*BAB8;
CPAB9 <= M*BAB9; CPAB10 <= M*BAB10; CPAB11 <= M*BAB11;

!Network B;
CPBA1 <= M*BBA1; CPBA2 <= M*BBA2; CPBA3 <= M*BBA3; CPBA4 <= M*BBA4;
CPBA5 <= M*BBA5; CPBA6 <= M*BBA6; CPBA7 <= M*BBA7; CPBA8 <= M*BBA8;
CPBA9 <= M*BBA9; CPBA10 <= M*BBA10; CPBA11 <= M*BBA11;
M = 1000;

!Defining binary variables;
!Network A;
@BIN (BAB1); @BIN (BAB2); @BIN (BAB3); @BIN (BAB4);
@BIN (BAB5);
@BIN (BAB6); @BIN (BAB7); @BIN (BAB8); @BIN (BAB9);
@BIN (BAB10); @BIN (BAB11);

!Network B;
@BIN (BBA1); @BIN (BBA2); @BIN (BBA3); @BIN
(BBA4); @BIN (BBA5);
@BIN (BBA6); @BIN (BBA7); @BIN (BBA8); @BIN (BBA9);
@BIN (BBA10); @BIN (BBA11);

! Constraints for CUF ;
! Interception unit I;
! Waste cascade;
S1I = S0I + TRDI - WWI - TRDIA - TRDIB ;
S2I = S1I + (WWA2I - TR1I) ;
S3I = S2I + (WWA1I + WWA3I + WWA4I + WWB1I - TR2I) ;
S4I = S3I + (WWA5I - TR3I) ;
S5I = S4I + (WWB2I - TR4I) ;

! No net waste flowrate enters the first and the last
levels;
S0I = 0 ;
S5I = 0 ;

! Flowrate balance(s) for treatment unit(s) ;
TRDI = TR1I + TR2I + TR3I + TR4I ;
```

```
! Flowrate constraints of treated wastewater source ;
TR1I <= WWA2I ;
TR2I <= WWA1I + WWA3I + WWA4I + WWB1I ;
TR3I <= WWA5I ;
TR4I <= WWB2I ;

! Waste load cascade (in kg/h) ;
J1I = 0 ;
J2I = J1I + S1I*(4.5-2)/1000 ;
J3I = J2I + S2I*(5-4.5)/1000 ;
J4I = J3I + S3I*(10-5)/1000 ;
J5I = J4I + S4I*(11-10)/1000 ;
J6I = J5I + S5I*(1000000-11)/1000 ;

! J2I-J6I should take non-negative values ;
J2I >= 0 ; J3I >= 0 ; J4I >= 0 ;
J5I >= 0 ; J6I >= 0 ;
! S1I-S4I may take negative values ;
@free (S1I) ; @free (S2I) ;
@free (S3I) ; @free (S4I) ;

! Interception unit II;
! Waste cascade;
S1II = S0II + TRDII - WWII - TRDIIA - TRDIIB ;
S2II = S1II + (WWA2II - TR1II) ;
S3II = S2II + (WWA1II + WWA3II + WWA4II + WWB1II - TR2II) ;
S4II = S3II + (WWA5II - TR3II) ;
S5II = S4II + (WWB2II - TR4II) ;

! No net waste flowrate enters the first and the last
levels;
S0II = 0 ;
S5II = 0 ;

! Flowrate balance(s) for treatment unit(s) ;
TRDII = TR1II + TR2II + TR3II + TR4II ;

! Flowrate constraints of treated wastewater source ;
TR1II <= WWA2II ;
TR2II <= WWA1II + WWA3II + WWA4II + WWB1II ;
TR3II <= WWA5II ;
TR4II <= WWB2II ;

! Waste load cascade (in kg/h) ;
J1II = 0 ;
J2II = J1II + S1II*(4.5-2)/1000 ;
J3II = J2II + S2II*(5-4.5)/1000 ;
J4II = J3II + S3II*(10-5)/1000 ;
J5II = J4II + S4II*(11-10)/1000 ;
J6II = J5II + S5II*(1000000-11)/1000 ;

! J2II-J6II should take non-negative values ;
J2II >= 0 ; J3II >= 0 ; J4II >= 0 ;
J5II >= 0 ; J6II >= 0 ;

! S1II-S4II may take negative values ;
@free (S1II) ; @free (S2II) ;
@free (S3II) ; @free (S4II) ;
END
```

The following is the solution report generated by LINGO:

```
Global optimal solution found.
Objective value:                 0.2621726E+08
Extended solver steps:                       0
Total solver iterations:                    45

              Variable        Value
              TAC          0.2621726E+08
              AOC          0.2620653E+08
              PC           46460.31
              UPWA         252.1614
              UPWB         221.1816
              TRDI         635.9163
              TRDII        1171.204
              PAB1          0.000000
              PAB2          0.000000
              PAB3          0.000000
              PAB4          0.000000
              PAB5          0.000000
              PAB6          0.000000
              PAB7          0.000000
              PAB8          0.000000
              PAB9          0.000000
              PAB10         0.000000
              PAB11         0.000000
              PBA1          0.000000
              PBA2          0.000000
              PBA3          0.000000
              PBA4         46460.31
              PBA5          0.000000
              PBA6          0.000000
              PBA7          0.000000
              PBA8          0.000000
              PBA9          0.000000
              PBA10         0.000000
              PBA11         0.000000
              D1A          252.1614
              D0A          252.1614
              CPBA1         0.000000
              CPAB1         0.000000
              D2A          252.1614
              CPBA2         0.000000
              CPAB2         0.000000
              D3A         -447.8386
              CPBA3         0.000000
              CPAB3         0.000000
              D4A          380.6667
              CPBA4        107.3016
              CPAB4         0.000000
              TRDIIA       1171.204
              D5A         -119.3333
              CPBA5         0.000000
              CPAB5         0.000000
              D6A           0.000000
              CPBA6         0.000000
```

CPAB6	0.000000
TRDIA	469.3333
D7A	0.000000
CPBA7	0.000000
CPAB7	0.000000
WWA1I	350.0000
WWA1II	0.000000
D8A	0.000000
CPBA8	0.000000
CPAB8	0.000000
WWA2I	58.79627
WWA2II	341.2037
D9A	0.000000
CPBA9	0.000000
CPAB9	0.000000
WWA3I	0.000000
WWA3II	250.0000
D10A	0.000000
CPBA10	0.000000
CPAB10	0.000000
WWA4I	0.000000
WWA4II	280.0000
D11A	0.000000
CPBA11	0.000000
CPAB11	0.000000
WWA5I	0.000000
WWA5II	300.0000
E1A	0.000000
E2A	1.739914
E3A	11.19597
E4A	0.000000
E5A	6.813933
E6A	0.000000
E7A	0.000000
E0A	0.000000
E9A	0.000000
E10A	0.000000
E11A	0.000000
E12A	0.000000
D1B	221.1816
D0B	221.1816
D2B	39.18156
D3B	−119.8184
D4B	0.000000
WWB1I	0.000000
WWB1II	0.000000
TRDIIB	0.000000
D5B	0.000000
D6B	0.000000
TRDIB	0.000000
D7B	0.000000
D8B	0.000000
WWB2I	227.1200
WWB2II	0.000000
D9B	0.000000
D10B	0.000000

```
D11B                    0.000000
E1B                     0.000000
E2B                     1.526153
E3B                     2.995461
E4B                     0.000000
E5B                     0.000000
E6B                     0.000000
E7B                     0.000000
E8B                     0.000000
E9B                     0.000000
E10B                    0.000000
E11B                    0.000000
E12B                    0.000000
BAB1                    0.000000
D                     100.0000
BAB2                    0.000000
BAB3                    0.000000
BAB4                    0.000000
BAB5                    0.000000
BAB6                    0.000000
BAB7                    0.000000
BAB8                    0.000000
BAB9                    0.000000
BAB10                   0.000000
BAB11                   0.000000
BBA1                    0.000000
BBA2                    0.000000
BBA3                    0.000000
BBA4                    1.000000
BBA5                    0.000000
BBA6                    0.000000
BBA7                    0.000000
BBA8                    0.000000
BBA9                    0.000000
BBA10                   0.000000
BBA11                   0.000000
M                    1000.000
S1I                     0.000000
S0I                     0.000000
WWI                   166.5829
S2I                     0.000000
TR1I                   58.79627
S3I                     0.000000
TR2I                  350.0000
S4I                     0.000000
TR3I                    0.000000
S5I                     0.000000
TR4I                  227.1200
J1I                     0.000000
J2I                     0.000000
J3I                     0.000000
J4I                     0.000000
J5I                     0.000000
J6I                     0.000000
S1II                    0.000000
```

S0II	0.000000
WWII	0.000000
S2II	0.000000
TR1II	341.2037
S3II	0.000000
TR2II	530.0000
S4II	0.000000
TR3II	300.0000
S5II	0.000000
TR4II	0.000000
J1II	0.000000
J2II	0.000000
J3II	0.000000
J4II	0.000000
J5II	0.000000
J6II	0.000000

The optimization result indicates that the TAC for the IPRCN in this case is \$26.2 million. The simplified cascade diagrams that consist of the material cascades of individual networks, as well as the waste cascades for both interception units, are shown in Figure 16.10 (load cascades are omitted for both cases). As shown, Networks A and B consume 252.16 ($F_{\text{UPW,A}}$) and 221.18 t/h ($F_{\text{UPW,B}}$) of UPW, respectively. An inter-plant connection is found at the operator level of 0.125 (MΩ m)$^{-1}$, which transfers 107.3 t/h of water from Network B to Network A. The model also determines the water source flow rates to be sent for interception in the CUF, for further recovery to Network A ($F_{\text{TRD,I,A}} = 1171.20$ t/h; $F_{\text{TRD,II,A}} = 469.33$ t/h), as well as for final waste discharge ($F_{\text{WW,I}} = 166.58$ t/h).

16.4 Insights from ATM for IPRCN Synthesis

In earlier chapters, it has been demonstrated how insights from ATM can be used for generating the network structure for an RCN. Similar approach can also be applied for IPRCN. However, care should be taken when more complicated cases are involved, such as those in IPRCN. In particular, when a negative net material flowrate is observed in the material cascade, it does not necessarily mean that the material flowrate is cascaded to the level immediately above the level where the source exists (but instead to the higher quality region; see Section 4.1 for a detailed discussion). Hence, it is always essential to do a cross-check to ensure that the sinks are allocated with sources that will fulfill both its flowrate and load constraints. This can be done using Equations 16.28 and 16.29 as follows:

$$F_{SRi,SKj} + F_{SRi+1,SKj} = F_{SKj} \tag{16.28}$$

$$F_{SRi,SKj}\, q_{SRi} + F_{SRi+1,SKj}\, q_{SRi+1} = F_{SKj}\, q_{SKj} \tag{16.29}$$

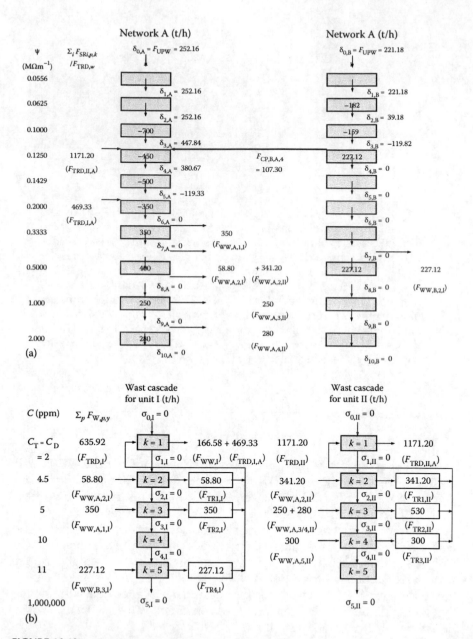

FIGURE 16.10
Result of ATM for Example 16.3: (a) material cascades for Networks A and B and (b) waste cascades for interception units I and II in CUF (load cascades are not displayed).

where $F_{SRi,SKj}$ is the allocation flowrate sent from SR_i to SK_j. Note that Equations 16.28 and 16.29 are essentially the same as those in *nearest neighbor algorithm* (NNA) for a detailed network synthesis (see Equations 7.1 and 7.2). However, for more complicated cases, the IPRCN should be synthesized with NNA completely. Insights from targeting can be incorporated to obtain an accurate solution (e.g., see Section 10.3 or for illustration).

Example 16.4 Generation of IPRCN structure from ATM insights

Rework the water minimization problem in Example 16.3. Generate the structure of the IPRCN using the insights of the ATM.

Solution

Since this example involves waste treatment section, it is more convenient to start the task from this section. From Figure 16.10a, we observe that all process sources of Network A are sent to either interception units I or II (between operation levels 0.3333–10 $(M\Omega\ m)^{-1}$). In other words, none of these sources is directly reused/recycled to the sinks in Network A. On the other hand, source SR8 of Network B is also sent for purification in interception unit I. Also, a portion of source SR7 is sent as a cross-plant stream to sink SK2 in Network A at the same operator level (i.e., 0.125 $(M\Omega\ m)^{-1}$ level). From Figure 16.10b, we then observe that all sources sent to the interception units are purified either for further recovery to Networks A or B ($F_{TRD,I,A}$ and $F_{TRD,II,A}$) or for final wastewater discharge ($F_{WW,I}$—from interception unit I). These matches are shown in Figure 16.11.

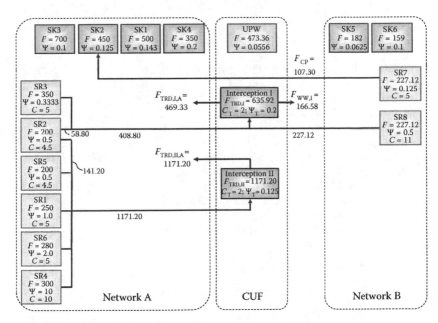

FIGURE 16.11
Preliminary network structure for IPRCN in Example 16.3.

Next, we focus on the material cascade of Networks A and B, as given in Figure 16.10a. For Network A, 252.16 t/h of UPW is cascaded from the highest operator level and is used by sink SK3 ($F_{SK3}=700$) at the operator level of 0.1 $(M\Omega\ m)^{-1}$. Due to the insufficiency of flowrate, a portion of the flowrate requirement of SK3 is to be fulfilled by flowrates (447.84 t/h) allocated from lower operator level, i.e., 0.125 $(M\Omega\ m)^{-1}$. This is indicated by the negative net material flowrate ($\delta_{3,A}$) between levels 0.1 and 0.125 $(M\Omega\ m)^{-1}$. Equations 16.28 and 16.29 are then used to ensure the allocated flowrate of the sources that will fulfill the flowrate and load requirements of SK3:

$$252.16 + 447.84 = 700\,t/h$$

$$252.16(0.0556) + 447.84(0.125) = 700\,t/h(0.1)$$

At the level of 0.125 $(M\Omega\ m)^{-1}$, where sink SK2 ($F_{SK2}=450$) is present, the purified source from interception unit II ($F_{TRD,II,A}$) as well as the cross plant stream from Network B are also found. Apart from allocating flowrate to sinks at a higher level (SK3, 0.1 $(M\Omega\ m)^{-1}$), these sources also fulfill the flowrate requirement of SK2 at the same level. Besides, the leftover flowrates (($\delta_{4,A}=380.67$ t/h) are cascaded to lower operator level, i.e., 0.1492 $(M\Omega\ m)^{-1}$). At the latter level, where SK1 exists, the cascade flowrate from 0.125 $(M\Omega\ m)^{-1}$ is insufficient for use and, hence, calls for 119.33 t/h of flowrate from lower level, i.e., 0.2 $(M\Omega\ m)^{-1}$, indicated by the negative net material flowrate ($\delta_{5,A}$) between levels 0.1429 and 0.2 $(M\Omega\ m)^{-1}$. Similarly, Equations 16.28 and 16.29 are used to ensure that the flowrate and load requirements of SK1 are fulfilled by the allocated sources:

$$380.67 + 119.33 = 500\,t/h$$

$$380.67(0.125) + 119.33(0.2) = 500(0.1429)\,t/h\ M\Omega\ m$$

At the operator level of 0.2 $(M\Omega\ m)^{-1}$, where SK4 ($\delta_{5,A}$) and the purified source from interception unit I ($F_{TRD,II,A}$) are present, the latter is used to fulfill the flowrate requirement of SK4, while its excessive flowrate is allocated to higher operator level (0.1492 $(M\Omega\ m)^{-1}$).

Similar exercises may also be carried out for Network B. The UPW ($F_{UPW,B}=221.18$ t/h) is cascaded to the operator level of 0.0625 $(M\Omega\ m)^{-1}$), where sink SK5 is present. Since this sink only requires 182 t/h of UPW, the leftover UPW flowrate ($\delta_{2,B}=39.18$ t/h) is then cascaded to SK6 at the level of 0.1 $(M\Omega\ m)^{-1}$). However, SK6 has a flowrate requirement of 159 t/h, which calls for flowrate allocation by source SR7 at lower operator level (indicated by the negative net material flowrate, $\delta_{3,B}$). However, the verification with Equations 16.28 and 16.29 indicates that the operator load balance does not hold:

$$39.18 + 119.82 = 159\,t/h$$

$$39.18(0.556) + 119.82(0.125) \neq 159(0.1)\,t/h\ M\Omega\ m$$

In order to restore flowrate and load feasibility of SK6, the allocated source flowrates from UPW and SR7 to sinks SK5 and SK6 are to be adjusted. One obvious solution is to allocate both SK5 and SK6 with UPW and SR7. Hence, solving Equations 16.28 and 16.29 yields the allocated flowrates as follows:

$$F_{UPW,SK5} = 163.9\,t/h;\; F_{SR7,SK5} = 18.1\,t/h$$

$$F_{UPW,SK6} = 57.28\,t/h;\; F_{SR7,SK6} = 101.72\,t/h$$

As described earlier, sources in both networks at an operator level lower than 0.3333 $(M\Omega\ m)^{-1}$ are sent for purification in the interception units.

A complete network structure for the IPRCN can then be developed based on the insights of ATM as described earlier and is shown in Figure 16.12. As shown, all sinks in Network A receive purified sources from both interception units, while those in Network B receive direct reuse/recycle water from SR7. Besides, a portion of the purified source is discharged as wastewater stream from interception unit I in the CUF (with $F_{WW,I} = 166.58\,t/h$; and $C_{WW} = C_T = 2\,ppm$). We also observe that this network structure is a hybrid form of direct (with cross plant stream) and indirect integration schemes (with the presence of a CUF).

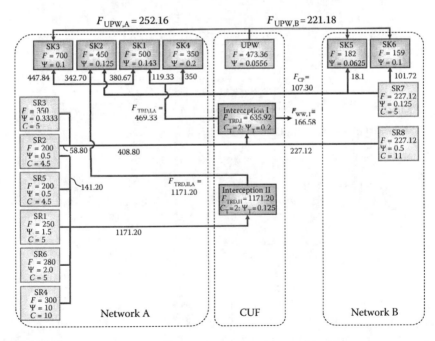

FIGURE 16.12
Complete network structure for IPRCN in Example 16.3. (From *Chem. Eng. J.*, 153(1–3), Chew, I.M.L. and Foo, D.C.Y., Automated targeting for inter-plant water integration, 23–36, Copyright 2009, with permission from Elsevier.)

16.5 Further Reading

Apart from ATM, several other mathematical optimization approaches have been proposed recently by various researchers. Most of these models are based on superstructural approaches (which has been discussed briefly in Chapter 12). These include the work of Chew et al. (2008), Lovelady and El-Halwagi (2009), Lovelady et al. (2009), Rubio-Castro et al. (2010, 2011), Chen et al. (2010), Aviso et al. (2010a,b, 2011).

Problems

Water Minimization Problems

16.1 An integrated iron and steel mill that consist of multiple water networks are analyzed for their water saving potential (Chew and Foo, 2009). The limiting water data for its individual network are shown in Table 16.5. As shown, there are five water networks in the mill, which

TABLE 16.5

Limiting Water Data for Problem 16.1

Sinks, SK$_j$	F_{SKj} (mil m³/y)	C_{SKj} (mg/L)	Sources, SR$_i$	F_{SRi} (mil m³/y)	C_{SRi} (mg/L)
Network A—Raw material storage yard					
SK1	10.00	20	SR1	9.00	23
Network B—Cooking plant					
SK2	12.29	20	SR2	11.92	23
SK3	12.29	19	SR3	11.92	23
Network C—Steel making plant					
SK4	59.60	75	SR4	57.81	100
SK5	39.73	80	SR5	38.54	100
Network D—Casting/rolling mills					
SK6	198.66	20	SR6	192.70	20.5
SK7	198.66	20	SR7	192.70	20.5
SK8	44.73	20	SR8	43.39	21
SK9	178.92	20	SR9	173.55	20.5
SK10	44.73	100	SR10	43.39	400
Network E—Indirect cooling					
SK11	468.55	20	SR11	459.18	20.2
$\Sigma_j F_{SKj}$	1268.16		$\Sigma_i F_{SRi}$	1234.10	

are segregated based on the different processing areas of the plant, i.e., raw material storage yard, cooking plant, steel making plant, casting/ rolling mills, and indirect cooling. The main contaminant of concern for water recovery is the ionic chloride content in the water. In this case, the freshwater supply contains a Cl⁻ level (C_{FW}) of 15 mg/L. Determine the minimum overall freshwater and cross-plant flowrates using the two-step optimization approach.

16.2 Rework the inter-plant water minimization problem in Example 10.2 (Olesen and Polley, 1996). Determine the minimum freshwater and cross-plant flowrates for direct integration scheme using the two-step optimization approach.

16.3 In an eco-industrial park, five process plants are involved in sharing water resources via an inter-plant water network (Lovelady and El-Halwagi, 2009). Limiting data for the water sinks and sources of these plants are given in Table 16.6. freshwater resource that is free from impurity content (i.e., $C_{FW} = 0$ ppm, unit cost $CT_{FW} = \$\,0.6/\text{ton}$) is available for use, when water sources are insufficient. A CUF is available for service, which consists of an interception unit that can regenerate water sources to 500 ppm. The unit cost of regenerated water sources is taken as \$0.05/kg impurity load removal (use Equation 14.37 for calculation of impurity load removal). The AOT is assumed to be 8760 h. Solve the following tasks:

a. Determine the MOC for this IPRCN, which is mainly contributed by freshwater and regeneration costs.

TABLE 16.6

Limiting Water Data for Problem 16.3

Sinks, SK$_j$	F_{SKj} (t/h)	C_{SKj} (ppm)	Sources, SR$_i$	F_{SRi} (t/h)	C_{SRi} (ppm)
Plant A					
SK1	3500	500	SR1	3000	50
Plant B					
SK2	1000	760	SR2	1000	400
Plant C					
SK3	2500	40	—	—	—
Plant D					
—	—	—	SR3	3000	1100
Plant E					
SK4	2000	225	SR4	1300	1600
—	—	—	SR5	200	1800
$\Sigma_j F_{SKj}$	9000		$\Sigma_i F_{SRi}$	8500	

b. What is the minimum cross-plant flowrate associated with the MOC in part (a) (hint: use the two-stage approach to identify the solution)?

c. Generate the network structure of IPRCN with insights gained from ATM. Does the IPRCN involve direct or indirect integration scheme?

16.4 Rework the inter-plant water network problem in Example 16.2. Determine the minimum TAC target for the IPRCN for the following cases:

a. The interception unit (with outlet concentration of 20 ppm) is installed in Network B instead of Network A for the purification of water sources. Determine the minimum TAC target for the IPRCN, assuming that all other parameters remain identical to those in Example 16.2.

b. Assume that a CUF is available for service. The CUF consists of an interception unit (with an outlet concentration of 50 ppm) that can regenerate water sources for recovery to both Networks A and B, as well as to comply with an environmental limit for final discharge (given as 100 ppm). The unit costs for freshwater (CT_{FW}) and regenerated source (CT_{TRD}) are given as \$1/ton and \$0.8/ton, respectively. Other parameters for capital and total cost calculation (i.e., piping cost, distance, and annualizing factor) are assumed to be identical to those in Example 16.2.

16.5 Generate the network structure of IPRCN for Examples 16.1* and 16.2 with insights gained from ATM. Take note of the load feasibility during flowrate allocations.

Utility Gas Recovery Problems

16.6 Rework the inter-plant hydrogen network case in Problem 10.3. Determine the MOC of the IPRCN, which is mainly contributed by the fresh hydrogen and fuel saving costs (from purge gas). The unit cost for fresh hydrogen (CT_{FH}) is given as \$0.002/mol, while fuel savings can be determined from the value of 0.00085/mol of hydrogen gas.

Property Integration Problems

16.7 Rework the IPRCN problem in Example 16.3. In this case, only consider direct integration and reuse/recycle scheme between the two wafer fabrication plants. Determine the minimum UPW and cross-plant flowrates using the two-step optimization approach. Generate the network structure of IPRCN with insights gained from the ATM.

---

* Compare the network structure with that synthesized by NNA in Example 10.3.

16.8 Rework the wafer fabrication problem in Problem 4.19, with the limiting data given in Table 4.32. In this case, assume that the plant is segregated into two different sections, i.e., the wafer fabrication section (FAB) and other processes (sinks SK5, SK6, and SK7). Carry out flowrate targeting for the following scenarios:

a. Target the minimum UPW flowrate needed for the FAB section alone (the formulation is similar to that given in Example 16.3). Also identify the flowrates of the reject streams from reverse osmosis (RO) and ultrafiltration (UF) units in the pretreatment system.

b. Target the minimum freshwater flowrates needed for other processes (SK5–SK7). Apart from the municipal freshwater feed, the UF and RO rejects from the pretreatment system may also be used here, in which they may be viewed as process sources for this section. However, they have limited flowrates (as a result of part (a)).

c. A membrane unit with a recovery factor of 0.95 (see Equation 11.3) is installed to purify sources in the FAB section for further recovery. The permeate stream ($15\,M\Omega$) of the unit is to be sent to the FAB section, while its reject stream is used for other processes. Determine the minimum fresh resource flowrates for both sections, as well as the minimum treatment flowrate (hint: determine the resistivity operator of the product and reject streams of the membrane unit using Equation 2.64).

References

Aviso, K. B., Tan, R. R., and Culaba, A. B. 2010a. Designing eco-industrial water exchange networks using fuzzy mathematical programming. *Clean Technologies and Environmental Policy*, 12, 353–363.

Aviso, K. B., Tan, R. R., Culaba, A. B., and Cruz, J. B. 2010b. Bi-level fuzzy optimization approach for water exchange in eco-industrial parks. *Process Safety and Environmental Protection*, 88, 31–40.

Aviso, K. B., Tan, R. R., Culaba, A. B., Foo, D. C. Y., and Hallale, N. 2011. Fuzzy optimization of topologically constrained eco-industrial resource conservation networks with incomplete information. *Engineering Optimisation*, 43(3), 257–279.

Chen, C.-L., Hung, S.-W., and Lee, J.-Y. 2010. Design of inter-plant water network with central and decentralized water mains. *Computers and Chemical Engineering*, 34, 1522–1531.

Chew, I. M. L. and Foo, D. C. Y. 2009. Automated targeting for inter-plant water integration. *Chemical Engineering Journal*, 153(1–3), 23–36.

Chew, I. M. L., Tan, R., Ng, D. K. S., Foo, D. C. Y., Majozi, T., and Gouws, J. 2008. Synthesis of direct and indirect interplant water network. *Industrial and Engineering Chemistry Research*, 47(23), 9485–9496.

Lovelady, E. M. and El-Halwagi, M. M. 2009. Design and integration of eco-industrial parks for managing water resources. *Environmental Progress and Sustainable Energy*, 28(2), 265–272.

Lovelady, E. M., El-Halwagi, M. M., Chew, I. M. L., Ng, D. K. S., Foo, D. C. Y., and Tan, R. R. 2009. A property-integration approach to the design and integration of eco-industrial parks. *7th International Conference on Foundations of Computer-Aided Process Design (FOCAPD)*, Breckenridge, CO, pp. 559–567, June 7–12.

Olesen, S. G. and Polley, G. T. 1996. Dealing with plant geography and piping constraints in water network design. *Process Safety and Environmental Protection*, 74, 273–276.

Rubio-Castro, E., Ponce-Ortega, J. M., Nápoles-Rivera, F., El-Halwagi, M. M., Serna-González, M., and Jiménez-Gutiérrez, A. 2010. Water integration of eco-industrial parks using a global optimization approach. *Industrial and Engineering Chemistry Research*, 49, 9945–9960.

Rubio-Castro, E., Ponce-Ortega, J. M., Serna-González, M., Jiménez-Gutiérrez, A., and El-Halwagi, M. M. 2011. A global optimal formulation for the water integration in eco-industrial parks considering multiple pollutants. *Computers and Chemical Engineering*, 35, 1558–1574.

17

Automated Targeting Model for Batch Material Networks

As discussed in Chapter 11, both quality and time constraints are to be considered in synthesizing a batch material network (BMN). To overcome time constraint, mass storage system is commonly employed to recover material across different time intervals. Automated targeting model (ATM) developed for resource conservation networks (RCNs) for continuous processes in the previous chapters are extended for BMNs in this chapter.

17.1 Basic ATM Procedure for Batch Material Networks

The basic procedure for performing flow/cost targeting in a BMN using the ATM is similar to that in Chapter 11 using insight-based methods, outlined as follows:

1. Define the time intervals for the BMN based on the start and end times of the process sinks and sources.
2. Locate the process sinks and sources in their respective time intervals, and determine their interval flows, based on the ratio of the duration of the time interval to that of the sink/source (see Example 11.1 for the details of this step).
3. Perform ATM for the BMN.

The following sections of this chapter outline the various ATMs for direct reuse/recycle, regeneration, as well as waste treatment for a BMN. Note that the various ATMs are essentially extensions of the variants found in Chapters 13 through 15 for continuous processes. Hence, the readers are advised to refer to these earlier chapters for details.

17.2 ATM for Direct Reuse/Recycle Network

The ATM framework for direct reuse/recycle scheme for a BMN is shown in Figure 17.1. Note that the framework resembles that of the direct reuse/ recycle scheme for a continuous process (Figure 13.1). However, the process sinks and sources now take the unit of material flow, instead of flowrates in continuous processes. Note also that a new subscript t is added to each flow term, denoting the time interval that it belongs to. Also, two new flow terms are added to the framework, i.e., stored material from earlier time interval $t-1$ (with flow $F_{ST,t-1,t}$), and the material to be stored for later time interval $t+1$ (with flow $F_{ST,t,t+1}$). As shown in Figure 17.1, the stored material from earlier time interval ($F_{ST,t-1,t}$) is added as a new source, along with the total process source flow in time interval t, on the left side of the material cascade. On the other hand, the material to be stored for later time interval ($F_{ST,t,t+1}$) is added as a new sink, along with the total process sink flow in interval t, on the right side of the material cascade. The stored material enables the process source to be recovered to the sinks at later time interval(s). In other words, the time constraint is overcome by this stored material. Note that an inherent assumption made here is that mass storage is potentially available at each quality level k. Their actual existence is to be decided upon solving the ATM.

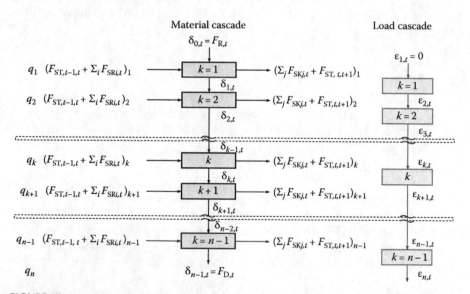

FIGURE 17.1
ATM framework for BMNs for direct reuse/recycle with mass storage. (From Foo, D.C.Y., Automated targeting technique for batch process integration, *Ind. Eng. Chem. Res.*, 49(20), 9899. Copyright 2010 American Chemical Society. With permission.)

The following constraints are imposed for all time intervals, based on the ATM framework in shown Figure 17.1:

$$\delta_{k,t} = \delta_{k-1,t} + (\Sigma_i \; F_{SRi} - \Sigma_j \; F_{SKj})_{k,t} + (F_{ST,t-1,t} - F_{ST,t,t+1})_k$$
$$k = 1, 2, \; \dots, n-1; \; \forall t \tag{17.1}$$

$$\delta_{k-1,t} \geq 0 \quad k = 1, \; n; \; \forall t \tag{17.2}$$

$$\varepsilon_{k,t} = \begin{cases} 0 & k = 1; \; \forall t \\ \varepsilon_{k-1,t} + \delta_{k-1,t} \left(q_{k,t} - q_{k-1,t} \right) & k = 2, 3, \dots, n; \; \forall t \end{cases} \tag{17.3}$$

$$\varepsilon_{k,t} \geq 0 \quad k = 2, 3, \dots, n; \; \forall t \tag{17.4}$$

Equation 17.1 describes the material cascades across all the quality levels for all time intervals. As shown, the *net material flow* of level k of interval t ($\delta_{k,t}$) is a result of the summation of the net material flow cascaded from an earlier quality level ($\delta_{k-1,t}$) with the flow balance at quality level k ($\Sigma_i F_{SRi} - \Sigma_j F_{SKj})_{k,t}$, as well as the net stored material flow at each level, ($F_{ST,t-1,t} - F_{ST,t,t+1})_k$. The next three constraints resemble those of the reuse/recycle scheme for continuous processes in Chapter 13 (Equations 13.4 through 13.6). In Equation 17.2, the net material flows entering the first ($\delta_{0,t}$) and last levels ($\delta_{n-1,t}$) of each interval t should take non-negative values. These variables correspond to the fresh resource ($F_{R,t}$) and waste discharge flow ($F_{D,t}$) of interval t. Next, Equation 17.3 states that the *residual impurity/property load* ($\varepsilon_{k,t}$) at the first quality level of each time interval t has zero value, i.e., $\varepsilon_{1,t} = 0$, while those in other quality levels are contributed by the residual load cascaded from previous level ($\varepsilon_{k-1,t}$), as well as the net material flow within each quality interval. The latter is given by the product of the net material flow from previous level ($\delta_{k-1,t}$) and the difference between two adjacent quality levels ($q_{k,t} - q_{k-1,t}$). All residual loads are set to take non-negative values following Equation 17.4, except at level 1 (which is set to zero according to Equation 17.3).

In addition, several constraints are needed to size the mass storage system. Equation 17.5 defines the cumulative content for a mass storage at quality level k for time interval t ($F_{MS,t,k}$), which is contributed by the cummulative content from previous time interval ($F_{MS,t-1,k}$), and the net flow between stored mass from earlier time interval ($F_{ST,t-1,t}$), and that to be recovered at later interval ($F_{ST,t,t+1}$). The size of the storage at each quality level k ($F_{STG,k}$) is then determined based on the maximum cumulative content across all time intervals, given by Equation 17.6. Equations 17.7 and 17.8 indicate that both the stored mass flow and the cumulative content of a mass storage should take positive values.

$$F_{MS,t,k} = (F_{MS,t-1} + F_{ST,t-1,t} + F_{ST,t,t+1})_k \quad \forall k, \forall t \tag{17.5}$$

$$F_{STG,k} \geq F_{MS,t,k} \quad \forall k, \forall t \tag{17.6}$$

$$F_{ST,t,t+1,k} \geq 0 \quad \forall k, \forall t \tag{17.7}$$

$$F_{MS,t+1,k} \geq 0 \quad \forall k \tag{17.8}$$

Optimization objective of this problem can be set to minimize the minimum fresh resource consumption and its associated minimum mass storage size or alternatively the overall cost of the BMN. For the former objective, the two-stage optimization approach introduced in Chapters 12 and 14 may be used, as it involves two optimization variables (i.e., fresh resource and mass storage). In the first stage, the total fresh resource across all time intervals ($\Sigma_t F_{R,t}$) is first minimized with the optimization objective in Equation 17.9, subject to the constraints in Equations 17.1 through 17.8. Next, the minimum mass storage size is determined using the optimization objective in Equation 17.10, with the minimum fresh resource determined in Stage 1 being added as a new constraint (Equation 17.11), along with other constraints in Stage 1 (Equations 17.1 through 17.8).

$$\text{Minimize } \Sigma_t F_{R,t} \tag{17.9}$$

$$\text{Minimize } \Sigma_k F_{STG,k} \tag{17.10}$$

$$F_{R,min} = \Sigma_t F_{R,t} \tag{17.11}$$

Alternatively, the optimization objective may also be set to determine the minimum cost of the BMN, which takes the form of the minimum operating cost (MOC), or the minimum total annualized cost (TAC), as given in Equation 17.12:

$$\text{Minimize cost} \tag{17.12}$$

Note that if we were to analyze a BMN without a mass storage system, the stored material terms in Equation 17.1 are set to zero, while the constraints in Equations 17.5 through 17.8 are excluded. The optimization objective may be set to determine the minimum flow/cost of fresh resource, using Equations 17.9 or 17.12, respectively.

Note that the ATM is linear and can be solved to achieve global optimum solution.

Example 17.1 Batch water network with direct reuse/recycle

The classical *batch water network* example in Example 11.1 (Wang and Smith, 1995) is revisited here. The limiting water data for the three semi-continuous* water-using processes (each consists of a pair of water sink and source) are given in Table 17.1 (reproduced from Table 11.1). The freshwater feed is available at 0 ppm. Determine the network targets for the following cases:

1. The minimum freshwater and wastewater flows for the batch water network, when no water storage is present.
2. For the batch water network with water storage tanks, use the two-stage optimization approach to determine the minimum freshwater and wastewater flows, as well as the minimum water storage capacity that achieves these flow targets. Develop the network structure based on the insights of ATM.

Solution

Following Steps 1 and 2 of the procedure, the interval flows of the water sinks and sources are located in their respective time intervals, i.e., 0–0.5 h (t=A), 0.5–1.0 h (t=B), and 1.0–1.5 h (t=C), as given in Table 17.2 (reproduced from Table 11.2—see Example 11.1 for detailed steps).

1. Case 1—Batch water network without water storage tank
   Based on the ATM framework in Figure 17.1, the cascade diagrams for all time intervals are shown in Figure 17.2. Note that since no water storage tank is present in this case, all stored water flows are set to zero:

$$F_{ST,t-1,t,k} = F_{ST,t,t+1,k} = 0 \quad k = 1, 2, 3, 4; \quad t = A, B, C \quad (17.13)$$

We then set the optimization objective to minimize the total freshwater flows across all time intervals, using the revised form of Equation 17.9:

TABLE 17.1

Limiting Water Data for Example 17.1

SK_j	F_{SKj} (tons)	C_{SKj} (ppm)	T^{STT} (h)	T^{END} (h)
SK1	100	100	0.5	1.5
SK2	40	0	0	0.5
SK3	25	100	0.5	1.0
SR_i	F_{SRi} (tons)	C_{SRi} (ppm)	T^{STT} (h)	T^{END} (h)
SR1	100	400	0.5	1.5
SR2	40	200	0	0.5
SR3	25	200	0.5	1.0

---

* See Section 11.1 for discussion of batch resource consumption unit.

TABLE 17.2

Interval Water Flows for Process Sinks and
Sources in Their Respective Time Intervals
(All Flow Terms in tons)

SK$_j$/SR$_i$	Time Interval, t		
	$t=A$ (0–0.5 h)	$t=B$ (0.5–1.0 h)	$t=C$ (1.0–1.5 h)
SK1		$F_{SK1,B}=50$	$F_{SK1,C}=50$
SK2	$F_{SK2,A}=40$		
SK3		$F_{SK3,B}=25$	
SR1		$F_{SR1,B}=50$	$F_{SR1,C}=50$
SR2	$F_{SR2,A}=40$		
SR3		$F_{SR3,B}=25$	

$$\text{Minimize } F_{FW,A} + F_{FW,B} + F_{FW,C} \qquad (17.14)$$

subject to the constraints given by Equations 17.1 through 17.4 for all time intervals. We first describe the constraints for time interval A (0–0.5 h).

Water cascade across all quality levels is given by Equation 17.1 (note that both $F_{ST,t-1,t,k}$ and $F_{ST,t,t+1,k}$ terms are set to zero):

$$\delta_{1,A} = \delta_{0,A} - 40$$

$$\delta_{2,A} = \delta_{1,A}$$

$$\delta_{3,A} = \delta_{2,A} + 40$$

$$\delta_{4,A} = \delta_{3,A}$$

Based on Equation 17.2, the net material flows that enter the first and last concentration levels should take non-negative values:

$$\delta_{0,A} \geq 0$$

$$\delta_{4,A} \geq 0$$

Note that these terms correspond to the minimum freshwater ($F_{FW,A}$) and wastewater flows ($F_{WW,A}$) for time interval A:

$$\delta_{0,A} = F_{FW,A}$$

$$\delta_{4,A} = F_{WW,A}$$

The impurity load cascades (in kg) across all quality levels are given by Equation 17.3:

$$\varepsilon_{1,A} = 0$$

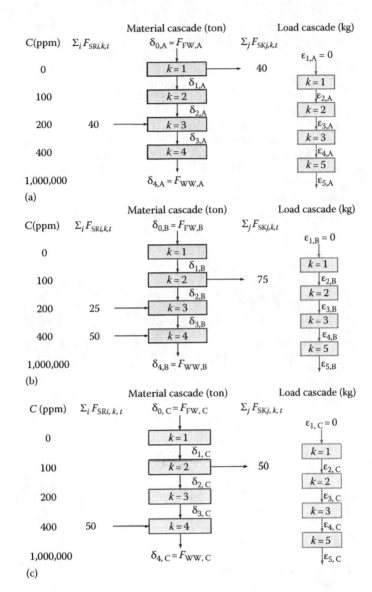

FIGURE 17.2
Cascade diagram for Example 17.1: Case 1 (without water storage): (a) time interval A, (b) time interval B, and (c) time interval C.

$$\varepsilon_{2,A} = \varepsilon_{1,A} + \delta_{1,A}(100 - 0)/1,000$$

$$\varepsilon_{3,A} = \varepsilon_{2,A} + \delta_{2,A}(200 - 100)/1,000$$

$$\varepsilon_{4,A} = \varepsilon_{3,A} + \delta_{3,A}(400 - 200)/1,000$$

$$\varepsilon_{5,A} = \varepsilon_{4,A} + \delta_{4,A}(1,000,000 - 400)/1,000$$

Equation 17.4 indicates that all residual loads (except at level 1) should take non-negative values:

$$\varepsilon_{2,A} \geq 0; \quad \varepsilon_{3,A} \geq 0$$

$$\varepsilon_{4,A} \geq 0; \quad \varepsilon_{5,A} \geq 0$$

Next, the constraints described in Equations 17.1 through 17.4 are also written for time interval B (0.5–1.0 h). For brevity, only the net material flow in Equation 17.1 is discussed here; the remaining constraints for intervals B and C are exactly the same as that in interval A (with subscripts for all variable terms changed from A to B or C, respectively). According to Equation 17.1, the water cascade for interval B is given as follows:

$$\delta_{1,B} = \delta_{0,B}$$

$$\delta_{2,B} = \delta_{1,B} - 75$$

$$\delta_{3,B} = \delta_{2,B} + 25$$

$$\delta_{4,B} = \delta_{3,B} + 50$$

Likewise, the water cascade for time interval C is given as follows:

$$\delta_{1,C} = \delta_{0,C}$$

$$\delta_{2,C} = \delta_{1,C} - 50$$

$$\delta_{3,C} = \delta_{2,C}$$

$$\delta_{4,C} = \delta_{3,C} + 50$$

The LINGO formulation for the linear program (LP) model given earlier can be written as follows:

Model:

```
! Optimisation objective: to minimise total freshwater flow
in time intervals A (FWA), B (FWB) and C (FWC);
min = FWA + FWB + FWC ;

! Constraints for interval A;
! Water cascade;
D1A = D0A - 40;
D2A = D1A;
D3A = D2A + 40;
D4A = D3A;

! Net material flows entering the first and the last
concentration levels should take non-negative values;
D0A >= 0;
D4A >= 0;
```

```
! These net material flows correspond to minimum freshwater
(FWt) and wastewater (WWt) flows for each time interval;
D0A = FWA;
D4A = WWA;

! Impurity load cascade (in kg) ;
E1A = 0 ;
E2A = E1A + D1A'(100 - 0)/1000;
E3A = E2A + D2A'(200 - 100)/1000;
E4A = E3A + D3A'(400 - 200)/1000;
E5A = E4A + D4A'(1000000 - 400)/1000;

! E2A-E5A should take non-negative values ;
E2A >= 0; E3A >= 0;
E4A >= 0; E5A >= 0;

! D1A-D3A may take negative values ;
@free (D1A); @free (D2A); @free (D3A);
```

! Constraints for interval B;
```
! Water cascade;
D1B = D0B;
D2B = D1B - 75;
D3B = D2B + 25;
D4B = D3B + 50;

! Net material flows entering the first and the last
concentration levels should take non-negative values;
D0B >= 0;
D4B >= 0;

! These net material flows correspond to minimum freshwater
(FWt) and wastewater (WWt) flows for each time interval;
D0B = FWB;
D4B = WWB;

! Impurity load cascade (in kg) ;
E1B = 0 ;
E2B = E1B + D1B'(100 - 0)/1000;
E3B = E2B + D2B'(200 - 100)/1000;
E4B = E3B + D3B'(400 - 200)/1000;
E5B = E4B + D4B'(1000000 - 400)/1000;

! E2B-E5B should take non-negative values ;
E2B >= 0; E3B >= 0;
E4B >= 0; E5B >= 0;

! D1B-D3B may take negative values ;
@free (D1B); @free (D2B); @free (D3B);
```

! Constraints for interval C;
```
! Water cascade;
D1C = D0C;
D2C = D1C - 50;
D3C = D2C;
D4C = D3C + 50;

! Net material flows entering the first and the last
concentration levels should take non-negative values;
D0C >= 0;
D4C >= 0;
```

```
! These net material flows correspond to minimum freshwater
(FWt) and wastewater (WWt) flows for each time interval;
D0C=FWC;
D4C=WWC;

! Impurity load cascade (in kg) ;
E1C=0 ;
E2C=E1C+D1C*(100 - 0)/1000;
E3C=E2C+D2C*(200 - 100)/1000;
E4C=E3C+D3C*(400 - 200)/1000;
E5C=E4C+D4C*(1000000 - 400)/1000;

! E2C-E5C should take non-negative values ;
E2C >= 0; E3C >= 0;
E4C >= 0; E5C >= 0;

! D1C-D3C may take negative values ;
@free (D1C); @free (D2C); @free (D3C);
END
```

Similar to the case in previous chapters, the last set of constraints are only necessary when LINGO is used to solve the optimization model, as the software assumes all variables to have non-negative values by default. For the same reason, the constraints for non-negativity (given by Equation 17.4) need not be coded in LINGO.

The following is the solution report generated by LINGO:

```
Global optimal solution found.
Objective value:                        121.2500
Total solver iterations:                       0

              Variable           Value
              FWA             40.00000
              FWB             43.75000
              FWC             37.50000
              D1A              0.000000
              D0A             40.00000
              D2A              0.000000
              D3A             40.00000
              D4A             40.00000
              WWA             40.00000
              E1A              0.000000
              E2A              0.000000
              E3A              0.000000
              E4A              8.000000
              E5A          39992.00
              D1B             43.75000
              D0B             43.75000
              D2B            -31.25000
              D3B             -6.250000
              D4B             43.75000
              WWB             43.75000
              E1B              0.000000
              E2B              4.375000
              E3B              1.250000
              E4B              0.000000
              E5B          43732.50
```

D1C	37.50000
D0C	37.50000
D2C	-12.50000
D3C	-12.50000
D4C	37.50000
WWC	37.50000
E1C	0.000000
E2C	3.750000
E3C	2.500000
E4C	0.000000
E5C	37485.00

The results indicate that the minimum freshwater flows for the BMN without water storage system are targeted at 121.25 tons. This is contributed by time intervals A (40 tons), B (43.75 tons), and C (37.5 tons). The revised cascade diagram for the BWN with all optimized variables is shown in Figure 17.3. The results are identical to those obtained using graphical and algebraic targeting techniques (see Examples 11.1 and 11.2—Case 1).

2. Case 2—Batch water network with water storage tanks
Since water storage tanks are available for use, the ATM framework in Figure 17.1 is revised as the cascade diagrams in Figure 17.4. In the first stage of optimization, the objective is set to minimize the total freshwater flows across all time intervals according to Equation 17.14, subject to the constraints in Equations 17.1 through 17.8. For brevity, we shall only illustrate the constraints that are different from Case 1 (without water storage tanks).

Water cascades across all quality levels are given by Equation 17.1:

$$\delta_{1,A} = \delta_{0,A} - 40 + (F_{ST,C,A,1} - F_{ST,A,B,1})$$
$$\delta_{2,A} = \delta_{1,A} + (F_{ST,C,A,2} - F_{ST,A,B,2})$$
$$\delta_{3,A} = \delta_{2,A} + 40 + (F_{ST,C,A,3} - F_{ST,A,B,3})$$
$$\delta_{4,A} = \delta_{3,A} + (F_{ST,C,A,4} - F_{ST,A,B,4})$$

Likewise, the water cascades for time intervals B and C are given as follows:

$$\delta_{1,B} = \delta_{0,B} + (F_{ST,A,B,1} - F_{ST,B,C,1})$$
$$\delta_{2,B} = \delta_{1,B} - 75 + (F_{ST,A,B,2} - F_{ST,B,C,2})$$
$$\delta_{3,B} = \delta_{2,B} + 25 + (F_{ST,A,B,3} - F_{ST,B,C,3})$$
$$\delta_{4,B} = \delta_{3,B} + 50 + (F_{ST,A,B,4} - F_{ST,B,C,4})$$
$$\delta_{1,C} = \delta_{0,C} + (F_{ST,B,C,1} - F_{ST,C,A,1})$$
$$\delta_{2,C} = \delta_{1,C} - 50 + (F_{ST,B,C,2} - F_{ST,C,A,2})$$
$$\delta_{3,C} = \delta_{2,C} + (F_{ST,B,C,3} - F_{ST,C,A,3})$$
$$\delta_{4,C} = \delta_{3,C} + 50 + (F_{ST,B,C,4} - F_{ST,C,A,4})$$

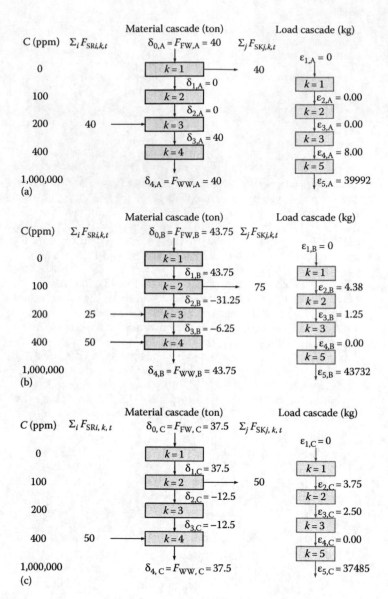

FIGURE 17.3
Result of ATM for Example 17.1: Case 1 (without water storage): (a) time interval A, (b) time interval B, and (c) time interval C.

Note that the constraints given by Equations 17.2 through 17.4 remain identical to those in Case 1. We next discuss the constraints for the water storage system. The cumulative content for the water storage at quality level 1 (i.e., 0 ppm) across time intervals A–C is described by Equation 17.5:

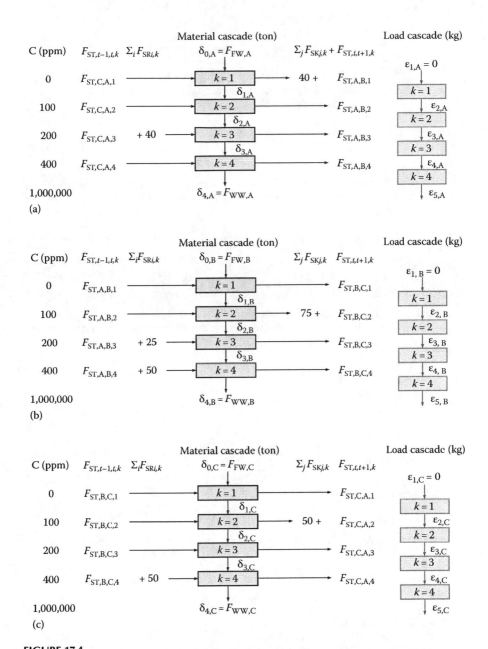

FIGURE 17.4
Cascade diagram for Example 17.1: Case 2 (with water storage): (a) time interval A, (b) time interval B, and (c) time interval C.

$$F_{MS,B,1} = F_{MS,A,1} - F_{ST,C,A,1} + F_{ST,A,B,1}$$

$$F_{MS,C,1} = F_{MS,B,1} - F_{ST,A,B,1} + F_{ST,B,C,1}$$

$$F_{MS,A,1} = F_{MS,C,1} - F_{ST,B,C,1} + F_{ST,C,A,1}$$

Equation 17.6 indicates that the size of the water storage at each quality level is determined based on the maximum cumulative content across all time intervals:

$$F_{STG,1} \geq F_{MS,A,1}; \quad F_{STG,1} \geq F_{MS,B,1}$$

$$F_{STG,1} \geq F_{MS,C,1}$$

Equations 17.7 and 17.8 indicate that both the stored water flow and the cumulative content for a mass storage should take positive values:

$$F_{ST,A,B,1} \geq 0; \quad F_{MS,A,1} \geq 0$$

$$F_{ST,B,C,1} \geq 0; \quad F_{MS,B,1} \geq 0$$

$$F_{ST,C,A,1} \geq 0; \quad F_{MS,C,1} \geq 0$$

Likewise, the constraints for quality levels 2–4 (100, 200, and 400 ppm) may be written as those for 0 ppm. The LINGO formulation for the LP problem may be written as follows:

```
Model:
min = FWA + FWB + FWC;

! Constraints for interval A;
! Water cascade;
D1A = D0A - 40 + (STCA1 - STAB1);
D2A = D1A + (STCA2 - STAB2);
D3A = D2A + 40 + (STCA3 - STAB3);
D4A = D3A + (STCA4 - STAB4);

! Net material flows entering the first and the last
concentration levels should take non-negative values;
D0A >= 0;
D4A >= 0;

! These net material flows correspond to minimum freshwater
(FWt) and wastewater (WWt) flows for each time interval;
D0A = FWA;
D4A = WWA;

! Impurity load cascade (in kg);
E1A = 0 ;
E2A = E1A + D1A*(100 - 0)/1000;
E3A = E2A + D2A*(200 - 100)/1000;
E4A = E3A + D3A*(400 - 200)/1000;
E5A = E4A + D4A*(1000000 - 400)/1000;

! E2A-E5A should take non-negative values ;
E2A >= 0; E3A >= 0;
E4A >= 0; E5A >= 0;
```

```
! D1A-D3A may take negative values;
@free (D1A); @free (D2A); @free (D3A);

! Constraints for interval B;
! Water cascade;
D1B=D0B+(STAB1 - STBC1);
D2B=D1B - 75+(STAB2 - STBC2);
D3B=D2B+25+(STAB3 - STBC3);
D4B=D3B+50+(STAB4 - STBC4);

! Net material flows entering the first and the last
concentration levels should take non-negative values;
D0B >= 0;
D4B >= 0;

! These net material flows correspond to minimum freshwater
(FWt) and wastewater (WWt) flows for each time interval;
D0B=FWB;
D4B=WWB;

! Impurity load cascade (in kg) ;
E1B=0 ;
E2B=E1B+D1B*(100 - 0)/1000;
E3B=E2B+D2B*(200 - 100)/1000;
E4B=E3B+D3B*(400 - 200)/1000;
E5B=E4B+D4B*(1000000 - 400)/1000;

! E2B-E5B should take non-negative values;
E2B >= 0; E3B >= 0;
E4B >= 0; E5B >= 0;

! D1B-D3B may take negative values;
@free (D1B); @free (D2B); @free (D3B);

! Constraints for interval C;
! Water cascade;
D1C=D0C+(STBC1 - STCA1);
D2C=D1C   50+(STBC2   STCA2);
D3C=D2C+(STBC3 - STCA3);
D4C=D3C+50+(STBC4 - STCA4);

! Net material flows entering the first and the last
concentration levels should take non-negative values;
D0C >= 0;
D4C >= 0;

! These net material flows correspond to minimum freshwater
(FWt) and wastewater (WWt) flows for each time interval;
D0C=FWC;
D4C=WWC;

! Impurity load cascade (in kg);
E1C=0 ;
E2C=E1C+D1C*(100 - 0)/1000;
E3C=E2C+D2C*(200 - 100)/1000;
E4C=E3C+D3C*(400 - 200)/1000;
E5C=E4C+D4C*(1000000 - 400)/1000;

! E2C-E5C should take non-negative values;
E2C >= 0; E3C >= 0;
E4C >= 0; E5C >= 0;
```

```
! D1C-D3C may take negative values;
@free (D1C); @free (D2C); @free (D3C);

! Constraints for water storage tanks;
! Cumulative content of water storage tanks;
MSB1=MSA1  - STCA1+STAB1;  !0ppm level;
MSC1=MSB1  - STAB1+STBC1;
MSA1=MSC1  - STBC1+STCA1;
MSB2=MSA2  - STCA2+STAB2;  !100ppm level;
MSC2=MSB2  - STAB2+STBC2;
MSA2=MSC2  - STBC2+STCA2;
MSB3=MSA3  - STCA3+STAB3;  !200ppm level;
MSC3=MSB3  - STAB3+STBC3;
MSA3=MSC3  - STBC3+STCA3;
MSB4=MSA4  - STCA4+STAB4;  !400ppm level;
MSC4=MSB4  - STAB4+STBC4;
MSA4=MSC4  - STBC4+STCA4;

! Sizing of water storage at each level;
STG1 >= MSA1; STG1 >= MSB1; STG1 >= MSC1;
STG2 >= MSA2; STG2 >= MSB2; STG2 >= MSC2;
STG3 >= MSA3; STG3 >= MSB3; STG3 >= MSC3;
STG4 >= MSA4; STG4 >= MSB4; STG4 >= MSC4;

! Stored water flows should take non-negative values;
STAB1 >= 0; STBC1 >= 0; STCA1 >= 0;
STAB2 >= 0; STBC2 >= 0; STCA2 >= 0;
STAB3 >= 0; STBC3 >= 0; STCA3 >= 0;
STAB4 >= 0; STBC4 >= 0; STCA4 >= 0;

! Cumulative content of water storage tanks should take
non-negative values;
MSA1 >= 0; MSB1 >= 0; MSC1 >= 0;
MSA2 >= 0; MSB2 >= 0; MSC2 >= 0;
MSA3 >= 0; MSB3 >= 0; MSC3 >= 0;
MSA4 >= 0; MSB4 >= 0; MSC4 >= 0;
END
```

The following is the solution report generated by LINGO (the results for optimization variables are omitted, as they are not necessary for now):

```
Global optimal solution found.
Objective value:                    102.5000
Total solver iterations:                  20
```

The optimization result indicates that the minimum freshwater flow for the batch water network is determined as 102.5 tons. This minimum flow is then included as a new constraint in Stage 2 of the optimization approach, based on the revised form of Equation 17.11:

$$F_{FW,A} + F_{FW,B} + F_{FW,C} = 102.5 \qquad (17.15)$$

while the optimization objective is set to minimize the total water storage size, given by the revised form of Equation 17.10:

$$\text{Minimize } F_{STG,1} + F_{STG,2} + F_{STG,3} + F_{STG,4} \qquad (17.16)$$

The LINGO formulation for the second optimization stage remains identical to that in the first stage, except for the new optimization objective (given by Equation 17.16) and the new sets of constraints for freshwater flow (Equation 17.15). In this case, the LINGO formulation for the Stage 2 optimization objective and the new constraints are given as follows:

```
min = STG1 + STG2 + STG3 + STG4 ;
FWA + FWB + FWC = 102.5;
```

The following is the solution report generated by LINGO for Stage 2 of the optimization model:

```
Global optimal solution found.
Objective value:                    37.50000
Total solver iterations:                  17
```

Variable	Value
STG1	0.000000
STG2	0.000000
STG3	37.50000
STG4	0.000000
FWA	40.00000
FWB	37.50000
FWC	25.00000
D1A	0.000000
D0A	40.00000
STCA1	0.000000
STAB1	0.000000
D2A	0.000000
STCA2	0.000000
STAB2	0.000000
D3A	2.500000
STCA3	0.000000
STAB3	37.50000
D4A	2.500000
STCA4	0.000000
STAB4	0.000000
WWA	2.500000
E1A	0.000000
E2A	0.000000
E3A	0.000000
E4A	0.5000000
E5A	2499.500
D1B	37.50000
D0B	37.50000
STBC1	0.000000
D2B	-37.50000
STBC2	0.000000
D3B	0.000000
STBC3	25.00000
D4B	50.00000

```
STBC4        0.000000
WWB         50.00000
E1B          0.000000
E2B          3.750000
E3B          0.000000
E4B          0.000000
E5B      49980.00
D1C         25.00000
D0C         25.00000
D2C        -25.00000
D3C          0.000000
D4C         50.00000
WWC         50.00000
E1C          0.000000
E2C          2.500000
E3C          0.000000
E4C          0.000000
E5C      49980.00
MSB1         0.000000
MSA1         0.000000
MSC1         0.000000
MSB2         0.000000
MSA2         0.000000
MSC2         0.000000
MSB3        37.50000
MSA3         0.000000
MSC3        25.00000
MSB4         0.000000
MSA4         0.000000
MSC4         0.000000
```

The optimization result indicates that the minimum water storage size is determined as 37.5 tons. The revised cascade diagram that incorporates the optimized variables is shown in Figure 17.5. Note that only the material cascade of all time intervals are shown; while the impurity/property

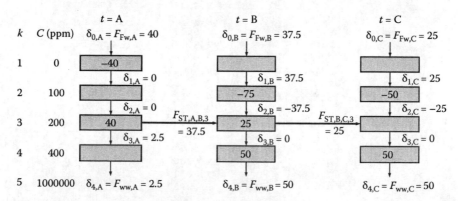

FIGURE 17.5
Result of ATM for Example 17.1: Case 2 (with water storage; flow in tons).

load cascade has been omitted in the diagram. Also, the values in boxes indicate the flow balances for a given quality level k within the respective time interval, i.e., $\Sigma_i F_{SRi,k,t} - \Sigma_j F_{SKj,k,t}$. It is also observed in Figure 17.5 that only one storage tank is needed to recover the water across different time intervals at 200 ppm level. The resulting freshwater flows are determined as 40, 37.5, and 25 tons for intervals A, B, and C, respectively, while the wastewater flows for these intervals are found as 2.5, 50, and 50 tons, respectively. Also note that the total flows for freshwater and wastewater, as well as the storage size, are identical to those obtained using the insight-based targeting technique (see Example 11.2).

For a simple case such as this one,[*] we can interpret the ATM results in Figure 17.5 and then develop a network structure based on the insights gained. From Figure 17.5, the flow balances in time interval A (0–0.5 h) indicate that a flow deficiet is observed at 0 ppm; while a flow surplus at 200 ppm. We can then identify from Tables 17.1 and 17.2 that these flow deficiet and surplus are contributed by water sink SK2 (0 ppm) and source SR2 (200 ppm), respectively. Next, Figure 17.5 also indicates that 40 tons of freshwater flow ($F_{FW,A}$) that is fed to time interval A is fully consumed by SK2 at 0 ppm level, since no net interval flow is leaving 0 ppm level ($\delta_{1,A} = 0$). On the other hand, at 200 ppm level where source SR2 exists ($k = 3$), it is observed that a major portion of this source is sent to water storage tank ($F_{ST,A,B,3} = 37.5$ tons), to be stored for later use in time interval B. The remaining water flow from this source, i.e., 2.5 tons ($F_{WW,A}$), is then discharged as wastewater flow from this interval.

In time interval B (0.5–1.0 h), the flow balances are contributed by the interval water flows of sinks SK1 and SK3 (a total of 75 tons, both at 100 ppm), as well as those of sources SR1 (50 tons, 400 ppm) and SR3 (25 tons, 200 ppm), respectively. It is then observed that 37.5 tons of freshwater flow ($F_{FW,A}$) is sent to this time interval. The freshwater flow is fully consumed by the process sink at 100 ppm ($k = 2$) that requires 75 tons of water flow. The insufficient flow is then supplied from a lower quality level, i.e., 200 ppm ($k = 3$), indicated by the negative value of net material flow, i.e., $\delta_{2,B} = -37.5$ tons. At 200 ppm level, apart from water source SR3 (25 tons), there are also 37.5 tons of stored water available for use in the storage tank, which is transferred from time interval A ($F_{ST,A,B,3}$). Hence, one option we have is to send the 37.5 tons of stored water to the 100 ppm sink, and then store 25 tons of water source SR3 for further recovery in time interval C ($F_{ST,B,C,3}$). Alternatively, we may also mix the stored water from interval A with source SR3 for reuse in the 100 ppm sink and for recovery at time interval C. At final concentration level, i.e., 400 ppm, the interval flow of source SR1 is discharged as wastewater flow from this interval ($F_{WW,B} = 50$ tons).

Similar insights may also be derived for time interval C (1.0–1.5 h). In this interval, the flow balances are contributed by the net interval flows of water sink SK1 (100 ppm) and source SR1 (400 ppm), respectively. The

---

[*] An inspection indicates that only one sink (of non-zero quality index) exists within the each pinch region in each time interval.

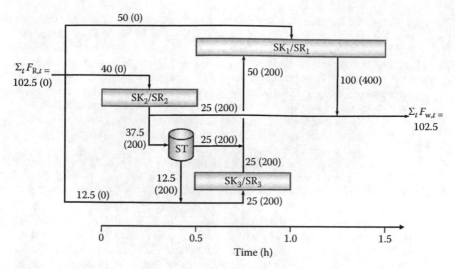

FIGURE 17.6
Batch water network for Example 17.1: Case 2. (From Foo, D.C.Y., Automated targeting technique for batch process integration, *Ind. Eng. Chem. Res.*, 49(20), 9899. Copyright 2010 American Chemical Society. With permission.)

freshwater flow of 25 tons is fully consumed by the process sink SK1 at 100 ppm ($k=2$) that requires 50 tons of water flow. The insufficient flow is then supplied from lower quality level, i.e., 200 ppm ($k=3$), which is available from the stored water from time interval B ($F_{ST,B,C,3}=25$ tons). Similar to the earlier time interval, the interval water flow of source SR1 at the final concentration level (400 ppm) of time interval C is discharged as wastewater from this interval ($F_{WW,C}=50$ tons).

Based on the aforementioned insights, the batch water network that fulfils the various network targets is shown in Figure 17.6. Note that one may also make use of the *nearest neighbor algorithm* (NNA) in synthesizing the network, as described in Chapter 11 (see Section 11.5 for detailed steps).

17.3 ATM for Batch Regeneration Network

As discussed in the previous chapters, interception units may be used to regenerate process sources for further recovery in an RCN. In this section, the ATM for a *batch regeneration network* is outlined. The ATM framework for material regeneration network with different types of interception unit for continuous processes in Chapter 14 may be extended for use in batch regeneration network. We discuss two such examples in the following section.

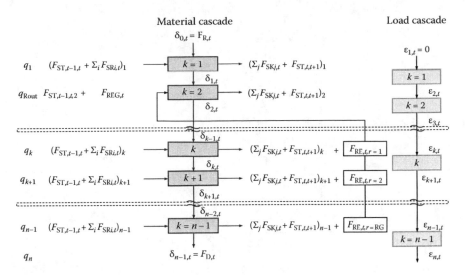

FIGURE 17.7
ATM framework for batch regeneration network with single pass interception unit of fixed outlet quality type. (From Foo, D.C.Y., Automated targeting technique for batch process integration, *Ind. Eng. Chem. Res.*, 49(20), 9899. Copyright 2010 American Chemical Society. With permission.)

When a *single pass* interception unit of the fixed outlet quality type is used, the ATM framework takes the form as outlined in Figure 17.7. It is observable that the framework in Figure 17.7 resembles the one in Figure 14.2. Note, however, that in the former, mass storage tanks are potentially present at each quality level k, which enable the transfer of stored material flows from earlier ($F_{ST,t-1,t,k}$) and to later time intervals ($F_{ST,t,t+1,k}$), similar to the case in Figure 17.1. Note also that, in some cases, the mass storage may be used to store the regenerated source for recovery to sinks at later time intervals.

Since the ATM framework in Figure 17.7 is an extended form of the one in Figure 14.2, their constraints should be similar. In other words, the regeneration flow terms are to be included in the constraints, given as follows:

$$\delta_{k,t} = \delta_{k-1,t} + (\Sigma_i\ F_{SRi} - \Sigma_j\ F_{SKj})_{k,t} + (F_{ST,t-1,t} - F_{ST,t,t+1})_k + F_{REG,t,r} - \Gamma_{RE,t,r}$$
$$k = 1,\ldots,n-1; \quad \forall t \tag{17.17}$$

$$F_{REG,t} = \Sigma_r F_{RE,r,t} \quad \forall t \tag{17.18}$$

$$F_{RE,r,t} \leq \Sigma_i F_{SRi,t} \quad r = 1,2,\ldots,RG; \ q_k < q_{Rout} \forall t \tag{17.19}$$

Equation 17.17 is the revised form of Equation 17.1, in which the regeneration flow terms are added to the net material flow calculation. The flow balance of the interception unit is given by Equation 17.18, while the flow availability for regeneration is given by Equation 17.19. Note that these constraints are extended

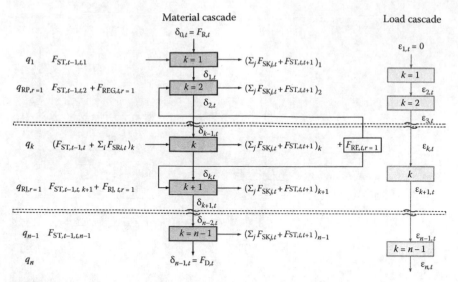

FIGURE 17.8
ATM framework for batch regeneration network with partitioning interception unit of fixed outlet quality type.

from Equations 14.1 through 14.3. In addition, the constraints in Equations 17.2 through 17.8 are also required for the ATM framework of this case.

The ATM framework for another variant of batch regeneration network is shown in Figure 17.8, where a *partitioning* interception unit of fixed outlet quality type is used to regenerate process sources. This framework resembles its counterpart for RCN in continuous processes (see Figure 14.11). Similar to the previous case, stored material flows from earlier ($F_{ST,t-1,t,k}$) and later time intervals ($F_{ST,t,t+1,k}$) are added at each level k of the framework (see Figure 17.8).

For the ATM framework in Figure 17.8, the net material flow is defined by Equation 17.20, which takes into account the purified ($F_{REG,t,r}$) and reject flow terms ($F_{RJ,t,r}$) of the partitioning interception unit. The flow balance of the latter is defined by Equation 17.21. Besides, an overall *product recovery factor* (RC) is used to define the recovery of the desired product from the inlet stream, given in Equations 17.22 and 17.23 for quality and concentration indexes, respectively (either one can be chosen):

$$\delta_{k,t} = \delta_{k-1,t} + (\Sigma_i\, F_{SRi} - \Sigma_j\, F_{SKj})_{k,t} + (F_{ST,t-1,t} - F_{ST,t,t+1})_k + F_{REG,t,r}$$
$$+ F_{RJ,t,r} - F_{RE,t,r} \quad k = 1,\ldots,n-1;\, \forall t \tag{17.20}$$

$$F_{REG,t,r} = F_{RJ,t,r} + F_{RE,t,r} \quad r = 1, 2, \ldots, RG;\, \forall t \tag{17.21}$$

$$F_{RE,t,r}\, q_{RE,r}\, RC = F_{REG,t,r}\, q_{REG,r} \quad r = 1, 2, \ldots, RG;\, \forall t \tag{17.22}$$

$$F_{RE,r}(C^{max} - C_{RE,r})RC = F_{REG,t,r}(C^{max} - C_{REG,r}) \quad r = 1, 2, \ldots, RG; \forall t \qquad (17.23)$$

where

C^{max} is the highest possible value for a given concentration index (e.g., 1,000,000 ppm, 100 wt%)

$q_{RE,r}$ and $C_{RE,r}$ are the quality and concentration of the inlet stream of interception unit r

$q_{REG,r}$ and $C_{REG,r}$ are the quality and concentration of the outlet stream of interception unit r

Note that the reject stream quality (q_{RJ}) of the interception unit should be determined and added in the ATM framework, using the method as described in Section 14.4 (see Example 14.3 for detailed step). Apart from the two variants in Figures 17.7 and 17.8, other types of interception units (e.g., with fixed removal ratio type) may also be used for a batch regeneration network. Their respective constraints may be derived similarly as in earlier cases.

Example 17.2 Batch water regeneration network

Example 17.1 is revisited here. In this case, a single pass interception unit with an outlet concentration of 20 ppm is used for regeneration. Determine the minimum TAC target for the batch water regeneration network. Additional information for cost calculation is given as follows:

1. Unit costs for freshwater (CT_{FW}), regeneration flow (CT_{RW}), and wastewater (CT_{WW}) are given as $1/ton, $0.8/ton, and $1/ton, respectively. Note that the unit cost for regeneration flow (CT_{RW}) includes capital and operating costs.
2. Capital cost for the network is mainly contributed by the water storage tanks (SC_{CP}, in USD), which is given by Equation 17.24 (Kim and Smith, 2004). Note that the cost for stream connections within the batch plant is omitted, as it is insignificant compared to water storage cost:

$$SC_{CP} = (235 \, F_{ST} + 20300) \qquad (17.24)$$

where F_{ST} is the stored water flows (tons) between time intervals.
3. For calculation of the TAC, the annualizing factor (AF) in Equation 12.15 may be used. The piping cost is annualized to a period of 5 years, with an interest rate of 5%. Assume an annual production of 5300 batches (i.e., an annual operating time of 7950 h, with a cycle time of 1.5 h).

Solution

The revised cascade diagram for all time intervals takes the form as in Figure 17.9. We first discuss the constraints needed for time interval A.

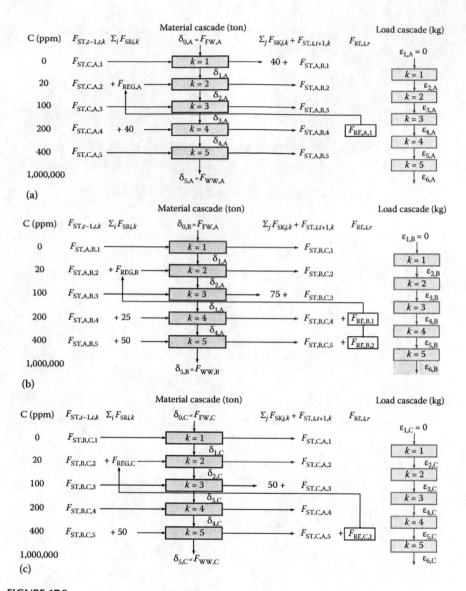

FIGURE 17.9
Cascade diagrams for Example 17.2: (a) time interval A, (b) time interval B, and (c) time interval C.

Water cascades across all quality levels are given by Equation 17.17:

$$\delta_{1,A} = \delta_{0,A} - 40 + (F_{ST,C,A,1} - F_{ST,A,B,1})$$

$$\delta_{2,A} = \delta_{1,A} + (F_{ST,C,A,2} - F_{ST,A,B,2}) + F_{REG,A}$$

$$\delta_{3,A} = \delta_{2,A} + (F_{ST,C,A,3} - F_{ST,A,B,3})$$

$$\delta_{4,A} = \delta_{3,A} + 40 + (F_{ST,C,A,4} - F_{ST,A,B,4}) - F_{RE,A,1}$$
$$\delta_{5,A} = \delta_{4,A} + (F_{ST,C,A,5} - F_{ST,A,B,5})$$

Next, flow balance and availability of the interception unit in this time interval are given by Equations 17.18 and 17.19, respectively:

$$F_{REG,A} = F_{RE,A,1}$$
$$F_{RE,A,1} \le 40$$

Similarly, water cascade for time interval B, as well as the flow balance and availability of the interception unit in this interval, is given as follows:

$$\delta_{1,B} = \delta_{0,B} + (F_{ST,A,B,1} - F_{ST,B,C,1})$$
$$\delta_{2,B} = \delta_{1,B} + (F_{ST,A,B,2} - F_{ST,B,C,2}) + F_{REG,B}$$
$$\delta_{3,B} = \delta_{2,B} - 75 + (F_{ST,A,B,3} - F_{ST,B,C,3})$$
$$\delta_{4,B} = \delta_{3,B} + 25 + (F_{ST,A,B,4} - F_{ST,B,C,4}) - F_{RE,B,1}$$
$$\delta_{5,B} = \delta_{4,B} + 50 + (F_{ST,A,B,5} - F_{ST,B,C,5}) - F_{RE,B,2}$$
$$F_{REG,B} = F_{RE,B,1} + F_{RE,B,2}$$
$$F_{RE,B,1} \le 25$$
$$F_{RE,B,2} \le 50$$

Likewise, water cascade, as well as the flow balance and availability of the interception unit, for time interval C is given as follows:

$$\delta_{1,C} = \delta_{0,C} + (F_{ST,B,C,1} - F_{ST,C,A,1})$$
$$\delta_{2,C} = \delta_{1,C} + (F_{ST,B,C,2} - F_{ST,C,A,2}) + F_{REG,C}$$
$$\delta_{3,C} = \delta_{2,C} - 50 + (F_{ST,B,C,3} - F_{ST,C,A,3})$$
$$\delta_{4,C} = \delta_{3,C} + (F_{ST,B,C,4} - F_{ST,C,A,4})$$
$$\delta_{5,C} = \delta_{4,C} + 50 + (F_{ST,B,C,5} - F_{ST,C,A,5}) - F_{RE,C,1}$$
$$F_{REG,C} = F_{RE,C,1}$$
$$F_{RE,C,1} \le 50$$

The remaining constraints (given by Equations 17.2 through 17.8) are identical to those in Example 17.1, Case 2, and, hence, are omitted here for brevity.

In this case, it is desired to determine the TAC target of the batch regeneration network. Hence, the optimization objective should take a revised form of Equation 17.12:

$$\text{Minimize TAC} \tag{17.25}$$

Various cost elements for the TAC are discussed next, which is similar to Example 12.3 (see Section 12.3 for a more detailed description). The operating cost (OC) of the network is contributed by the various water costs:

$$OC \, (\$/h) = \Sigma_t \, F_{FW,t} \, CT_{FW} + \Sigma_r \, F_{RE,A,r} \, CT_{RW} + \Sigma_t \, F_{WW,t} \, CT_{WW}$$

$$= (F_{FW,A} + F_{FW,B} + F_{FW,C} + F_{WW,A} + F_{WW,B} + F_{WW,C}) \, (\$1/ton)$$

$$+ (F_{RE,A,1} + F_{RE,B,1} + F_{RE,B,2} + F_{RE,C,1}) \, (\$0.8/ton)$$

Next, the AOC is determined from the product of OC and the annual production, given as follows:

$$AOC(\$/year) = OC(5300)$$

The total storage (SC) and, hence, the capital cost (CC) for the network are contributed by the storage cost at each concentration level:

$$CC = SC = SC_{STG,1} + SC_{STG,2} + SC_{STG,3} + SC_{STG,4} + SC_{STG,5}$$

To determine the storage cost, Equation 17.24 is revised as Equation 17.26, where a binary term ($B_{STG,k}$) is added to indicate the presence of a water storage tank at level k:

$$SC_{STG,k} = (235 \, F_{STG,k} + 20,300 \, B_{STG,k}) \quad k = 1, \ldots, n-1 \qquad (17.26)$$

Equation 17.26 is used in conjunction with Equation 17.27 to examine the presence/absence of a water storage tank at level k:

$$\frac{F_{STG,k}}{M} \leq B_{STG,k} \quad k = 1, \ldots, n-1 \qquad (17.27)$$

where M is an arbitrary large number. Note that similar forms of Equations 17.26 and 17.27 are used for piping cost calculations for an RCN in continuous process (see Section 12.3 and Example 12.3 for detailed example). In this case, the value of M is taken as 1000. Equations 17.26 and 17.27 are applied for water storage for all concentration levels, given as follows:

$$SC_{STG,1} = (235 \, F_{STG,1} + 20,300 \, B_{STG,1}); \quad F_{STG,1} \leq 1,000 \, B_{STG,1}$$

$$SC_{STG,2} = (235 \, F_{STG,2} + 20,300 \, B_{STG,2}); \quad F_{STG,2} \leq 1,000 \, B_{STG,2}$$

$$SC_{STG,3} = (235 \, F_{STG,3} + 20,300 \, B_{STG,3}); \quad F_{STG,3} \leq 1,000 \, B_{STG,3}$$

$$SC_{STG,4} = (235 \, F_{STG,4} + 20,300 \, B_{STG,4}); \quad F_{STG,4} \leq 1,000 \, B_{STG,4}$$

$$SC_{STG,5} = (235 \, F_{STG,5} + 20,300 \, B_{STG,5}); \quad F_{STG,5} \leq 1,000 \, B_{STG,5}$$

The annualizing factor in Equation 12.15 is calculated as follows:

$$AF = \frac{0.05(1+0.05)^5}{(1+0.05)^5 - 1} = 0.231$$

Hence, the TAC in Equation 17.25 is given as follows:

$$TAC = AOC + CC\,(0.231)$$

Due to the existence of the binary variable, the earlier problem is a mixed integer linear program (MILP). The LINGO formulation for the earlier problem may be written as follows:

Model:
```
min = TAC;
TAC = AOC + SC*0.231;
OC = (FWA + FWB + FWC + WWA + WWB + WWC)*1 + (REA1 + REB1 + REB2
+ REC1)*0.8;
AOC = OC*5300;
SC = CSTG1 + CSTG2 + CSTG3 + CSTG4 + CSTG5;

! Constraints for interval A;
! Water cascade;
D1A = D0A - 40 + (STCA1 - STAB1);
D2A = D1A + (STCA2 - STAB2) + REGA;
D3A = D2A + (STCA3 - STAB3);
D4A = D3A + 40 + (STCA4 - STAB4) - REA1;
D5A = D4A + (STCA5 - STAB5);

! Net material flows entering the first and the last
concentration levels should take non-negative values;
D0A >= 0;
D5A >= 0;

! These net material flows correspond to minimum freshwater
(FWt) and wastewater (WWt) flows for each time interval;
D0A = FWA;
D5A = WWA;

! Flow balance and availability of the interception unit;
REGA = REA1;
REA1 <= 40;

! Impurity load cascade (in kg);
E1A = 0 ;
E2A = E1A + D1A*(20 - 0)/1000;
E3A = E2A + D2A*(100 - 20)/1000;
E4A = E3A + D3A*(200 - 100)/1000;
E5A = E4A + D4A*(400 - 200)/1000;
E6A = E5A + D5A*(1000000 - 400)/1000;

! E2A-E6A should take non-negative values ;
E2A >= 0; E3A >= 0; E4A >= 0;
E5A >= 0; E6A >= 0;

! D1A-D4A may take negative values;
@free (D1A); @free (D2A);
@free (D3A); @free (D4A);
```

```
! Constraints for interval B;
! Water cascade;
D1B=D0B+(STAB1 - STBC1);
D2B=D1B+(STAB2 - STBC2)+REGB;
D3B=D2B - 75+(STAB3 - STBC3);
D4B=D3B+25+(STAB4 - STBC4) - REB1;
D5B=D4B+50+(STAB5 - STBC5) - REB2;

! Net material flows entering the first and the last
concentration levels should take non-negative values;
D0B >= 0;
D5B >= 0;

! These net material flows correspond to minimum freshwater
(FWt) and wastewater (WWt) flows for each time interval;
D0B=FWB;
D5B=WWB;

! Flow balance and availability of the interception unit;
REGB=REB1+REB2;
REB1 <= 25;
REB2 <= 50;

! Impurity load cascade (in kg) ;
E1B=0 ;
E2B=E1B+D1B*(20 - 0)/1000;
E3B=E2B+D2B*(100 - 20)/1000;
E4B=E3B+D3B*(200 - 100)/1000;
E5B=E4B+D4B*(400 - 200)/1000;
E6B=E5B+D5B*(1000000 - 400)/1000;

! E2B-E6B should take non-negative values;
E2B >= 0; E3B >= 0; E4B >= 0;
E5B >= 0; E6B >= 0;

! D1B-D4B may take negative values;
@free (D1B); @free (D2B);
@free (D3B); @free (D4B);

! Constraints for interval C;
! Water cascade;
D1C=D0C+(STBC1 - STCA1);
D2C=D1C+(STBC2 - STCA2)+REGC;
D3C=D2C - 50+(STBC3 - STCA3);
D4C=D3C+(STBC4 - STCA4);
D5C=D4C+50+(STBC5 - STCA5) - REC1;

! Net material flows entering the first and the last
concentration levels should take non-negative values;
D0C >= 0;
D5C >= 0;

! These net material flows correspond to minimum freshwater
(FWt) and wastewater (WWt) flows for each time interval;
D0C=FWC;
D5C=WWC;

! Flow balance and availability of the interception unit;
REGC=REC1;
REC1 <= 50;
```

```
! Impurity load cascade (in kg);
E1C = 0 ;
E2C = E1C + D1C* (20 - 0)/1000;
E3C = E2C + D2C* (100 - 20)/1000;
E4C = E3C + D3C* (200 - 100)/1000;
E5C = E4C + D4C* (400 - 200)/1000;
E6C = E5C + D5C* (1000000 - 400)/1000;

! E2C-E6C should take non-negative values;
E2C >= 0; E3C >= 0; E4C >= 0;
E5C >= 0; E6C >= 0;

! D1C-D4C may take negative values;
@free (D1C); @free (D2C);
@free (D3C); @free (D4C);

! Constraints for water storage tanks;
! Cumulative content of water storage tanks;
MSB1 = MSA1 - STCA1 + STAB1; !0 ppm level;
MSC1 = MSB1 - STAB1 + STBC1;
MSA1 = MSC1 - STBC1 + STCA1;
MSB2 = MSA2 - STCA2 + STAB2; !20 ppm level;
MSC2 = MSB2 - STAB2 + STBC2;
MSA2 = MSC2 - STBC2 + STCA2;
MSB3 = MSA3 - STCA3 + STAB3; !100 ppm level;
MSC3 = MSB3 - STAB3 + STBC3;
MSA3 = MSC3 - STBC3 + STCA3;
MSB4 = MSA4 - STCA4 + STAB4; !200 ppm level;
MSC4 = MSB4 - STAB4 + STBC4;
MSA4 = MSC4 - STBC4 + STCA4;
MSB5 = MSA5 - STCA5 + STAB5; !400 ppm level;
MSC5 = MSB5 - STAB5 + STBC5;
MSA5 = MSC5 - STBC5 + STCA5;

! Sizing of water storage at each level;
STG1 >= MSA1; STG1 >= MSB1; STG1 >= MSC1;
STG2 >= MSA2; STG2 >= MSB2; STG2 >= MSC2;
STG3 >= MSA3; STG3 >= MSB3; STG3 >= MSC3;
STG4 >= MSA4; STG4 >= MSB4; STG4 >= MSC4;
STG5 >= MSA5; STG5 >= MSB5; STG5 >= MSC5;

! Stored water flows should take non-negative values;
STAB1 >= 0; STBC1 >= 0; STCA1 >= 0;
STAB2 >= 0; STBC2 >= 0; STCA2 >= 0;
STAB3 >= 0; STBC3 >= 0; STCA3 >= 0;
STAB4 >= 0; STBC4 >= 0; STCA4 >= 0;
STAB5 >= 0; STBC5 >= 0; STCA5 >= 0;

! Cumulative content of water storage tanks should take
non-negative values;
MSA1 >= 0; MSB1 >= 0; MSC1 >= 0;
MSA2 >= 0; MSB2 >= 0; MSC2 >= 0;
MSA3 >= 0; MSB3 >= 0; MSC3 >= 0;
MSA4 >= 0; MSB4 >= 0; MSC4 >= 0;
MSA5 >= 0; MSB5 >= 0; MSC5 >= 0;

! Cost estimation for water storage tank;
!Capital cost estimation;
CSTG1 = (235*STG1 + 20300*BSTG1);
```

```
CSTG2 = (235*STG2 + 20300*BSTG2);
CSTG3 = (235*STG3 + 20300*BSTG3);
CSTG4 = (235*STG4 + 20300*BSTG4);
CSTG5 = (235*STG5 + 20300*BSTG5);

!Binary variables that represent present/absent of a tank;
STG1 <= M*BSTG1; STG2 <= M*BSTG2; STG3 <= M*BSTG3;
STG4 <= M*BSTG4; STG5 <= M*BSTG5;
M = 1000;

!Defining binary variables;
@BIN (BSTG1); @BIN (BSTG2); @BIN (BSTG3);
@BIN (BSTG4); @BIN (BSTG5);
END
```

The following is the solution report generated by LINGO:

```
Global optimal solution found.
Objective value:                     724792.5
Extended solver steps:                      3
Total solver iterations:                   30


                  Variable          Value
                  TAC            724792.5
                  AOC            718444.4
                  SC             27480.56
                  OC             135.5556
                  FWA            40.00000
                  FWB            0.000000
                  FWC            0.000000
                  WWA            9.444444
                  WWB            8.333333
                  WWC            22.22222
                  REA1           0.000000
                  REB1           0.000000
                  REB2           41.66667
                  REC1           27.77778
                  CSTG1          0.000000
                  CSTG2          0.000000
                  CSTG3          0.000000
                  CSTG4          27480.56
                  CSTG5          0.000000
                  D1A            0.000000
                  D0A            40.00000
                  STCA1          0.000000
                  STAB1          0.000000
                  D2A            0.000000
                  STCA2          0.000000
                  STAB2          0.000000
                  REGA           0.000000
                  D3A            0.000000
                  STCA3          0.000000
                  STAB3          0.000000
                  D4A            9.444444
                  STCA4          0.000000
```

STAB4	30.55556
D5A	9.444444
STCA5	0.000000
STAB5	0.000000
E1A	0.000000
E2A	0.000000
E3A	0.000000
E4A	0.000000
E5A	1.888889
E6A	9442.556
D1B	0.000000
D0B	0.000000
STBC1	0.000000
D2B	41.66667
STBC2	0.000000
REGB	41.66667
D3B	−33.33333
STBC3	0.000000
D4B	0.000000
STBC4	22.22222
D5B	8.333333
STBC5	0.000000
E1B	0.000000
E2B	0.000000
E3B	3.333333
E4B	0.000000
E5B	0.000000
E6B	8330.000
D1C	0.000000
D0C	0.000000
D2C	27.77778
REGC	27.77778
D3C	−22.22222
D4C	0.000000
D5C	22.22222
E1C	0.000000
E2C	0.000000
E3C	2.222222
E4C	0.000000
E5C	0.000000
E6C	22213.33
MSB1	0.000000
MSA1	0.000000
MSC1	0.000000
MSB2	0.000000
MSA2	0.000000
MSC2	0.000000
MSB3	0.000000
MSA3	0.000000
MSC3	0.000000
MSB4	30.55556
MSA4	0.000000
MSC4	22.22222
MSB5	0.000000
MSA5	0.000000

MSC5	0.000000
STG1	0.000000
STG2	0.000000
STG3	0.000000
STG4	30.55556
STG5	0.000000
BSTG1	0.000000
BSTG2	0.000000
BSTG3	0.000000
BSTG4	1.000000
BSTG5	0.000000
M	1000.000

The optimization result indicates that the batch regeneration network has a TAC of $725,000. The revised cascade diagrams for all time intervals are shown in Figure 17.10 (impurity load cascades are omitted). As shown, freshwater is only needed in time interval A, with a flow of 40 tons. On the other hand, the wastewater flows are determined as 9.44, 8.33, and 22.22 tons for time intervals A, B, and C, respectively. The model also determines that water sources at 400 ppm in time intervals B (41.67 tons) and C (27.78 tons) are sent for regeneration within the time intervals. This leads to complete removal of freshwater flows in these time intervals. The model also determines that a water storage tank of 30.56 tons is required for water recovery between time intervals. The results are identical to those obtained using pinch-based targeting technique (see Example 11.3).

Similar to the case in Example 17.1, one may utilize the insights of the ATM results in Figure 17.10 to develop the structure of the batch regeneration network. Alternatively, we may also make use of the NNA, which will result in a network structure like that in Figure 11.7 (see Example 11.4 for detailed steps).

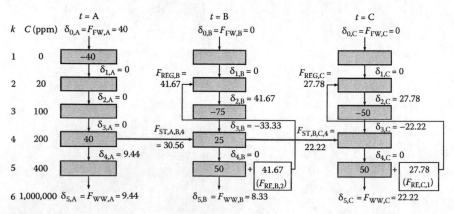

FIGURE 17.10
Result of ATM for Example 17.2 (flow in tons).

17.4 ATM for Batch Total Network

ATM for *total material networks* that employs different type of regeneration and waste treatment units have been discussed in Chapter 15. The models are now extended for use in *batch total networks*. The ATM framework for one of the variants is shown in Figure 17.11, which is a revised form of that in Figure 15.9. In this case, process sources may be sent for regeneration in a fixed outlet quality type interception unit (with flow terms $F_{RE,t,r}$) or treated in waste treatment units (with flow terms $F_{TR,t,w}$) prior to final environmental discharge.

For the case in Figure 17.11, the effluent from the waste treatment units will not be recycled to the BMN. Hence, the waste treatment network (WTN) has a separate ATM framework. Similar to the cases in batch regeneration networks, the constraints for the ATM framework in Figure 17.11 are extended from those in Figure 15.9 (Equations 15.10 through 15.15), given as follows:

$$\delta_{k,t} = \delta_{k-1,t} + (\Sigma_i\, F_{SRi} - \Sigma_j\, F_{SKj})_{k,t} + (F_{ST,t-1,t} - F_{ST,t,t+1})_k$$

$$+ F_{REG,t,r} - \Sigma_r F_{RE,t,r} - F_{TR,t,w} \quad k = 1,\, \ldots,\, n-1;\, \forall t;\, \forall r;\, \forall w \quad (17.28)$$

$$\delta_{k-1,t} = \begin{cases} \geq 0 & k = 1;\, \forall t \\ 0 & k = n;\, \forall t \end{cases} \quad (17.29)$$

$$F_{RE,r,t} + F_{TR,t,w} \leq F_{SRi,t} \quad r = 1, 2, \ldots, RG;\, q_k < q_{Rout}\ \forall t \quad (17.30)$$

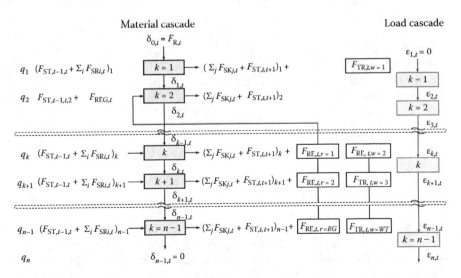

FIGURE 17.11
ATM framework for batch total network without waste recycling.

Equation 17.28 is an extended form of net material flow in Equation 17.17, to which the waste treatment flow terms ($F_{TR,t,w}$)are added. According to Equation 17.29, the net material flow that enters the first quality level (δ_0) should take non-negative values, while that to the last levels (δ_{n-1}) is set to zero. The constraint in Equation 17.30 indicates the flow availability of source to be sent for regeneration and waste treatment. In addition, the constraints in Equations 17.2, 17.4 through 17.8, and 17.18 are also needed for the ATM framework in Figure 17.11.

As mentioned earlier, a separate ATM framework is needed for the WTN, as the effluent of the WTN cannot be recycled within the BMN. Similar to the earlier cases, there exist two variants of ATM frameworks for the WTN, due to the characteristics of the waste treatment units, i.e., fixed outlet quality or removal ratio types. Both of these variants have been discussed in Chapter 15 and, hence, will not be further elaborated here (see Sections 15.1 and 15.2 for details). In principle, different sets of constraints are to be included when different types of waste treatment unit are used in the WTN. For instance, when the WTN employs a fixed outlet quality type treatment unit, the ATM framework is given by Figure 15.1 and the constraints in Equations 15.1 through 15.6 are to be used (see Example 17.3 for detailed steps). Alternatively, when a removal ratio type treatment unit is used, the ATM framework is represented by that in Figure 15.2, with the constraints given by Equations 15.2 and 15.4 through 15.8. Note however that the main assumption for the adaption of this model is that, the WTN is operated in a pseudo-continuous mode of operation. Due to the waste streams that are emitted from different time in interval of the BMN, we will have to make use of mass storage tank to mix the treated and bypassed waste streams in order to meet the environmental limit for final discharge. This will be shown in Example 17.3.

In another variant, where the effluent of the WTN is allowed for recycling in the BMN, the corresponding ATM framework is shown in Figure 17.12 (a revised form of Figure 15.19 for continuous RCN). For this case, apart from Equations 17.2, 17.4 through 17.8, 17.18, 17.29, and 17.30, two other new constraints are also needed:

$$\delta_{k,t} = \delta_{k-1,t} + (\Sigma_i F_{SRi} - \Sigma_j F_{SKj})_{k,t} + (F_{ST,t-1,t} - F_{ST,t,t+1})_k + F_{REG,t,r} - \Sigma_r F_{RE,t,r}$$
$$+ F_{TRD,t,w} - \Sigma_r F_{TR,t,w} - F_{D,t} \quad k = 1, \ldots, n-1; \forall t; \forall r; \forall w \tag{17.31}$$

$$F_{TRD,t} = \Sigma_r F_{TR,t,w} \quad \forall t \tag{17.32}$$

In Equation 17.31, the treated waste ($F_{TRD,t,w}$) and the final discharge flows ($F_{D,t}$) are both included in calculating the net material flow. Equation 17.32 describes the flow balance of the waste treatment unit. Note that one may also view the waste treatment unit as an interception unit that regenerate sources for further recovery (see Example 15.4 for such a case). The final

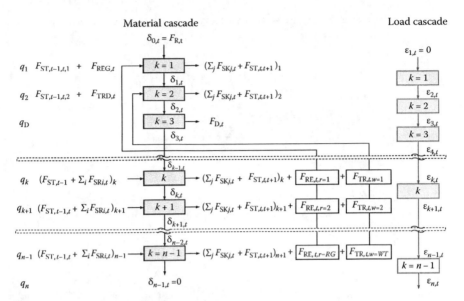

FIGURE 17.12
ATM framework for batch total network with waste recycling.

selection of sources for regeneration or waste treatment will rely on cost consideration, which is the result of the ATM.

Apart from the two variants given in Figures 17.11 and 17.12, different variants of the batch total network introduced in Chapter 15 may be extended here, e.g., those with partitioning interception unit for regeneration. Similarly, the constraints that correspond to each variant may be extended according to those in Chapter 15.

Example 17.3 Batch total water network

Example 17.2 is revisited. In this case, apart from the interception unit that is used for regeneration, a waste treatment unit with outlet concentration (C_T) of 20 ppm is used for treating terminal wastewater streams before they are discharged to the environment. The environment discharge limit (C_D) is set to 50 ppm in this case. The treated effluent from the wastewater treatment unit will not be recycled in the batch water network. Determine the minimum TAC target for the batch water regeneration network. All cost elements remain identical to that of Example 17.2, except that the cost for wastewater treatment is given as a function of impurity load removal, i.e., $0.5/kg impurity load.

Solution

The revised cascade diagrams for all time intervals are shown in Figure 17.13a through c, while that for the pseudo-continuous WTN is shown in Figure 17.13d.

For time interval A, the water cascade across all quality levels is given by Equation 17.28:

$$\delta_{1,A} = \delta_{0,A} - 40 + (F_{ST,C,A,1} - F_{ST,A,B,1})$$

$$\delta_{2,A} = \delta_{1,A} + (F_{ST,C,A,2} - F_{ST,A,B,2}) + F_{REG,A}$$

$$\delta_{3,A} = \delta_{2,A} + (F_{ST,C,A,3} - F_{ST,A,B,3})$$

$$\delta_{4,A} = \delta_{3,A} + 40 + (F_{ST,C,A,4} - F_{ST,A,B,4}) - F_{RE,A,1} - F_{WW,A,1}$$

$$\delta_{5,A} = \delta_{4,A} + (F_{ST,C,A,5} - F_{ST,A,B,5})$$

Equation 17.29 indicates that net material flows that enter the first concentration level should take non-negative values, while no flow should enter the last level:

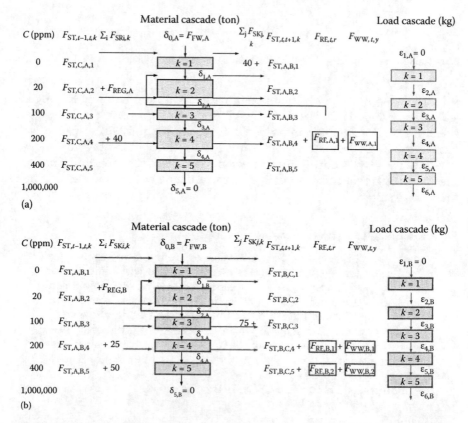

(a)

(b)

FIGURE 17.13
Cascade diagrams for Example 17.3: (a) time interval A, (b) time interval B, (c) time interval C, and (d) WTN.

(continued)

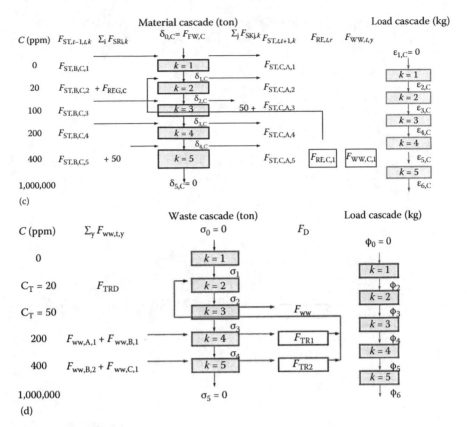

FIGURE 17.13 (continued)

$$\delta_{0,A} \geq 0$$

$$\delta_{5,A} = 0$$

Next, flow balance and availability of the interception unit in this time interval are given by Equations 17.18 and 17.30, respectively:

$$F_{REG,A} = F_{RE,A,1}$$

$$F_{RE,A,1} + F_{WW,A,1} \leq 40$$

Similarly, water cascade, as well as the flow balance and availability of the interception unit, for time interval B is given by Equations 17.18, 17.28, 17.29, and 17.30:

$$\delta_{1,B} = \delta_{0,B} + (F_{ST,A,B,1} - F_{ST,B,C,1})$$

$$\delta_{2,B} = \delta_{1,B} + (F_{ST,A,B,2} - F_{ST,B,C,2}) + F_{REG,B}$$

$$\delta_{3,B} = \delta_{2,B} - 75 + (F_{ST,A,B,3} - F_{ST,B,C,3})$$

$$\delta_{4,B} = \delta_{3,B} + 25 + (F_{ST,A,B,4} - F_{ST,B,C,4}) - F_{RE,B,1} - F_{WW,B,1}$$

$$\delta_{5,B} = \delta_{4,B} + 50 + (F_{ST,A,B,5} - F_{ST,B,C,5}) - F_{RE,B,2} - F_{WW,B,2}$$

$$\delta_{0,B} \geq 0$$

$$\delta_{5,B} = 0$$

$$F_{REG,B} = F_{RE,B,1} + F_{RE,B,2}$$

$$F_{RE,B,1} + F_{WW,B,1} \leq 25$$

$$F_{RE,B,2} + F_{WW,B,2} \leq 50$$

Similarly, the water cascade equations as well as the flow balance and availability of the interception unit for time interval C are given as follows:

$$\delta_{1,C} = \delta_{0,C} + (F_{ST,B,C,1} - F_{ST,C,A,1})$$

$$\delta_{2,C} = \delta_{1,C} + (F_{ST,B,C,2} - F_{ST,C,A,2}) + F_{REG,C}$$

$$\delta_{3,C} = \delta_{2,C} - 50 + (F_{ST,B,C,3} - F_{ST,C,A,3})$$

$$\delta_{4,C} = \delta_{3,C} + (F_{ST,B,C,4} - F_{ST,C,A,4})$$

$$\delta_{5,C} = \delta_{4,C} + 50 + (F_{ST,B,C,5} - F_{ST,C,A,5}) - F_{RE,C,1} - F_{WW,C,1}$$

$$\delta_{0,C} \geq 0$$

$$\delta_{5,C} = 0$$

$$F_{REG,C} = F_{RE,C,1}$$

$$F_{RE,C,1} + F_{WW,C,1} \leq 50$$

The rest of the constraints for impurity load and mass storage system (given by Equations 17.2, 17.4 through 17.8) for each time interval are identical to those in Examples 17.1 (Case 2) and 17.2, and, hence, are omitted here for brevity.

On the other hand, for the pseudo-continous WTN represented by the ATM framework in Figure 17.13d, the constraints are given by Equations 15.1 through 15.6, which resemble the case in Example 15.2. Equation 15.1 describes the waste cascade across all quality levels:

$$\sigma_1 = \sigma_0$$

$$\sigma_2 = \sigma_1 + F_{TRD}$$

$$\sigma_3 = \sigma_2 - F_{WW}$$

$$\sigma_4 = \sigma_3 + (F_{WW,A,1} + F_{WW,B,1} - F_{TR1})$$

$$\sigma_5 = \sigma_4 + (F_{WW,B,2} + F_{WW,C,1} - F_{TR2})$$

No net wastewater flow enters the first and the last levels of the waste cascade, based on Equation 15.2:

$$\sigma_0 = 0$$

$$\sigma_5 = 0$$

Equation 15.3 describes the flow balance of the wastewater treatment unit:

$$F_{TRD} = F_{TR1} + F_{TR2}$$

Next, the wastewater source to be treated should not be more than what is available, based on Equation 15.4:

$$F_{TR1} \leq F_{WW,A,1} + F_{WW,B,1}$$

$$F_{TR2} \leq F_{WW,B,2} + F_{WW,C,1}$$

Equation 15.5 describes the waste load cascades (in kg) across all concentration levels:

$$\varphi_1 = 0$$

$$\varphi_2 = \varphi_1 + \delta_1(20 - 0)/1,000$$

$$\varphi_3 = \varphi_2 + \delta_2(50 - 20)/1,000$$

$$\varphi_4 = \varphi_3 + \delta_3(200 - 50)/1,000$$

$$\varphi_5 = \varphi_4 + \delta_4(400 - 200)/1,000$$

$$\varphi_6 = \varphi_5 + \delta_5(1,000,000 - 400)/1,000$$

According to Equation 15.6, all residual loads (except at level 1) should take non-negative values:

$$\varphi_2 \geq 0; \quad \varphi_3 \geq 0$$

$$\varphi_4 \geq 0; \quad \varphi_5 \geq 0$$

$$\varphi_6 \geq 0$$

For the optimization objective, the various cost elements are identical to that in Example 17.2, except that the operating cost (OC) is revised as follows:

$$OC \ (\$/h) = \Sigma_t F_{FW,t} \ CT_{FW} + \Sigma_r F_{RE,A,r} \ CT_{RW} + \Sigma_w F_{TR,w} \ (C_{SRi} - C_T) \ CT_{TRD}$$

$$= (F_{FW,A} + F_{FW,B} + F_{FW,C}) \ (\$1/\text{ton})$$

$$+ (F_{RE,A,1} + F_{RE,B,1} + F_{RE,B,2} + F_{RE,C,1}) \ (\$0.8/\text{ton})$$

$$+ F_{TR,1} \ (200 - 20)/1000 \ (\$0.5/\text{kg})$$

$$+ F_{TR,2} \ (400 - 20)/1000 \ (\$0.5/\text{kg})$$

The LINGO formulation for the MILP model discussed earlier can be written as follows:

```
Model:
min = TAC;
TAC = AOC + SC'0.231;
OC = (FWA + FWB + FWC)'1 + (REA1 + REB1 + REB2 + REC1)'0.8 +
```

```
    (TR1'(200 - 20)/1000+TR2'(400 - 20)/1000)'0.5;
AOC=OC'5300;
SC= (CSTG1+CSTG2+CSTG3+CSTG4+CSTG5) ;
```

```
! Constraints for interval A;
! Water cascade;
D1A=D0A - 40+(STCA1 - STAB1);
D2A=D1A+(STCA2 - STAB2)+REGA;
D3A=D2A+(STCA3 - STAB3);
D4A=D3A+40+(STCA4 - STAB4) - REA1 - WWA1;
D5A=D4A+(STCA5 - STAB5);
```

```
! Net material flow that enters the first concentration
level should take non-negative values, while that to the
last level should take zero value;
D0A >= 0;
D5A=0;
```

```
! The net material flow entering the first concentration
level correspond to minimum freshwater (FWt) flow for the
time interval;
D0A=FWA;
```

```
! Flow balance and availability of the interception units;
REGA=REA1;
REA1+WWA1 <= 40;
```

```
! Impurity load cascade (in kg);
E1A=0 ;
E2A=E1A+D1A'(20 - 0)/1000;
E3A=E2A+D2A'(100 - 20)/1000;
E4A=E3A+D3A'(200 - 100)/1000;
E5A=E4A+D4A'(400 - 200)/1000;
E6A=E5A+D5A'(1000000 - 400)/1000;
```

```
! E2A-E6A should take non-negative values ;
E2A >= 0; E3A >= 0; E4A >= 0;
E5A >= 0; E6A >= 0;
```

```
! D1A-D4A may take negative values;
@free (D1A); @free (D2A);
@free (D3A); @free (D4A);
```

```
! Constraints for interval B;
! Water cascade;
D1B=D0B+(STAB1 - STBC1);
D2B=D1B+(STAB2 - STBC2)+REGB;
D3B=D2B - 75+(STAB3 - STBC3);
D4B=D3B+25+(STAB4 - STBC4) - REB1 - WWB1;
D5B=D4B+50+(STAB5 - STBC5) - REB2 - WWB2;
```

```
! Net material flow that enters the first concentration
level should take non-negative values, while that to the
last level should take zero value;
D0B >= 0;
D5B=0;
```

```
! The net material flow entering the first concentration
level correspond to minimum freshwater (FWt) flow for the
time interval;
D0B = FWB;

! Flow balance and availability of the interception units;
REGB = REB1 + REB2;
REB1 + WWB1 <= 25;
REB2 + WWB2 <= 50;

! Impurity load cascade (in kg) ;
E1B = 0 ;
E2B = E1B + D1B*(20 - 0)/1000;
E3B = E2B + D2B*(100 - 20)/1000;
E4B = E3B + D3B*(200 - 100)/1000;
E5B = E4B + D4B*(400 - 200)/1000;
E6B = E5B + D5B*(1000000 - 400)/1000;

! E2B-E6B should take non-negative values;
E2B >= 0; E3B >= 0; E4B >= 0;
E5B >= 0; E6B >= 0;

! D1B-D4B may take negative values;
@free (D1B); @free (D2B);
@free (D3B); @free (D4B);

! Constraints for interval C;
! Water cascade;
D1C = D0C + (STBC1 - STCA1);
D2C = D1C + (STBC2 - STCA2) + REGC;
D3C = D2C - 50 + (STBC3 - STCA3);
D4C = D3C + (STBC4 - STCA4);
D5C = D4C + 50 + (STBC5 - STCA5) - REC1 - WWC1;

! Net material flow that enters the first concentration
level should take non-negative values, while that to the
last level should take zero value;
D0C >= 0;
D5C = 0;

! The net material flow entering the first concentration
level correspond to minimum freshwater (FWt) flow for the
time interval;
D0C = FWC;

! Flow balance and availability of the interception units;
REGC = REC1;
REC1 + WWC1 <= 50;

! Impurity load cascade (in kg);
E1C = 0 ;
E2C = E1C + D1C*(20 - 0)/1000;
E3C = E2C + D2C*(100 - 20)/1000;
E4C = E3C + D3C*(200 - 100)/1000;
E5C = E4C + D4C*(400 - 200)/1000;
E6C = E5C + D5C*(1000000 - 400)/1000;

! E2C-E6C should take non-negative values;
E2C >= 0; E3C >= 0; E4C >= 0;
E5C >= 0; E6C >= 0;
```

```
! D1C-D4C may take negative values;
@free (D1C); @free (D2C); @free (D3C); @free (D4C);

! Constraints for water storage tanks;
! Cumulative content of water storage tanks;
MSB1 = MSA1 - STCA1 + STAB1; !0 ppm level;
MSC1 = MSB1 - STAB1 + STBC1;
MSA1 = MSC1 - STBC1 + STCA1;
MSB2 = MSA2 - STCA2 + STAB2; !20 ppm level;
MSC2 = MSB2 - STAB2 + STBC2;
MSA2 = MSC2 - STBC2 + STCA2;
MSB3 = MSA3 - STCA3 + STAB3; !100 ppm level;
MSC3 = MSB3 - STAB3 + STBC3;
MSA3 = MSC3 - STBC3 + STCA3;
MSB4 = MSA4 - STCA4 + STAB4; !200 ppm level;
MSC4 = MSB4 - STAB4 + STBC4;
MSA4 = MSC4 - STBC4 + STCA4;
MSB5 = MSA5 - STCA5 + STAB5; !400 ppm level;
MSC5 = MSB5 - STAB5 + STBC5;
MSA5 = MSC5 - STBC5 + STCA5;

! Sizing of water storage at each level;
STG1 >= MSA1; STG1 >= MSB1; STG1 >= MSC1;
STG2 >= MSA2; STG2 >= MSB2; STG2 >= MSC2;
STG3 >= MSA3; STG3 >= MSB3; STG3 >= MSC3;
STG4 >= MSA4; STG4 >= MSB4; STG4 >= MSC4;
STG5 >= MSA5; STG5 >= MSB5; STG5 >= MSC5;

! Stored water flows should take non-negative values;
STAB1 >= 0; STBC1 >= 0; STCA1 >= 0;
STAB2 >= 0; STBC2 >= 0; STCA2 >= 0;
STAB3 >= 0; STBC3 >= 0; STCA3 >= 0;
STAB4 >= 0; STBC4 >= 0; STCA4 >= 0;
STAB5 >= 0; STBC5 >= 0; STCA5 >= 0;

! Cumulative content of water storage tanks should take
non-negative values;
MSA1 >= 0; MSB1 >= 0; MSC1 >= 0;
MSA2 >= 0; MSB2 >= 0; MSC2 >= 0;
MSA3 >= 0; MSB3 >= 0; MSC3 >= 0;
MSA4 >= 0; MSB4 >= 0; MSC4 >= 0;
MSA5 >= 0; MSB5 >= 0; MSC5 >= 0;

! Cost estimation for water storage tank;
!Capital cost estimation;
CSTG1 = (235*STG1 + 20300*BSTG1);
CSTG2 = (235*STG2 + 20300*BSTG2);
CSTG3 = (235*STG3 + 20300*BSTG3);
CSTG4 = (235*STG4 + 20300*BSTG4);
CSTG5 = (235*STG5 + 20300*BSTG5);

!Binary variables that represent present/absent of a tank;
STG1 <= M*BSTG1; STG2 <= M*BSTG2; STG3 <= M*BSTG3;
STG4 <= M*BSTG4; STG5 <= M*BSTG5;
M = 1000;

!Defining binary variables;
@BIN (BSTG1); @BIN (BSTG2); @BIN (BSTG3);
```

```
@BIN (BSTG4); @BIN (BSTG5);
```

! Constraints for WTN ;
```
! Waste cascade;
S1 = S0 ;
S2 = S1 + TRD ;
S3 = S2 - WW ;
S4 = S3 + ( WWA1 + WWB1 - TR1 ) ;
S5 = S4 + ( WWB2 + WWC1 - TR2 ) ;

! No net waste flow enters the first and the last levels;
S0 = 0 ;
S5 = 0 ;

! Flow balance(s) for treatment unit(s) ;
TRD = TR1 + TR2 ;

! Flow constraints of treated wastewater source ;
TR1 <= WWA1 + WWB1 ;
TR2 <= WWB2 + WWC1 ;

! Waste load cascade (in kg) ;
J1 = 0 ;
J2 = J1 + S1*(20 - 0)/1000 ;
J3 = J2 + S2*(50 - 20)/1000 ;
J4 = J3 + S3*(200 - 50)/1000 ;
J5 = J4 + S4*(400 - 200)/1000 ;
J6 = J5 + S5*(1000000 - 400)/1000 ;

! J2-J6 should take non-negative values ;
J2 >= 0 ; J3 >= 0 ; J4 >= 0 ;
J5 >= 0 ; J6 >= 0 ;

! S1-S4 may take negative values ;
@free (S1) ; @free (S2) ;
@free (S3) ; @free (S4) ;
END
```

The following is the solution report generated by LINGO:

```
Global optimal solution found.
Objective value:                    544886.9
Extended solver steps:                     3
Total solver iterations:                  79

                Variable          Value
                TAC             544886.9
                AOC             538538.9
                SC              27480.56
                OC              101.6111
                FWA             40.00000
                FWB             0.000000
                FWC             0.000000
                REA1            0.000000
                REB1            0.000000
                REB2            41.66667
                REC1            27.77778
                TR1             2.777778
```

TR2	30.55556
CSTG1	0.000000
CSTG2	0.000000
CSTG3	0.000000
CSTG4	27480.56
CSTG5	0.000000
D1A	0.000000
D0A	40.00000
STCA1	0.000000
STAB1	0.000000
D2A	0.000000
STCA2	0.000000
STAB2	0.000000
REGA	0.000000
D3A	0.000000
STCA3	0.000000
STAB3	0.000000
D4A	0.000000
STCA4	0.000000
STAB4	30.55556
WWA1	9.444444
D5A	0.000000
STCA5	0.000000
STAB5	0.000000
E1A	0.000000
E2A	0.000000
E3A	0.000000
E4A	0.000000
E5A	0.000000
E6A	0.000000
D1B	0.000000
D0B	0.000000
STBC1	0.000000
D2B	41.66667
STBC2	0.000000
REGB	41.66667
D3B	-33.33333
STBC3	0.000000
D4B	0.000000
STBC4	22.22222
WWB1	0.000000
D5B	0.000000
STBC5	0.000000
WWB2	8.333333
E1B	0.000000
E2B	0.000000
E3B	3.333333
E4B	0.000000
E5B	0.000000
E6B	0.000000
D1C	0.000000
D0C	0.000000
D2C	27.77778
REGC	27.77778
D3C	-22.22222
D4C	0.000000

```
D5C               0.000000
WWC1             22.22222
E1C               0.000000
E2C               0.000000
E3C               2.222222
E4C               0.000000
E5C               0.000000
E6C               0.000000
MSB1              0.000000
MSA1              0.000000
MSC1              0.000000
MSB2              0.000000
MSA2              0.000000
MSC2              0.000000
MSB3              0.000000
MSA3              0.000000
MSC3              0.000000
MSB4             30.55556
MSA4              0.000000
MSC4             22.22222
MSB5              0.000000
MSA5              0.000000
MSC5              0.000000
STG1              0.000000
STG2              0.000000
STG3              0.000000
STG4             30.55556
STG5              0.000000
BSTG1             0.000000
BSTG2             0.000000
BSTG3             0.000000
BSTG4             1.000000
BSTG5             0.000000
M              1000.000
S1                0.000000
S0                0.000000
S2               33.33333
TRD              33.33333
S3               -6.666667
WW               40.00000
S4                0.000000
S5                0.000000
J1                0.000000
J2                0.000000
J3                1.000000
J4                0.000000
J5                0.000000
J6                0.000000
```

The optimization result indicates that the batch total network has a TAC of $545,000. The revised cascade diagrams for all time intervals as well as that for the WTN are shown in Figure 17.14 (impurity load cascades are omitted for the former). As observed in Figure 17.14a, the results are quite similar to the case in Example 17.2 (see Figure 17.10), except that the wastewater flows are now emitted from the concentration levels where

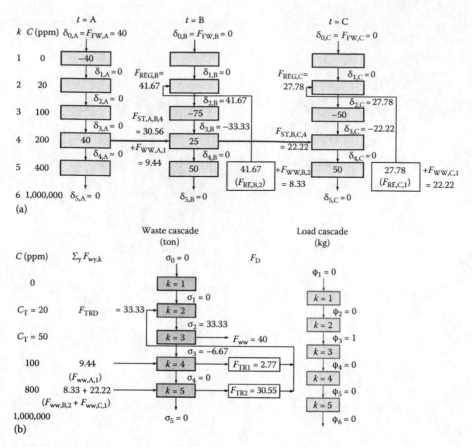

FIGURE 17.14
Result of ATM for Example 17.3: (a) water cascade for time intervals A–C (flow in tons) and (b) cascade diagram for WTN.

the sources exist, i.e., 200 ppm (time interval A—$F_{WW,A,1} = 9.44$ tons) and 400 ppm (time intervals B and C—$F_{WW,B,2} = 8.33$ tons; $F_{WW,C,1} = 22.22$ tons), respectively. Similar to the earlier cases, we may make use of the insights in Figure 17.14a or the NNA to design the batch total network. For this case, the network structure should take the same form as in Example 17.2.

The wastewater flows emitted from the water network are sent to the WTN for end-of-pipe treatment before final discharge. The ATM result in Figure 17.14b indicates that the wastewater sources at 400 ppm from time intervals B and C ($F_{TR2} = 30.56$ tons) are fully treated. On the other hand, only 2.77 tons of wastewater source at 200 ppm (F_{TR1} from time interval A) is treated, while 6.67 tons (= 9.44 − 2.77 tons) of this wastewater source will bypass the treatment unit. In other words, only 33.33 tons (F_{TRD}) of the total wastewater flow ($F_{WW} = 40$ tons) will be treated prior to environmental discharge. As mentioned earlier, a water storage tank is needed to mix the treated and bypassed wastewater streams in order to meet the environmental discharge limit of 50 ppm. This is similar to the case in Figure 11.9. Hence, the network structure of the batch total network takes the same form as that in Figure 11.10.

17.5 Further Reading

In this chapter, the ATM has been utilized to optimize for minimum water flows/storage size, as well as the minimum cost solution for a BMN. In some cases, space requirement is a critical issue in the batch process plants. For such cases, we may incorporate fuzzy optimization techniques within the ATM for trade-off analysis among the minimum flows of freshwater and regeneration, as well as the mass storage size. This is reported in the work of Nun et al. (2011).

Apart from ATM, many other mathematical optimization approaches have also been recently developed. Most of these models are based on super-structural approaches (similar to the basic framework in Chapter 12). These include the works of Almató et al. (1999), Kim and Smith (2004), Li and Chang (2006); Ng et al. (2008), Chen et al. (2008, 2009, 2010), Majozi (2005, 2006), Cheng and Chang (2007), Gouws and Majozi (2008), and Majozi and Gouws (2009). A comparison for these techniques may be found in the review paper by Gouws et al. (2010). A useful text book for this technique is also available (Majazi, 2010).

Problems

Water Minimization Problems

17.1 For the batch water network in Problem 11.2 (Kim and Smith, 2004), use ATM to determine the minimum freshwater and wastewater flows for a direct reuse/recycle scheme for the case in tasks: (a) without water storage and (b) with water storage system.

17.2 Use ATM to determine the minimum freshwater and wastewater flows for the batch water network in Problem 11.3 (Chen and Lee, 2008), operated in repeated batch mode. Also, determine the number and sizes of water storage tanks needed.

17.3 Revisit Problem 11.4, where water is used extensively as a washing agent and reaction solvent in an agrochemical manufacturing plant (Majozi et al., 2006). Determine the minimum freshwater and wastewater flows for direct reuse/recycle scheme for repeated batch processes with a water storage tank. Also determine the size of the water storage tank. How is the solution different from the one obtained with the insight-based technique (Task (b) in Problem 11.4)?

17.4 Rework the water recovery problem for the batch polyvinyl chloride (PVC) resins manufacturing plant in Problem 11.5 (Chan et al., 2008). Determine the minimum TAC for the water network, assuming that the process has a cycle time of 14.78 h, and an annual operating time of 7982 h. The unit costs of freshwater (CT_{FW}) and wastewater (CT_{WW}) are given as \$1/ton each. The capital cost for water storage tanks is given in Equation 17.24, while piping cost may be ignored. The storage cost is annualized to a period of 5 years, with an interest rate of 5%. What is the storage size for this case and does it match with that determine using the insight-based approach in Problem 11.5? (note: beware of using a single water source that is available as a continuous stream; the parameter M in Equation 17.27 should take a much larger number; a licensed LINGO software is needed to solve this case, due to the large number of constraints and variables in the ATM).

17.5 Rework the water minimization problem in Example 17.3 (Wang and Smith, 1995). In this case, the treated effluent from the wastewater treatment unit may be recycled in the water network. All process and economic data remain identical to that in Example 17.3. Determine the minimum TAC for the water network.

17.6 Table 17.3 shows the limiting water data for a water minimization study in a fruit juice production plant (Almató et al., 1999; Li and Chang, 2006; Chen and Lee, 2010; Foo, 2010). Solve the following cases assuming that the plant is operated in repeated batch mode with a water storage system:

TABLE 17.3

Limiting Water Data for Problem 17.4

SK_j	F_{SKj} (tons)	C_{SKj} (ppm)	T^{STT} (h)	T^{END} (h)
SK1	20	0	0.5	2.5
SK2	20	6	5	7
SK3	20	15	9.5	11.5
SK4	16	5	17	19
SK5	20	7	6	8
SR_i	F_{SRi} (tons)	C_{SRi} (ppm)	T^{STT} (h)	T^{END} (h)
SR1	20	5	2.5	4.5
SR2	20	14	7	9
SR3	20	20	11.5	13.5
SR4	8	25	17	19
SR5	16	10	10.5	14.5

a. Determine the minimum freshwater and wastewater flow targets for a direct reuse/recycle scheme, as well as the minimum size of the water storage tanks.

b. Determine the minimum TAC target for a direct reuse/recycle scheme. The unit costs of freshwater (CT_{FW}) and wastewater (CT_{WW}) are given as $1/ton each. Capital cost for water storage tanks is given in Equation 17.24, while piping cost may be ignored. The storage cost is annualized to a period of 5 years, with an interest rate of 5%. Assume an annual production of 420 batches.

c. Determine the minimum TAC target for a water regeneration network. In this case, a partitioning interception unit is used to regenerate process sources. The unit has a recovery factor (RC)* of 0.8 and an outlet concentration of 5 ppm. The unit cost for water regeneration (CT_{RW}) is assumed to be $0.8/ton, which includes capital and operating costs. Other cost parameters remain unchanged as in (b). Identify also the time intervals where the interception unit will be in operation.

d. Determine the minimum TAC target for a batch total water network. In this case, a single pass interception unit with outlet concentration of 5 ppm is used for source purification. The purified source may be sent to the process sinks for further reuse/recycle, or for final waste discharge. The environmental discharge limit is set to 5 ppm. The unit cost for water regeneration (CT_{RW}) is assumed to be $0.8/ton, which includes capital and operating costs (this includes the cost of wastewater treatment). Other cost parameters remain unchanged as in (b). Identify the minimum TAC target for this case. Identify also the time intervals where the interception unit will be in operation. Also identify when the storage tank will be used to store the regenerated water source for recovery at later time intervals.

e. Synthesize the batch water network for case (d), either with the insights gained from the ATM or using the NNA (refer to Section 11.5 for detailed steps).

Property Integration Problems

17.7 Revisit the property-based batch water network in Problem 11.6 (Ng et al., 2008), where pH is the main water stream quality to be considered for water recovery (see mixing rule in Equation 11.1). Two fresh resources are available for use without any flow constraint, i.e., acid and freshwater, with unit costs of $0.20/kg ($CT_{AC}$) and $0.01/kg ($CT_{FW}$), respectively. The total annualized cost of a storage tank (including pumping and piping) is given as $15,000/year. Limiting data are given in Table 11.22. Determine the minimum TAC target for this BMN, assuming an annual operating time (AOT) of 8760 h.

---

* Refer to Example 14.3 for calculation of a partitioning interception unit.

References

Almató, M., Espuña, A., and Puigjaner, L. 1999. Optimisation of water use in batch process industries. *Computers and Chemical Engineering*, 23,1427–1437.

Chan, J. H., Foo, D. C. Y., Kumaresan, S., Aziz, R. A., and Hassan, M. A. A. 2008. An integrated approach for water minimisation in a PVC manufacturing process. *Clean Technologies and Environmental Policy*, 10(1), 67–79.

Chen, C.-L., Chang, C.-Y., and Lee, J.-Y. 2008. Continuous-time formulation for the synthesis of water-using networks in batch plants. *Industrial and Engineering Chemistry Research*, 47, 7818–7832.

Chen, C. L. and Lee, J. Y. 2008. A graphical technique for the design of water-using networks in batch processes. *Chemical Engineering Science*, 63, 3740–3754.

Chen, C. L. and Lee, J. Y. 2010. On the use of graphical analysis for the design of batch water networks. *Clean Technologies and Environmental Policy*, 12, 117–123.

Chen, C.-L., Lee, J.-Y., Ng, D. K. S., and Foo D. C. Y. 2010. A unified model of property integration for batch and continuous processes. *AIChE J*, 56, 1845–1858.

Chen, C.-L., Lee, J.-Y., Tang, J.-W., and Ciou, Y.-J. 2009. Synthesis of water-using network with central reusable storage in batch processes. *Computers and Chemical Engineering*, 33, 267–276.

Cheng, K. F. and Chang, C. T. 2007. Integrated water network designs for batch processes. *Industrial and Engineering Chemistry Research*, 46,1241–1253.

Foo, D. C. Y. 2010. Automated targeting technique for batch process integration. *Industrial and Engineering Chemistry Research*, 49(20), 9899–9916.

Gouws, J. F. and Majozi, T. 2008. Impact of multiple storage in wastewater minimisation for multi-contaminant batch plants: Towards zero effluent. *Industrial and Engineering Chemistry Research*, 47, 369–379.

Gouws, J., Majozi, T., Foo, D. C. Y., Chen, C. L., and Lee, J.-Y. 2010. Water minimisation techniques for batch processes. *Industrial and Engineering Chemistry Research*, 49(19), 8877–8893.

Kim, J.-K. and Smith, R. 2004. Automated design of discontinuous water systems. *Process Safety and Environmental Protection*, 82(B3), 238–248.

Li, B. H. and Chang, C. T. 2006. A mathematical programming model for discontinuous water-reuse system design. *Industrial and Engineering Chemistry Research*, 45, 5027–5036.

Majozi, T. 2005. Wastewater minimization using central reusable storage in batch plants. *Computers and Chemical Engineering*, 29, 1631–1646.

Majozi, T. 2006. Storage design for maximum wastewater reuse in multipurpose batch plants. *Industrial and Engineering Chemistry Research*, 45, 5936–5943.

Majozi, T., Brouckaert, C. J., and Buckley, C. A. 2006. A graphical technique for wastewater minimization in batch processes. *Journal of Environmental Management*, 78, 317–329.

Majozi, T. and Gouws, J. 2009. A mathematical optimization approach for wastewater minimization in multiple contaminant batch plants. *Computers and Chemical Engineering*, 33, 1826–1840.

Majozi, T. 2010. Batch Chemical Process Integration: Analysis, Synthesis and Optimization. Springer.

Ng, D. K. S., Foo, D. C. Y., Rabie, A., and El-Halwagi, M. M. 2008. Simultaneous synthesis of property-based water reuse/recycle and interception networks for batch processes. *AIChE Journal*, 54(10), 2624–2632.

Nun, S., Tan, R. R., Razon, L. F., Foo, D. C. Y., and Egashirad, R. 2011. Fuzzy automated targeting for trade-off analysis in batch water networks. *Asia-Pacific Journal of Chemical Engineering*, 6(3), 537–551.

Wang, Y. P. and Smith, R. 1995. Time pinch analysis. *Chemical Engineering Research and Design*, 73, 905–914.

Appendix: Case Studies and Examples

Throughout this book, various hypothetical and industrial examples have been utilized for methodology illustration, as well as problems for exercises. A complete list of all examples and problems in this book is given here for better reference for the readers of specific interest. An example is indicated as "E," while a problem is indicated as "P."

Water Minimization Problems

Wang and Smith (1994): P3.3, P6.2, P7.16, E8.1, E8.2, E8.3, E8.4, E10.1, P12.1, P13.7, E14.1, E14.2, P14.1, E15.1, E15.2, E15.4

Polley and Polley (2000): P3.4, P4.11, E6.1, E6.2, E7.1, E7.3, E7.4, P7.1, P7.13, P7.14, P8.1, E10.1, P12.2, P13.6

Savelski and Bagajewicz (2001): P4.4, P7.17

Wang and Smith (1995): P4.12

Jacob et al. (2002): P4.3, P7.10

Sorin and Bédard (1999): P6.4, P7.12, P8.4, P12.7, P13.3, P14.3, P15.1

Gabriel and El-Halwagi (2005): P14.2

Zero fresh resource network: E3.5, P3.7, E4.3, P4.7, P7.9

Zero waste discharge network (Foo, 2008): E3.6, P3.7, E4.4, P4.7, P7.8, P13.2

Acrylonitrile (AN) production case study: E2.1, E2.2, E3.1, E3.2, E3.4, E4.1, E5.1, E6.3, P7.2, E13.1, E13.2, P13.5, P14.5

Tire-to-fuel process (El-Halwagi, 1997; Noureldin and El-Halwagi, 1999): P2.1, P3.1, P7.3, P13.1

Kraft pulping process (El-Halwagi, 1997): P2.2, P4.1, P6.6, P7.15, P8.3, P14.6, P15.2

Tricresyl phosphate process (El-Halwagi, 1997): P2.3, P4.2

Specialty chemical production process (Wang and Smith, 1995): P4.6, P6.3, P7.7, P8.2, P12.5

Bulk chemical production (Foo, 2012): P3.5, P7.6, P12.4, P13.4

Organic chemical production (Hall, 1997; Foo, 2008): P3.6, P7.11

Paper milling process (Foo et al., 2006b): P4.5, P12.8, P14.4

Textile plant (Ujang et al., 2002): P2.4, P3.2, P12.3

Palm oil mills (Chungsiriporn et al., 2006): P4.8, P7.4, P12.6

Steel plant (Tian et al., 2008): P4.9, P7.5

Kraft pulping process (Parthasarathy and Krishnagopalan, 2001): P4.10

Integrated pulp mill and bleached paper plant: P12.15

Utility Gas Recovery Problems

Refinery hydrogen network (Hallale and Liu, 2001): E2.4, E3.3, P4.13, P7.20, P12.10, P13.8, E14.4, E15.3

Refinery hydrogen network (Alves and Towler, 2002): P2.6, P4.15, P7.21, P12.11, P13.9, P14.7, P15.3

Refinery hydrogen network (Foo and Manan, 2006): P4.16, P7.22, P8.5

Magnetic tape manufacturing process (El-Halwagi, 1997): P2.5, P3.8, P4.14, P7.18, P14.8

Oxygen-consuming process (Foo and Manan, 2006): P3.9, P7.19, P12.9

Property Integration Problems

Metal degreasing process (Kazantzi and El-Halwagi, 2005): E2.5, E3.7, P4.17, E5.2, E7.2, P7.23, P13.10

Kraft papermaking process (Kazantzi and El-Halwagi, 2005): E2.6, E3.8, P3.10, E4.5, P7.24, P12.12, E13.3

Microelectronics manufacturing (Gabriel et al., 2003; El-Halwagi, 2006): P3.11, P7.25, P8.6, P12.13

Wafer fabrication process (Ng et al., 2009, 2010): P2.7, P4.19, P7.26, P8.7, P13.11, P14.9, P15.4

Palm oil mill (Ng et al., 2009): E13.4, P14.10, P15.5

Vinyl acetate manufacturing (El-Halwagi, 2006): P2.8, P3.10, P4.18

Paper recovery study (Wan Alwi et al., 2010): P3.12

Paper recovery study (Wan Alwi et al., 2010): P4.20

Pre-Treatment Networks

Tan et al. (2010): Case 1 and 2a: E9.1, E9.2, E14.5
Tan et al. (2010): Case 2b: P9.1, P14.11
Tan et al. (2010): Case 3: P9.2, P14.12

Inter-Plant Resource Conservation Networks

Inter-plant water networks (Bandyopadhyay et al., 2010): E10.1, E10.3, E10.4, P12.14, E16.1, E16.2, P16.4, P16.5

Inter-plant water networks (Olesen and Polley, 1996): E10.2, P10.2, P12.18, P16.2

Inter-plant water networks in catalyst plant (Feng et al., 2006): P10.1

Inter-plant water networks in integrated pulp mill and a bleached paper plant (Lovelady et al., 2007; Chew et al., 2008): P12.15

Inter-plant water networks in integrated iron and steel mill (Chew and Foo, 2009): P16.1

Inter-plant water networks in eco-industrial park (Lovelady and El-Halwagi, 2009): P16.3

Inter-plant hydrogen networks (Chew et al., 2010a): P10.3, P12.16, P16.6

Inter-plant property-based water networks in wafer fabrication plants (Chew and Foo, 2009; Chew et al., 2010): P10.4, P12.17, E16.3, E16.4, P16.7, P16.8

Batch Material Networks

Wang and Smith (2005): E11.1 through 11.5, P11.1, E17.1, E17.2, E17.3, P17.5
Kim and Smith (2004): P11.2, P17.1
Chen and Lee (2008): P11.3, P17.2
Majozi et al. (2006): P11.4, P17.3
Chan et al. (2008): P11.5, P17.4
Almató et al. (1999): P17.6
Property network (Ng et al., 2008; Foo, 2010): P17.7

Index